AF293989

Die Grundlehren der mathematischen Wissenschaften

in Einzeldarstellungen
mit besonderer Berücksichtigung
der Anwendungsgebiete

Band 138

Herausgegeben von

J. L. Doob · E. Heinz · F. Hirzebruch · E. Hopf · H. Hopf
W. Maak · S. Mac Lane · W. Magnus · D. Mumford
M. M. Postnikov · F. K. Schmidt · D. S. Scott · K. Stein

Geschäftsführende Herausgeber

B. Eckmann und B. L. van der Waerden

Wolfgang Hahn

Stability of Motion

Translated by

Arne P. Baartz

With 63 Figures

Springer-Verlag New York Inc. 1967

Professor Dr. phil. Wolfgang Hahn
Technische Hochschule Graz
Graz (Austria)

Professor Arne P. Baartz, Ph. D.
University of Victoria
Department of Mathematics, Victoria (British Columbia)

Geschäftsführende Herausgeber:

Professor Dr. B. Eckmann
Eidgenössische Technische Hochschule Zürich

Professor Dr. B. L. van der Waerden
Mathematisches Institut der Universität Zürich

ISBN 978-3-642-50087-9   ISBN 978-3-642-50085-5 (eBook)
DOI 10.1007/978-3-642-50085-5

Preface

The theory of the stability of motion has gained increasing significance in the last decades as is apparent from the large number of publications on the subject. A considerable part of this work is concerned with practical problems, especially problems from the area of controls and servo-mechanisms, and concrete problems from engineering were the ones which first gave the decisive impetus for the expansion and modern development of stability theory.

In comparison with the many single publications, which are numbered in the thousands, the number of books on stability theory, and especially books not written in Russian, is extraordinarily small. Books which give the student a complete introduction into the topic and which simultaneously familiarize him with the newer results of the theory and their applications to practical questions are completely lacking. I hope that the book which I hereby present will to some extent do justice to this double task. I have endeavored to treat stability theory as a mathematical discipline, to characterize its methods, and to prove its theorems rigorously and completely as mathematical theorems. Still I always strove to make reference to applications, to illustrate the arguments with examples, and to stress the interaction between theory and practice.

The mathematical preparation of the reader should consist of about two to three years of university mathematics. Here and there a few fundamental concepts of the theory of metric spaces are needed, but I have formulated the arguments in such a way that the reader can usually find an interpretation in n-dimensional Euclidean space. On the whole I limited the selection of materials mainly to the stability of motions in Euclidean space, particularly since the majority of applications are concerned with such motions. But I have stated the basic definitions of stability and proved a number of criteria in a general form, and pointed out take-off points for further investigations, as for instance in the theory of differential equations and difference equations.

Even when limited to Euclidean space (*i.e.* to common differential and difference equations) a complete presentation of the field is not possible in an introduction. Many fine isolated results had to be left out. But I also had to omit several larger topics. Among them was the stability of random processes (*cf.* the remark in sec. 36) as well as the method of "harmonic balance" in the section on periodic motions, which although mathematically suspect is indispensible to the practical man. To include the work done to give a rigorous foundation to such methods (BOGOLYUBOV and MITROPOLSKII in Russia, CESARI, HALE, and co-workers in the USA) would have gone beyond the limits of this book.

It is only natural that the section on periodic motions is rather short compared to the other sections. After all, the subject matter of stability theory involves primarily assertions about the stability of the equilibrium, whereas the numerous contributions to periodic motions are mainly concerned with existence questions.

It was not necessary to treat in more detail individual investigations on second-order differential equations since only quite recently an excellent monograph by REISSIG, SANSONE, and CONTI has appeared, which reports on the newest developments.

As I intended to write a text book and not a handbook, the bibliography is by no means complete. It comprises those publications which I actually used (in the text I frequently referred to sources) and several works of interest for further study.

In preparing the work I received very valuable help from many sources. Dr. KAPPEL read the manuscript and the galley-proofs and sketched the figures. Dr. W. MÜLLER, Prof. REISSIG, Dr. STETTNER, and Dr. INGE TROCH read the galley-proofs and made many critical remarks. I wish to express my sincerest gratitude to all the above. I especially thank Prof. ARNE P. BAARTZ for translating the German manuscript into English and also for making a number of useful suggestions.

A major part of the manuscript was written during my stay at the Mathematics Research Center, Madison, Wis. I am very much obliged to its Directors, Prof. R. E. LANGER and Prof. J. B. ROSSER, for having arranged that stay. Finally, I wish to acknowledge the cooperation of the Springer-Verlag and to appreciate the consideration given to this book by the Editors of the Yellow Series.

Graz, July 1967

WOLFGANG HAHN

Contents

Notations and Formulas

1. The Definition Symbol. The symbol $:=$ or $=:$ defines the variable standing next to the colon. For example, $c := f(a, b)$. c is being introduced, the right side is known.

2. End of Proof. In case of a longer proof, the end of the proof is indicated by the symbol $\bullet$. It corresponds to the classical q.e.d. Sometimes, it denotes the end of the discussion of an example.

3. Vectors. Vectors in n-dimensional Euclidean space R_n are denoted by lower case Latin, and occasionally Greek letters in semi-bold face. Their components have indices. All vectors are column vectors

$$x = \operatorname{col}(x_1, \ldots, x_n).$$

x^T is the transpose of x and thus is a row vector; $x^T y$ is the inner product of x and y. The norm $|x|$ is the Euclidean norm

$$|x|^2 := x_1^2 + \cdots + x_n^2 = x^T x.$$

The zero vector $\operatorname{col}(0, \ldots, 0)$ is simply denoted by 0. The inequality

$$|x| < a$$

defines an open ball in R_n with radius a and center at the origin; it is denoted by K_a. The "half cylinder" defined by $|x| < a, t \geq t_0$, is denoted by K_{a,t_0}.

4. Matrices[1]) are denoted by capital Latin letters. Their elements are represented by lower case Latin letters with double indices, *e.g.*

$$A = (a_{ik}), \quad i, k = 1, 2, \ldots, n.$$

The determinant of A is denoted by $\det A$, its trace by $\operatorname{Tr} A$,

$$\operatorname{Tr} A := \sum_{i=1}^{n} a_{ii}.$$

If $a_{ij} = 0$, $i \neq j$, we write

$$A = \operatorname{diag}(a_{11}, \ldots, a_{nn}).$$

A^T is the transpose of the matrix A, A^I is the inverse of A in case $\det A \neq 0$. E is the unit matrix. Bilinear forms are written in vector notation

$$\sum_{i,k=1}^{n} a_{ik} x_i y_k =: x^T A y.$$

---

[1]) *Cf.*, for instance, BELLMAN [2], SCHMEIDLER [1].

If the quadratic form $x^T Bx$, B symmetric, is positive definite we write $B > 0$.

The norm of A is the square root of the largest characteristic root of $A^T A$, or equivalently,

$$||A|| := \sup \left\{ \frac{|A x|}{|x|} \quad x \in R_n \right\}.$$

5. Differentiation. For a scalar function f depending on a variable vector x (a mapping of a domain in R_n into the real line) we write

$$f_x = \frac{\partial f}{\partial x} := \operatorname{grad} f = \operatorname{col} \left(\frac{\partial f}{\partial x_1}, \ldots, \frac{\partial f}{\partial x_n} \right).$$

The derivative of a vector $g(x)$ with respect to x is the Jacobian or functional matrix

$$\frac{\partial g}{\partial x} := \left(\frac{\partial g_i}{\partial x_k} \right), \quad i = 1, 2, \ldots, m; \quad k = 1, 2, \ldots, n.$$

The derivative with respect to the "time" variable t is denoted by a raised dot, $\dot{x} := \frac{dx}{dt}$.

The differential equation

$$\dot{x} = f(x, t)$$

replaces the n scalar differential equations

$$\dot{x}_i = f_i(x_1, \ldots, x_n, t), \quad i = 1, 2, \ldots, n.$$

In general we shall assume without further mention that the equation possesses solutions in a certain neighborhood K_h of the origin which are uniquely determined by the *initial time* t_0 and the *initial point* x_0. The solution determined by t_0 and x_0 is denoted by $p(t, x_0, t_0)$, so that $p(t_0, x_0, t_0) = x_0$.

Uniqueness is assured if $f(x, t)$ satisfies a Lipschitz condition ($f \in C_0$), *i.e.* if there exists a number L such that

$$|f(x, t) - f(\bar{x}, t)| < L |x - \bar{x}|$$

for all $\bar{x}, x \in K_h$, $t_0 \leq t \leq t_1$. In that case there is an estimate

$$|x_0 - \bar{x}_0| e^{-nL|t-t_0|} \leq |p(t, x_0, t_0) - p(t, \bar{x}_0, t_0)| \leq |x_0 - \bar{x}_0| e^{+nL|t-t_0|}.$$

If $f(x, t)$ satisfies this condition with the same L for all $\bar{x}, x$ in the closed ball $|x| \leq h$ and for all $t \geq t_0$, we say that f satisfies a Lipschitz condition with a uniform Lipschitz constant ($f \in \bar{C}_0$) (*cf.* also CODDINGTON and LEVINSON [1], KAMKE [1], among others).

6. Point Sets. If G is a point set in R_n, $\bar{G}$ denotes the closure, $b(G)$ its boundary. If $G_2 < G_1$, $G_1 \, G_2$ denotes the difference.

Chapter I

Generalities

1. The Stability Concept in Mechanics

The word "stability" originates in mechanics where it characterizes the equilibrium of a rigid body. The equilibrium is called *stable* if the body returns to its original position, having been "disturbed" by being moved slightly from its position of rest. If the body after a slight displacement tends toward a new position its equilibrium is called *unstable*. If one point of the rigid body, e.g. of a compound pendulum, is fixed and the body is subjected to gravity, a well-known geometric criterion for the stability of its equilibrium can be given: it is stable if the fixed point lies above the center of gravity and unstable if the fixed point lies below the center of gravity. If the center of gravity and the point of suspension (point of support) coincide we have the limiting case of a *neutral* equilibrium. The body does not return to its old position but an arbitrarily small displacement causes it to remain "arbitrarily close" to its original position.

The geometric criterion just mentioned has very limited application. It already fails, for instance, to characterize the equilibrium of a ball lying on a horizontal plane, because as the ball moves none of its points remain fixed. However, it is possible to describe the stability of the equilibrium of a ball as follows: The equilibrium of a ball resting on a (not necessarily level) base is stable if its center of gravity is raised by the initial displacement, unstable if it is lowered. The equilibrium of a ball on a level base must thus be called neutral because arbitrary shifting neither raises nor lowers the center of gravity.

Using this formulation we are less concerned with the position of the center of gravity than with the change in its position and we compare the position of the ball not with its final position but rather with the behavior of the ball in motion (strictly speaking, this is already true for the compound pendulum). Thus already for the simple task of discussing the equilibrium of a ball at rest it is necessary to consider the *moving* body: We compare the motion of the ball with its position at rest which can be considered as a special type of motion, or rather as a limiting case of a motion. By the *motion* of a point we essentially mean here the change of its position in the course of time; it is known if the coordinates of the point are given as functions of time. These coordinates are, in this context, not only the coordinates in space but quite generally the quantities which determine the state of the point in question (Lagrangian coordinates); so we might for instance include velocity and acceleration components among others. The number of general coordinates and consequently the dimension of the motion depend on the particular stability problem at hand. So the space in which the motion is interpreted is not 3-dimensional Euclidean space but a *phase space* of higher dimension. As usually, we write the n coordinates $x_1, \ldots, x_n$ of the moving point as an n-dimensional column vector

$$(1.1) \qquad\qquad \boldsymbol{x} = \mathrm{col}(x_1, \ldots, x_n)$$

and regard the x_i as scalar functions of t; then the variable position vector $\boldsymbol{x} = \boldsymbol{x}(t)$ describes the motion. If the motion is along a straight line then $\boldsymbol{x}$ has only one component and actually is a scalar. It is convenient to consider this case as a special case of (1.1) with $n = 1$. If we define the *norm* or *length* of $\boldsymbol{x}$ by

$$(1.2) \qquad\qquad |\boldsymbol{x}| := \sqrt{x_1^2 + \cdots + x_n^2}$$

the phase space obtains the usual Euclidean metric.

Since stability investigations involve the comparison of several motions the vector function $\boldsymbol{x}$ of a stability problem must range over a *family* of motions, *i.e.* it must depend on a parameter $\boldsymbol{a}$ which, of course, may itself be a vector. A motion is then characterized by a vector $\boldsymbol{x}(t, \boldsymbol{a}, t_0)$; the third variable expresses the fact that the motion involves times $t \geq t_0$. In many cases, particularly in practical applications, the motion is defined by letting the vector $\boldsymbol{x}(t, \boldsymbol{a}, t_0)$ be a solution of a differential equation $\dot{\boldsymbol{x}} = \boldsymbol{f}(\boldsymbol{x}, t)$; frequently, but not always, the parameter $\boldsymbol{a}$ belongs to the set of initial values

$$(1.3) \qquad\qquad \boldsymbol{a} = \boldsymbol{x}(t_0).$$

If we are to examine equilibrium stability the family must contain the constant motion $\boldsymbol{x} = \boldsymbol{x}_0$, its equilibrium; *i.e.* there must exist a parameter $\boldsymbol{a}_0$ such that

$$\boldsymbol{x}(t, \boldsymbol{a}_0, t_0) \equiv \boldsymbol{x}_0 \qquad (t \geq t_0).$$

Apart from this obvious stipulation the motion whose stability behavior we study may be defined in a great number of different ways. We give a few simple examples.

1. Let $x(t, a, 0)$ be the particular solution of the scalar differential equation

$$(1.4) \qquad\qquad \ddot{x} + c_1 \dot{x} + c_2 x = a \qquad (t_0 = 0, \quad c_2 \neq 0),$$

which has initial values $x(0) = \dot{x}(0) = 0$. Denoting the zeros of the characteristic polynomial

$$f(s) := s^2 + c_1 s + c_2$$

by α_1 and α_2, it is well-known that the general solution of (1.4) is

$$x(t, a, 0) = \frac{a}{c_2} + k_1 e^{\alpha_1 t} + k_2 e^{\alpha_2 t},$$

where k_1 and k_2 are constants of integration (we assume $\alpha_1 \neq \alpha_2$). Introducing the initial conditions we obtain

$$(1.5) \qquad x(t, a, 0) = \frac{a}{c_2}\left(1 + \alpha_2 \frac{e^{\alpha_1 t}}{\alpha_1 - \alpha_2} - \alpha_1 \frac{e^{\alpha_2 t}}{\alpha_1 - \alpha_2}\right)$$

as an analytic expression for the motion to be studied. Since the second order scalar equation (1.4) is equivalent to a system of two first order equations we are induced to consider in addition to (1.5) the derivative

$$(1.6) \qquad \dot{x}(t, a, 0) = \frac{a\,\alpha_1\,\alpha_2}{c_2(\alpha_1 - \alpha_2)}\left(e^{\alpha_1 t} - e^{\alpha_2 t}\right)$$

and to interpret the motion in an $(x, \dot{x})$-space. The equilibrium $x = \dot{x} = 0$ corresponds to the parameter $a = 0$. If one of the numbers α_1 or α_2 has a positive real part then the right sides of (1.5) and (1.6) become arbitrarily large as t increases, regardless how small a is. If both real parts are non-positive the exponential functions are bounded; (1.5) and (1.6) can be made arbitrarily small by an appropriate choice of a. The equilibrium of the family (1.5), (1.6) is thus stable for $\operatorname{Re} \alpha_i \leq 0$ $(i = 1, 2)$; otherwise it is unstable.

Equation (1.4) has various well-known physical interpretations. It describes for instance the (transient) oscillation in a mechanical or electrical oscillatory system under the influence of a constant external force (for example in a mass-spring system under the influence of gravity or in an electric oscillatory circuit under constant DC voltage, fig. 1.1, 1.2). The assertion "the system is stable" can be interpreted by stating that to a small change in the external force there corresponds a small change in the reaction.

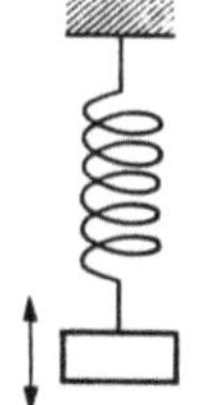

Fig. 1.1. Mechanical oscillating system

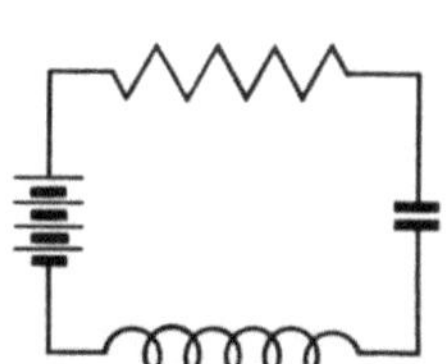

Fig. 1.2. Electrical oscillating system

1*

2. Let the motions be solutions of the linear homogeneous difference equation

$$(1.7) \qquad x(t + 2h) = c_1 x(t + h) + c_2 x(t); \quad t = 0, h, 2h, \ldots .$$

The parameter vector $\boldsymbol{a}$ is defined in terms of the initial values:

$$x(0) = x_0, \quad x(h) = x_1, \quad \boldsymbol{a} = \mathrm{col}(x_0, x_1).$$

If the roots ϱ_1 and ϱ_2 of the so-called characteristic equation

$$p^2 = c_1 p + c_2$$

are distinct the family of motions can be given analytically in the form

$$x(nh, \boldsymbol{a}, 0) = \frac{1}{\varrho_2 - \varrho_1} \left(x_0 (\varrho_2 \varrho_1^n - \varrho_1 \varrho_2^n) + x_1 (\varrho_2^n - \varrho_1^n) \right).$$

The equilibrium $x(t) = x(t + h) = 0$ corresponds to the initial values $x_0 = x_1 = 0$. As in example (1.4) we recognize the following: If one of the numbers ϱ_1, ϱ_2 has absolute value greater than 1 the equilibrium is clearly unstable. If $|\varrho_i| \leq 1$, $i = 1, 2$, then the equilibrium is stable, *i.e.* $x(t)$ remains under a predetermined bound if we choose x_0 and x_1 sufficiently small. This can be expressed by the inequality

$$(1.8) \qquad |x(nh, \boldsymbol{a}, 0)| \leq \frac{2}{|\varrho_1 - \varrho_2|} (|x_0| + |x_1|) \leq \frac{2\sqrt{2}\,|\boldsymbol{a}|}{|\varrho_1 - \varrho_2|}.$$

If $|\varrho_i| < 1$, $i = 1, 2$, we even have

$$(1.9) \qquad |x(nh, \boldsymbol{a}, 0)| \leq \frac{2\sqrt{2}}{|\varrho_1 - \varrho_2|} |\boldsymbol{a}| \cdot \gamma^n,$$

where $\gamma = \max(|\varrho_1|, |\varrho_2|)$. In this case the equilibrium is not only stable but *attractive*: All solutions tend to zero as the independent variable $t = nh$ increases without bound.

3. Let the motions be the particular solutions of the scalar differential equation

$$(1.10) \qquad \ddot{x} + x = a \sin t,$$

which have vanishing initial values for $t = t_0$; a is the parameter. As we did in example (1.4) we find for this family an analytic representation

$$x(t, a, t_0) = \frac{1}{2} a (\cos t_0 \sin(t - t_0) - (t - t_0) \cos t).$$

Since for arbitrarily small non-zero a the right side becomes arbitrarily large as t increases, we have instability here.

Equation (1.10) describes an oscillatory undamped circuit tuned to resonance and the instability we determined corresponds to a well-known physical fact.

4. Let $k \neq 0$ be a constant. Define a family of motions by

$$x(t) = k \sin t.$$

Let the parameter of the family be the initial value $a = x(t_0) = k \sin t_0$. In case $t_0 \neq n\pi$ we have

$$x(t) = (a \sin t) / \sin t_0.$$

For sufficiently small a $x(t)$ remains small; we thus have stability with respect to a. If, on the other hand, $t_0 = n\pi$ then the motion is independent of the value $x(t_0)$ and our present concepts do not apply.

2. Stability in the Sense of Liapunov

The few examples given in sec. 1 and the great variety of possibilities that appeared show that exact definitions are needed. The basic concepts will be defined in sec. 35 in complete generality. For now we limit ourselves to the most important special case of families of motions defined by a differential equation

$$(2.1) \qquad \dot{x} = f(x, t), \quad t \geq t_0.$$

This equation shall admit the solution $x = 0$, the so called *trivial solution*, which we will also call the *equilibrium*. It is expressed by

$$f(0, t) = 0, \quad t \geq t_0.$$

Further we assume that the solutions belonging to the initial points x_0 in a certain neighborhood $|x_0| < h$ of the origin exist for all $t \geq t_0$ and are uniquely determined by the initial values x_0, t_0[1]. The solution which at the initial time t_0 passes through the initial point x_0 is denoted by $p(t, x_0, t_0)$, so that we have

$$p(t_0, x_0, t_0) = x_0.$$

This notation suggests that we intend to examine stability with respect to the initial values [*cf.* (1.3)].

Def. 2.1. The equilibrium of the differential equation (2.1) is called *stable* (in the sense of Liapunov) if for each $\varepsilon > 0$ there exists a $\delta > 0$ such that

$$(2.2) \qquad |p(t, x_0, t_0)| < \varepsilon \quad \text{for} \quad t \geq t_0$$

is valid whenever

$$(2.3) \qquad |x_0| < \delta.$$

The property defined in Def. 2.1 can be visualized geometrically as follows. Consider the solutions of (2.1) as curves in a $(n + 1)$-dimensional

---

[1] Existence and uniqueness theorems can be found in every textbook on differential equations. See for instance CODDINGTON and LEVINSON [1], KAMKE [1], as well as the compilation by CESARI [1].

(x, t)-space. Then inequality (2.2) determines a cylinder of radius ε about the t-axis, and inequality (2.3) defines a ball in the hyperplane $t = t_0$. By choosing the initial points in a sufficiently small ball we can force the graph of the solution (for $t \geq t_0$) to lie entirely inside a given cylinder (fig. 2.1). This consideration also yields an important property of stability for differential equations: If the equilibrium of a differential equation (2.1) is stable at an initial time $t = t_0$ then it is also stable at every initial time $t_1 > t_0$, possibly requiring a different value for δ. For the

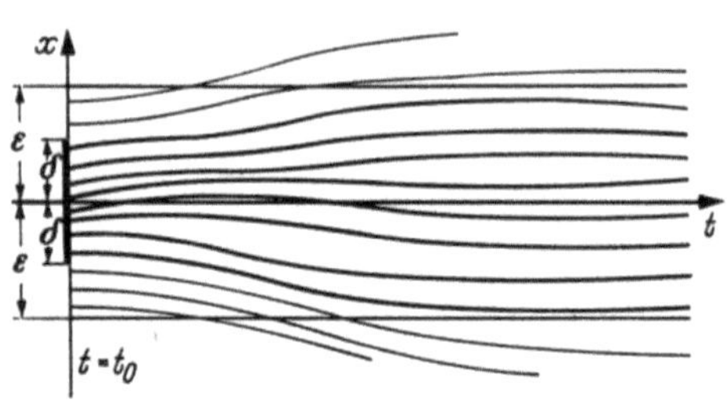

Fig. 2.1. Trajectories in the (t, x)-plane

spherical neighborhood of zero at t_0 is mapped by the graphs of the solutions onto a neighborhood of zero at t_1 which of course contains a certain ball $|x| < \delta_1$ entirely in its interior. If we choose $|x_1| < \delta_1$, (2.2) implies that $|p(t, x_1, t_1)| < \varepsilon$ for $t \geq t_1$. It is therefore unnecessary to require in Def. 2.1 that the desired property hold for all t_0. We must however keep in mind that the number δ depends on t_0 (*cf.* sec. 36, uniform stability).

Def. 2.2. The equilibrium of the differential equation (2.1) is called *attractive* if there exists a number $\eta > 0$ having the property:

$$(2.4) \qquad \lim_{t \to \infty} p(t, x_0, t_0) = 0 \quad \text{whenever} \quad |x_0| < \eta.$$

We are assuming here that the equilibrium $x = 0$ is *isolated*, *i.e.* the neighborhood $|x_0| < h$ contains no point x' other than the origin such that $f(x', t) = 0$ for all t. The concepts *stable* and *attractive* are independent of each other (see sec. 40).

Def. 2.3. The equilibrium of the differential equation (2.1) is called *asymptotically stable* (in the sense of Liapunov) if it is both stable and attractive.

For many practical problems asymptotic stability is of special significance; this explains why a special term was coined, particularly in view of the fact that the independence of stability and attractivity was discovered quite late.

Stability and attractivity were defined as *local* properties at the origin. At first we will consider only solutions with initial values in a sufficiently small neighborhood of the origin. If the relation (2.4) holds for all initial values we speak of asymptotic stability *in the whole*; we are then dealing with a *global* property.

The concepts *stable* and *asymptotically stable* as given in Defs. 2.1 and 2.3 were introduced by LIAPUNOV [1]. In addition, several other definitions are being used which deviate somewhat from the given ones, so that in case of doubt we have to say "stable, resp. asymptotically stable, in the sense of Liapunov". We will at first consider no other kind of stability and will hence omit the qualifier.

Def. 2.4. The equilibrium of the differential equation (2.1) is called *unstable* if it is not stable.

It is useful to state what exactly happens in this case. There exists an ε, in fact an arbitrarily small one, for which no δ can be found; *i.e.* one can choose a sequence $x_{0n} \to 0$ of initial points and a sequence t_n in such a way that

$$|p(t_0 + t_n, x_{0n}, t_0)| \geq \varepsilon \text{ for all } n.$$

It can still happen that all the solutions tend to zero with increasing t: instability and attractivity are compatible, as we already noted (*cf.* sec. 39). The equilibrium is necessarily unstable if every neighborhood of the origin contains initial points for unbounded solutions. It does not follow from the last definition that instability for the initial time t_0 implies instability for any later initial time.

In example (1.7), inequalities (1.8) and (1.9) implied the properties of stability and attractivity of the equilibrium. A similar principle applies quite generally. Below we redefine the terms stable and attractive in a way equivalent to Defs. 2.1 and 2.2, but we will postpone the proof of the equivalence to a later section (sec. 25). For linear differential equations with constant coefficients, which we will consider presently, the equivalence is obvious at any rate. Before stating the definitions we introduce the following notation.

Def. 2.5. A real-valued function $\varphi(r)$ *belongs to class* K $(\varphi \in K)$ if it is defined, continuous, and strictly increasing on $0 \leq r \leq r_1$, resp. $0 \leq r < \infty$, and if it vanishes at $r = 0$: $\varphi(0) = 0$.

Def. 2.6. A real-valued function $\sigma(s)$ *belongs to class* L if it is defined, continuous, and strictly decreasing on $0 \leq s_1 \leq s < \infty$ and if $\lim \sigma(s) = 0$ $(s \to \infty)$.

Def. 2.7. The equilibrium of the differential equation (2.1) is *stable* in case there exists a function φ of class K such that

$$(2.5) \qquad |p(t, x_0, t_0)| \leq \varphi(|x_0|), \quad \text{for} \quad t \geq t_0.$$

Def. 2.8. The equilibrium of the differential equation (2.1) is *attractive* in case there exists an $\eta > 0$ and for each x_0 satisfying $|x_0| < \eta$ there exists a function σ of class L such that

$$(2.6) \qquad |p(t, x_0, t_0)| < \sigma(t - t_0), \quad \text{for} \quad t > t_0.$$

In general, the comparison functions depend on the secondary variables x_0 and t_0. Since this dependence has no significance in the immediately following considerations it was not especially mentioned in (2.5) and (2.6).

Def. 2.9. The equilibrium of the differential equation (2.1) is *asymptotically stable* in case there exists a function $\varphi \in K$, a number $\eta > 0$, and for each x_0 satisfying $|x_0| < \eta$, there exists a function $\sigma \in L$ such that

$$|p(t, x_0, t_0)| \leq \varphi(|x_0|)\, \sigma(t - t_0), \quad \text{for} \quad t \geq t_0.$$

Again the dependence of the comparison functions on the initial values was ignored.

The comparison functions used in example (1.7) were obviously $\varphi(r) = r$ and $\sigma(s) = \gamma^s$. Since by hypothesis $\gamma < 1$, we have $\sigma \in L$. It is apparent that Defs. 2.7 and 2.9 also make sense when applied to difference equations.

In the next few sections we will deal with a number of stability problems which are connected with more or less concrete situations. From a mathematical point of view we will encounter mainly the following type of questions.

a) A family of motions is given (usually by means of a differential equation). We wish to know the stability behavior of the equilibrium, an estimate of the number η in Def. 2.2, resp. 2.8, and the nature of the comparison functions φ and σ in Defs. 2.7 and 2.9.

b) A family of motions is given which depends on certain parameters. We seek sufficient conditions on the parameters for asymptotic stability of the equilibrium. We wish to emphasise especially the significance of the estimate of the number η mentioned under a). For if it is very small and if the solutions with initial points outside the sphere $|x_0| = \eta$ tend away from the origin, then the concrete system described by the equation is *practically unstable* despite its asymptotic stability. Fig. 2.2 and 2.3

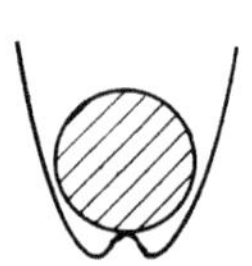

Fig. 2.2. "Practically stable" equilibrium

Fig. 2.3. "Practically unstable" equilibrium

illustrate this: The equilibrium of the ball in fig. 2.3 is practically unstable, whereas the unstable equilibrium of the ball in fig. 2.2 can be considered as *practically stable*. A fully satisfactory definition of the concept of *practical stability* has not yet been found; we shall mention suggested definitions in the appropriate place (*cf.* sec. 56).

Chapter II

Linear Functional Equations with Constant Coefficients

3. Transfer Units

We begin with an interpretation of the solution of a differential equation which deviates from the usual mathematical explanation but which is useful for the understanding of stability considerations.

An electric motor can be thought of as a device which transforms a physical quantity which varies with time, the magnitude of the incoming direct or alternating current, into another quantity which varies with time, the torque of the motor shaft. Many other machines and instruments act in a similar manner: The thermometer transforms temperature into a geometric quantity, the length of the mercury thread, a clock transforms the potential energy of a taut spring into a circular motion, *i.e.* into an angle which changes with time, etc. In all of these cases there are two quantities which are functions of time, the *input* $x_I(t)$ and the *output* $x_O(t)$, and a functional relation (fig. 3.1):

$$(3.1) \qquad x_O(t) = \Re\left(x_I(t)\right),$$

Fig. 3.1. Transfer unit (block diagram)

where $\Re$ denotes an operator characterizing the particular transfer unit. Often the connection between input and output is given by a differential equation, e.g. by

$$(3.2) \qquad \ddot{x}_O(t) + c_1 \dot{x}_O(t) + c_2 x_O(t) = x_I(t).$$

In this simple case we can write down the general solution immediately. Using the notation of (1.5) with $t_0 = 0$ we have

$$(3.3) \quad x_O(t) = k_1 e^{\alpha_1 t} + k_2 e^{\alpha_2 t} + \frac{1}{\alpha_1 - \alpha_2} \int_0^t \left(e^{\alpha_1(t-u)} - e^{\alpha_2(t-u)}\right) x_I(u)\, du.$$

The constants k_1 and k_2 depend on the initial values. In trying to study the response of the transfer unit described by (3.2) to the input function x_I (this could for instance be of the nature of a periodic disturbance), we are interested in that part of the output quantities which is independent of the initial conditions: We wish to know the so-called *steady state* which is assumed after sufficient time has elapsed. However, such a state can develop only if the part of (3.3) which depends on the initial values tends to zero as time increases, *i.e.* if the equilibrium of the homogeneous differential equation is asymptotically stable. In this case all the functions (3.3) tend toward the particular solution of (3.2) which belongs to vanishing initial values, and the operator in (3.1) has then been uniquely defined in a reasonable manner. We see from these considerations how important the concept of asymptotic stability is in practice: Only if the homogeneous differential equation describing a transfer unit has an asymptotically stable equilibrium can such a transfer unit assign a steady output to a steady input. Of course, in our argument we make considerable use of linearity, which allows us to split the response to the input x_I, that is to split the general solution of the nonhomogeneous equation, into two parts, one of which is independent of the initial values, the other of the right side x_I. In the nonlinear case this is not possible, and our reasoning can thus not be immediately transferred to that case.

4. Linear Differential Equations with Constant Coefficients

Equation (3.2) is a special case of the nonhomogeneous n^{th} order differential equation

$$(4.1) \quad a_0 x^{(n)} + a_1 x^{(n-1)} + \cdots + a_{n-1}\dot{x} + a_n x = z(t) \quad (a_0 \neq 0).$$

The solution $x = x(t)$ is a scalar quantity which is uniquely determined only after the initial values have been fixed. The scalar equation (4.1) is equivalent to a system of n equations of the first order which can be constructed in the following manner. We set

$$(4.2) \qquad x_1 = x, \quad x_2 = \dot{x}, \ldots, x_n = x^{(n-1)},$$

introduce the vectors

$$\boldsymbol{x} := \mathrm{col}(x_1, \ldots, x_n), \quad \boldsymbol{z} := \mathrm{col}(0, 0, \ldots, 0, z),$$

and the matrix

$$(4.3) \qquad A := \begin{bmatrix} 0 & 1 & 0 & \cdots & 0 \\ 0 & 0 & 1 & \cdots & 0 \\ \cdots\cdots\cdots\cdots\cdots\cdots\cdots\cdots \\ 0 & 0 & 0 & \cdots & 1 \\ -\dfrac{a_n}{a_0} & -\dfrac{a_{n-1}}{a_0} & -\dfrac{a_{n-2}}{a_0} & \cdots & -\dfrac{a_1}{a_0} \end{bmatrix}$$

and obtain the vector equation

$$(4.4) \qquad \dot{\boldsymbol{x}} = A\boldsymbol{x} + \boldsymbol{z},$$

which is obviously equivalent to (4.1): to each solution vector $\boldsymbol{x}(t)$ of (4.4) there corresponds a solution $x(t)$ of (4.1) whose derivatives are equal to the components of $\boldsymbol{x}$. We shall throughout write equations of the type (4.1) in the simpler form (4.4) and call them differential equations, and only when there is danger of confusion shall we mention whether we are dealing with vector or scalar equations. To convert a given vector equation (4.4) into the scalar form we proceed as follows: we differentiate (4.4), substitute the result in (4.4), and obtain

$$\ddot{\boldsymbol{x}} = A\dot{\boldsymbol{x}} + \dot{\boldsymbol{z}} = A^2\boldsymbol{x} + A\boldsymbol{z} + \dot{\boldsymbol{z}}$$

and repeating the process we finally obtain

$$(4.5) \quad \boldsymbol{x}^{(h)} = A^h\boldsymbol{x} + A^{h-1}\boldsymbol{z} + A^{h-2}\dot{\boldsymbol{z}} + \cdots + \boldsymbol{z}^{(h-1)}, \quad h = 1, 2, \ldots, n,$$

assuming of course that $\boldsymbol{z}$ has derivatives of sufficiently high orders.

The equations in (4.5) are valid for each individual component of $\boldsymbol{x}$ resp. $\boldsymbol{z}$, so that we can write down a total of n^2 scalar equations of the form

$$(4.6) \qquad x_i^{(h)} = \sum_{j=1}^{n} a_{ij}^{(h)} x_j + f_{ih}(\boldsymbol{z}); \quad \begin{array}{l} h = 1, 2, \ldots, n \\ i = 1, 2, \ldots, n. \end{array}$$

The expressions $f_{ih}(\boldsymbol{z})$ are linear combinations of components of $\boldsymbol{z}$ and their derivatives up to and including the $(h-1)$st. For fixed i we can formally solve the system of equations (4.6) for $x_1, \ldots, x_n$ obtaining the expressions

$$(4.7) \qquad x_j = \sum_{k=1}^{n} c_{ik}^{(j)} x_i^{(k)} + g_{ij}(\boldsymbol{z}), \quad j = 1, 2, \ldots, n; \quad i \text{ fixed.}$$

The g_{ij} are of the same type as the f_{ij}. For fixed i one of the n equations (4.7), namely the one in which $j = i$, is a nonhomogeneous scalar differential equation of nth order of the form (4.1) for the component x_i of the vector $\boldsymbol{x}$.

In addition to the differentiability assumption on $\boldsymbol{z}$ we must further require that at least one of the systems (4.6) has a nonzero determinant. This requirement is not generally satisfied. Starting for instance from the two scalar equations

$$(4.8) \qquad \dot{x}_1 = x_1 + z_1, \quad \dot{x}_2 = x_2 + z_2$$

the procedure above does not produce a second order scalar differential equation. The reason for this is easy to see: The physical system described by (4.8) consists of two completely independent first order transfer units. Physically it makes no sense to combine them into a second order transfer unit. We are here concerned with a simple special case of a more general situation which has given rise to an extended theory[1]). The purpose of this discussion was only to show in what sense (4.1) and (4.4) can be considered as equivalent. The discussion of vector equations which follows is in any case applicable to scalar equations as well.

As is well known, the general solution of (4.4) can be given in closed form (*cf.* also sec. 58). We have using the notation defined in sec. 2

$$(4.9) \qquad \boldsymbol{p}(t, \boldsymbol{x}_0, t_0) = e^{A(t-t_0)} \left(\boldsymbol{x}_0 + \int_{t_0}^{t} e^{-A(n-t_0)} \boldsymbol{z}(n) \, dn \right).$$

The dependence on t_0 is not essential here and for the sake of simplicity we shall assume that $t_0 = 0$. The general solution of the homogeneous equation

$$(4.10) \qquad \dot{\boldsymbol{x}} = A \boldsymbol{x},$$

which determines the stability behavior has the form

$$\boldsymbol{p}(t, \boldsymbol{x}_0, 0) = e^{At} \boldsymbol{x}_0,$$

In order to investigate the stability behavior of the equilibrium of (4.10) we make the substitution

$$(4.11) \qquad \boldsymbol{x} = S \boldsymbol{y}$$

and thereby introduce new coordinates; then

$$(4.12) \qquad \dot{\boldsymbol{y}} = S^I A S \boldsymbol{y}$$

and using the abreviation

$$J := S^I A S$$

we obtain

$$(4.13) \qquad \boldsymbol{y}(t) = e^{Jt} \boldsymbol{y}_0.$$

Now a linear nonsingular substitution using a constant matrix has no effect on the stability behavior of the equilibrium. For we have (*cf.* the definition of the norm of a matrix on p. X).

$$|\boldsymbol{x}| \leq \|S\| \, |\boldsymbol{y}|$$

and if there exists for $\boldsymbol{y}(t)$ an estimate as in Def. 2.9,

$$|\boldsymbol{y}(t)| \leq \varphi(|\boldsymbol{y}_0|) \, \sigma(t - t_0),$$

then

$$|\boldsymbol{x}(t)| \leq \|S\| \, \varphi(|S^I \boldsymbol{x}_0|) \, \sigma(t - t_0) \leq \|S\| \, \varphi(\|S^I\| \, |\boldsymbol{x}_0|) \, \sigma(t - t_0).$$

---

[1]) *cf.* for instance KALMAN [2] (Observability and Controllability).

The function $\varphi(\|S^I\|\,|x_0|)$ considered as a function of $|x_0|$ is again a member of class K; and so there exists an estimate of the desired type for x also. We also say that the two differential equations (4.10) and (4.12) are equivalent with respect to the stability of the equilibrium.

The matrix S in (4.11) is so chosen that J becomes a Jordan normal form[1]. If we denote the distinct roots of the characteristic equation

$$(4.14) \qquad \det(A - \lambda E) = 0$$

by $\lambda_1, \lambda_2, \ldots, \lambda_m$ $(m \leq n)$ then J is partitioned into blocks

$$J = \mathrm{diag}(K_1, \ldots, K_r), \qquad m \leq r \leq n,$$

and each block has the form

$$K = \begin{bmatrix} \lambda & 0 & 0 & \cdots & 0 \\ 1 & \lambda & 0 & \cdots & 0 \\ 0 & 1 & \lambda & \cdots & 0 \\ & & \cdots & & \\ 0 & 0 & 0 & \cdots & \lambda \end{bmatrix}.$$

The order l of K is equal to the order of the elementary divisor belonging to λ (if we are dealing with a scalar equation (4.1) then the characteristic equation has the form

$$(4.15) \qquad f(\lambda) := a_0 \lambda^n + a_1 \lambda^{n-1} + \cdots + a_{n-1}\lambda + a_n = 0$$

and l is always equal to the multiplicity of the corresponding root). The matrix appearing in (4.13) satisfies

$$e^{Jt} = \mathrm{diag}(e^{K_1 t}, \ldots, e^{K_m t}).$$

A direct calculation shows that K^r is a triangular matrix with λ^r in its main diagonal while on the subdiagonal the elements λ^{r-1}, $\binom{r}{2}\lambda^{r-2}$, $\binom{r}{3}\lambda^{r-3}$, ... appear in turn. We therefore have

$$e^{Kt} = e^{\lambda t}\begin{bmatrix} 1 & 0 & 0 & \cdots 0 \\ t & 1 & 0 & \cdots 0 \\ \dfrac{t^2}{2!} & t & 1 & \cdots 0 \\ & & \cdots & \\ \dfrac{t^{l-1}}{(l-1)!} & \dfrac{t^{l-2}}{(l-2)!} & \dfrac{t^{l-3}}{(l-3)!} & \cdots 1 \end{bmatrix}.$$

---

[1] *cf.* for instance Bellman [2], Schmeidler [1].

From (4.13) arïd (4.11) we see that:

a) The components of the solution are linear combinations of terms of the form $t^{p_i} e^{\lambda_j t}$ with constant coefficients. The numbers p_i are at most equal to $n-1$ and are equal to 0 if λ_i belongs to a simple elementary divisor and therefore in particular if λ_i is a simple characteristic root.

b) The equilibrium is clearly attractive if all the roots of (4.14), resp. of (4.15), have negative real parts. For if $\lambda = \lambda' + i\lambda''$ then

$$\left| t^p e^{\lambda t} \right| = t^p e^{\lambda' t}.$$

If now $\lambda' < 0$ and ε is so small that $\lambda' + \varepsilon < 0$ then a t_1 can be found such that

$$t^p e^{\lambda' t} < e^{(\lambda' + \varepsilon) t} \quad \text{for } t > t_1.$$

Hence in case $\operatorname{Re} \lambda_i < 0$ $(i = 1, \ldots, m)$, all solutions decrease exponentially; the comparison function σ of Def. 2.9 is of the form $\sigma(s) = e^{-\mu s}$, $\mu > 0$.

c) Even if only one of the real parts $\operatorname{Re} \lambda_i$ is positive then some terms are unbounded and the equilibrium is unstable.

d) If all real parts are non-positive and some of them are actually equal to zero, then there are two cases. If a root $i\omega$ has zero real part and a simple elementary divisor then according to a) only terms of the form $e^{i\omega t}$ belong to it; these are bounded. If such a root has an elementary divisor of higher order then unbounded terms of the form $t^p e^{i\omega t}$ appear. If all terms are bounded then the norm of the matrix e^{Jt} is clearly bounded and then

$$\left| p(t, x_0, 0) \right| \leq \| S \| \, \| e^{Jt} \| \, \| S^I \| \, | x_0 |,$$

i.e. the equilibrium is stable according to Def. 2.7.

Summarizing the above we obtain

Theorem 4.1. The equilibrium of the homogeneous equation $\dot{x} = A x$ is asymptotically stable if all the characteristic roots of the matrix A have negative real parts. If at least one of the characteristic roots has a positive real part then the equilibrium is unstable. If all the characteristic roots have non-positive real parts and some of them actually have a zero real part then the equilibrium is stable, but not asymptotically stable, in case the elementary divisors belonging to the characteristic roots with vanishing real parts are all of the first order. Otherwise the equilibrium is again unstable.

In order to simplify this somewhat involved formulation we introduce the following terminology.

Def. 4.1. A matrix A with real coefficients is called *stable* if all its characteristic roots have negative real parts. If at least one of the characteristic roots has a positive real part the matrix is called *unstable*. A matrix which is neither stable nor unstable is called *critical*; in this case the characteristic roots with vanishing real parts are called *critical characteristic roots*.

This definition which does not concern a special stability concept but simply gives a suitable nomenclature permits a somewhat more concise formulation for Theorem 4.1:

Theorem 4.1a. The equilibrium of the equation $\dot{x} = A\,x$ is asymptotically stable or unstable according as the matrix A is stable or unstable. If A is critical then the equilibrium is stable, but not asymptotically stable, in case the elementary divisors belonging to the critical characteristic roots are all simple. Otherwise the equilibrium is again unstable.

For the sake of completeness we formulate the corresponding theorem for the homogeneous equation (4.1).

Theorem 4.2. The equilibrium of a homogeneous linear equation of nth order with constant coefficients is asymptotically stable if all the roots of the characteristic equation (4.15) have negative real parts. If at least one of the roots has a positive real part then the equilibrium is unstable. If all the roots have non-positive real parts and there exist roots with zero real parts then the equilibrium is stable (but not asymptotically stable) in case the roots whose real parts vanish are simple. If there exist multiple roots with vanishing real parts then the equilibrium is again unstable.

Here again the formulation can be simplified if we call the zeros of a polynomial *stable*, *unstable* or *critical* according as their real parts are negative, positive or equal to zero. A polynomial which has only stable zeros is often called a *Hurwitz polynomial*, for reasons apparent from Theorem 6.3.

Theorems 4.1 and 4.2 are applicable to equations with complex coefficients as well, since in their proofs we never used the fact that the coefficients were real. However, a differential equation with complex coefficients has no immediate interpretation as a transfer unit.

It follows from our decomposition of the general solution into terms of the form $t^p e^{\lambda t}$ that the instability caused by the roots with positive real part is of a different character from that caused by multiple roots with vanishing real parts. In the first case there exist solutions which grow exponentially whereas in the second case the unbounded solutions only grow like powers.

In formula (4.9) we recognize the so-called *forced* component of the solution of the nonhomogeneous equation. In the notation of (3.1) we have

$$\Re\big(z(t)\big) = e^{At} \int_0^t e^{-An}\,z(n)\,dn.$$

In case of stability we obtain

$$(4.16) \quad |\Re(z(t))| \leq \int_0^t \|e^{A(t-n)}\| \, |z(n)| \, dn \leq \int_0^t e^{-\mu(t-n)} \, |z(n)| \, dn$$

with μ positive. If $z(t)$ is bounded then the integral on the right is bounded, implying

Theorem 4.3. If the equilibrium of the homogeneous equation is asymptotically stable then all the solutions of the nonhomogeneous equation corresponding to a bounded right side are bounded.

Calling a transfer unit *stable* if it is described by a homogeneous differential equation with asymptotically stable equilibrium, we have: A stable transfer unit transforms a bounded input into a bounded output.

Expressions like "a stable transfer unit", etc., are found frequently in the technical literature, but we are almost always concerned with *asymptotic* stability of the equilibrium of the underlying differential equation. With a stable but not asymptotically stable equilibrium a bounded input can lead to an unbounded output as equation (1.10) shows. This is well known as *resonance*.

5. Geometrical Criteria for Stability

The results of sec. 4 have reduced the stability problem for linear differential equations with constant coefficients to a purely algebraic problem. Essentially we must determine whether a given polynomial

$$(5.1) \qquad f(s) = a_0 s^n + a_1 s^{n-1} + \cdots + a_{n-1} s + a_n$$

is a Hurwitz polynomial (sec. 4), that is whether its zeros all have negative, resp. non-positive real parts. In case zeros with vanishing real parts exist it is further necessary to determine their multiplicity (resp., in case we are dealing with the characteristic polynomial of a matrix, to determine the order of the corresponding elementary divisors). Various procedures are available for this task. We first discuss the graphical *criterion of Leonhard-Mikhailov*. It applies only to polynomials with real coefficients; we shall therefore suppose from now on that the differential equation under consideration and accordingly the polynomial (5.1) have real coefficients. We further assume that the polynomial (5.1) has p zeros in the right and $n - p$ zeros in the left half of the s-plane. There shall be no zeros on the imaginary axis. Let C be the contour formed by a semi-circle C' with radius r, centered at the origin, together with its diameter on the imaginary axis; and let r be chosen so large that the p zeros in the right half plane lie in the interior of the circle. By a well-known theorem from the elementary theory of functions

$$(5.2) \qquad p = \frac{1}{2\pi i} \int\limits_C \frac{f'(s)}{f(s)} \, ds = \frac{1}{2\pi i} \, \Delta_C \ln f(s),$$

where $\Delta_C \ln f(s)$ is the increment of $\ln f(s)$ along the contour C. Let $\Delta_{C'} \ln f(s)$ be the increment of $\ln f(s)$ along the semi-circular arc C': $s = r e^{i\varphi}$. For s on C',

$$f(s) = a_0 r^n e^{ni\varphi} \left(1 + O(r^{-1})\right),$$

and

$$\ln f(s) = \ln a_0 + n \ln r + ni\varphi + O(r^{-1}).$$

Hence

$$\Delta_{C'} \ln f(s) = ni(\pi/2 + \pi/2) + O(r^{-1}) = n\pi i + O(r^{-1}).$$

Now we let $r \to \infty$ and conclude that

$$(5.3) \qquad p = n/2 + 1/(2\pi i) \, \Delta_I \ln f(s),$$

where $\Delta_I \ln f(s)$ is the increment of the logarithm of $f(s)$ along the imaginary axis I, from $-\infty$ to $+\infty$. To determine this increment we set $s = i\omega$,

$$(5.4) \qquad f(s) = f(i\omega) =: R(\omega) \, e^{\theta(\omega)} =: U(\omega) + i V(\omega),$$

and consider R, θ, resp. U, V, as coordinates in a complex plane. As the real parameter ω ranges from $-\infty$ to $+\infty$ the point $f(i\omega)$ in (5.4) describes a curve which is called the *Leonhard response diagram*[1]) of the polynomial (5.1). Since $f(s)$ is real we have $R(\omega) = R(-\omega)$ and $\theta(\omega) = -\theta(-\omega)$. It suffices to consider the part of the response curve belonging to the positive values of the parameter ω. Hence (5.3) implies

$$(5.5) \qquad p = \frac{n}{2} - \frac{\theta(\infty)}{\pi} = \frac{1}{2}\left(n - \frac{\theta(\infty)}{\pi/2}\right),$$

where $\theta(\infty)$ is the limit to which the polar angle $\theta(\omega) = \arctan \frac{V(\omega)}{U(\omega)}$ of the response diagram tends as ω becomes unbounded. Since $U(\omega)$ and $V(\omega)$ are polynomials of different degrees, $|V(\omega)/U(\omega)|$ tends toward zero or becomes infinite: $\theta(\infty)$ is an integral multiple of $\pi/2$. In case of asymptotic stability $p = 0$ of necessity, *i.e.* $\theta(\infty) = n\frac{\pi}{2}$ and we have (fig. 5.1 and 5.2)

Theorem 5.1. The polynomial $f(s)$ has only zeros with negative real parts if and only if its Leonhard response diagram $f(i\omega)$, for $0 < \omega < \infty$, passes through exactly n quadrants in the positive sense.

---

[1]) In the Russian literature: *Mikhailov response diagram*.

Since $f(s)$ is real we can write

(5.6)
$$f(i\,\omega) = f_1(\omega^2) + i\,\omega f_2(\omega^2)\,,$$

where

$$\deg f_1(u) = \left[\frac{n}{2}\right]\,, \qquad \deg f_2(u) = \left[\frac{n-1}{2}\right]\,.$$

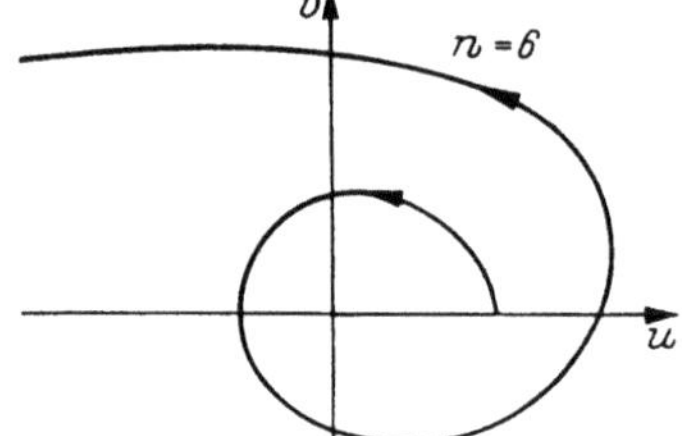

Fig. 5.1. Leonhard plot (stable case) Fig. 5.2. Leonhard plot (unstable case)

If $n = 2k$ then the degrees are k and $k - 1$; if $n = 2k + 1$ then they are both equal to k. To the zeros u_{ij}, $i = 1, 2$; $j = 1, 2, \ldots$ of the polynomials $f_i(u)$ correspond those values of ω^2 at which the response diagram intersects the axes. In case of stability these values of ω^2 must of course be real and increasing, *i.e.* the zeros u_{ij} must all be positive and alternate:

(5.7)
$$0 < u_{11} < u_{21} < u_{12} < u_{22} < \cdots$$

since otherwise the response curve does not make the proper number of turns. These considerations lead to the so-called *gap and position criterion* (fig. 5.3 and 5.4):

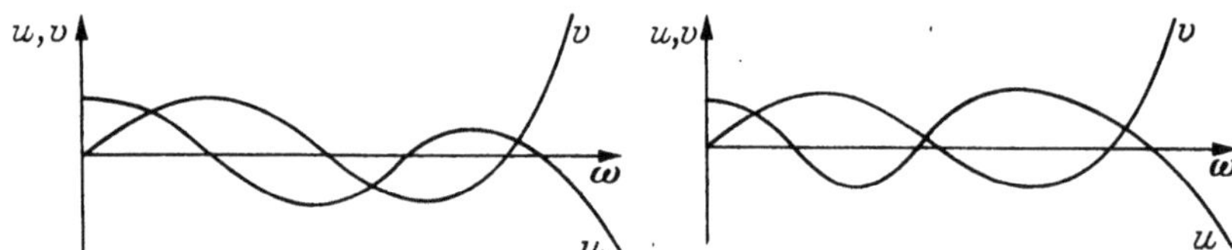

Fig. 5.3. Position criterion (stable case) Fig. 5.4. Position criterion (unstable case)

Theorem 5.2. The polynomial $f(s)$ has only zeros with negative real parts if and only if the zeros of the polynomials defined in (5.6) are real and satisfy the inequalities (5.7).

The Leonhard response diagram is unbounded and hence can never be graphed completely. We therefore frequently work with the reciprocal response diagram i.e. the response diagram of the function $\frac{1}{f(i\,\omega)}$. It approaches zero asymptotically and in case of stability rotates exactly through n quadrants in the *negative* sense. However, in applications of Theorem 5.1 the entire curve is not needed; rather it suffices to consider the interval $0 \leq \omega \leq \omega_0$ of the parameter, where

$$(5.8) \qquad \omega_0 := 1 + 3\mu, \qquad \mu := \max_i \left\{ \left| \frac{a_i}{a_0} \right| \right\}.$$

The reason for this is that we can choose for the radius r of the semi-circle C in (5.2) the value $r_0 = 1 + 3\mu$. For if $f(\lambda) = 0$ and $|\lambda| > 1$, then

$$1 = - \sum_{i=1}^{n} \frac{a_i}{a_0} \lambda^{-i},$$

$$1 \leq \mu \sum_{i=1}^{n} |\lambda|^{-i} \leq \frac{\mu}{|\lambda| - 1}, \qquad |\lambda| \leq \mu + 1 < 1 + 3\mu.$$

Therefore no zero can lie outside of the circle with radius r_0. For the circle with radius r_0 the error term in (5.2) which was majorized by the estimate $O(r^{-1})$ is exactly equal to

$$\ln \left(1 + \frac{a_1}{a_0 r_0} + \cdots + \frac{a_n}{a_0 r_0^n} \right),$$

and since

$$\sum_{i=1}^{n} \left| \frac{a_i}{a_0 r_0^i} \right| \leq \frac{\mu}{r_0 - 1} = \frac{1}{3}$$

it follows that

$$\arg \left(1 + \sum_{i=1}^{n} \frac{a_i}{a_0 r_0^i} \right) \leq \arcsin \frac{1}{3} < \frac{\pi}{4}.$$

At the value $\omega = \omega_0$ of the parameter the response curve is already in its last quadrant, since $\theta(\omega_0)$ differs from $\theta(\infty)$ by less than $\pi/4$.[1]

6. Algebraic Criteria for Stability

1. In applying the so-called algebraic criteria we compute from the coefficients of the polynomial (5.1) by rational steps certain numbers and test their signs. For this purpose the characteristic polynomial of a matrix A must first be put in the form (5.1) as was necessary if we wanted to apply the response diagram criterion.

Theorem 6.1. For (5.1) to be a Hurwitz polynomial it is necessary that the inequalities

$$(6.1) \qquad \frac{a_1}{a_0} > 0, \qquad \frac{a_2}{a_0} > 0, \ldots, \frac{a_n}{a_0} > 0$$

hold.

---

[1] *cf.* KAPLAN [1].

2*

Proof. Let $s_1, \ldots, s_n$ be the zeros and in particular let s_j' be the real and s_k'' the complex roots. Then

$$f(s) = a_0 \prod_j (s - s_j') \prod_k (s - s_k'')$$

and combining the complex conjugate factors we obtain an expression of the form

$$f(s) = a_0 \prod_j (s - s_j') \prod_k (s^2 - 2 \operatorname{Re} s_k'' + |s_k''|^2).$$

If now all the numbers s' and all the $\operatorname{Re} s_k''$ are negative we can obtain only positive coefficients for the powers of s when we multiply the product out.

2. We next assume (6.1), and in addition we may assume without loss of generality that a_0 is positive. A necessary and sufficient condition for $f(s)$ to be a Hurwitz polynomial is given by the *Routh criterion*. To use it we form the following scheme of numbers.

$$c_{10} = a_0, \quad c_{20} = a_2, \quad c_{30} = a_4, \quad c_{40} = a_6, \ldots$$

$$c_{11} = a_1, \quad c_{21} = a_3, \quad c_{31} = a_5, \quad c_{41} = a_7, \ldots$$

$$r_2 = \frac{a_0}{a_1} \; \Big| \; c_{12} = a_2 - r_2 a_3, \quad c_{22} = a_4 - r_2 a_5, \quad c_{32} = a_6 - r_2 a_7, \ldots$$

$$r_3 = \frac{c_{11}}{c_{12}} \; c_{13} = c_{21} - r_3 c_{22}, \quad c_{23} = c_{31} - r_3 c_{32}, \quad c_{33} = c_{41} - r_3 c_{42}, \ldots$$

$$\cdots \cdots \cdots \cdots \cdots \cdots$$

$$r_j = \frac{c_{1,j-2}}{c_{1,j-1}} \; c_{ij} = c_{i+1,j-2} - r_j c_{i+1,j-1}; \quad \begin{array}{l} i = 1, 2, \ldots \\ j = 2, 3, \ldots \end{array}$$

$$\cdots \cdots \cdots \cdots \cdots \cdots$$

$$c_{1n} = a_n.$$

If $n = 2m$, we have $c_{m+1,0} = c_{m+1,2} = a_n, c_{m+1,1} = c_{m+1,3} = 0$.

If $n = 2m - 1$, we have $c_{m0} = a_{n-1}, c_{m1} = a_n, c_{m2} = c_{m3} = 0$.

This scheme terminates after $n - 1$ steps in case all the numbers c_{ij} are different from zero; the last line defines c_{1n}.

Theorem 6.2. A polynomial $f(s)$ is a Hurwitz polynomial if and only if the inequalities (6.1) hold in conjunction with the inequalities

(6.2) $$c_{11} > 0, \quad c_{12} > 0, \ldots, c_{1n} > 0.$$

Proof. First let $n = 2m$. We define the polynomials

(6.3) $$h_1(s) := \frac{1}{2}\left(f(s) + f(-s)\right); \quad h_2(s) := \frac{1}{2}\left(f(s) - f(-s)\right)$$

and apply the Euclidean algorithm to determine the greatest common divisor of $h_1(s)$ and $h_2(s)$. This process yields the sequence of equations

$$
\begin{aligned}
h_1(s) &= r_2' s\, h_2(s) - h_3(s), \\
h_2(s) &= r_3' s\, h_3(s) - h_4(s),
\end{aligned}
$$
(6.4)

$$\cdots\cdots\cdots\cdots\cdots\cdots$$

The linear factors arising in the division have no constant term. The remainders have been written with negative signs. The numbers r_i' are, if we disregard their sign, equal to the multipliers of the Routh scheme; in fact $r_i' = (-1)^i r_i$. We now define a further sequence of polynomials by

$$
h_{2i-1}(s) =: g_{2i-1}(s^2), \quad h_{2i}(s) =: s\, g_{2i}(s^2), \quad i = 1, \ldots, m.
$$

They are related by the recursion formulae

$$
\begin{aligned}
g_{2i+1}(z) &= r_{2i}' z\, g_{2i}(z) - g_{2i-1}(z), \\
g_{2i+2}(z) &= r_{2i+1}' g_{2i+1}(z) - g_{2i}(z)
\end{aligned}
$$
(6.5)

which follow from (6.4). The first two polynomials of the sequence are

$$
\begin{aligned}
g_1(z) &= a_0 z^m + a_2 z^{m-1} + \cdots + a_{2m}, \\
g_2(z) &= a_1 z^{m-1} + a_3 z^{m-2} + \cdots + a_{2m-1},
\end{aligned}
$$

and they agree, except for the sign of the coefficients, with the polynomials f_i defined in (5.6). In fact,

$$
f_i(u) = g_i(-u).
$$
(6.6)

If we construct the Routh scheme, resp. the sequence of polynomials $g_i(z)$, from an arbitrary polynomial and encounter a zero row, resp. an identically vanishing polynomial g_j, then h_1 and h_2 resp. $f(s)$ and $f(-s)$ have a common divisor. In this case $f(s)$ possesses a divisor of the form $s^2 + \alpha$ with real α and certainly is not a Hurwitz polynomial.

We now assume that the hypotheses of Theorem 6.2 are satisfied. Then the numbers r_i' have alternating signs. The signs of the leading coefficients of the polynomials g_i, $i = 1, 2, \ldots$, are

$$+, +, -, -, +, +, -, -, \text{etc.}$$

The degrees of the polynomials are

$$m,\, m-1,\, m-1,\, m-2,\, m-2,\, \ldots,\, 1,\, 1,\, 0.$$

For a fixed z, $-\infty < z < +\infty$, consider the sequence of numbers

$$
g_1(z), g_2(z), \ldots, g_{2m}(z).
$$
(6.7)

The last polynomial $g_{2m}(z)$ is constant. Let $W(z)$ denote the number of sign changes in this sequence. If $z > 0$ and is very large then the signs of the sequence (6.7) correspond to those of the leading coefficients. For large negative values of z the signs alternate. The difference $W(-\infty) - W(+\infty)$ is always equal to m: as z changes from large negative values to large positive values m sign changes are lost. Such a loss can only occur at a zero of $g_1(z)$; for if z passes through a zero z' of $g_i(z)$, $1 < i < 2m$, no loss in the number of sign changes occurs since by (6.5) sgn $g_{i-1}(z') \neq$ sgn $g_{i+1}(z')$ (the polynomials g_i form a Sturmian chain). Consequently $g_1(z)$ has exactly m real zeros, and since by hypothesis all the coefficients are positive no positive zeros can occur. A similar argument shows that g_2 has exactly $m - 1$ real negative zeros. Since in the sequence (6.7) the largest possible loss of sign changes occurs, a loss must actually occur each time z passes through a zero of g_1. This however is only possible if g_2 in turn changes sign between every two zeros of g_1; otherwise we would have an additional change of signs. Hence the zeros of g_2 separate those of g_1. Because of the relationship (6.6), however, the statement about the zeros of g_1 and g_2 which we just derived is equivalent with inequality (5.7). The gap and position criterion applies. The condition that in (6.2) $c_{1i} > 0$ is thus sufficient for $f(s)$ to be a Hurwitz polynomial. At the same time it is also necessary since otherwise the count of the sign changes is too small and the condition of Theorem 5.2 is not satisfied: either g_1 has too few zeros or the polynomial does not satisfy the *gap condition* (5.7).

If the degree n of $f(s)$ is odd, $n = 2m + 1$, we interchange the definitions of $h_1(s)$ and $h_2(s)$ in (6.3) and furthermore set

$$h_{2i+1}(s) := s\,g_{2i+1}(s^2), \quad h_{2i}(s) := g_{2i}(s^2).$$

The degrees of the polynomials formed in this manner are m, m, $m - 1$, $m - 1, \ldots, 1, 1, 0$. The numbers (6.7) all have the same sign for large negative z. As above we conclude that the polynomials $g_1(z)$ and $g_2(z)$ each have m negative zeros, and that the gap criterion applies.

If one of the numbers c_{1j} vanishes the Routh scheme cannot be set up. But in this case $f(s)$ is certainly not a Hurwitz polynomial. For then the sequence $h_1, h_2, \ldots$ has fewer than $2m$ terms and the sequence (6.7) does not furnish enough sign changes. Since together with $f(s)$ the polynomial $s^n f(s^{-1})$, whose coefficients are $a_n, \ldots, a_0$, is also a Hurwitz polynomial (because sgn Re $s =$ sgn Re s^{-1}), Theorem 6.2 is also valid for the Routh scheme formed from the sequence $a_n, a_{n-1}, \ldots, a_1, a_0$.

Example. The polynomial

$$f(s) = 4s^5 + 2s^4 + 9s^3 + 4s^2 + 5s + 1$$

satisfies the necessary condition of Theorem 6.1. The Routh scheme is

$$
\begin{array}{c|ccc}
 & 4 & 9 & 5 \\
 & 2 & 4 & 1 \\
\hline
2 & 1 & 3 \\
2 & -2 & 1 \\
-0.5 & 3.5
\end{array}
$$

Since $c_{13} < 0$, the polynomial is not a Hurwitz polynomial.

In practical calculations it is often not necessary to determine the numbers in the scheme exactly; an approximate calculation is often sufficient. We must of course make sure that in rounding off the signs are not affected.

A further sequence of numbers can be applied to test for stability. For this purpose we use the coefficients of the polynomial to form the matrix

$$
(6.8) \qquad
\begin{array}{cccccccc}
a_1 & a_0 & 0 & 0 & \cdots & 0 & 0 \\
a_3 & a_2 & a_1 & a_0 & \cdots & 0 & 0 \\
a_5 & a_4 & a_3 & a_2 & \cdots & 0 & 0 \\
\multicolumn{8}{c}{\cdots\cdots\cdots\cdots\cdots\cdots\cdots\cdots\cdots} \\
0 & 0 & 0 & 0 & \cdots & a_{n-1} & a_{n-2} \\
0 & 0 & 0 & 0 & \cdots & 0 & a_n
\end{array}
$$

For $0 < 2i - j \leq n$, the general element $a_{ij} = a_{2i-j}$, otherwise $a_{ij} = 0$. Then we form the sequence of principal subdeterminants

$$H_1 = a_1, \quad H_2 = a_1 a_2 - a_0 a_3, \quad \ldots, \quad H_{n-1}, \quad H_n = a_n H_{n-1},$$

the so-called *Hurwitz determinants*, and we have the following theorem due to Hurwitz:

Theorem 6.3. A polynomial $f(s)$ is a Hurwitz polynomial if and only if the inequalities (6.1) and the inequalities

$$(6.9) \qquad H_1 > 0, \quad H_2 > 0, \ldots, H_n > 0$$

hold.

Proof. By a method similar to the so-called Gaussian algorithm we convert the matrix to a triangular matrix. For this purpose we first subtract r_2 times columns $1, 3, \ldots$ from columns $2, 4, \ldots$ resp., where $r_2 = a_0/a_1$ is the multiplier of the Routh scheme. Then we subtract r_3 times the $2k$th column from the $(2k + 1)$th column and continue in this manner until there are only zeros above the main diagonal. These calculations are exactly those which lead to the Routh scheme, so that after we have reduced the matrix, the numbers $c_{11} = a_1, c_{12}, \ldots, c_{1n} = a_n$ appear on the main diagonal. The last element of the main diagonal is not affected by this transformation. Since the transformation leaves the value of the principal subdeterminants unchanged we have

$$H_1 = a_1, \ H_2 = a_1 c_{12}, \ldots, H_i = c_{11} c_{12} \cdots c_{1i}, \ldots$$

and

$$a_1 = c_{11} = H_1, \quad c_{12} = \frac{H_2}{H_1}, \ldots, c_{1i} = \frac{H_i}{H_{i-1}}, \ldots.$$

Thus the Hurwitz determinants are all positive if and only if the same is true for the Routh numbers (6.2).

If the necessary conditions of Theorem 6.1 are satisfied (and $a_0 > 0$) we have $H_1 = a_1 > 0$, and H_n has the same sign as H_{n-1}; it is therefore unnecessary to check the condition for H_1 and H_n. The essential Hurwitz conditions are

$$(6.10) \quad \begin{aligned} &\text{for } n = 3: \quad a_1 a_2 - a_0 a_3 > 0, \\ &\text{for } n = 4: \quad a_1 a_2 - a_0 a_3 > 0; \ a_1 a_2 a_3 - a_0 a_3^2 - a_1^2 a_4 > 0. \end{aligned}$$

In case $n = 2$ the necessary condition (all coefficients have the same sign) is at the same time sufficient.

The $2n$ conditions in (6.1) together with (6.2), resp. (6.9), are dependent on each other. One can show that any n suitably chosen inequalities are implied by the remaining inequalities, so that n inequalities, which are necessary and sufficient, are all that is needed. So it suffices for instance that $a_{2i} > 0, H_{2i+1} > 0$. There is, however, little value in this for practical calculations since it is still necessary to compute the Hurwitz determinants one after another[1]).

A concrete problem induced Hurwitz to investigate and develop the criterion of determinants named after him: the problem was to find conditions for the stable behavior of a centrifugal steam engine governor. In contrast to the proof given above, Hurwitz's considerations can also be applied to polynomials with complex coefficients. Hurwitz was not aware of Routh's work. The close connection between the two criteria was discovered relatively late[1]).

For checking a polynomial with given numerical coefficients the Routh scheme is preferable in general because of its relative perspicuity. Investigation of the Hurwitz determinants is of special advantage if the influence of the coefficients on the stability is to be examined (*cf.* secs. 9, 11).

---

[1]) *cf.* CREMER and EFFERTZ [1].

Neither the Routh nor the Hurwitz criterion can be applied immediately if the polynomial under investigation is the characteristic polynomial of a matrix of the form

$$(6.11) \qquad \det(A - sE) = 0.$$

A practically feasible procedure applicable to (6.11) is based on the following considerations[1]).

The transformation of variables

$$(6.12) \qquad w = 1 + \frac{2}{s-1}$$

takes the left half of the s-plane onto the interior of the unit circle in the w-plane. It transforms a Hurwitz polynomial $f(s)$ into a rational function all of whose zeros lie in the interior of the unit circle. Now a matrix B has all of its characteristic roots in the interior of the unit circle if and only if the powers B^k converge to zero as k increases. For if ϱ is a zero of the characteristic polynomial

$$\det(B - wE) = 0$$

that is, if ϱ is a characteristic root of B, then ϱ^k is a characteristic root of B^k, and if $|\varrho| < 1$, then ϱ^k tends to zero. However, the transformation (6.12) corresponds to the matrix transformation

$$(6.13) \qquad B = E + 2(A - E)^I$$

in the sense that the characteristic roots of B are the image of the characteristic roots of A under the transformation (6.12). We have proved
Theorem 6.4. All of the characteristic roots of a matrix A have negative real parts if and only if the matrix B defined by (6.13) satisfies the condition $B^k \to 0$ $(k = 1, 2, 3, \ldots)$.

A different criterion immediately applicable to the matrix was given by H. R. Schwarz [1].

7. Orlando's Formula [2])

Let $s_1, s_2, \ldots, s_n$ be the zeros of the polynomial (5.1), as above.

Theorem 7.1. The following formula obtains:

$$(7.1) \qquad H_{n-1} = (-1)^{\frac{n(n-1)}{2}} a_0^{n-1} \prod_{i<k} (s_i + s_k).$$

---

[1]) Zubov [4].

[2]) Orlando [1]. The following proof differs from the original one.

Since $H_n = a_n H_{n-1}$ and $a_n = (-1)^n a_0 s_1 s_2 \cdots s_n$, we can write instead of (7.1) the equivalent formula

$$H_n = (-1)^{\frac{n(n+1)}{2}} a_0^n 2^{-n} \prod_{i < k} (s_i + s_k).$$

Proof. The resultant of two polynomials

$$P(s) = p_0 s^n + p_1 s^{n-1} + \cdots + p_n,$$
$$Q(s) = q_0 s^m + q_1 s^{m-1} + \cdots + q_m$$

with zeros $\alpha_1, \ldots, \alpha_n$, resp. $\beta_1, \ldots, \beta_m$ is defined by

$$(7.2) \qquad R(P, Q) := p_0^m q_0^n \prod (\alpha_i - \beta_j), \qquad \begin{matrix} i = 1, 2, \ldots, n, \\ j = 1, 2, \ldots, m, \end{matrix}$$

and may be written as a determinant with $m + n$ rows:

$$R(P, Q) = \begin{vmatrix} p_0 & p_1 & \cdots & p_n & 0 & \cdots & 0 \\ 0 & p_0 & \cdots & p_{n-1} & p_n & \cdots & 0 \\ \cdots & \cdots & \cdots & \cdots & \cdots & \cdots & \cdots \\ 0 & 0 & \cdots & \cdots & \cdots & \cdots & p_n \\ q_0 & q_1 & \cdots & \cdots & \cdots & \cdots & 0 \\ 0 & q_0 & \cdots & \cdots & \cdots & \cdots & 0 \\ \cdots & \cdots & \cdots & \cdots & \cdots & \cdots & \cdots \\ 0 & 0 & \cdots & \cdots & \cdots & \cdots & q_m \end{vmatrix} \begin{matrix} \left. \vphantom{\begin{matrix}a\\b\\c\end{matrix}} \right\} m \text{ rows} \\ \\ \left. \vphantom{\begin{matrix}a\\b\\c\end{matrix}} \right\} n \text{ rows} \end{matrix}$$

Starting with the polynomials $f(s)$ and $f(-s)$ we obtain a determinant of $2n$ rows and $2n$ columns. The kth and the $(m + k)^{th}$ rows are given by

$$(7.3) \qquad \begin{matrix} 0 \cdots 0 & a_0 & a_1 \cdots a_n \cdots 0, \\ 0 \cdots 0 & (-1)^n a_0 & (-1)^{n-1} a_1 \cdots a_n \cdots 0. \end{matrix}$$

Without changing the value of the determinant we can add $(-1)^n$ times the second row in (7.3) to the first row, factor out the scalar 2 and then add $(-1)^{n-1}$ times the new first row to the second. Apart from a factor of $(-1)^{n-1} \cdot 2$ the two lines then go over into

$$0 \ldots 0 \; a_0 \; 0 \; a_2 \; 0 \ldots \quad 0 \ldots 0$$
$$0 \ldots 0 \; 0 \; a_1 \; 0 \; a_3 \ldots \quad 0 \ldots 0.$$

We repeat this operation for all pairs of rows k, $n + k$. In the first and in the last column of the resulting determinant only the element a_0, resp. a_n, differs from 0. The expansion by minors contains the factor $a_0 a_n$ as well as a $(2n-2)$-row determinant which starts with the array

$$
\begin{array}{cccc}
a_1 & 0 & a_3 & 0 \ldots \\
a_0 & 0 & a_2 & 0 \ldots \\
0 & a_1 & 0 & a_3 \ldots \\
0 & a_0 & 0 & a_2 \ldots .
\end{array}
$$

The rows and columns are then so rearranged that the determinant starts with

$$
\begin{array}{ccc}
a_1 & a_3 & a_5 \ldots \\
a_0 & a_2 & a_4 \ldots \\
0 & a_1 & a_3 \ldots .
\end{array}
$$

The zeros are then distributed in such a way that the determinant consists of two blocks which both agree with H_{n-1}. We therefore have

$$
R\big(f(s), f(-s)\big) = (-1)^{n^2-n}\, a_0 a_n \cdot 2^n H_{n-1}^2 .
$$

On the other hand we have according to (7.2)

$$
R\big(f(s), f(-s)\big) = (-1)^{n^2} a_0^{2n} \prod_{i=1}^{n} (s_i + s_i)\, \prod_{i<j} (s_i + s_j)^2 .
$$

Furthermore, applying a relation which we used once before,

$$
a_0 \prod_{i=1}^{n} (s_i + s_i) = 2^n a_0 s_1 s_2 \cdots s_n = (-1)^n\, 2^n a_n ,
$$

we obtain the relation

$$
H_{n-1}^2 = a_0^{2n-2} \prod_{i<j} (s_i + s_j)^2 .
$$

(7.1) follows except for the sign. The sign is determined by comparing coefficients. The determinant for H_{n-1} involves the term $+ a_1 a_2 \cdots a_{n-1}$. We express the a_i by means of the elementary symmetric functions in terms of the roots s_i and compare the signs of the term $a_0^{n-1} s_1^{n-1} s_2^{n-2} \cdots$ $\cdots s_{n-1}$ on the two sides of (7.1). The validity of Orlando's formula then becomes apparent.

An immediate consequence is

Theorem 7.2. The polynomial $f(s)$ has a pair of opposite but equal zeros if and only if H_{n-1} is zero.

8. Linear Transfer Systems

In sec. 3 a transfer unit was defined as a device implementing a functional transformation of the form

$$
(8.1) \qquad\qquad x_O(t) = \Re\big(x_I(t)\big).
$$

We already mentioned that especially those operators are of practical importance which can be expressed by means of a linear differential equation with constant coefficients. The *solution* is then the output variable; the input $x_I(t)$ appears on the right side of the differential equation. It is convenient to express the operation d/dt by the symbol D and to express the left side of the differential equation symbolically as a polynomial in D

$$(8.2) \qquad a_0 x^{(n)} + a_1 x^{(n-1)} + \cdots + a_{n-1} \dot{x} + a_n x =: f(D) x,$$

$$(8.3) \qquad f(s) := a_0 s^n + a_1 s^{n-1} + \cdots + a_{n-1} s + a_n.$$

The relation (8.1) then becomes

$$(8.4) \qquad\qquad f(D) x_O = x_I.$$

In many cases not only the input $x_I(t)$ itself is effective but also its first, second, ... derivative, occasionally even its time integral: the transfer unit measures all of these quantities, multiplies them with the appropriate factors and forms the sum, and it is this sum which effectively acts as the input quantity. In place of (8.4) we have in this case the more general equation

$$(8.5) \qquad\qquad f(D) x_O = g(D) x_I.$$

The operator on the right has the form

$$(8.6) \qquad g(D) = b_{-1} D^{-1} + b_0 + b_1 D + b_2 D^2 + \cdots$$

and

$$D^{-1} x := \int\limits_{t_0}^{t} x(\tau)\, d\tau.$$

From a purely mathematical point of view there is no difference between (8.4) and (8.5): on the right we have in both cases a function $z(t)$. For applications however and especially for stability considerations the form (8.5) is important.

The appropriate tool for the mathematical treatment of (8.5) is the Laplace transformation[1]. We shall denote the variable of the transform by s, the transformed function by a bar over the original letter. So for instance

$$\mathfrak{L}\{x(t)\} := \int\limits_{0}^{\infty} e^{-st} x(t)\, dt =: \bar{x}(s).$$

At first we apply Laplace transform to (8.5) quite formally and obtain

$$f(s)\bar{x}_O = g(s)\bar{x}_I + P\big(s; x_O(0), x_I(0)\big),$$

---

[1] *cf.* for instance DOETSCH [1], KAPLAN [1].

where P denotes a polynomial in s whose coefficients depend in a well defined manner on the initial values of x_0 and x_I and which vanishes if these initial values are zero. Neglecting this term we have

$$(8.7) \qquad \bar{x}_0 = \frac{g(s)}{f(s)} \bar{x}_I$$

as the relation between the transforms in the steady state. As we explained in sec. 3 such a steady state prevails after the influence of the initial values has faded away, provided the transfer unit is stable. The rational function

$$(8.8) \qquad F(s) := \frac{g(s)}{f(s)}$$

is called the *transfer function* of the transfer unit described in (8.5). This function can of course also be given for an unstable transfer unit. In accordance with the well known rules of the Laplace transformation the transformed equation (8.7) represents a reasonable relation between the original functions only in case the degree of the numerator of (8.8) is at least one less than the degree of the denominator. We must also take into consideration that $g(s)$ possibly involves negative powers of s and therefore contributes to the degree of the denominator. Keeping this in mind and using the results of sec. 4 we have: The rational function (8.8) is the transfer function of a stable unit if and only if the degree of the denominator exceeds the degree of the numerator and all of the poles of the function have negative real parts.

In that case the right side of (8.8) admits an expansion into partial fractions of the form

$$(8.9) \qquad \sum_{i,j} \frac{c_{ij}}{(s - s_j)^{p_{ij}}} \qquad (p_{ij} \text{ integers})$$

and the corresponding original function is a linear combination of terms of the form $t^{p_{ij}-1} e^{s_j t}$ (see also the remark in sec. 4 on the decomposition of $e^{At} x_0$). Although this original function is a solution of the homogeneous equation it can also be interpreted as a solution of the nonhomogeneous equation, *i.e.* as an output quantity, if we admit the unit impulse function (the Dirac delta function) as an input. This solution (the response to the unit impulse) is also called the *weighting function* of the unit. It must be distinguished from the *step response*. The step response is the output whose input is the unit step function

$$(8.10) \qquad x_u(t) = \begin{matrix} 0, t < 0 \\ 1, t \geq 0 \end{matrix} \quad \text{resp.} \quad x_u(t) = \frac{1}{2}(1 + \mathrm{sgn}\, t),$$

that is, it is the solution of the nonhomogeneous equation (8.4) which has vanishing initial values and (8.10) as its right side. The Laplace transform of the step response is $\dfrac{g(s)}{s f(s)}$. Choosing for s in (8.8) a purely imaginary variable $s = i\omega$ we obtain a quantity

$$(8.11) \qquad F(i\omega) = \frac{g(i\omega)}{f(i\omega)}$$

which is known as the (complex) *frequency response function* of the transfer unit. It indicates how the transfer unit, which was assumed stable, reacts upon a periodic input $x_I = a e^{i\omega t}$. For, taking the inverse transform of (8.7), we obtain

$$(8.12) \qquad x_O(t) = a F(i\omega) e^{i\omega t}.$$

Thus the operator $\Re$ is applied to $a e^{i\omega t}$ by simply multiplying by the frequency response. Writing the complex number (8.11) in the form $R(\omega) e^{i\theta(\omega)}$, $R(\omega)$ indicates the amplitude distortion and $\theta(\omega)$ the phase shift. Interpreting this number in a complex plane and allowing ω to range from 0 to $+\infty$ yields the frequency response diagram. In case $g(s) = 1$ it agrees with the reciprocal response diagram of the polynomial $f(s)$ which was mentioned in sec. 5. For an nth order stable transfer unit it must accordingly pass through n quadrants in the negative sense (*cf.* fig. 8.1). It is therefore possible to characterize the stability of the transfer unit with the help of the frequency response diagram (*cf.* Theorem 5.1). Since, as we mentioned, the frequency response function is the response of the transfer unit to a very special input we see that the stability behavior is already completely characterized by a single special response. Of course this is typically a *linear* property: the superposition principle applies here and if an arbitrary disturbance is represented by a Fourier series, resp. a Fourier integral, then the corresponding output can be obtained accordingly by superposition.

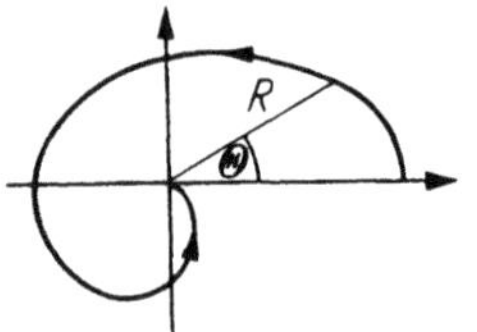

Fig. 8.1. Frequency response (unstable)

If the output equation is given in matrix form

$$(8.13) \qquad \dot{x}_O = A x_O + B x_I$$

then (8.7) corresponds to the equation

$$(s E - A) \bar{x}_O = B \bar{x}_I$$

and the transfer function is meaningfully defined as the determinant of the transfer matrix $(s E - A)^I B$. Its poles are the characteristic roots of A.

Example. The transfer function of the transfer unit described by

$$\dot{x}_0 + x_0 = \int_0^t x_I(\tau)\, d\tau$$

is

$$F(s) = 1/s(s+1).$$

The unit is thus unstable. The characterization for stable behavior given at the end of sec. 4 is violated here: the bounded input (8.10) results in the unbounded output

$$x_0(t) = t - 1 + e^{-t}.$$

By applying a transfer unit to the output of another transfer unit we generate a *transfer system* which can, of course, be considered as a single transfer unit of higher order. Conversely, many transfer units can in practice be decomposed into component units of lower order, often even in several ways. Such a decomposition is in general carried out less according to mathematical than physical principles. From a mathematical point of view however we are interested in the question: "What can be said about the stability behavior of the composite system if the behavior of the individual components is known?"

There are three basic types of compositions, or connections, of transfer units which we must study.

1. Connection in series (fig. 8.2). The output of the first unit is the input of the second. The defining equations are, according (8.2),

$$f_1(D)\,x_{01} = g_1(D)\,x_{I1}; \quad f_2(D)\,x_{02} = g_2(D)\,x_{I2}; \quad x_{01} = x_{I2}.$$

If we introduce the transfer functions this becomes

$$\bar{x}_{01} = F_1(s)\,\bar{x}_{I1}; \quad \bar{x}_{02} = F_2(s)\,\bar{x}_{I2} = F_2(s)\,\bar{x}_{01} = F_2(s)\,F_1(s)\,\bar{x}_{I1}.$$

Thus the transfer function of the combinations of the two unit is

(8.14) $F(s) = F_1(s)\,F_2(s)$,

the product of the transfer functions of the component units.

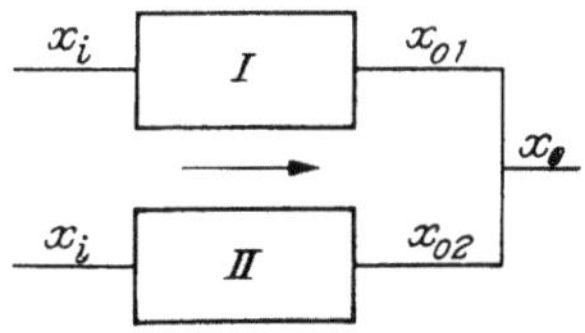

Fig. 8.2. Connection in series

2. Parallel connection (fig. 8.3). The two units have the same input. The outputs are added (superimposed). In this case we have the equations

$$f_1(D)\,x_{01} = g_1(D)\,x_I, \quad f_2(D)\,x_{02} = g_2(D)\,x_I;$$

$$x_0 = x_{01} + x_{02}$$

Fig. 8.3. Parallel connection

and we find as above that

$$\bar{x}_O = \big(F_1(s) + F_2(s)\big)\bar{x}_I,$$

which yields the theorem: In a parallel connection the transfer function of the total system equals the sum of the transfer functions of the components.

3. Feedback connection (fig. 8.4). The units are connected in series; but in such a way that the output of the second unit and the input of the first unit are superimposed; this superposition is usually considered as a subtraction. The equations are

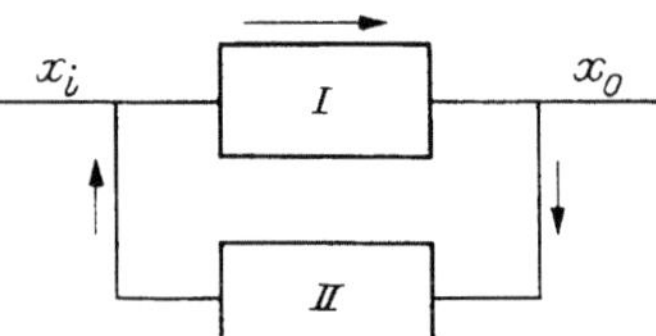

Fig. 8.4. Closed loop (feedback)

$$f_1(D)\,x_{O1} = g_1(D)\,(x_I - x_{O2}),$$

$$f_2(D)\,x_{O2} = g_2(D)\,x_{O1}.$$

Taking Laplace transforms we obtain after a short calculation

$$(8.15) \qquad \bar{x}_{O1} = \frac{F_1(s)}{1 + F_1(s)\,F_2(s)}\,\bar{x}_I; \qquad \bar{x}_{O2} = \frac{F_1(s)\,F_2(s)}{1 + F_1(s)\,F_2(s)}\,\bar{x}_I.$$

Here we obtain two transfer functions which characterize the relation between x_I and x_{O1}, resp. x_{O2}.

The main area of applications for systems with feedback is that of automatic controls. The part of the system described by the first equation, resp. by the transfer function $F_1(s)$, is called the *controlled system* (plant). x_{O1} is the quantity to be controlled, *i.e.* to be kept free from undesired disturbances. The equation for x_{O2} describes the action of the controller itself and the first equation in (8.15) shows what effect the control has. The controlled quantity $\bar{x}_{O1}$ is the output of (8.15) in place of the uncontrolled quantity $F_1(s)\,\bar{x}_I$. Controlled system and controller feedback form a closed loop; controlled system and controller in series, that is before x_{O2} and x_{I1} are superimposed, form an open loop (fig. 8.5).

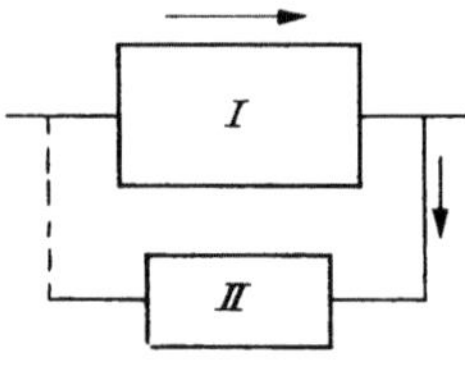

Fig. 8.5. Open loop

Formulas (8.13) and (8.14) yield immediately: for series and parallel arrangements the combined system is stable if all the components are stable. However, this condition is in general not necessary; it may happen that a pole of a transfer function is cancelled out so that the total system may be stable even if it contains unstable components (see also sec. 11). For a feedback the situation is much more complicated. The stability depends on the position of the zeros of the function $1 + F_1(s)\,F_2(s)$ for which there is no general rule.

Example. Let
$$F_1(s) = 1/(s + 1)(s + 2), \qquad F_2(s) = k/(s + 3),$$
so that the numerator of $1 + F_1(s) F_2(s)$ is
$$s^3 + 6s^2 + 11s + 6 + k.$$

In this case condition (6.10) says $66 - (6 + k) > 0$ and we see that stability obtains only for $k < 60$. If the parameter k which can physically be interpreted as an amplification factor (gain) (*cf.* sec. 10) is larger than 60 we have instability.

9. An Example

A classical example for the considerations of secs. 4 and 8 is the speed control by means of a centrifugal pendulum (fig. 9.1). Hurwitz started from such a concrete system when he developed his determinant criterion. We shall not discuss the physical process but limit ourselves to writing down the equations symbolized in the block diagram, fig. 9.2.

Controlled system
$$\dot{x} + \varepsilon a x = - b y,$$

Measuring unit
$$\alpha \ddot{z} + \beta \dot{z} + \gamma z = \delta x,$$

Governor slide
$$u = c_1 z - c_2 y,$$

Motor
$$y = \int^{t} c_3 u \, dt.$$

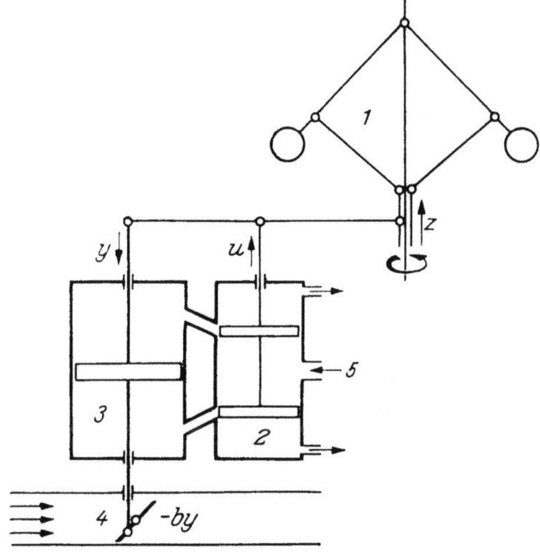
Fig. 9.1. The classical governor
1 Centrifugal pendulum; *2* Steering piston; *3* Actuator; *4* Steam valve; *5* Oil supply

The right side of the first equation should actually be $b(x_I - y)$ where x_I represents an external force (compulsive force). However, for stability considerations the input variable x_I may be neglected. The constants are all positive, $\varepsilon = +1, 0,$ or -1, depending on the type of controlled system. The system involves a mechanical

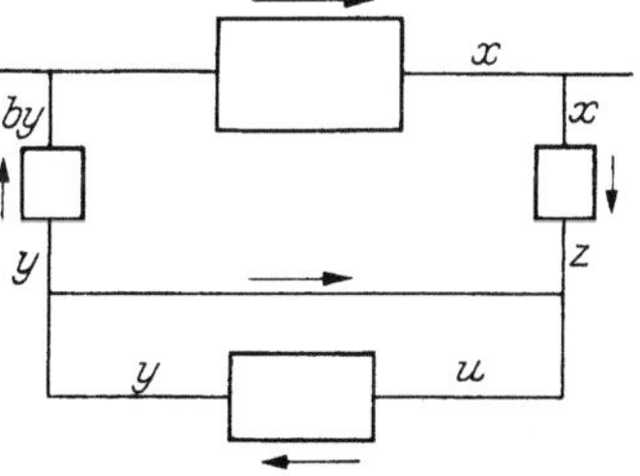
Fig. 9.2. Block diagram of Fig. 9.1

feedback which is indicated in the third equation by the term $-c_2 y$. The transfer function of the closed loop described by the last two equations is

$$F_3 = \frac{c_1 c_3}{s + c_2 c_3}.$$

The transfer functions of the first and second equations are, resp.

$$F_1 = \frac{b}{s + \varepsilon a}, \qquad F_2 = \frac{\delta}{x s^2 + \beta s + \gamma}.$$

The transfer function for the controller proper is therefore $F_2 F_3$, and according to sec. 8, the closed loop consisting of controlled system and controller feedback has for its transfer function

$$F(s) = \frac{F_1(s)}{1 + F_1(s) F_2(s) F_3(s)}.$$

The poles of this function are the zeros of the polynomial

$$(9.1) \qquad (s + \varepsilon a)(x s^2 + \beta s + \gamma)(s + c_2 c_3) + b c_1 c_3 \delta,$$

which we will discuss next. The coefficients of this polynomial are

$$a_0 = \alpha, \quad a_1 = \varepsilon a \alpha + \beta + \alpha c_2 c_3, \quad a_2 = (\beta + \varepsilon a x) c_2 c_3 + \gamma + \varepsilon a \beta,$$

$$a_3 = \varepsilon a \gamma + c_2 c_3 (\gamma + \varepsilon a \beta), \quad a_4 = \varepsilon a c_2 c_3 \gamma + b c_1 c_3 \delta.$$

In order to have stability these must be positive which is clearly the case for $\varepsilon = 0$ and $\varepsilon = 1$; in addition (6.10) must be satisfied.

Frequently the constants in a concrete system, for instance in a control circuit are not given as fixed numbers but rather as lying in a certain interval, and we must find out in which intervals stability is assured. For example, let c_1 and c_2 be variable in (9.1), let the other constants be fixed, and let $\varepsilon = +1$. Then the third Hurwitz determinant is an expression of the form

$$H_3 = h_0 c_2^3 + h_1 c_2^2 + h_2 c_2 + h_3 + g_0 c_1 + g_1 c_1 c_2 + g_2 c_1 c_2^2$$

and a_4 has the form

$$a_4 = k_1 c_1 + k_2 c_2.$$

The condition

$$H_3 > 0, \quad a_4 > 0$$

determines a well-defined domain in a (c_1, c_2)-plane. The coordinates of points in this domain are pairs of values c_1, c_2 belonging to a stable system. The domain is bounded by the line $a_4 = 0$ and the curve $H_3 = 0$; on the curve $H_3 = 0$, c_1 is a rational function of c_2.

10. The Nyquist Criterion

The application of stability criteria in practice frequently meets basic difficulties stemming from the fact that the defining differential equations are not exactly known. These equations are determined on the basis of general physical laws; however frequently idealizations and simplifications are necessary whose full impact is sometimes not easily assessed. In contrast, the response diagram criteria have the advantage that they can be based directly on observations and measurements: The frequency response function of the transfer unit may be determined with the help of (8.12), by observing $x_0(t)$ for a periodic input of varying frequency and then applying Theorem 5.1 to the reciprocal response diagram. To test for the stability of a closed control loop, however, we cannot proceed in this manner: for obvious reasons the system can be operated only if it is known to be stable. It is thus impossible to measure the transfer function of a closed loop in order to test its stability. The so-called *Nyquist criterion*, on the other hand, allows us to use the transfer function of the open loop to make inferences about the stability of the closed loop. And the open circuit is much more accessible to measurements.

Henceforth let $F(s)$ be the transfer function of the open loop which can be determined in accordance with sec. 8 from the transfer function of the individual components. As in sec. 5 we consider the frequency response diagram, *i.e.* the curve

$$F(i\,\omega) = u(\omega) + i\,v(\omega) = R(\omega)\,e^{i\theta(\omega)}$$

in the complex plane and observe that this curve can be considered as the image of the contour denoted by C in sec. 5 (semi-circle and imaginary axis) under the transformation $z = F(s)$. Since the response diagram gives us directly the increment of the logarithm of $F(s)$ along this contour we conclude: As s moves around the right half plane along the path C and in the process encircles n_1 poles and n_2 zeros of $F(s)$, the response diagram circles the origin n_1 times in one direction and n_2 times in the opposite direction, all together $n_1 - n_2$ times. Now at each zero of the function $1 + F(s)$ which we actually need to consider, the function $F(s)$ takes on the value -1. Both functions have the same poles. We now apply our rule of revolutions to the function $1 + F$ and obtain: as s describes the curve C the frequency response diagram circles the point $-1 + 0i$ exactly k times, where $k = n_1 - n_2$ is the number of poles less the number of zeros of $1 + F$ in the right half plane.

If $F(s)$ has no poles in the right half plane (the open loop is then stable) we have $n_1 = 0$. For the closed loop to be stable it is necessary that $n_2 = 0$ and accordingly $k = 0$. The Nyquist criterion follows:

3*

Theorem 10.1. A closed loop is stable if the open loop is stable and if its response diagram traversed in the direction of increasing ω has the "critical" point $(-1, 0)$ on its left.

In case $n_1 = 0$ this rule has an even simpler proof as follows: $F(s)$ must never equal -1 in the right half-plane, *i.e.* the image of the half-plane $\operatorname{Re} s > 0$ under the transformation $z = F(s)$ must not contain the point $z = -1$. The frequency response diagram is the image of the imaginary axis which, directed upwards, has the half-plane $\operatorname{Re} s > 0$ on its right. The point $z = -1$ does not lie in the image of this half-plane and hence cannot lie to the right of the image of the imaginary axis, since the mapping is conformal (fig. 10.1, 10.2).

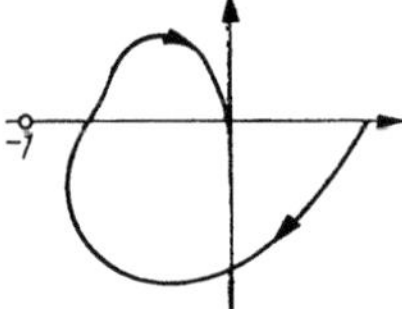

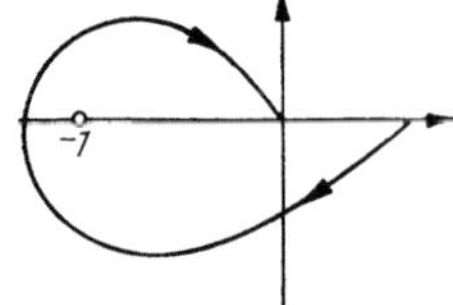

Fig. 10.1. Nyquist diagram (stable case)   Fig. 10.2. Nyquist diagram (unstable case)

If the transfer function has poles on the imaginary axis the criterion fails in the present form[1]).

The condition "the critical point lies to the left of the response diagram" can also be formulated as follows: the response diagram cuts the real interval $(-\infty, -1)$ either not at all or an even number of times. In fact it meets the interval as often from below as from above.

Let us now consider a special case. Let the transfer function of the open loop be $kf(s)$, where k is a positive parameter and $f(s)$ a polynomial. If the open circuit is stable and the degree of $f(s)$ is larger than 2 then according to sec. 5, the reciprocal Leonhard response diagram of $f(s)$ must at least pass through the fourth, third and second quadrants. Thus it cuts the negative real axis and it is certainly possible to choose k so large that the response diagram of kF intersects the interval $(-\infty, -1)$, *i.e.* the closed loop is unstable. Usually k is an amplification factor and the value of k at which the system ceases to be stable is the critical amplification factor. If the degree of the denominator is larger than 2 we know therefore that a critical amplification factor exists (see also the second example in sec. 8). We further note that $(-1, 0)$ is the critical point only if the equations of the feedback system have been written as in sec. 8, *i.e.* if x_{02} is *subtracted* in the first equation. Occasionally these equations are written with a $+$ sign; in this case $(+1, 0)$ is the critical point and the criterion must be reformulated accordingly.

---

[1]) *cf.* for instance LEHNIGK [3], SOLODOVNIKOV [1].

11. The Boundary of Stability

The coefficients a_i of the characteristic polynomial of the differential equation are, in systems with a physical interpretation, functions of certain parameters accessible to measurements, for example masses, spring tensions, capacities, etc. The obvious question arises how the stability of the physical system is influenced if the parameters are changed. In attacking this question we first of all ignore the physical interpretation entirely and consider the coefficients a_i as single-valued continuous functions of parameters $\alpha_1, \alpha_2, \ldots, \alpha_k$ which can be considered as the coordinates of a point in a k-dimensional space. The case in which the coefficients themselves are the parameters is included. To each point (α_j) of the parameter space (or at least of a certain domain in this space) there corresponds then a characteristic polynomial resp. a differential equation, of the type under consideration. The totality of all points corresponding to systems with an asymptotically stable equilibrium forms a well-defined domain, the *domain of stability* in the parameter space. According to sec. 6 this domain is characterized by the inequalities

$$(11.1) \qquad a_i > 0, \; H_i > 0 \qquad (i = 1, 2, \ldots, n)$$

(possibly by a subcollection of these inequalities). The a_i and H_i must be considered as functions of the parameters α_j. If the parameters are changed so that the system passes from a stable to an unstable state then at least one of the *stable* roots (*i.e.* one with a negative real part) must be replaced by an *unstable* root (with positive real part). Thus either a zero root or a pair of conjugate imaginary roots $\pm i\omega$ occur since the parameters vary continuously. For polynomials with two roots $+i\omega$, $-i\omega$, we have $H_{n-1} = 0$ in accordance with sec. 7; if one of the roots is zero then $a_n = 0$. It follows that the points of the parameter space which form the so-called *stability boundary i.e.* the boundary of the stability domain satisfy the equation

$$(11.2) \qquad a_n = 0, \quad H_{n-1} = 0$$

(*cf.* the example in sec. 9). Since H_{n-1} may also vanish at points of the domain of instability the stability boundary is in many cases only a subset of the manifold defined in (11.2).

Example. The system of equations

$$\dot{x}_1 = a x_1 + b x_2, \quad \dot{x}_2 = c x_1 + d x_2$$

is equivalent to a second order scalar equation with characteristic polynomial

$$(11.3) \qquad s^2 - (a + d) s + (a d - b c).$$

The domain of stability in the 4-dimensional parameter space a, b, c, d is given by

$$a + d < 0, \quad ad - bc > 0$$

and the stability boundary is defined by the equations

$$a + d = 0, \quad ad - bc = 0.$$

With each point P of the parameter space at which the polynomial $f(s)$ is defined we can associate the number $N = N(P)$ of the stable roots: that is, $N(P)$ indicates how many roots with a negative real part the polynomial belonging to P has. Obviously $0 \leq N \leq n$, and the domain of stability is characterized by $N = n$. The number N can change only at such points at which $f(s)$ has roots with vanishing real part. The set of all those points is called the *boundary of the D-decomposition* in the parameter space. In general it consists of at least $n - 1$ separate manifolds each of which separates two domains with different values of N. The stability boundary is a subset of the boundary of the D-decomposition, and an analytic expression for the boundary of the D-decomposition is obtained by setting $s = i\omega$ and considering the equations

$$(11.4) \qquad \mathrm{Re}\, f(i\omega) = 0, \quad \mathrm{Im}\, f(i\omega) = 0, \quad 0 \leq \omega < \infty$$

as a parametric representation of the boundary.

If the boundary of the D-decomposition and the value of $N(P)$ for a given fixed point P of the parameter space are known then $N(Q)$ can be obtained for each other point Q: We connect P with Q by means of a continuous path, mark the points of intersection of this path with the boundary and observe how N changes at these points of intersection. Of course we need to know in addition how N changes as the path goes through a segment of the boundary, whether by one or two units, increasing or decreasing.

In certain, simple but practically important cases the boundary and the rule of "crossing points" can be given explicitly, especially in case the polynomial $f(s)$ depends, linearly, on no more than two parameters. We then have

$$(11.5) \qquad\qquad f(s) = \alpha\, p(s) + \beta\, q(s) + r(s)$$

where p, q, r are polynomials. If we set $s = i\omega$ and separate the real and imaginary parts

$$p(i\omega) = p_1(\omega) + i\, p_2(\omega) \quad \text{etc.}$$

then it follows from (11.4) that

$$u_j(\omega) := \alpha\, p_j(\omega) + \beta\, q_j(\omega) + r_j(\omega) = 0 \quad (j = 1, 2)$$

and solving the equations we obtain the parametric representation

$$(11.6) \qquad \alpha = \frac{r_2 q_1 - r_1 q_2}{p_1 q_2 - p_2 q_1} ; \quad \beta = \frac{r_1 p_2 - r_2 p_1}{p_1 q_2 - p_2 q_1}$$

provided the equations are linearly independent, *i.e.* numerator and denominator do not vanish simultaneously. Since $u_1(\omega)$ is an even, $u_2(\omega)$ an odd function of ω, α and β are even functions. Each point on the boundary corresponds to two values of ω with opposite sign. If u_1 and u_2 are linearly dependent for a certain value of ω then numerator and denominator vanish together. To such a value ω_0 there corresponds a straight line $u_1(\omega_0) = 0$ in the (α, β)-plane, a so-called *singular line*. $\omega = 0$ is always one of these exceptional values. Similarly $\omega = \infty$ is exceptional and must be considered separately. We now imagine the boundary

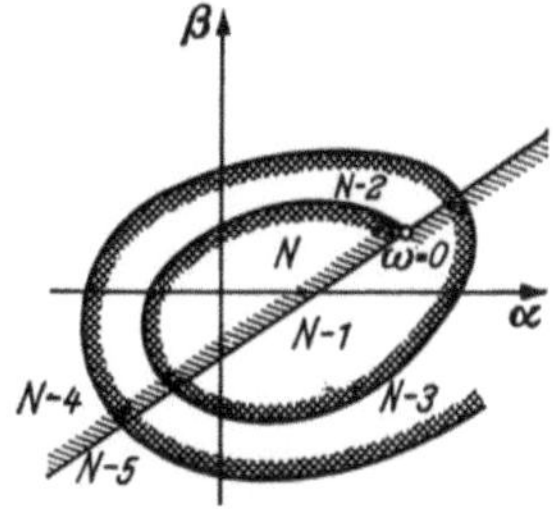

Fig. 11.1. *D*-Decomposition

graphed, inclusive of the singular lines, and the half-plane in which N assumes the larger value designated by shading. As $s = i\omega$ moves along the imaginary axis with increasing ω so that $f'(s) \neq 0$, the mapping of the s-plane onto the (u_1, u_2)-plane is locally single-valued in a neighborhood of the imaginary axis; similarly for the linear mapping of the (u_1, u_2)-plane onto the (α, β)-plane. We then have

$$\Delta := \frac{\partial(u_1, u_2)}{\partial(\alpha, \beta)} = p_1 q_2 - p_2 q_1 \neq 0 \quad \text{if} \quad \text{Re } s = 0.$$

Hence the half-plane Re $s < 0$ corresponds to the region left of the boundary in case $p_1 q_2 - p_2 q_1 > 0$. We therefore pass through the boundary in such a way that this determinant is positive and cross-hatch the left region since N increases by two each time we cross the imaginary axis, resp. the boundary.

A singular line caused by an exceptional value of ω, $0 < \omega < \infty$, intersects the curve (11.6). If the sign of Δ remains unchanged as we pass through the point of intersection, crossing the singular line has no influence on the value of N; the line may therefore be ignored. If the sign changes we cross-hatch the side of the line directed toward the hatch marks of the curve since only with this hatching the value of N remains the same after a complete rotation about the point of intersection. At the point of the curve corresponding to the value $\omega = 0$ the curve ends and forms a T with the singular line. This line is provided with simple shading. The same holds for $\omega = \infty$; the singular line is asymptotic to the curve here.

Example. The polynomial (9.1) is linear in α and β. Introducing the abbreviations

$$P(s) := (s + \varepsilon a)\,(s + c_2 c_3), \quad k := b c_1 c_3 \delta,$$

we have

$$f(s) = s^2 P(s)\,\alpha + s P(s)\,\beta + \gamma P(s) + k.$$

Let

$$P(i\omega) = \lambda(\omega) + i\mu(\omega).$$

Then using the notation introduced above,

$$p_1 = -\lambda\omega^2, \quad q_1 = -\mu\omega, \quad r_1 = \gamma\lambda,$$
$$p_2 = -\mu\omega^2, \quad q_2 = \lambda\omega, \quad r_2 = \gamma\mu + k,$$
$$\Delta = -(\lambda^2 + \mu^2)\,\omega^3,$$
$$\alpha = \frac{-\gamma(\lambda^2 + \mu^2) - k\mu}{\omega^2(\lambda^2 + \mu^2)}, \quad \beta = \frac{k\lambda}{\lambda^2 + \mu^2}.$$

The determinant Δ vanishes only for $\omega = 0$. This value has no significance here since it corresponds to the infinitely distant point $\alpha = \beta = \infty$. The value $\omega = \infty$ corresponds to the origin $\alpha = \beta = 0$. Since $\Delta < 0$, we put hatch marks on the right (as ω grows) of the boundary which is easy to draw if the constants are given numerically. The stability domain, *i.e.* the domain for $N = 4$ is known from the considerations of sec. 9.

Occasionally the D-decomposition with respect to a complex parameter plays a part. Let z be a complex parameter, $p(s)$ and $q(s)$ real polynomials, and

$$F(s, z) = p(s) - z q(s).$$

Setting $F(s, z) = 0$, z becomes a function of s,

$$(11.7) \qquad\qquad z = \frac{p(s)}{q(s)}.$$

As s moves along the imaginary axis z describes the image of this axis under the mapping (11.7). This however is the boundary of the D-decomposition in the plane of the complex parameter z. The half-plane $\mathrm{Re}\,s < 0$ is mapped onto the region to the right of this boundary. For the real polynomial $F(s, -1)$ to be a Hurwitz polynomial the point $z = -1$ must lie to the left of the boundary of the D-decomposition; for the zeros of the polynomial $F(s, -1)$ correspond to that point. In this way we have returned to the Nyquist criteron.

It is conceivable that a given system has no stability domain at all so that the D-decomposition does not furnish a domain in which N equals the degree of the characteristic polynomial. Such a system is called *structurally unstable*. We can define this important concept more precisely, while stressing its concrete significance, as follows. Let a transfer system be given which is composed of a number of transfer units of the first and second order. As characteristic polynomials of the components only the types

$$s, \quad cs + 1, \quad c's - 1, \quad as^2 + bs + 1,$$
$$a's^2 - b's + 1, \quad a''s^2 + 1$$

occur; the coefficients a, b, a', etc., are positive real numbers. If the characteristic polynomial is of the form $as^2 \pm bs - 1$ then it admits a real factorization $(\alpha s + 1)(\beta s - 1)$. Thus the corresponding transfer unit can be replaced by two transfer units of the first order connected in series. The transfer system is called *structurally unstable* if there exists no choice of the (by hypothesis positive) coefficients a, a', etc., which makes it stable. For example, a transfer system whose characteristic polynomial is

$$k + (cs - 1)(as^2 + 1) \quad \text{or} \quad k + k_1 s^2 (as^2 + bs + 1)$$

si structurally unstable whereas for the case

$$k + (cs - 1)(as^2 + bs + 1)$$

a domain of stability exists. A concrete arrangement which is structurally unstable occurs for instance in the regulation of the water level in a container[1]). The equation of the water level is

$$T_1 \dot{x}_1 = - x_3$$

and the one of the float is

$$T_2^2 \ddot{x}_2 + T_2' \dot{x}_2 + x_2 = k_2 x_1.$$

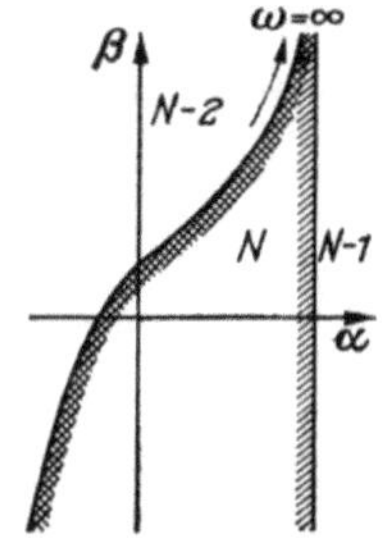

Fig. 11.2. *D*-Decomposition
($\omega = \infty$ is exceptional value)

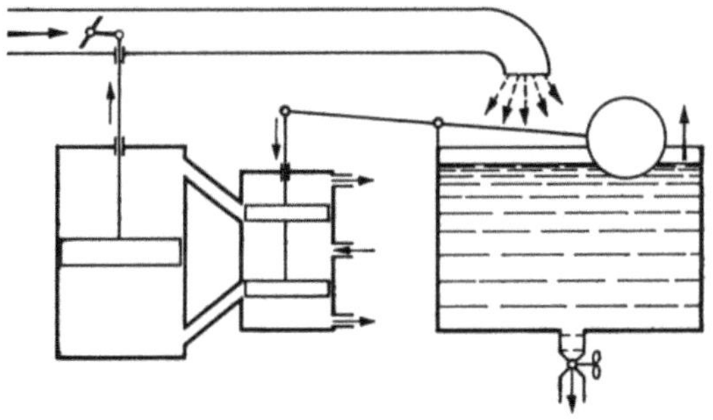

Fig. 11.3. Structurally unstable system (Aizerman [3])

[1]) AIZERMAN [3].

The equation of the servomotor is

$$T_3 \dot{x}_3 = x_2$$

and hence the characteristic polynomial is

$$T_1 T_3 s^2 (T_2^2 s^2 + T_2' s + 1) + k_2 .$$

It appears that structural instability is primarily ascribable to unstable terms (with $c's - 1$). If such terms appear their influence must be compensated for by a certain minimal number of stable terms. A number of isolated results have been published relating to simple systems[1]. They give the minimum and maximum number of terms permissible for the components of the various types. Since in practice the total number of components may not exceed a certain bound and the maximum number of units of a certain construction is prescribed by structural necessities, the criteria mentioned have considerable practical importance.

A general mathematical formulation of the problem of structural stability of linear systems is as follows: let

$$\dot{\boldsymbol{x}} = A \boldsymbol{x}$$

be given. Suppose we know for each element a_{ik} of the matrix whether it is positive, negative or zero. The system is called structurally unstable if for no choice of the values a_{ik}, holding the signs fixed, it is possible to obtain stability. We ask for which combination of signs structural instability obtains. The problem has never yet been dealt with in this general form.

The structural instability of linear transfer systems is unrelated to the concept of structural stability used for non-linear differential equations (see sec. 21).

12. Linear Differential Difference Equations [2]

Let a transfer system be given which consists of two transfer units connected in series. The individual equations are

$$f_1(D) x_{O1} = g_1(D) x_{I1} ,$$

$$f_2(D) x_{O2} = g_2(D) x_{I2} .$$

The output of the first unit is the input of the second unit but the connection is such that the second unit receives its input not immediately but

---

[1] *cf*. LEHNIGK [1] and especially LEHNIGK [3], where a detailed synopsis is given.

[2] *cf*. for instance BELLMAN and COOKE [1], HALANAY [1], PINNEY [1].

after a certain time, τ, the *delay period* or *time lag*. The relation connecting the two units is therefore

$$(12.1) \qquad x_{I2}(t) = x_{01}(t - \tau).$$

Passing as in sec. 8 to the transfer functions and observing that (in the notation of sec. 8)

$$(12.2) \qquad \mathfrak{L}\{x(t - \tau)\} = e^{-s\tau}\,\overline{x}(s),$$

we obtain — at first quite formally — the transfer function of the combined system

$$F_1(s)\,F_2(s)\,e^{-s\tau}.$$

The analogous treatment of two units with feedback yields the transfer function

$$(12.3) \qquad F(s) = \frac{F_1(s)}{1 + e^{-s\tau}\,F_1(s)\,F_2(s)}.$$

This is no longer a rational function as in sec. 8, but because of its exponential factor is a meromorphic function possessing in general infinitely many poles (see also sec. 13).

We next wish to obtain from the equations of the closed loop an equation for the original function $x_{01}(t)$, rather than its transform. For this purpose we eliminate $x_{02}(t)$, utilizing (12.1), and obtain a functional equation whose left side has the form

$$f_2(D)\,f_1(D)\,x_{01}(t) + g_2(D)\,g_1(D)\,x_{01}(t - \tau).$$

On the right side we have an expression which depends on x_{I1}. We are therefore dealing with a *differential difference equation*: It relates values of the unknown function and its derivatives at a point t with the corresponding values at the point $t - \tau$. The simplest case of such an equation is

$$(12.4) \qquad \dot{x}(t) + a\,x(t) + b\,x(t - \tau) + c\,\dot{x}(t - \tau) = 0.$$

In the general case where there are several delay periods, resp. delay terms, the homogeneous equation has the form

$$(12.5) \qquad \sum_{i=0}^{n} \sum_{k=0}^{m} a_{ik}\,x^{(i)}(t - \tau_k) = 0, \qquad 0 = \tau_0 < \tau_1 < \cdots < \tau_m,$$

which can of course be written in vector form as in (4.2):

$$(12.6) \qquad \sum_{k=0}^{m} \left(A_k\,x(t - \tau_k) + B_k\,\dot{x}(t - \tau_k)\right) = 0.$$

In addition there is an even more general form. Assuming that the process under investigation depends on all the values which the function $x(t)$ and its derivatives assume over a certain t-interval (and not just on their values at the discrete times $t, t - \tau_1, \ldots, t - \tau_m$) we obtain a functional equation whose left side is

$$\text{(12.7)} \qquad \sum_{i=0}^{n} \int_{0}^{\infty} x^{(i)}(t - \tau)\, d\alpha_i(\tau),$$

resp. in matrix form

$$\text{(12.8)} \qquad \int_{0}^{\infty} dG(\tau)\, \boldsymbol{x}(t - \tau) + \int_{0}^{\infty} dH(\tau)\, \dot{\boldsymbol{x}}(t - \tau).$$

These integrals are Stieltjes integrals having τ as their variable of integration and it is assumed that the functions $\alpha_i(\tau)$, resp. $g_{ij}(\tau)$ and $h_{ij}(\tau)$, are constant for $\tau \geq \tau_m$ so that we are actually integrating over finite intervals. If the functions $g_{ij}(\tau)$ and $h_{ij}(\tau)$ are step functions then (12.7) and (12.8) become (12.5), resp. (12.6).

The general solution of a differential difference equation depends on arbitrary functions and not only, as is the case with differential equations, on arbitrary constants. This is seen in the example of equation (12.4). To find the solution $x(t)$ for $t > t_0$, x and $\dot{x}$ must be known in the interval $t_0 - \tau \leq t < t_0$, or must be given a priori (these are the so-called *initial functions*):

$$x(t) = \lambda(t), \qquad \qquad t_0 - \tau \leq t < t_0,$$
$$\dot{x}(t) = \mu(t).$$

Here it is not necessary that $\mu(t) = \dot{\lambda}(t)$; for the functional equation is effective only for values of $t > t_0$. On the basis of the initial functions we obtain for $x(t)$ on the interval $t_0 < t < t_0 + \tau$ a linear nonhomogeneous differential equation

$$\dot{x} + ax + b\lambda(t - \tau) + c\mu(t - \tau) = 0.$$

The initial value $x(t_0)$ may also be predetermined or it may be given by the assumption that $x(t)$ connects at the point $t = t_0$ continuously with the initial function. In the interval $t_0 + \tau < t < t_0 + 2\tau$ a different linear differential equation is valid; it is obtained by entering in (12.4) the functional values from the first interval. In this manner we progress from interval to interval. By construction the solution is everywhere continuous and is continuously differentiable except at the junctions $t = t_0 + \tau, t_0 + 2\tau, \ldots$. At these places the derivative has in general a jump discontinuity, given that $c \neq 0$. In this case the equation is of the *neutral type*. If $c = 0$, *i.e.* if the highest derivative does not have a delayed argument then the solution is continuously differentiable at the junctions as well. The same is true in the general case: The solution is uniquely determined by the initial function, resp. the initial function vector, *i.e.* by the values of the components of $x(t)$ in the interval $t_0 - \tau_m < t \leq t_0$ and is continuously differentiable except at the junctions. If the equation is not neutral then the solution is continuously differentiable at the junctions as well. Of course, the initial function vector must then also have the appropriate properties.

If we wish to determine the general solution of the non-homogeneous equation

$$(12.9) \qquad \sum_{k=0}^{m} \left(A_k x(t - \tau_k) + B_k \dot{x}(t - \tau_k) \right) = z(t), \quad t > 0,$$

by means of the Laplace transformation we must take into consideration that (12.2) is valid only for functions which vanish for $t < 0$. For arbitrary functions we have the relations

$$(12.10) \qquad \mathfrak{L}\{x(t - \tau)\} = e^{-\tau s} \, \mathfrak{L}\{x(t)\} + e^{-\tau s} \int_{-\tau}^{0} e^{-su} x(u) \, du,$$

$$\mathfrak{L}\{\dot{x}(t - \tau)\} = e^{-\tau s} s \, \mathfrak{L}\{x(t)\} - e^{-\tau s} x(0) + e^{-\tau s} \int_{-\tau}^{0} e^{-su} \dot{x}(u) \, du.$$

The integrands $x(u)$ and $\dot{x}(u)$ on the right are the initial vectors. Applying (12.10) we obtain

$$\sum_{k=0}^{m} (e^{-\tau_k s} A_k \bar{x} + s \, e^{-\tau_k s} B_k \bar{x}) - \sum_{k=0}^{m} e^{-\tau_k s} B_k x(0)$$

$$(12.11) \qquad + \sum_{k=0}^{m} e^{-\tau_k s} \int_{-\tau_k}^{0} e^{-us} \left(A_k x(u) + B_k \dot{x}(u) \right) du = \bar{z},$$

$$\left[\sum_{k=0}^{m} e^{-\tau_k s} (A_k + B_k s) \right] \bar{x}(s) = \bar{z}(s) + \bar{g}(s).$$

The function $\bar{g}(s)$ combines the terms depending on the initial values and those depending on the initial functions. Now let $K(t)$ be the matrix defined by

$$\mathfrak{L}\{K(t)\} = \left[\sum_{k=0}^{m} e^{-\tau_k s} (A_k + B_k s) \right]^I$$

and $g(t)$, a function of time, be the function whose Laplace transform is $\bar{g}(s)$. Then the solution is given in closed form by inverting (12.11):

$$x(t) = K(t) * g(t) + K(t) * z(t)$$

$$(12.12)$$

$$[g(t) \neq 0 \text{ only for } -\tau_k \leq t \leq 0].$$

As usual, the star denotes convolution. Formula (12.12) is the exact analogue to (4.9). For the functional equation belonging to (12.8) we obtain a corresponding formula; $K(t)$ is the inverse transform of

$$\left[\int_0^\infty \left(dG(\tau) + s\,dH(\tau)\right) e^{-s\tau}\right]^I.$$

One can show by substitution that (12.12) is actually the "correct" solution.

13. Stability for Linear Differential Difference Equations with Constant Coefficients

The results of the last section show that the general solution of a differential difference equation depends on an arbitrary initial function, resp. an arbitrary initial function vector. So it cannot be interpreted, like the general solution of an nth order differential equation, in the Euclidean space R_n. Henceforth we shall denote the initial vector by $a(\theta)$, $-\tau_k \leq \theta \leq 0$, and assume for the sake of simplicity that $a(\theta)$ is continuous on its domain of definition and that its value for $\theta = 0$ equals the initial value of the solution: $a(0) = x(0)$. This assumption does not limit the generality of the argument significantly, yet it simplifies our analysis. The solution is considered as an element in a normed linear space with norm

$$(13.1) \qquad \| x(t) \| = \sup_\theta | x(t + \theta) |, \quad -\tau_k \leq \theta \leq 0.$$

Accordingly, the norm of the initial vector is defined by

$$(13.2) \qquad \| a(0) \| = \sup_\theta | a(\theta) |, \quad -\tau_k \leq \theta \leq 0.$$

The definitions of sec. 2 are now directly applicable. We simply replace all the absolute values by norms and note that the parameter a in sec. 2 is now the initial vector $a(\theta)$, also an element of a normed linear space. We only state one form of the definitions, and do so for the homogeneous equation (12.6), resp. (12.8), now written in the form

$$(13.3) \qquad L x = 0.$$

L denotes the linear differential difference operator acting on x.

Def. 13.1. The equilibrium of the equation (13.3) is called *stable* if there exists a function φ of class K (*cf.* sec. 2) such that

$$(13.4) \qquad \| x(t) \| \leq \varphi(\| a(0) \|), \quad t \geq 0.$$

Def. 13.2. The equilibrium of the equation (13.3) is called *attractive* if there exists a function σ of class L such that

$$\| x(t) \| \leq \sigma(t), \quad t \geq 0.$$

Def. 13.3. The equilibrium of the equation (13.3) is called *asymptotically stable* if it is simultaneously stable and attractive.

Since (13.3) is linear and autonomous we may without loss of generality set $t_0 = 0$ and need not incorporate a bound for the initial values into the definition, as was necessary in Def. 2.8. The function $g(t)$ in (12.12) is clearly bounded whenever the initial vector is bounded. (12.12) implies: If the norm of the matrix $K(t)$ is bounded then the equilibrium is stable. If the norm of the matrix $K(t)$ tends to zero as t increases then the equilibrium is attractive. Thus the stability behavior is determined by the properties of the matrix $K(t)$.

By definition, the Laplace transform of the matrix $K(t)$ is a matrix whose elements are meromorphic functions of s. Each such function can be written as a quotient of two *exponential polynomials* (linear combinations of terms of the form $s^p e^{-qs}$). The numbers p are integers between 0 and n. The numbers q are linear combinations of the delays $\tau_0, \tau_1, \ldots, \tau_m$. To calculate the function of time (inverse transform) belonging to such a quotient, that is to obtain an element of the matrix $K(t)$, we must expand the function into a partial fraction series analogous to (8.9). It contains summands of the form

$$(13.5) \qquad \frac{c_{i,r_i}}{(s - \lambda_i)^{r_i}}, \quad r_i \geq 0 \text{ an integer},$$

where λ_i ranges over all the zeros of the function

$$f(s) := \det \sum_{k=0}^{m} e^{-\tau_k s}(A_k + s B_k).$$

(13.5) is the transform of the function

$$(13.6) \qquad c_{i,r_i} t^{r_i - 1} e^{\lambda_i t}.$$

Hence the general element of the matrix $K(t)$ is an infinite series formed from the terms (13.6). This formal discussion must now be substantiated and in particular it must be shown that the partial fraction decomposition is admissible, in the given form, that the series of partial fractions converges uniformly, and that the transition to the t-functions can be made term-wise. For this purpose we must carefully study the zeros of functions of the type under consideration. We only assert here that the various steps can be justified and that the proofs can be found in the literature[1].

The elements of $K(t)$ can be represented as infinite series of terms of the form (13.6). The exponents r_i are bounded by a fixed finite number. The asymptotic behavior of these series for large t depends on the zeros λ_i of the characteristic function $f(s)$, as was the case for linear differential equations, more precisely on the signs of their real parts. If all the real

---

[1] *cf.* for instance PINNEY [1].

parts are smaller than a fixed negative number $-\delta$ then there is an exponential estimate for the series similar to the one in sec. 4,

$$\| K(t) \| \leq c_1 e^{-\mu t}, \quad 0 < \mu < \delta.$$

An immediate consequence is

Theorem 13.1. The equilibrium of a linear differential difference equation with constant coefficients is asymptotically stable if the real parts of the roots of the characteristic equation are smaller than a fixed negative number. The comparison function of Def. 13.2 is in this case an exponential function with negative exponent.

Clearly the equilibrium is unstable if at least one of the roots has a positive real part.

Given conditions under which the general solutions decrease exponentially, we have an estimate similar to the one in (4.16) for the expression

$$K(t) * z(t)$$

appearing in (12.12). The rule stated at the end of sec. 4 thus applies also for stable "transfer units with delay period (time lag)"; they also transform a bounded input into a bounded output. We are therefore justified in associating a transfer function with such a stable unit and the introduction of (12.3) which had only been stated formally, has now been justified. The denominator of (12.3) is an exponential polynomial of the form

$$(13.7) \qquad\qquad h(s) = p(s) + e^{-\tau s} q(s),$$

where p and q are polynomials and the degree of p is greater than the degree of q, assuming that we are dealing with the composition of two stable transfer units (sec. 8). It has been tried to derive for functions of the type (13.7) stability criteria similar to those in secs. 5 and 6 but these attempts have so far been successful only in very special cases. We mention without proof a relatively general theorem due to PONTRYAGIN [1]. Let $s = i\omega$ and

$$h(i\omega) = u(\omega) + iv(\omega).$$

A sufficient condition for $h(s)$ to have all of its zeros in the left half-plane is as follows: The real zeros of $u(\omega)$ and $v(\omega)$ alternate along the real axis and there exists at least one value $\omega = \omega_0$ such that

$$u'(\omega_0) v(\omega_0) - v'(\omega_0) u(\omega_0) < 0.$$

In comparing this theorem with the gap and position criterion we must note that $u(\omega)$ and $v(\omega)$ are complicated transcendental functions and not polynomials.

Since the characteristic function is not a polynomial and has infinitely many zeros the statement $\mathrm{Re}\,\lambda_i < 0$ does not imply the statement $\mathrm{Re}\,\lambda_i \leq -\delta < 0$. It can happen (however only in the neutral case, see above, sec. 12) that the characteristic equation has a set of zeros in the left half-plane which come arbitrarily close to the imaginary axis. For example, there exist characteristic functions with zeros satisfying an asymptotic formula of the kind

$$\mathrm{Re}\,\lambda_k = -\,a\,k^{-b} + o(k^{-b}), \quad a > 0, \quad b \geq 1,$$

and which are all double zeros. It can be shown[1]) that even in this case the equilibrium of the corresponding differential difference equation is attractive provided no zero with non-negative real part exists. But this equation admits a partial solution of the form $te^{\lambda_k t}$. Such a solution attains its maximum in absolute value at

$$t = t_k' = -\,\frac{1}{\mathrm{Re}\,\lambda_k}.$$

The maximum itself is $e^{-1}t_k' = O(k^b)$ and hence the general solution is unbounded. Thus the equilibrium is attractive but unstable.

14. Linear Difference Equations with Constant Coefficients

If we set the matrices B_k in (12.9) equal to zero we obtain from the differential difference equation a pure difference equation. The characteristic function involves exponential functions only and no powers. The general solution is derived in the same way and the stability discussion is similar. Still it is useful to treat at least the simplest case of a difference equation with a constant delay separately.

$$(14.1) \qquad\qquad \theta x = A x + z$$

is such an equation in vector form. In this formula θ denotes the shift operator, $\theta x(t) = x(t + 1)$. It is easy to put the more general equation

$$x(t + h) = A x(t) + z(t)$$

or the equation

$$\Delta x = B x + y, \quad \Delta x := \frac{1}{h}\left(x(t + h) - x(t)\right)$$

written as a difference equation, into the form (14.1). Generally we consider the equation (14.1) and correspondingly its solution only for discrete values $t_0, t_0 + 1, t_0 + 2, \ldots$ of the independent variable and we denote these frequently by use of the letter n which traditionally suggests an integer. A difference equation also assigns an output $x(n)$ to an input $z(n)$. In practice difference equations apply above all to the treatment of pulse systems (scanning systems) whose peculiarity it is that they

---

[1]) HAHN [3].

do not work continuously but that the incoming pulse is operative at regular intervals for a short while, quiescent in between.

Such a system (see fig. 14.1) has in addition to a continuously operating part which may or may not be linear, a so-called *pulse unit* whose action can for instance be described by an equation of the type

$$(14.2) \qquad x_O(t) = \begin{cases} k\,x_I(n\,T), & n\,T \le t < (n + \gamma)\,T, \\ 0, & (n + \gamma)\,T \le t < (n + 1)\,T, \end{cases}$$

$$\gamma \text{ fixed, } 0 < \gamma < 1.$$

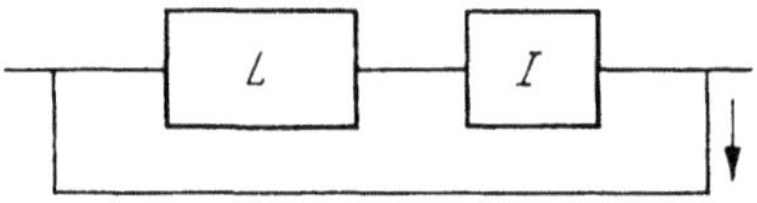

Fig. 14.1. Pulse system. L linear part, I pulse unit

The output variable thus depends only on the value of the input at the times $t = n\,T$, the *scanning times*. Another possibility is

$$(14.3) \qquad x_O(t) = \begin{cases} k\,\operatorname{sgn}\,x_I(n\,T), & n\,T \le t < (n + \gamma)\,T, \\ 0, & (n + \gamma)\,T \le t < (n + 1)\,T, \end{cases}$$

where this time the duration of the impulse γT is proportional to the quantity $|x_I(n\,T)|$. A closed loop with a linear component and a pulse component is described by means of a difference equation. This can be seen as follows.

We may assume that the linear part is described by an rth order scalar differential equation whose form has no further significance. We only need the step response $u_0(t)$ (see sec. 8) and the fundamental solutions $u_1(t), \ldots, u_r(t)$ of the homogeneous equation. The output variable corresponding to an input step c [*cf.* (8.10)] is c times the step response plus a quantity which depends on the initial values at time $t = t_0$, and it can be written in the form (*cf.* sec. 4)

$$x(t) = c\,u_0(t - t_0) + \sum_{i=1}^{r} v_i(t - t_0)\,x^{(i-1)}(t_0).$$

The functions $v_i(t)$ are certain linear combinations of the $u_i(t)$, $i = 1, 2, \ldots, n$. We set up the last equation for the times $t, t - T, t - 2T, \ldots, \ldots, t - (r - 1)\,T$, set $t = t_0$, and consider the resulting system of equations as determining for the initial values $x^{(i-1)}(t_0)$, $i = 1, 2, \ldots, r$. We substitute these initial values, which are linear combinations of the functional values $x(t_0), x(t_0 - T), \ldots, x(t_0 - (r - 1)\,T)$ and the constant c, in the expression for the solution $x(t)$ given above and obtain

$$x(t) = c\,\psi_0(t - t_0) + \sum_{i=1}^{r} \psi_i(t - t_0)\,x\big(_0 - (i - 1)\,T\big).$$

The functions $\psi_i(t)$ are again linear combinations of the $u_i(t)$. In a closed loop the jump c equals the value of the output of the pulse unit with $x(t)$ as input. c can hence be computed from (14.2). In a closed loop we have therefore

$$x(t) = k\,x(n\,T)\,\psi_0(t - t_0) + \sum_{i=1}^{r} \psi_i(t - t_0)\,x\big((n - i + 1)\,T\big),$$

$$n\,T \le t < (n + \gamma)\,T,$$

$$x(t) = \sum_{i=1}^{r} \psi_i(t - t_0)\,x\big((n - i + 1 + \gamma)\,T\big),$$

$$(n + \gamma)\,T \le t < (n + 1)\,T,$$

and because of the continuity of x the first equation is valid for $t = (n + \gamma)\,T$ and the second for $t = (n + 1)\,T$, so that for $t_0 = n\,T$ we have

$$x\big((n + \gamma)\,T\big) = k\,x(n\,T)\,\psi_0(\gamma\,T) + \sum_{i=1}^{r} \psi_i(\gamma\,T)\,x\big((n - i + 1)\,T\big),$$

$$x\big((n + 1)\,T\big) = \sum_{i=1}^{r} \psi_i(T)\,x\big((n - i + 1 + \gamma)\,T\big).$$

The first equation can be set up for $n,\ n - 1, \ldots, n - r + 1$ and the quantities $x\big((n - j + \gamma)\,T\big)$, $j = 0, 1, \ldots, r - 1$ can be eliminated from the second equation. In this manner a linear relation is obtained between the quantities

$$x\big((n + 1)\,T\big), \quad x(n\,T), \ldots, \quad x\big((n - r + 1)\,T\big),$$

i.e. a difference equation. The coefficients depend on $\psi_i(T)$ and $\psi_i(\gamma\,T)$; in the case (14.2) they are constant. For the impulse width modulation (14.3) a non-linear difference equation is obtained in a similar manner, since in this case the values $\psi_i(T)$ depend on the values $x\big((n - j)\,T\big)$ also.

The scalar form of (14.1) is

$$(14.4) \qquad a_0\,\theta^m x + a_1\,\theta^{m-1} x + \cdots + a_{m-1}\,\theta x + a_m\,x = f(t).$$

The equivalence of the two forms can be established as in sec. 4. (14.1) yields

$$\boldsymbol{x}(n + r) = \theta^n \boldsymbol{x}(r) = A^n \boldsymbol{x}(r) + A^{n-1} \boldsymbol{z}(r) + A^{n-2} \theta \boldsymbol{z}(r)$$

$$+ A^{n-3} \theta^2 \boldsymbol{z}(r) + \cdots + A\,\theta^{n-2} \boldsymbol{z}(r) + \theta^{n-1} \boldsymbol{z}(r),$$

$$(14.5) \qquad \boldsymbol{x}(n + r) = A^n \boldsymbol{x}(r) + \sum_{l=0}^{n-1} A^l\,\theta^{n-l-1} \boldsymbol{z}(r).$$

Hence we can immediately write down the solution in closed form. We investigate the asymptotic behavior of the solutions of the homogeneous equation in the same way as in the case of differential equations. We

4*

make the substitution $x = Sy$ [see (4.11)] in such a manner that the difference equation

$$(14.6) \qquad\qquad \theta y = J y$$

involves a matrix in Jordan canonical form. It is established immediately that for each block

$$
\begin{array}{ccccccc}
\varrho & 1 & 0 & 0 & \cdots & 0 \\
0 & \varrho & 1 & 0 & \cdots & 0 \\
0 & 0 & \varrho & 1 & \cdots & 0 \\
& & \cdots\cdots\cdots & & \\
0 & 0 & 0 & 0 & \cdots & \varrho
\end{array}
$$

with m rows there are the m linearly independent solutions

$$\mathrm{col}\,(\varrho^t, 0, 0, 0, \ldots, 0),$$

$$\mathrm{col}\,(t\varrho^{t-1}, \varrho^t, 0, 0, \ldots, 0),$$

$$\mathrm{col}\left(\binom{t}{2}\varrho^{t-2}, t\varrho^{t-1}, \varrho^t, 0, \ldots, 0\right),$$

$$\cdots\cdots\cdots\cdots\cdots\cdots\cdots\cdots\cdots$$

$$\mathrm{col}\left(\binom{t}{m-1}\varrho^{t-m+1}, \binom{t}{m-2}\varrho^{t-m+2}, \ldots, t\varrho^{t-1}, \varrho^t\right).$$

Inverting the linear transformation $x = Sy$ we recognize that the general solution, denoted as in sec. 2 by $p(t, x_0, t_0)$, resp. by $p(n+r, x_0, r)$, is a linear combination of terms of the form $t^k \varrho^t$ resp. $n^k \varrho^n$ with constant (independent of t resp. n) coefficients. The numbers ϱ are the roots of the characteristic equation

$$\det(A - \varrho E) = 0.$$

This representation of the solution allows us to read off the stability behavior; the concepts *stable* and *attractive* are defined in the same way as for linear differential equations (sec. 4) or for linear differential difference equations (sec. 13).

Theorem 14.1. The equilibrium of the difference equation

$$\theta x = A x$$

is asymptotically stable if and only if all the characteristic roots of the matrix A are smaller than one in absolute value. It is stable but not asymptotically stable if in addition to characteristic roots of absolute value less than one there exist characteristic roots of absolute value equal to one which belong to elementary divisors of the first order. The equilibrium is unstable if there exists a characteristic root with absolute value greater than one or if a characteristic root of absolute value one belongs to an elementary divisor of higher order.

Theorem 14.2. The equilibrium of the homogeneous difference equation (14.4) is asymptotically stable if all the zeros of the characteristic equation

$$a_0 \varrho^m + a_1 \varrho^{m-1} + \cdots + a_{m-1} \varrho + a_m = 0$$

are smaller than one in absolute value. It is stable but not asymptotically stable if all the roots have absolute value less than or equal one but if there are only simple roots of absolute value one. It is unstable if at least one root has absolute value greater than one or if there exist multiple roots of absolute value one.

If the equilibrium of the homogeneous equation is asymptotically stable then all the solutions of the non-homogeneous equation (14.1) are bounded provided the disturbance z is bounded. This follows from (14.5), for

$$|x(n + r)| \le |A^n x(r)| + \sum_{l=0}^{n-1} |A^l \theta^{n-l-1} z(r)|.$$

For bounded z, $|\theta^j z(r)|$ is smaller than a constant c so that

$$|x(n + r)| \le \|A^n\| \, |x(r)| + c \sum_{l=0}^{n-1} \|A^l\|.$$

The series on the right is bounded if the absolute values of the characteristic roots of A are smaller than one:

$$|x(n + r)| \le \|A^n\| \, |x(r)| + c \, \frac{1}{1 - \|A\|}.$$

15. Linear Operators

The similarity of the stability theorems of secs. 4, 13 and 14 is based on the fact that the three types of equations (4.4), (12.6), and (14.1), resp. the corresponding scalar equations, have similar interpretations. We have already pointed out such an interpretation: The equation assigns (in the stable case) to an input $x_I(t)$, resp. $z(t)$, a well defined output. This was already expressed in (3.1). The solution of the homogeneous equation can also be interpreted as a mapping of the space of initial values to itself: The solution maps the initial point a to a point $p(t, a)$, depending on t, where without loss of generality we can set the initial instant $t_0 = 0$ since we are dealing with autonomous equations. In the case of a differential, resp. difference, equation a and p are points in n-dimensional Euclidean space. In the case of a differential difference equation $a = a(\theta)$ is a point of the following vector space: The components of each vector are continuous functions on the interval $-\tau_k \le \theta \le 0$. The solution belongs to the same space and its variable ranges over the interval $(t - \tau_k, t)$. For this interpretation the solution of the differential difference equation must always be thought of as defined on an interval of length

τ_k. The relation between solution and initial value is expressed by the notation

$$(15.1) \qquad p(t, a) = \mathfrak{F}_t a.$$

The exact analytic expression for the operator which acts on the space of initial values and which depends on the parameter t (again $t_0 = 0$) is

$$\mathfrak{F}_t a = e^{At} a \quad \text{for differential equations,}$$

$$\mathfrak{F}_t a = A^t a \quad \text{for difference equations.}$$

For differential difference equations $\mathfrak{F}_t$ can either be represented by means of the matrix K in (12.12) or one can utilize the progressive continuation from interval to interval which was discussed in conjunction with (12.4). For differential difference equations whose left side is (12.8), for instance, we define an operator

$$\psi(v(t)) := \int_{-\tau_k}^{0} dG(\theta)\, v(t - \theta) + \int_{-\tau_k}^{0} dH(\theta)\, \dot{v}(t - \theta)$$

and write the solution in the form

$$\mathfrak{F}_t a(\theta) = \begin{cases} a(t + \theta), & t + \theta \leq 0, \\[2mm] a(0) + \int_0^{t+\theta} \left[\psi(\mathfrak{F}_u a(\theta))\right] du, & \theta + t > 0,\ -\tau_k \leq \theta \leq 0. \end{cases}$$

An easy calculation shows that this expression actually satisfies the differential difference equation.

The norm of a, resp. p, is the Euclidean vector norm in R_n. In dealing with differential difference equations the norm (13.1) is usually used,

$$\|a(t)\| := \max |a(t + \theta)|, \qquad -\tau_k \leq \theta \leq 0,$$

but occasionally other norms are suitable, for instance

$$(15.2) \qquad \left[\int_0^{\tau_k} |a(t - u)|^2\, du\right]^{1/2}.$$

To define the norm of the linear operator $\mathfrak{F}_t$ we use the norm of a and set

$$(15.3) \qquad \|\mathfrak{F}_t\| = \sup_{a \neq 0} \frac{\|\mathfrak{F}_t a(\theta)\|}{\|a\|}.$$

If a belongs to a Euclidean space this is the usual matrix norm. The norm $\|\mathfrak{F}_t\|$ depends of course on the parameter t. If it is finite then $\mathfrak{F}_t$ is called bounded.

Equation (15.1) implies

$$\|p(t, a)\| \leq \|\mathfrak{F}_t\| \cdot \|a\|$$

and we recognize:

1. The equilibrium of the functional equation under consideration is stable if $\|\mathfrak{F}_t\|$ is uniformly bounded with respect to t,

$$\|\mathfrak{F}_t\| \leq c.$$

2. The equilibrium is attractive if $\|\mathfrak{F}_t\|$ tends to zero with increasing t.

Thus the stability problem reduces to finding the norm of the linear operator $\mathfrak{F}_t$ or at least to estimating it. Since in the cases we considered this norm was determined by the roots of the characteristic equation the formal similarity of the stability theorems is not surprising.

Chapter III

The Equilibrium of Autonomous Differential Equations

16. Fundamental Concepts, Definitions and Notations [1])

As we remarked earlier the great majority of all statements about stability is related to motions definable by a differential equation

$$\dot{x} = f(x, t), \tag{16.1}$$

where $x = x(t)$ is an n-dimensional vector. It is advisable to introduce the following notations. Let K_h be the set of points x such that $|x| < h$, *i.e.* an open spherical neighborhood of the origin in R_n. The corresponding closed set $|x| \leq h$, is denoted by $\bar{K}_h$. The symbol K_{h,t_0} denotes the "half cylinder" of points (x, t) of a $(n + 1)$-dimensional space, $R_n \times I$, defined by

$$|x| < h, \qquad t \geq t_0.$$

Again $\bar{K}_{h,t_0}$ is the closure. We make the following assumptions on the differential equation:

1) It admits the trivial solution $x = 0$, *i.e.* $f(0, t) = 0$ for $t \geq t_0$.

2) There exists an $h > 0$ and a $\tau > 0$ such that, for each initial value $(x_0, t_0) \in K_{h,\tau}$, there exists exactly one solution $p(t, x_0, t_0)$ for $t \geq t_0$.

3) The origin is an *isolated* singular point, *i.e.* there exists an h' such that the equation $f(x', t) \equiv 0$, for $t \geq t_0$, holds for no $x' \neq 0$, $x' \in K_{h'}$.

We shall say "f belongs to class E" ($f \in E$) to indicate that f satisfies the three conditions above. Occasionally condition 3 will not be required.

As is well known, a sufficient condition for 2) is the Lipschitz condition; *i.e.* the inequality

$$|f(x', t) - f(x'', t)| < L|x' - x''| \tag{16.2}$$

for all (x', t), (x'', t) in K_{h,t_0}. The functions f satisfying a Lipschitz condition form a set C_0:

$$f \in C_0.$$

---

[1]) References for this section are, for instance, CODDINGTON and LEVINSON [1], KAMKE [1], LEFSCHETZ [1], NEMYTSKII and STEPANOV [1].

If the constant L can be chosen independently of t (uniform Lipschitz condition) we usually write

$$f \in \overline{C}_0.$$

The Lipschitz condition is clearly satisfied if the components of the right hand side have bounded first order partial derivatives with respect to $x_1, \ldots, x_n$. This is expressed in the notation,

$$f \in C_1;$$

$f \in \overline{C}_1$ means that the partial derivatives are uniformly bounded with respect to t. Analogously we use for functions with (bounded) higher partial derivatives the notations

$$f \in C_r, \quad f \in \overline{C}_r.$$

For other uniqueness conditions we refer the reader to the literature but we wish to mention here that any condition which guarantees the uniqueness of the solution simultaneously assures that the solution depends continuously on the initial values[1]).

Def. 2.1 which defines the stability concept presupposes the existence of a number $h > 0$ and a number t such that for $|x_0| < h$ all solutions $p(t, x_0, t_0)$ exist for $t \geq t_0$. On the other hand it makes sense to call the equilibrium unstable even if the neighboring solutions do not exist for all future times t.

Example. The general solution of the scalar differential equation

$$\dot{x} = x^3$$

is

$$p(t, x_0, t_0) = [x_0^{-2} - 2(t - t_0)]^{-1/2},$$

which becomes infinite for $t - t_0 = \frac{1}{2} x_0^{-2}$; hence for $x_0 \neq 0$ there exists no solution defined on an infinite time interval. None the less we are going to call the equilibrium unstable; the characterization of the concept *unstable* given immediately after Def. 2.4 applies here.

If the solution is uniquely determined and exists for all t, $t_0 \leq t < \infty$, we have the relation

$$(16.3) \qquad p(t_1 + t_2, p(t_1, x_0, t_0), t_1) = p(t_1 + t_2, x_0, t_0)$$

for $x_0 \in K_h$.

There exist conditions which guarantee for arbitrary x_0 the existence of the solutions $p(t, x_0, t_0)$ for *all* $t \geq t_0$. Sufficient for this is the continuity

---

[1]) *cf.* CESARI [1].

of $f(x, t)$ and the existence of an estimate

$$(16.4) \qquad |f(x, t)| \leq L|x|, \quad x \in R_n$$

with fixed L.[1]

The discussion of a differential equation (16.1) is often simplified if a first integral is known.

Def. 16.1. A function $h(x, t)$ which becomes constant if a solution $x(t)$ of the differential equation is substituted for x, is called a *first integral of the differential equation*. Thus we have

$$h\big(p(t, x_0, t_0), t\big) = \text{constant}.$$

The constant depends of course on x_0 and t_0. Obviously the total derivative of the left side with respect to t, *i.e.* the expression

$$(16.5) \qquad \frac{dh}{dt} := \frac{\partial h}{\partial x_1} \dot{x}_1 + \frac{\partial h}{\partial x_2} \dot{x}_2 + \cdots + \frac{\partial h}{\partial x_n} \dot{x}_n + \frac{\partial h}{\partial t}$$

is identically 0. This expression is called the *derivative of the function $h(x, t)$ for the differential equation* (16.1) [or with regard to the differential equation (16.1)]. It is again a function of x and t.

A differential equation

$$(16.6) \qquad \dot{x} = f(x)$$

in which the independent variable t does not appear explicitly is called *autonomous* in contrast to the *heteronomous* differential equation in (16.1). A very special case of equation (16.6) is the linear differential equation with constant coefficients. In the present chapter we are dealing with equation (16.6) and we shall assume $f \in E$ unless the contrary is explicitly mentioned. Because of autonomy we have

$$(16.7) \qquad p(t + c, x_0, t_0 + c) = p(t, x_0, t_0).$$

The dependence of the solution on the initial time is therefore not essential. Frequently we can assume without loss of generality that $t_0 = 0$ and write the general solution in the simplified form $p(t, x_0)$, insisting that $p(0, x_0) = x_0$. Relation (16.3) then becomes

$$(16.8) \qquad p(t_1 + t_2, x_0) = p(t_2, p(t_1, x_0)).$$

It is often necessary not only to consider solutions for $t \geq 0$ but also for the interval $-\infty < t \leq 0$, provided that the solution exists on that interval. This is for instance not the case for $f(x) = -\frac{1}{2} x^3$, for which the general solution is

$$p(t, x_0) = (t + x_0^{-2})^{-1/2}.$$

---

[1] CESARI [1], p. 3.

The existence of the solution for $t \geq t_0$ is clearly assured whenever $f(x)$ is bounded [cf. (16.4)]. If this is not the case we introduce arc length as the new independent variable using the transformation

$$ds = dt \sqrt{1 + |f(x)|^2}$$

and thus produce an equation which is equivalent to the original equation with respect to stability of the origin. This equation is

$$(16.9) \qquad \frac{dx}{ds} = \frac{f(x)}{\sqrt{1 + |f(x)|^2}} .$$

The right side is bounded and solutions exist for $-\infty < s < +\infty$.

The vector variable $p(t, x_0)$ can be interpreted as a curve in an $(n + 1)$-dimensional space $R_n \times I$, but frequently it is better to interpret $p(t, x_0)$ in R_n and to consider t as the curve parameter. This curve is called the *phase trajectory* or also the *phase curve*; it is the projection of the *motion* $p(t, x_0)$ into the phase space R_n. By a phase trajectory we simply mean the whole curve for $-\infty < t < +\infty$. Each point x_0 splits the trajectory $p(t, x_0)$ into two *half-trajectories*; they are denoted by $p_+(t, x_0)$, resp. $p_-(t, x_0)$. This decomposition depends of course on the choice of the point x_0. To visualize this we imagine that a movable point, the *phase point*, moves along the trajectory. A simple example for real trajectories is given by the linear equation $\dot{x} = A x$, if the matrix A has a real characteristic root λ. Then there exists a real solution (cf. sec. 4)

$$p(t, x_0) = e^{\lambda t} x_0 ,$$

which is defined for all finite t. The trajectory is the ray passing from the origin through x_0. It is cut into two half-trajectories by the point x_0. Since $-p(t, x_0)$ is also a solution the ray in the opposite direction is also a trajectory. The straight line through 0 and x_0 consists thus of two trajectories, actually of three since the origin itself must also be considered as a trajectory.

A point $Q \in R_n$ is called an *ω-limit point* for the trajectory $p(t, x_0)$, if there exists a sequence $t_n \to \infty$ such that

$$p(t_n, x_0) \to Q .$$

Analogously Q' is an α-limit point if there exists a sequence $t_n \to \infty$ such that $p(-t_n, x_0) \to Q'$. The totality of all ω-limit points is called the *ω-limit set Ω^+*; the α-limit points form the α-limit set Ω^-, the union of these two sets is the limit set Ω. We state without proof the following theorems[1].

---

[1] cf. for instance NEMYTSKII and STEPANOV [1], REISSIG, SANSONE and CONTI [1].

Theorem 16.1. The limit set is a union of phase trajectories.
Theorem 16.2. A trajectory which has a point in common with a limit set is contained in the limit set.

We further mention

Def. 16.2. A trajectory which is contained in a limit set is called *Poisson stable*.

Besides the interpretation of the expression $p(t, x)$ as a phase trajectory passing through the point x there is another interpretation: we map the point x to the point $p(t, x)$ in R_n. $p(t, x)$ forms thus a one-parameter mapping of R_n into itself (we mentioned this already in sec. 15 for the linear equation). (16.8) implies that these mappings form a group. Because $p(0, x) = x$ the parameter $t = 0$ corresponds to the identity map, resp. the unit element of the group. The interpretation of $p(t, x)$ as a mapping leads to

Def. 16.3. A set M is called *invariant* with respect to $p(t, x)$ [resp. with respect to the differential equation (16.6)] if $x \in M$ implies that

$$p(t, x) \in M, \quad -\infty < t < +\infty.$$

An invariant set is therefore always made up of whole trajectories.

Theorem 16.3. The closure of an invariant set is also invariant.

Proof. Let $\overline{M}$ be the closure of M, that is the union of M and the set of all accumulation points of M, and let $x \in \overline{M}$. If x is also a point of M then by the definition of an invariant set, $p(t, x) \in M$ for all t, and hence $p(t, x) \in \overline{M}$. If on the other hand $x \notin M$, choose a sequence $x_n \in M$, which converges to x. Then the sequence $p(t, x_n)$ converges to $p(t, x)$ which implies $p(t, x) \in \overline{M}$ for all t.

In particular, the topological boundary of an open invariant set is made up of whole trajectories.

It should be noted that these considerations and definitions are based only on the fact that the mapping $p(t, x)$ exists for all t, is continuous with respect to x and has the group property (16.8). It is immaterial that $p(t, x)$ happens to be defined by means of a differential equation. A one parameter set of mappings with the group property (16.8) is also called a *dynamical system*.

A point x_1 which satisfies the equation $f(x_1) = 0$ is called a *singular point* of the differential equation (16.6). To it corresponds a constant solution which can be thought of as an "equilibrium", in case (16.6) describes a concrete physical system. A point which is not singular is called *regular*. (16.6) assigns to each regular point a tangential direction (linear element)

$$dx_1 : dx_2 : \cdots : dx_n = f_1 : f_2 : \cdots : f_n.$$

Hence the differential equation defines a directional field (oriented field of linear elements) undefined only at the singular points. Exactly one trajectory passes through each regular point (sufficiently close to the origin). Since the solutions depend continuously on the initial values we have

Theorem 16.4. Consider the normal plane to a trajectory at a regular point. All trajectories which meet this plane in a sufficiently small neighborhood of the point pass through it in the same direction.

Assuming that $f \in E$, the origin has a neighborhood in which it is the only singular point. In this case it can for instance never happen that the components of $f(x)$ have a common factor $\gamma(x)$ which vanishes on a hypersurface passing through the origin.

By the *distance* $\varrho(p, M)$ from a point to a set we mean the expression

$$\varrho(p, M): = \inf \varrho(p, q), \qquad q \in M,$$

where q ranges over all points of M. $\varrho(p, q)$ is the distance $\overline{pq}$, also written as $|p - q|$.

With the help of the distance concept we can define the concept of an attractive set which generalizes Def. 2.2.

Def. 16.4. An invariant set is called *attractive* if

$$\lim \varrho\big(p(t, x_0), M\big) = 0, \quad t \to \infty,$$

for all x_0 belonging to a certain neighborhood of the set M, *i.e.* such that $\varrho(x_0, M) < h$.

Besides the differential equations whose right sides belong to class E we also occasionally need in applications differential equations (16.6) whose right sides have discontinuities. In this case there are certain known *surfaces of discontinuity*

$$F_j(x) = 0, \qquad j = 1, 2, \ldots.$$

The components $f_i(x)$ of $f(x)$ belong to class E in the interior of each of the domains bounded by the surfaces $F_j(x) = 0$ but they may have a jump discontinuity as they pass through one of the surfaces. The statement of the differential equation must be augmented by requirements on the behavior of the solutions as these pass through one of the surfaces of discontinuity. Continuity will always be required but usually further assumptions are needed. We note that we have somewhat generalized the concept of a "solution" since the existence of the derivatives of the "solution" on the surfaces of discontinuity is in general no longer assured. It can also happen that a solution cannot be continued across a surface of discontinuity. This occurs if two trajectories are defined in adjacent domains in such a way that at the surface of discontinuity separating these domains, they abut with opposite orientations. In this case we speak of *end points* of a trajectory, resp. of the corresponding solution. As we change from t to $-t$ they become *initial points*. We shall further be concerned with differential equations with discontinuities in secs. 22 and 74.

17. Homogeneous Right Side

We are again concerned with equation (16.6), $\dot{x} = f(x)$, and we assume that the right side is a homogeneous function:

$$(17.1) \qquad f(c\,x) = c^k f(x).$$

We shall require that the degree k of homogeneity be a positive number; for if $k < 0$ the equation has a discontinuity at $x = 0$. Further restrictions for k result from the following considerations. For $0 < k < 1$ the Lipschitz condition (16.2) is not satisfied at the point $x = 0$. If k is not a rational number of the special form p/q, where p and q are relatively prime and q is odd then $f(x)$ is not real in a neighborhood of the origin. This is seen by setting $c = -1$ in (17.1). We shall therefore usually have to assume $k = p/(2r + 1) > 1$. Some of the considerations below however apply in the general case. A special case of the homogeneous functions with $k = 1$ are the linear functions; but there also exist nonlinear functions with $k = 1$, for instance

$$f(x_1, x_2) = \frac{x_1^2 + x_2^2}{x_1 + x_2}.$$

The substitution

$$x = c\,y, \qquad s = c^{k-1}t, \qquad ds = c^{k-1}dt$$

transforms the equation (16.6) into

$$\frac{dy}{ds} = f(y).$$

The relation $p(t, x_0) = p(t, cy_0) = cp(s, y_0)$ follows, and changing the notation we have

$$(17.2) \qquad c\,p\,(c^{k-1}t, x_0) = p(t, c\,x_0).$$

This means that a change of scale on the t axis can be compensated for by a homogeneous transformation on the phase space. From (17.2) we obtain (see Def. 2.3):

Theorem 17.1. If the equilibrium of an autonomous differential equation with homogeneous right side is asymptotically stable than it is asymptotically stable in the whole.

Let a trajectory have the property that two of its points x', x_0'' lie on a straight line through the origin. Then $x_0'' = a x_0'$ with real a. If we choose x_0' as starting point the trajectory is given by $p(t, x_0')$ and there exists a value t' such that

$$(17.3) \qquad p(t', x_0') = a\,x_0'.$$

All trajectories which meet the straight line through x_0' and x_0'' have the same property. For if in (17.2) we set

$$c x_0 = x_0', \quad t = t',$$

then

$$c \, \boldsymbol{p} \, (c^{k-1} t', \, c^{-1} x_0') = \boldsymbol{p} \, (t', \, x_0')$$

results, and applying (17.3)

$$\boldsymbol{p} \, (c^{k-1} t', \, c^{-1} x_0') = a \, c^{-1} x_0'$$

is obtained. Thus since c is arbitrary, $c^{-1} x_0'$ is an arbitrary point of our straight line and the last equation can be written

$$\boldsymbol{p} \, (t'', \, x_0) = a x_0,$$

that is it is of the form (17.3).

Let x be a solution of (16.6) and $r = |x|$ its absolute value. Furthermore let y be the unit vector obtained from x,

$$(17.4) \qquad y = \frac{x}{|x|}.$$

By differentiating the identity

$$r^2 = x^T x$$

we obtain

$$2 r \dot{r} = \dot{x}^T x + x^T \dot{x} = r \left(f(ry)^T y + y^T f(ry) \right) = 2 r^{k+1} y^T f(y)$$

$$(17.5) \qquad \dot{r} = r^k y^T f(y)$$

and furthermore

$$(17.6) \qquad \dot{y} = \frac{d}{dt} \left(\frac{x}{r} \right) = \frac{\dot{x}}{r} - \frac{x \dot{r}}{r^2} = r^{k-1} \left(f(y) - \left(y^T f(y) \right) y \right).$$

The transformation (17.4) projects the trajectories of (16.6) onto the unit sphere. A constant solution y_0 of (17.6) satisfies the vector equation

$$(17.7) \qquad f(y_0) = \left(y_0^T f(y_0) \right) y_0$$

and corresponds to a point on the unit sphere. The corresponding trajectory of (16.6) is the ray going from the origin through the point y_0. Along this ray the scalar differential equation

$$\dot{r} = \beta r^k, \quad \beta := y_0^T f(y_0)$$

which follows from (17.5) and which can be directly integrated, is valid. For $k \neq 1$ and $t_0 = 0$ we have

$$r^{1-k} - r_0^{1-k} = (1 - k) \beta t,$$

$$(17.8) \qquad r(t) = \left[r_0^{1-k} + (1 - k) \beta t \right]^{\frac{1}{1-k}}.$$

In case $k = 1$,

$$r(t) = r_0 e^{\beta t}.$$

The behavior of the motion depends on the sign of the quantity $(1 - k)\beta$, respectively in case $k = 1$, on the sign of β. If $\beta < 0$, and $k > 1$, then $1/(1 - k) < 0$ and r tends to zero as t increases. The same is true if $k = 1, \beta < 0$. If $\beta < 0, 0 < k < 1$ then the value $r = 0$ is attained after a finite time interval. The solution is therefore not uniquely determined by the initial value $r = 0$. This is possible since the Lipschitz condition is not satisfied. For $\beta > 0$ r grows without bounds; in case $k > 1$ this happens on a finite time interval. Then there exists no solution on an infinite time interval (see sec. 16). We therefore have the necessary condition for stability that

$$(17.9) \qquad\qquad y_0^T f(y_0) < 0$$

for all the constant solutions of (17.6). Clearly it is not satisfied for an even k nor for $k = 2p/(2q + 1)$. For together with y_0, $-y_0$ is also a constant solution of (17.7) and since

$$-y_0^T f(-y_0) = (-1)^{k+1} y_0^T f(y_0) = -y_0^T f(y_0),$$

$y_0^T f(y_0)$ and $(-y_0^T f(-y_0))$ have opposite signs.

If $k = 1$ and $f(x) = A x$ is linear then equation (17.6) becomes

$$\dot{y} = A y - y(y^T A y).$$

A constant solution of this equation must be a characteristic vector belonging to the characteristic root $(y_0^T A y_0)$. Condition (17.9) implies the familiar result that in case of stability the real characteristic roots of A must be negative.

Let $\hat{y}$ be a periodic solution of (17.6) with period T, $i.e.$

$$\hat{y}(t + T) = \hat{y}(t).$$

The geometric image of this solution is a closed curve on the unit sphere. Let $\hat{y}_0$ be a point on this curve. Since the projection of the trajectory $p(t, \hat{y}_0)$ of (16.6) onto the unit sphere is periodic the trajectory itself has the property (17.3). To determine the proportionality factor a we use the relation which follows (17.4), and applying (17.5),

$$\ln|a| = \ln r(t') = \int_0^{t'} \frac{\dot{r}}{r}\, dt = \int_0^{t'} \hat{y}^T f(\hat{y})\, r^{k-1}\, dt$$

results.

We introduce by means of $d\tau = r^{k-1}\, dt$ the time scale of the motions on the unit sphere. Then the interval of integration becomes equal to the period T and

$$(17.10) \qquad\qquad \ln|a| = \int_0^T \hat{y}^T f(\hat{y})\, d\tau.$$

We conclude: If the integral (17.10) is negative for the periodic solution $\hat{y}$ then $|a| < 1$ and the trajectory $p(t, \hat{y}_0)$ tends to zero as t increases.

A further necessary condition for asymptotic stability has been obtained. Unfortunately it is in general quite difficult to deal with the differential equation (17.6). If $n = 2$ then each of these two necessary conditions is also sufficient as we shall see in the following sections. They are also sufficient for the case $n = 3$ as COLEMAN [1] has shown by discussing the trajectories of (17.6) on the unit sphere (see also sec. 57).

18. General Systems of the Second Order

For systems of the second order the phase space is a Euclidean plane. In studying the behavior of the phase curves the properties of the plane can be utilized and thus the considerations of sec. 16 can be carried considerably further. It is useful here to write the differential equation in scalar rather than vector form,

$$(18.1) \qquad \dot{x} = f(x, y), \quad \dot{y} = g(x, y).$$

We assume that the right sides are such that (18.1) belongs to class E and only finitely many singular points exist. At the regular points the system is equivalent to the ordinary first order differential equation

$$(18.2) \qquad \frac{dy}{dx} = \frac{g(x, y)}{f(x, y)}.$$

Many nonlinear oscillatory equations can be written in the form (18.1). For instance the so-called (generalized) *Liénard equation*

$$(18.3) \qquad \ddot{x} + \dot{x} f(x) + g(x) = 0$$

assumes the form (18.1) if we make the substitution

$$\dot{x} = y, \quad \dot{y} = -y f(x) - g(x).$$

Or we could also set

$$(18.4) \qquad \dot{u} = -F(u) + v, \quad \dot{v} = -g(u), \quad F(u) := \int^{u} f(x)\, dx.$$

A periodic solution of (18.1) is represented by a closed phase curve which in the limiting case can degenerate into a singular point. A closed trajectory is also called a cycle. If the cycle is isolated it is called a *limit cycle*. If several singular points are connected by phase trajectories a closed curve is also obtained but it is not a cycle since it does not represent a periodic solution but a *phase polygon*. Special cases are the phase-2-gon

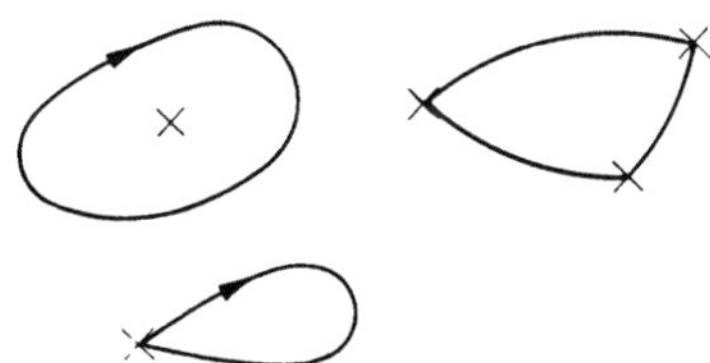
Fig. 18.1. Phase plane with cycle and phase polygons ($\times$ singular points)

5 Hahn, Stability

(spindle) and the phase-1-gon (loop) (fig. 18.1). In general the phase plane decomposes into a number of regions within each of which the phase curves may show a different type of behavior. For example, they might approach different singular points. The curves separating these regions are called *separatrices* (for examples see sec. 22).

The important theorem of BENDIXSON which we state here without proof[1]) says essentially that the limit sets of an equation (18.1) are points or phase polygons or cycles.

Theorem 18.1. If a half trajectory γ_+ of (18.1) lies in a bounded region of the plane then only the following cases are possible: a) γ_+ is a singular point, b) γ_+ is a cycle, c) γ_+ approaches a singular point, d) γ_+ approaches a cycle or a phase polygon spirally.

An analogous statement is of course valid for negative half trajectories. The following theorem is also due to BENDIXSON[1]).

Theorem 18.2. The region interior to a cycle always contains at least one singular point.

An immediate consequence of Theorem 18.1 is

Theorem 18.3. Necessary and sufficient for the existence of constant or periodic solutions is the existence of at least one bounded half trajectory.

This half trajectory is either itself *stationary i.e.* constant or periodic or it tends toward a stationary solution or toward a phase polygon on which there is at least one singular point.

The following criterion which excludes the existence of cycles and thereby of periodic solutions is due to Bendixson also. It applies whenever the right sides in (18.1) have continuous partial derivatives of the first order in a connected region D of the plane, so that we can form the divergence of the vector col (f, g),

$$(18.5) \qquad \frac{\partial f}{\partial x} + \frac{\partial g}{\partial y}.$$

Theorem 18.4. If the expression (18.5) does not change sign in the region D then D contains no cycle, not even a 1-gon or a phase polygon with finitely many sides all of which are traversed in the same sense.

Proof: Let Γ be a cycle in D and let G be the region bounded by it. According to Green's theorem

$$(18.6) \qquad \iint\limits_{G} \left(\frac{\partial f}{\partial x} + \frac{\partial g}{\partial y} \right) dx\, dy = \int\limits_{\Gamma} (f\, dy - g\, dx) = 0.$$

The integral on the right is zero since Γ is a cycle. Equality can obtain only if the integrand on the left changes its sign.

If Γ is a polygon we "around off" the corners, *i.e.* the trajectories forming the polygon are replaced in the immediate vicinity of singular

---

[1]) *cf.* CESARI [1], LEFSCHETZ [1], REISSIG, SANSONE and CONTI [1].

points by arcs of arbitrarily small length so that at the points where the arcs meet the trajectories we have a continuous tangent. Let

$$a := \sup \sqrt{f^2 + g^2}, \quad b := \inf \left| \frac{\partial f}{\partial x} + \frac{\partial g}{\partial y} \right|, \quad (x, y) \in D.$$

Also let σ be the total length of the arcs denoted by Γ'', and let G and G_1 be the regions bounded by Γ, resp. by the substitute curve. In place of (18.6) we now have

$$\iint\limits_{G_1} \left(\frac{\partial f}{\partial x} + \frac{\partial g}{\partial y} \right) dx \, dy = \int\limits_{\Gamma - \Gamma'} (f \, dy - g \, dx) + \int\limits_{\Gamma'} (f \, dy - g \, dx).$$

If we assume that the integrand on the left is positive then the area integral approaches an expression which is larger than $b \, |G_1|$ while the right integral is smaller in absolute value than $a\sigma$ and becomes arbitrarily small. This contradiction implies the theorem.

Theorem (18.4) can be generalized somewhat: In place of (18.5) we can use the divergence

$$(18.7) \qquad\qquad \frac{\partial(hf)}{\partial x} + \frac{\partial(hg)}{\partial y},$$

where $h = h(x, y)$ denotes a continuous scalar function with continuous partial first order derivatives (*Dulac's criterion*)[1].

Examples. The singular points of the scalar system

$$\dot{x} = x - y^2, \quad \dot{y} = -y^3 + x^2$$

are (0, 0) and (1, 1). The point (1, 1) is asymptotically stable. This follows from Theorem 21.1 if we make the substitution $x = \xi + 1$, $y = \eta + 1$ and thereby change the system to the form

$$\dot{\xi} = \xi - \eta^2 - 2\eta, \quad \dot{\eta} = -3\eta + 2\xi + \xi^2 - 3\eta^2 - \eta^3.$$

The expression (18.5) is $-1 - 3y^2$, and thus is always negative. Therefore the system has no periodic solutions and we can further conclude that (0, 0) must be unstable.

The scalar system

$$\dot{x} = x(ax + by + c),$$

$$\dot{y} = y(\alpha x + \beta y + \gamma),$$

where $a \neq 0$, $\beta \neq 0$, $a\beta - b\alpha \neq 0$, has four singular points: $(0, 0)$, $(0, -\gamma/\beta)$, $(-c/a, 0)$, (x_0, y_0) with $ax_0 + by_0 + c = 0$ and $\alpha x_0 + \beta y_0 + \gamma = 0$. Since the axes are trajectories a limit cycle, if it exists, must be located so that it does not cut the axes and that it encircles the point (x_0, y_0). To apply the Bendixson criterion in the generalized form we

---

[1] Andronov, Witt and Khaikin [1].

choose $h(x, y) = x^{k-1} y^{g-1}$, where the constants k, g remain undetermined for now. The expression (18.7) then becomes

$$x^{k-1} y^{g-1} \big((a + ka + g\alpha)\, x + (kb + g\beta + \beta)\, y + kc + g\gamma\big),$$

and if we choose k, g so that the coefficients of x, y vanish, then an expression $\delta x^{k-1} y^{g-1}$ results for (18.7). Therefore no limit cycle exists since this expression does not change sign within a quadrant.

It follows from these considerations how important it is for studying (18.1) and the concrete processes described by it, to know the singular points and the cycles, *i.e.* to know the equilibria and the periodic solutions. But we need to know more: Periodic motions and equilibria can be realized physically only if they are stable.

We shall put off the discussion of the stability of a cycle until a later section (sec. 81) and in the next section deal only with the stability of a singular point for equations of the form (18.1).

19. Second Order Systems with Homogeneous Right Sides

In this section we assume that (18.1) belongs to class E and $f(x, y)$ and $g(x, y)$ are relatively prime homogeneous polynomials of a degree $k \geq 1$. The results of sec. 17 apply. The transformation (17.4) amounts to an introduction of polar coordinates. We set

$$(19.1) \qquad\qquad x = r \cos \varphi; \quad y = r \sin \varphi,$$

$$(19.2) \qquad \dot{x} = \dot{r} \cos \varphi - r \dot{\varphi} \sin \varphi; \quad \dot{y} = \dot{r} \sin \varphi + r \dot{\varphi} \cos \varphi$$

and obtain

$$(19.3) \qquad \dot{r} = f \cos \varphi + g \sin \varphi; \quad \dot{\varphi} = \frac{1}{r} (g \cos \varphi - f \sin \varphi).$$

In the last two equations the arguments of f and g must be expressed in terms of r and φ. Since f and g are homogeneous we can put (19.3) into the form

$$(19.4) \qquad\qquad \dot{r} = r^k P(\varphi), \quad \dot{\varphi} = r^{k-1} Q(\varphi),$$

respectively

$$(19.5) \qquad\qquad \frac{dr}{d\varphi} = r \frac{P(\varphi)}{Q(\varphi)}.$$

P and Q denote certain polynomials in $\sin \varphi$ and $\cos \varphi$, which are homogeneous of degree $k + 1$. Equation (19.4) corresponds to equations (17.5) and (17.6). P and Q do not have a common zero since by hypothesis f and g have no common divisor. If the equation

$$(19.6) \qquad\qquad Q(\varphi) = 0$$

has real roots then there exist trajectories of (18.1) with constant polar angles; these are rays through the origin. If $\tilde{\varphi}$ is such a root (19.4) yields [*cf.* (17.8)]

$$(19.7) \qquad r = [r_0^{1-k} - (k-1)\,P(\tilde{\varphi})\,t]^{\frac{1}{1-k}}, \qquad k > 1,$$

$$(19.8) \qquad r = r_0 \exp\left(t\,P(\tilde{\varphi})\right), \qquad k = 1.$$

Obviously r tends to zero as t increases if $P(\tilde{\varphi}) < 0$. If $\tilde{\varphi}$ is a root of (19.6) then so is $\pi + \tilde{\varphi}$; the ray in the opposite direction is therefore also a phase curve. Since the degree of Q is $k+1$ there exist at most $2k+2$ such rays. They decompose the plane into sectors such that any trajectory other than a ray remains entirely inside a given sector. Since the origin is the only singular point, Theorem 18.2 implies that there are no cycles. Let φ_1, φ_2 be the polar angles of two adjacent phase rays. As φ approaches one of these two values the right side of (19.5) becomes unbounded but (19.5) implies

$$(19.9) \qquad r(\varphi) = r_0 \exp\left(\int_{\varphi_0}^{\varphi} \frac{P(u)}{Q(u)}\,du\right).$$

Hence, depending on the sign of the integrand, r approaches zero or infinity according as φ approaches φ_1 or φ_2. The phase trajectories are either tangent to the phase ray at the origin or asymptotic to it. The following possibilities arise (fig. 19.1 to 19.3): a) Both rays of the sector are tangents at the origin. The phase curves start at the singular point and return to it.

Fig. 19.1. Elliptic sector (nested ovals)

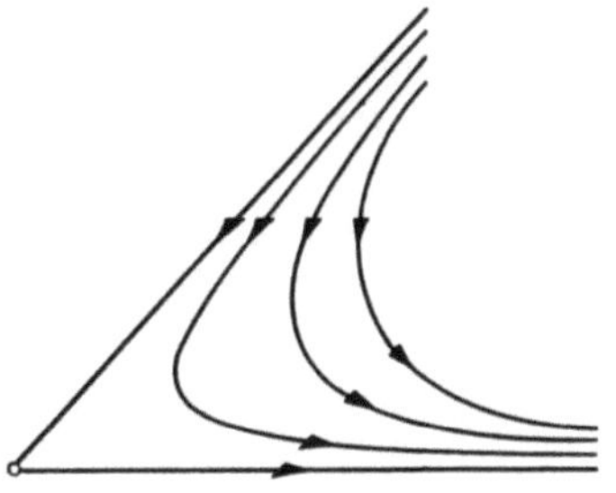

Fig. 19.2. Hyperbolic sector (saddle)

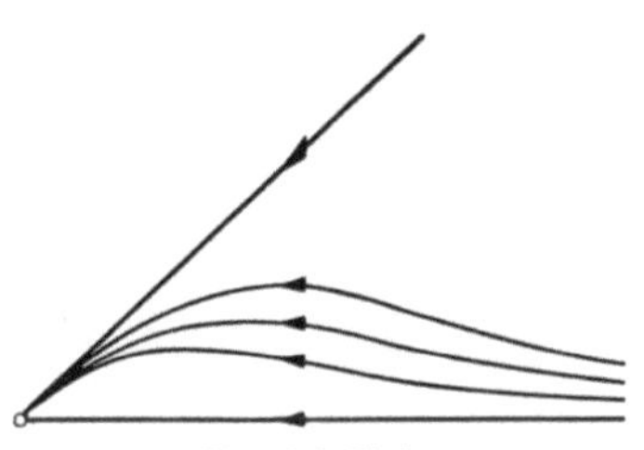

Fig. 19.3. Node

Such a sector is called *elliptic*. b) Both rays are asymptotes. The phase curves are of hyperbolic form and the sector is called *hyperbolic* (*saddle* type). c) One ray is a tangent at the origin, the direction of the other is the asymptotic direction. The sector is then called *nodal* (parabolic).

If $Q(\varphi)$ is identically zero then the trajectories of the equation are a pencil of rays through the origin.

If (19.6) has no real roots then there is no phase curve which approaches the origin in a fixed tangential direction. The integrand of (19.9) is periodic of period 2π in this case. We make use of

Theorem 19.1. If $q(u)$ is an integrable periodic function of period T then
$\int\limits^{v} q(u)\, du - hv$ is a periodic function of v with period T where

$$h := \frac{1}{T} \int\limits_0^T q(u)\, du.$$

Proof. The function

$$p(v) := \int\limits_0^v q(u)\, du$$

satisfies the functional equation

$$p(v + T) = p(v) + \int\limits_0^T q(u)\, du.$$

The function $p(v) - hv$ is therefore periodic.

For the integral (19.9) this yields the representation

$$(19.10) \qquad h\varphi + \text{periodic function}, \qquad h := \frac{1}{2\pi} \int\limits_0^{2\pi} \frac{P(\varphi)}{Q(\varphi)}\, d\varphi.$$

Hence

$$r(\varphi) = r_0\, e^{h\varphi} H(\varphi), \qquad H(\varphi) \text{ periodic of period } 2\pi,$$

and the character of the origin is determined by the number h. If $h = 0$ then $r(\varphi)$ is periodic. The origin is surrounded by closed phase curves and is called a *center*. The equation has a family of cycles in this case. If $h \neq 0$ then the phase curves are spiral shaped in the vicinity of the origin: The origin is a *focus* (vortex or spiral point). Its stability depends on whether r increases or decreases with increasing t. The second case occurs when $h < 0$ and φ increases with increasing t or when $h > 0$ and φ approaches $-\infty$ as t increases. The condition for this case is $\operatorname{sgn} h \neq \operatorname{sgn} Q(\varphi)$.

Asymptotic stability occurs in the following cases:

a) The origin is a stable focus (Q has no zeros, $\operatorname{sgn} h \neq \operatorname{sgn} Q$). b) The origin is a stable node (for all the zeros $\tilde{\varphi}$ of Q, $P(\tilde{\varphi}) < 0$); this includes the special case $Q = 0$, $P < 0$. (The nomenclature used here is not quite consistent: We should properly speak of an asymptotically stable focus, resp. node.)

The equilibrium is stable but not asymptotically stable if the origin is a center ($h = 0$). In all other cases the origin is unstable.

Comparing with sec. 17 we see that the stability conditions given there are equivalent to the present ones; they are therefore necessary and sufficient for $n = 2$.

The terms "saddle point", "node", etc., are also used in the general case of the system of equations (18.1) to describe the behavior of the trajectories in a neighborhood of the singular point. Apart from those mentioned many other types of singularities can appear. For example the system

$$\dot{x} = \alpha x + y - x^2, \quad \dot{y} = y - x^3$$

has three singular points $(0, 0)$, (β, β^3), (γ, γ^3), where β and γ are the roots of the equation $x^2 - x + \alpha = 0$. As α approaches zero one of these roots tends to 1, the other to zero. Accordingly, for $\alpha = 0$, the origin must be considered as a *double* singularity of the system[1]).

Next consider the **example**

$$\dot{x} = cxy, \quad \dot{y} = ax^2 + by^2.$$

In this case

$$P(\varphi) = (c + a) \cos^2 \varphi \sin \varphi + b \sin^3 \varphi,$$

$$Q(\varphi) = a \cos^3 \varphi + (b - c) \cos \varphi \sin^2 \varphi.$$

The condition $Q(\varphi) = 0$ implies $\cos \varphi = 0$ or $\tan \varphi = \sqrt{a/(c - b)}$. There exist therefore trajectories which are rays through the origin and since the order $k = 2$ is even, the origin is unstable according to the discussion following (17.8). In order to sketch the phase picture a more exact discussion of the expressions $P(\varphi)$ and $Q(\varphi)$ is required. We must distinguish between the cases $(c - b)/a < 0$ and $(c - b)/a > 0$.

In the special case $a = -1, b = +1, c = 2$ the system is equivalent to

$$(19.11) \qquad \dot{r} = r^2 \sin \varphi, \quad \dot{\varphi} = -r \cos \varphi,$$

$$\frac{dr}{d\varphi} = -r \tan \varphi \qquad \left(\varphi \neq \frac{\pi}{2}, \ \varphi \neq \frac{3\pi}{2}\right).$$

The solution is

$$r = r_0 \, |\cos \varphi|, \qquad \varphi \neq \frac{\pi}{2}, \ \varphi \neq \frac{3\pi}{2},$$

and

$$\varphi = \frac{\pi}{2}, \quad \varphi = \frac{3\pi}{2}.$$

---

[1]) A detailed classification of the singular points is found in Nemytskii and Stepanov [1], Chap. II; see also Forster [1], Frommer [1], Keil [1].

The phase curves (fig. 19.4) form two families of circles whose midpoints lie on the x-axis and which are tangent to the y-axis at the origin. The

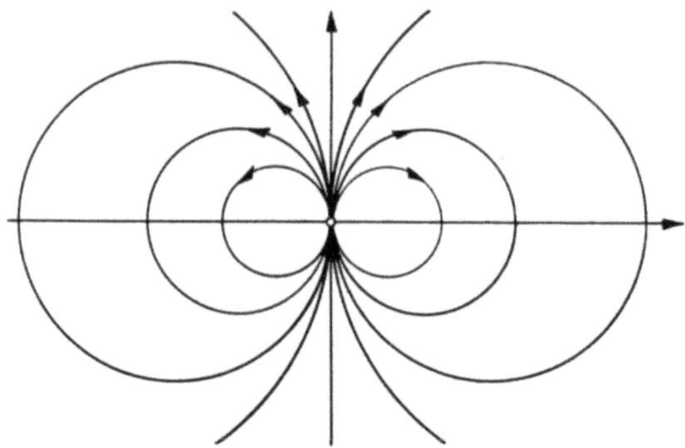

Fig. 19.4. Phase portrait of (19.11)

two halves of the y-axis are also trajectories; they decompose the plane into two elliptic sectors. Along the y-axis we have

$$\dot{r} = \pm r^2, \quad r(t) = \left(\frac{1}{r_0} - t\right)^{-1} \text{ or } = \left(\frac{1}{r_0} + t\right)^{-1}.$$

One of the solutions "explodes". The remaining motions tend toward the origin. The following state of affairs was already described in conjunction with Def. 2.4: Arbitrarily close to the origin there exist starting points such that the phase curve passing through them moves further than an arbitrarily pre-assigned distance R away from the origin. These starting points lie between the circles with radius R centered at $\left(+\frac{R}{2}, 0\right)$ and $\left(-\frac{R}{2}, 0\right)$. The larger R becomes, the narrower is the intersection of the two circular discs. This can be made analytically more precise if we introduce the concept of probability of stability due to STEPANOFF [1]. Let $E(R, r)$ denote the measure of the set of points x_0 in the ball K_r for which the maximal distance of the motion $p(t, x_0)$ from the origin is no larger than R:

$$\sup |p(t, x_0)| \leq R,$$

and let $S(r)$ be the volume of the ball K_r itself. Form the quotient $Q(R, r) = E(R, r)/S(r)$. Stepanoff calls the limit

$$\lim_{R \to 0} \lim_{r \to 0} Q(R, r)$$

in case it exists, the *probability of stability* of the trivial solution.

In the example above $S(r) = \pi r^2$. To determine $E(R, r)$ we compute the area I of the intersection of the circular disc about the origin with radius r and the disc about $\left(\dfrac{R}{2}, 0\right)$ with radius $\dfrac{R}{2}$ (fig. 19.5):

$$I = 2\left(\frac{R^2 \alpha}{8} - \frac{(R/2 - x)\, y}{2}\right)$$
$$+ \left(\frac{r^2 \beta}{2} - \frac{x\,y}{2}\right)$$
$$= \frac{R^2}{4}\alpha + r^2\beta - \frac{R}{2}y.$$

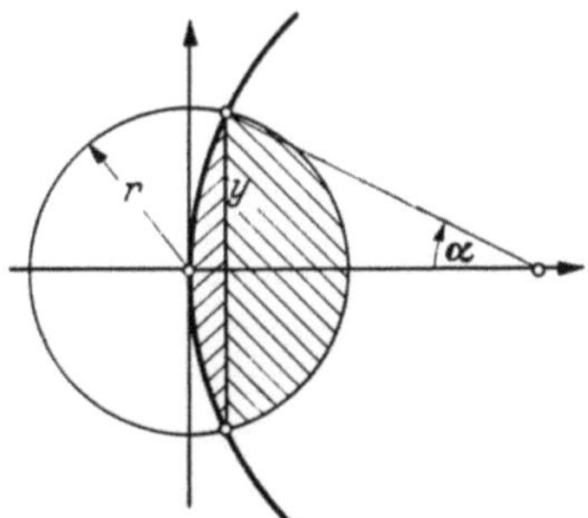

Fig. 19.5. Stability probability (*cf.* text)
$(\beta = \measuredangle\,(r,\,x))$

Since $\sin \alpha = (2r/R)\sin \beta$ and $x^2 = r^2 - y^2$ we obtain for small r approximately

$$\alpha \sim \frac{2r}{R}, \quad y \sim r, \quad \beta \sim \frac{\pi}{2}, \quad I \sim \frac{R^2 r}{2R} + \frac{r^2 \pi}{2} - \frac{R r}{2} = \frac{r^2 \pi}{2}.$$

Now $E(R, r) = 2I$ and therefore

$$\frac{E(R, r)}{S(r)} \sim \frac{r^2 \pi}{r^2 \pi} = 1$$

independent of R. So the probability of stability is equal to one.

We consider another **example:**

$$\dot{x} = y^2, \quad \dot{y} = -x^2,$$

$$P(\varphi) = \sin \varphi \cos \varphi (\sin \varphi - \cos \varphi),$$

$$Q(\varphi) = -(\sin^3 \varphi + \cos^3 \varphi) = (\cos \varphi + \sin \varphi)(\sin \varphi \cos \varphi - 1).$$

The only zeros of Q are $\tilde{\varphi}_0 = \dfrac{3\pi}{4}$ and $\tilde{\varphi}_1 = \dfrac{7\pi}{4}$, and we have $P(\tilde{\varphi}_0) = -P(\tilde{\varphi}_1)$. Since P does not have the same sign at these two points the equilibrium is unstable. A first integral exists (*cf.* sec. 16) and we see that the phase curves satisfy the equation

$$x^3 + y^3 = \text{const.}$$

Here the singular point is not distinguished geometrically in any way. The member of the family which passes through the origin has equation $x + y = 0$. The parametrical representation in the sense of the differential equation is

$$x = (x_0^{-1} - t)^{-1}, \quad y = (t - x_0^{-1})^{-1}.$$

It shows that the solutions for $x_0 = -y_0 > 0$ do not exist on an infinite time interval. The line $x + y = 0$ is therefore not a phase trajectory.

This example makes it clear that the qualitative inspection of the phase curves occasionally gives no information on essential properties of the solution.

We shall give some criteria to distinguish between unstable and stable equilibrium as well as between focus and center in a different place (*cf.* secs. 79 ff.).

In concrete systems the nature of the singular point alone gives many insights into the general behavior of the motions. If the origin is a center the system shows undamped oscillations about the equilibrium; in the case of a focus the oscillations are damped, resp. anti-damped. A node indicates the occurence of aperiodically fading resp. swelling motions.

Without giving a detailed discussion we mention further that the method of this section applies also to difference equations

$$x_{n+1} = f(x_n, y_n), \quad y_{n+1} = g(x_n, y_n), \quad x_n = x(t_0 + n) \text{ etc.}$$

provided the functions f and g are homogeneous polynomials[1]).

20. Second Order Linear Systems

If in place of (18.1) we have a linear system

$$(20.1) \qquad \dot{x} = ax + by, \quad \dot{y} = cx + dy$$

the calculations for the last section can explicitly and completely be carried out. With the previously introduced notation we have

$$(20.2) \qquad P(\varphi) = a \cos^2\varphi + (b + c) \cos\varphi \sin\varphi + d \sin^2\varphi,$$

$$(20.3) \qquad Q(\varphi) = c \cos^2\varphi + (d - a) \cos\varphi \sin\varphi - b \sin^2\varphi.$$

The polynomial $Q(\varphi)$ has no real zeros if and only if the discriminant

$$(20.4) \qquad D := (a - d)^2 + 4bc < 0.$$

This condition is therefore characteristic for a focus or a center.

The characteristic equation

$$(20.5) \qquad \lambda^2 - (a + d)\lambda + ad - bc = 0$$

has complex zeros in this case. According to sec. 6 the inequalities

$$(20.6) \qquad p := -(a + d) > 0, \quad q := ad - bc > 0$$

are necessary and sufficient for asymptotic stability of the origin whereas in case

$$(20.7) \qquad p = 0, \quad q > 0$$

----

[1]) *cf.* PANOV [2].

non-asymptotic stability prevails. Thus (20.4) together with (20.6) form the condition for a stable focus and (20.4) together with (20.7) characterize a center. Because of

$$(20.8) \qquad\qquad D = p^2 - 4q,$$

$q < 0$ implies $D > 0$ (fig. 20.1, 20.2).

If $D \geq 0$ then equation (20.5) has two real zeros and we can be guided by the discussion following (19.6). If the two zeros are simple there exist two pairs of phase rays. If a real double root exists there is only one pair of phase rays. If the two zeros λ_1, λ_2 have opposite signs — the condition for this is $q < 0$ — then all the sectors are hyperbolic. The origin is a saddle point in this case (fig. 20.3). If $q > 0$ then $\operatorname{sgn} \lambda_1 = \operatorname{sgn} \lambda_2$; the sectors are all nodal and the origin is a node (fig. 20.4). It is stable if p is negative, otherwise unstable.

If D vanishes and $b^2 + c^2 > 0$ then there are only two phase rays and they separate the plane into two half planes. If $D = 0$, $b = c = 0$, then the characteristic

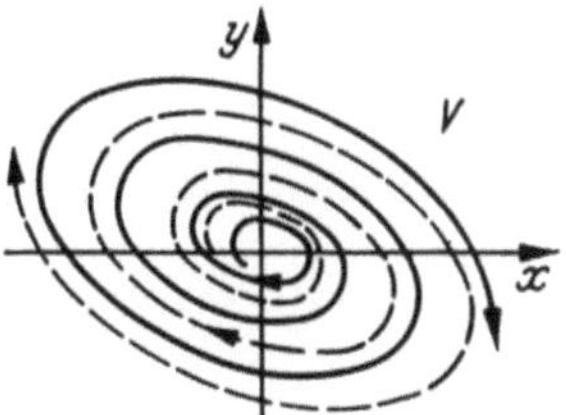
Fig. 20.1. Focus (unstable)

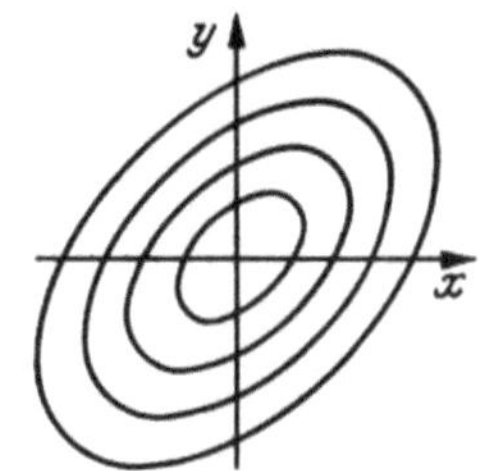
Fig. 20.2. Center

Fig. 20.3. Saddle

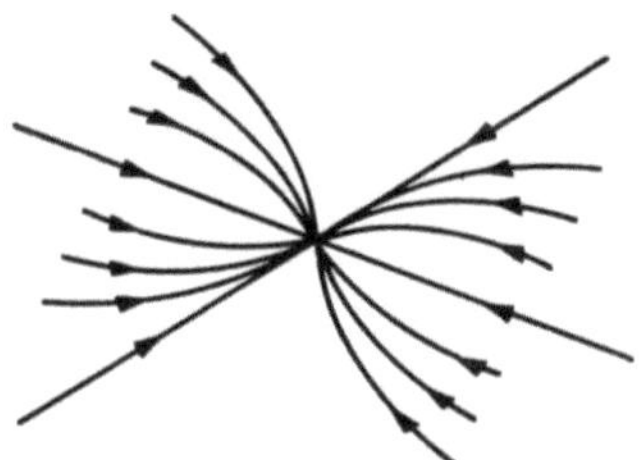
Fig. 20.4. Node (stable)

equation has a double root belonging to two elementary divisors of the matrix of the system. The trajectories then form a pencil of rays through the origin. In these last two cases we also speak of a (degenerate) node. (fig. 20.5) Elliptic sectors do not occur.

In case $\lambda_1 \neq \lambda_2$ the general solution of (20.1) is

$$x = c_1 x_1 e^{\lambda_1 t} + c_2 x_2 e^{\lambda_2 t}, \qquad y = c_1 y_1 e^{\lambda_1 t} + c_2 y_2 e^{\lambda_2 t},$$

where the x_i, y_i are the components of the characteristic vectors of the matrix of the equation and the c_i are arbitrary constants. Suppose

$0 < |\lambda_1| < |\lambda_2|$. The directional factor of the tangent to the phase trajectory, *i.e.* the expression $\dot{y}(t)/\dot{x}(t)$ tends to the value y_1/x_1 as the phase point approaches the origin (*i.e.* as $t \to \infty$ resp. $t \to -\infty$), provided $c_1 \neq 0$. The phase curves for $c_1 \neq 0$ have as their common tangent the ray belonging to the root with the smaller absolute value. Fig. 20.6 shows the stability situation in the (p, q)-plane. The stability boundary (see sec. 11) is formed by the positive half-axes.

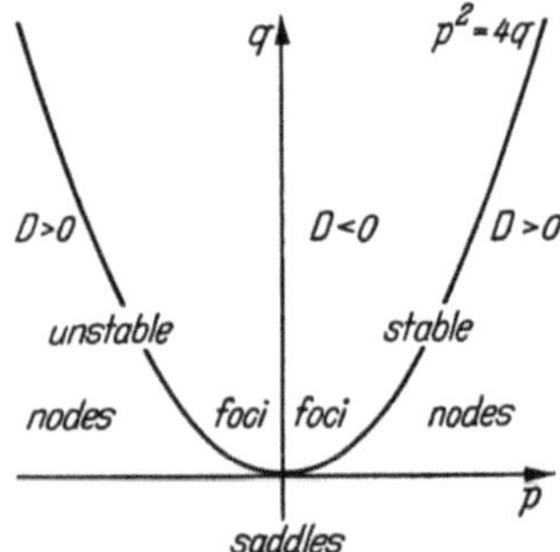

Fig. 20.5. Degenerated node (unstable)

Fig. 20.6. Stability domains in the (p, q)-plane

21. Perturbed Second Order Linear Systems

A system

(21.1)
$$\dot{x} = ax + by + h(x, y),$$
$$\dot{y} = cx + dy + k(x, y)$$

is called a *perturbed linear system* if the conditions

(21.2)
$$h(x, y) = o\left(\sqrt{x^2 + y^2}\right), \quad k(x, y) = o\left(\sqrt{x^2 + y^2}\right)$$
$$\text{or } h(x, y) = o(|x| + |y|), \quad k(x, y) = o(|x| + |y|)$$

are given.

Most concrete systems are of type (21.1) since the so-called linear laws of physics which are involved in deriving the equations of motions are, in fact, not linear. The linear formulation of Hooke's or Ohm's law for instance is an idealization. The derivations from the linear model are in general "small" and therefore the assumption (21.2) is justified within certain bounds. Of course we must then also in general restrict the dependent variable to "small" values. Accordingly the first question which presents itself is whether the stability behavior of the linear system implicitly defined in (21.1) is "sensitive", *i.e.* whether and to what extent it is affected by the nonlinear terms. To answer this question we usually introduce a further idealization namely, we assume that h and k have Taylor series expansions beginning with second or higher degree terms.

Relation (21.2) guarantees immediately that the origin is an isolated singular point. For if we assume that $(x_0, y_0) \neq (0, 0)$ is also a singular point, then a substitution and some calculation show [*cf.* (20.6)] that (21.1) implies

$$q x_0 = - dh(x_0, y_0) + bk(x_0, y_0),$$

$$q y_0 = c h(x_0, y_0) - a k(x_0, y_0)$$

and furthermore

$$(21.3) \quad |q| (|x_0| + |y_0|) \le (|a| + |b|) |k(x_0, y_0)| + (|c| + |d|) |h(x_0, y_0)|;$$

since by (21.2) the right side is $o(|x_0| + |y_0|)$, (21.3) cannot be valid for arbitrarily small x_0, y_0.

As already indicated, we shall compare (21.1) with the reduced or linearized system (20.1). Under the assumption (21.2) we have

Theorem 21.1. If the origin of the reduced system is asymptotically stable then the origin of the complete system is also asymptotically stable. If the origin of (20.1) is a saddle point, an unstable node, or an unstable focus (*i.e.* if the characteristic equation has at least one root with positive real part) then the origin for (21.1) is unstable.

This theorem is a special case of a general *theorem on stability in the first approximation* which shall be proved in sec. 28. There is no theorem on the behavior of the origin of (21.1) in case the origin of (20.1) is a center. For in that case the stability behavior is determined by the non-linear terms, as the following **example** shows. The system

$$(21.4) \quad \dot{x} = - y + \beta x(x^2 + y^2), \quad \dot{y} = x + \beta y(x^2 + y^2)$$

written in polar coordinate (19.1) has the form

$$\dot{r} = \beta r^3, \quad \dot{\varphi} = 1.$$

The trajectories are spirals. So the origin is a focus whose stability depends on the sign of β. But the number β appears in (21.4) for the first time in the third degree terms.

Theorem (21.1) makes no assertion on whether the behavior of the trajectories of 21.1 is the same as that of the curves of (20.1). In fact, under certain conditions there may be a difference; this is shown by the system

$$(21.5) \quad \dot{x} = - x + \frac{2y}{\ln(x^2 + y^2)}, \quad \dot{y} = - y - \frac{2x}{\ln(x^2 + y^2)}.$$

The origin is a focus. For the solution of the system, written in polar coordinates, is

$$r = r_0 e^{-(t - t_0)}, \quad \varphi = \varphi_0 + \ln(t_0' - t + \ln r_0) - \ln \ln r_0.$$

Its phase curves are spirals but the origin of the reduced system is a degenerate node.

We next give some sufficient conditions under which the general character of the phase curves in the vicinity of the origin is not altered or affected by the additional terms[1]). We make the additional assumption here that in the vicinity of the origin the four difference quotients

$$(21.6) \qquad \frac{h(x_1, y) - h(x_2, y)}{x_1 - x_2} , \ldots, \frac{k(x, y_1) - k(x, y_2)}{y_1 - y_2}$$

are bounded. This guarantees the uniqueness of the solution of (21.1) because for sufficiently small $(x, y) \neq (0, 0)$ at least one of the two equations

$$\frac{dy}{dx} = \frac{cx + dy + k(x, y)}{ax + by + h(x, y)}, \qquad \frac{dx}{dy} = \frac{ax + by + h(x, y)}{cx + dy + k(x, y)},$$

which have been constructed in accordance with (18.2), makes sense since the origin is an isolated singular point. (21.6) guarantees that the Lipschitz condition holds for these equations. (21.6) is satisfied for example if

$$(21.7) \qquad h(x, y) = O(r^{1+\delta}), \qquad k(x, y) = O(r^{1+\delta}), \qquad \delta > 0.$$

We change (21.1) to polar coordinates:

$$(21.8) \qquad \dot{r} = rP(\varphi) + rp(r, \varphi); \qquad \dot{\varphi} = Q(\varphi) + q(r, \varphi),$$

$$p := \frac{1}{r}(h \cos\varphi + k \sin\varphi); \qquad q := \frac{1}{r}(k \cos\varphi - h \sin\varphi).$$

$P(\varphi)$ and $Q(\varphi)$ are defined in (20.2), (20.3). Under the hypothesis of (21.7), $p = O(r^\delta)$, $q = O(r^\delta)$. If $Q(\varphi)$ has no real zeros we can write

$$\frac{dr}{d\varphi} = r\frac{P(\varphi)}{Q(\varphi)} + rs(r, \varphi), \qquad s = O(r^\delta),$$

and furthermore

$$\ln r = \int_0^\varphi \frac{P(u)}{Q(u)} du + \int_0^\varphi s(r(u), u) du.$$

If the origin for (20.1) is a focus then the number h defined by (19.10) is different from zero. Choosing r so small that

$$|s(r, \varphi)| < \frac{|h|}{2}$$

and setting $\varphi = 2\pi m$, we obtain, observing the signs,

$$|\ln r| \geq |h| 2\pi m - \frac{|h|}{2} 2\pi m = |h| \pi m.$$

As $h\varphi \to -\infty$, r decreases exponentially. So again the origin is a focus.

If $h = 0$ this manner of reasoning fails. If the origin of the reduced system is a saddle point then there exist exactly two trajectories which approach the origin in a definite tangential direction, namely the two

---

[1]) PERRON [1], cf. also SANSONE and CONTI [1].

straight lines. We can show that (21.1) also has this property. We can assume here that by means of a linear transformation the equation has already been changed to the form

$$(21.9) \quad \dot{x} = a(x + h(x, y)), \quad \dot{y} = -y + k(x, y), \quad a > 0.$$

We shall assume that the additional terms have continuous first order partial derivatives vanishing at the origin; then the expressions (21.6) are clearly bounded. Next choose a square

$$S: |x| + |y| > 0, \quad |x| \leq \eta, \quad |y| \leq \eta,$$

and pick the number η so small that in S

$$|h(x, y)| + |k(x, y)| \leq \frac{1}{3} (|x| + |y|).$$

Furthermore, for $0 \leq x \leq \eta$, let, for $k(x, y) \not\equiv 0$,

$$\alpha(x) := \sup_{\substack{0 \leq |y| \leq |x| \\ 0 \leq |\xi| \leq |x|}} \frac{k(\xi, y)}{|\xi| + |y|},$$

$$\beta(x) := \frac{\alpha(x)}{1 - \alpha(x)} x.$$

The function $\beta(x)$ is continuous and monotone increasing on the interval $0 \leq x \leq \eta$; $\beta(0) = 0$ $\beta(x) \leq \frac{x}{3}$, and $\beta(x) = o(x)$. Consider the differential equation obtained from (21.9)

$$(21.10) \qquad \frac{dy}{dx} = \frac{-y + k(x, y)}{a(x + h(x, y))} = : F(x, y)$$

defined in the domain K_x: $0 \leq x \leq \eta$, $|y| \leq \beta(x)$. The function $F(x, y)$ is continuous in this domain if we set $F(0, 0) = 0$, for we have

$$|x + h(x, y)| \geq x - \frac{1}{3} (|x| + |y|) \geq \frac{2}{3} x - \frac{1}{3} \beta(x) \geq \frac{1}{2} x,$$

$$|-y + k(x, y)| \leq |y| + |k(x, y)| \leq \beta(x) + \alpha(x) (x + \beta(x)) = 2\beta(x)$$

there. Hence (21.10) has a well defined solution $y = \varphi(x)$ in K_x, with $\varphi(0) = 0$ (because of the Lipschitz condition), and

$$\varphi'(0) = \lim_{x \to 0} \frac{y}{x} = 0$$

since in K_x, $\frac{y}{x} \leq \frac{\beta(x)}{x}$. This solution $y = \varphi(x)$ is a trajectory of (21.9) which has a definite tangent as it approaches the origin. The existence of a trajectory of the same kind for $x < 0$ is shown in the same way and likewise the existence of a similar trajectory defined by $x = \psi(y)$. We then introduce new variables by

$$(21.11) \qquad x - \psi(y) = u, \quad y - \varphi(x) = v.$$

Since the Jacobian

$$\frac{\partial(u, v)}{\partial(x, y)} = 1 - \psi'(y)\,\varphi'(x)$$

has the value 1 at the origin, the system (21.11) can be solved for x and y. (21.9) can therefore be put into the form

$$
\begin{aligned}
(21.12) \qquad & \left[a\big(x + h(x, y)\big) + \psi'(y)\big(y - k(x, y)\big)\right] dv \\
& + \left[a\big(x + h(x, y)\big)\,\varphi'(x) + y - k(x, y)\right] du = 0.
\end{aligned}
$$

In accordance with (21.11), x and y are expressed in terms of u and v. Now $x = \psi(y)$ is a solution of the differential equation (21.10). Hence we have the identity

$$a\big(\psi(y)\big) + h\big(\psi(y), y\big) + \big(y - k\big(\psi(y), y\big)\big)\,\psi'(y) = 0$$

which implies

$$a\big(x - u + h(x - u, y)\big) + \big(y - k(x - u, y)\big)\,\psi'(y) = 0.$$

Similarly we obtain the equation

$$a\big(x + h(x, y - v)\big)\,\varphi'(x) + \big(y - v - k(x, y - v)\big) = 0.$$

If we multiply the last two equations by dv resp. du and add them to (21.12) then this differential equation takes on the form

$$
\begin{aligned}
& \left[a\big(u + h(x, y) - h(x - u, y)\big) - \big(k(x, y) - k(x - u, y)\big)\,\psi'(y)\right] dv \\
& + \left[a\big(h(x, y) - h(x, y - v)\big)\,\varphi'(x) + v - \big(k(x, y) - k(x, y - v)\big)\right] du = 0.
\end{aligned}
$$

We may apply the mean value theorem to the differences appearing in this expression:

$$h(x, y) - h(x - u, y) = u\,h_x(x - \vartheta u, y), \qquad 0 < \vartheta < 1,$$

etc. Expressing x and y in terms of u and v, we see that the quantities h_x, h_y, considered as functions of u and v, approach zero as $(u, v) \to (0, 0)$. Therefore the differential equation (21.10) has the form

$$a\,u\big(1 + h_1(u, v)\big)\,dv + v\big(1 + k_1(u, v)\big)\,du = 0,$$

$$h_1(0, 0) = k_1(0, 0) = 0,$$

and is equivalent to the system

$$\dot{u} = a\,u\big(1 + h_1(u, v)\big), \qquad \dot{v} = -v\big(1 + k_1(u, v)\big).$$

The trajectories $u = 0$, $v = 0$ correspond to the trajectories $x = \psi(y)$ and $y = \varphi(x)$ of (21.9). There are no further trajectories which approach the origin; for in the vicinity of the origin

$$\operatorname{sgn}(u\,\dot{u}) \neq \operatorname{sgn}(v\,\dot{v})$$

so that u^2 and v^2 cannot decrease simultaneously. This shows that the system (21.9) also has exactly two trajectories approaching the origin; the saddle point character has been preserved.

If the origin of the reduced system is a non-degenerate node we can change the system to the form (possibly after substituting $t' = -t$)

$$\dot{x} = -\lambda x + h(x, y), \quad \dot{y} = -\mu y + k(x, y), \quad \lambda > \mu > 0,$$

respectively (*cf.* (21.8))

$$\dot{r} = -r(\lambda \cos^2 \varphi + \mu \sin^2 \varphi) + r p(r, \varphi),$$

$$\dot{\varphi} = \frac{1}{2}(\lambda - \mu) \sin 2\varphi + q(r, \varphi).$$

Applying Theorem 21.1 and the equation for $\dot{r}$, we conclude that r decreases steadily in the vicinity of the origin and because of the condition on the order of magnitude of $q(r, \varphi)$ we can arrange for q to be smaller than $c(\lambda - \mu)/2$ in absolute value, for an arbitrary given c. If at the beginning of the motion

$$\sin 2\varphi_0 > c$$

then φ increases but at most to the value $\frac{1}{2}(\pi + \arcsin c)$. Later on the motion φ decreases but it cannot fall below the value $\frac{1}{2}(\pi - \arcsin c)$. Eventually φ remains therefore in a sector whose central angle is $\arcsin c$, and since c can be chosen arbitrarily small we recognize that the trajectory approaches the ray $\varphi = \frac{\pi}{2}$ as r decreases, *i.e.* as t increases. From the equation

$$\frac{dy}{dx} = \frac{\mu \sin \varphi - \dfrac{1}{r} k}{\lambda \cos \varphi - \dfrac{1}{r} h}$$

we see that y' behaves like $\frac{\mu}{\lambda} \tan \varphi$ for small values of r; so as $r \to 0$, y' grows without bound. The origin of the complete system proves to be a node.

We have thus shown that under the given conditions a singular point of the type of a focus, a saddle point, or a non-degenerate node does not change its character if additional terms are added. These types of singular points are *insensitive* in the sense of the following

Def. 21.1. A property which the motions defined by a differential equation

$$\dot{x} = f(x, t)$$

possess in a certain neighborhood K_{h, t_0} of the equilibrium is called *insensitive (rough)* if there exists a function $\varphi(r)$ in K such that the motions defined by the perturbed differential equation

$$\dot{x} = f(x, t) + g(x, t)$$

have the same property, provided only that

(21.13) $$|g(x, t)| < \varphi(|x|).$$

If there exists at least one unperturbed differential equation which has the property while for each $\varphi \in K$ a disturbance $g(x, t)$ can be found such that (21.13) holds and the perturbed system does not have the property, then the property is called *sensitive*.

For the origin the property of being a center is sensitive. This can be seen from the system

$$\dot{x} = -y + \beta x f(r), \quad \dot{y} = x + \beta y f(r), \quad r^2 = x^2 + y^2,$$

resp.

$$\dot{r} = \beta r f(r), \quad \dot{\varphi} = 1$$

which is analogous to example (21.4), and in which the perturbation terms can be of arbitrarily high degree.

The degenerate node on the other hand is insensitive if we choose $\varphi(r) = r^2$: As PERRON [1] showed, the conditions [cf. (21.1)] $h = O(r^2)$, $k = O(r^2)$ are sufficient for the degenerate node to maintain its character.

A center may go over into a focus and one other possibility exists; this is shown by the system

$$\dot{x} = -y + rx \sin \frac{1}{r}, \quad \dot{y} = x + ry \sin \frac{1}{r},$$

resp.

(21.14) $$\dot{r} = r^2 \sin \frac{1}{r}, \quad \dot{\varphi} = 1.$$

It has a countably infinite number of closed phase curves $r = 1/\pi k$, $k = 1, 2, \ldots$. Between any two of these circles there are non-closed curves given by the equations

$$\varphi = \varphi_0 + \int_{r_0}^{r} \frac{dr}{r^2 \sin \dfrac{1}{r}}$$

which approach the containing circles spirally. In this case the origin is called a *center-focus*. This type never occurs if, in (21.1), the expression $x h(x, y) + y k(x, y)$ is different from zero in a neighborhood of the origin. For then the function $p(r, \varphi)$ in (21.8) is different from zero throughout and since $P(\varphi)$ vanishes identically, $\dot{r}$ has always the same sign. Hence r approaches a constant value c_0 if we rotate in the appropriate direction. If we had $c_0 \neq 0$ then $r = c_0$ would be a trajectory, and then $p(c_0, \varphi) = 0$, contrary to hypothesis. Therefore $\lim r = 0$ and the origin is a focus.

The actual reason why the center is sensitive, as was seen in examples (21.4) and (21.14), is that the center is representing a limiting case: The real parts of the characteristic roots are zero.

If the given system of equations has singular points other than the origin we can translate them one by one to the origin by a translation of the coordinate axes and analyze their character as above. Furthermore the overall phase picture can be observed to study its sensitivity. A system is called *structurally stable* if, roughly speaking, the topological behavior of the phase picture is not affected by small changes in the equations, especially if each singular point maintains its character and if the cycles of the new system correspond to those of the old system. It follows from the above that a structurally stable system cannot have a center. A further necessary condition which we mention without proof, says that no two saddle points may be joined by a trajectory. Furthermore only finitely many closed curves may exist.

An exact definition of the concept of structural stability, which must not be confused with the concept mentioned in sec. 11, is found in LEFSCHETZ [1][1].

For the sake of completeness we will shortly mention the rôle played by the infinitely distant. We study the behavior of the trajectories of

$$\dot{x} = f(x, y), \quad \dot{y} = g(x, y)$$

for very large values of x and y by performing the transformation

$$x = x_1/x_3, \quad y = x_2/x_3,$$

thereby introducing homogeneous coordinates. We have

$$\frac{dy}{dx} = \frac{x_3\, dx_2 - x_2\, dx_3}{x_3\, dx_1 - x_1\, dx_3} = \frac{g\,(x_1/x_3, x_2/x_3)}{f\,(x_1/x_3, x_2/x_3)}$$

which implies a relation

$$(21.15) \quad p\,(x_1, x_2, x_3)\, dx_1 + q\,(x_1, x_2, x_3)\, dx_2 + r\,(x_1, x_2, x_3)\, dx_3 = 0.$$

If f and g are polynomials we can multiply the equation with an appropriate power of x_3 and thereby arrange for p, q, r to be homogeneous polynomials in x_1, x_2, x_3.

On the finite plane we may set $x_3 = 1$, $dx_3 = 0$. Then the equation corresponds with the original system and furnishes finite singular points. If $dx_3 \neq 0$, then $r\,(x_1, x_2, x_3)$ must vanish. The singular structures — not necessarily "points" — are the common solutions of

$$(21.16) \quad \begin{aligned} p\,(x_1, x_2, x_3)\, dx_1 &= 0, \quad q\,(x_1, x_2, x_3)\, dx_2 = 0, \\ r\,(x_1, x_2, x_3)\, dx_3 &= 0. \end{aligned}$$

---

[1] *cf.* also DE BAGGIS [1], PEIXOTO [1].

6*

Example. Consider

$$\dot{x} = x^2 + y^2 - 1, \quad \dot{y} = 5(xy - 1).$$

Multiplying by x_3 we obtain

$$p = -5x_3(x_1 x_2 - x_3^2), \quad q = x_3(x_1^2 + x_2^2 - x_3^2),$$
$$r = 4x_1^2 x_2 - x_2^3 - 5x_1 x_3^2 + x_2 x_3^2.$$

The common solutions of (21.16) are given by

$$x_3 = 0, \quad x_2(4x_1^2 - x_2^2) = 0.$$

They are the three points

$$A = (1, 0, 0), \quad B = (1, 2, 0), \quad C = (1, -2, 0).$$

To determine the character of these points we have to return to the Cartesian coordinates. To study the first point we set $u = x_2/x_1$, $v = x_3/x_1$. Then the coordinates of the singular point are $u = v = 0$. Applying (21.15) with $x_1 = 1$, $dx_1 = 0$ we obtain

$$v(1 + u^2 - v^2)\, du + (4u - u^3 - 5v^2 + uv^2)\, dv = 0,$$

and we find that the origin of the (u, v)-plane is a saddle point. To the infinite point A there corresponds therefore a singular point of the system of equations of the character of a saddle point. To study B we first perform the transformation $x_1 = x_1'$, $x_2 = 2x_1' + x_2'$, $x_3 = x_3'$ which assigns B the coordinates $(1, 0, 0)$. Then again we go over to the coordinates u and v, and as above we obtain the result that the point B corresponds to a singular point of nodal type. This calculation however does not enable us to decide whether this node is stable or unstable, but this cannot be expected: If we assign to the points of the projective plane with $x_3 = 0$ a direction from the Cartesian plane we recognize that the same direction corresponds to a pair of points on the line at infinity: These two points are so to speak "diametrically opposed". So actually two singular points of the saddle type belong to A whereas two nodes each belong to B and C, one of which is stable and the other unstable. Thus the system of equations has a total of six singular points on the line at infinity. Since equation (21.15) holds identically for $x_3 = 0$, $dx_3 = 0$, $x_3 = 0$ is a trajectory, resp. consists of trajectories.[1]

Example

(21.17) $$\dot{x} = -x + y^2, \quad \dot{y} = -y + x^2.$$

Equation (21.15) here has the form

$$x_3(x_2 x_3 - x_1^2)\, dx_1 + x_3(x_2^2 - x_1 x_3)\, dx_2 + (x_1^3 - x_2^3)\, dx_3 = 0.$$

---

[1] *cf.* LEFSCHETZ [1].

The common solutions of (21.16) are determined by

$$x_3 = 0, \quad x_1 = x_2.$$

The "direction" $(1, 1, 0)$ corresponds to them. Shifting the point $(1, 1, 0)$ by means of a linear transformation to the point $(1, 0, 0)$ and again going to the (u, v)-plane yields

$$-v\big(v - (1 + u)^2\big)\, du + \big(1 - (1 + u)^3\big)\, dv = 0,$$

and the singular point proves to be a node. We therefore have a stable and an unstable node on the line at infinity. In addition we have in the finite plane the stable node $(0, 0)$ and the saddle point $(1, 1)$. Fig. 21.1 portrays the curves in the projective plane.

Example. Consider

$$\dot x = y,$$

$$\dot y = -x + a\big((1 + x^2)^{-1} - b\big)\, y.$$

This time

$$p = x_1 x_3 - a x_2 x_3 \left(\frac{x_3^2}{x_1^2 + x_3^2} - b\right),$$

$$q = x_2 x_3,$$

$$r = -\,(x_1^2 + x_2^2) + a x_1 x_2 \left(\frac{x_3^2}{x_1^2 + x_3^2} - b\right).$$

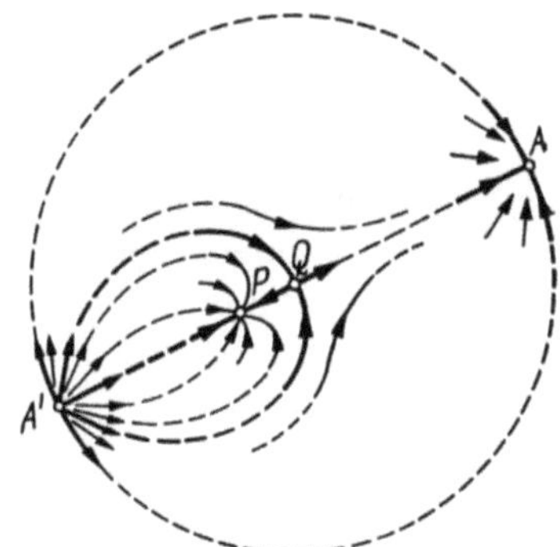

Fig. 21.1. Phase portrait of (21.17) in the projective plane

$x_1 = x_2 = 0$, $x_3 = 1$ yields the origin. Equations (21.16) are also satisfied for $x_3 \equiv 0$. If $|ab| > 2$, the coefficient of dx_3 is zero for $x_1^2 + x_2^2 + ab x_1 x_2 = 0$; this yields four more singular points on $x_3 = 0$. If $|ab| < 2$, then there is no singular point on $x_3 = 0$. The line at infinity is then a trajectory, so to speak an infinitely distant limit cycle[1]).

22. Conservative Second Order Systems

As an example for the discussion of the last section we consider the scalar second order differential equation

$$(22.1) \qquad\qquad \ddot x + f(x) = 0,$$

resp. the system of equations

$$(22.2) \qquad\qquad \dot x = y, \quad \dot y = - f(x).$$

---

[1]) OBMORSHEV [1]. This system of equations describes the oscillations in a generator-motor-system.

(22.1) describes the behavior of a physical system in which there is no energy loss due to friction, current heat etc. The first integral

$$(22.3) \qquad \frac{1}{2}\left(\dot{x}^2 + F(x)\right) = \text{constant}, \qquad F(x) := 2 \int^x f(n)\, dn,$$

is found immediately, the left side is known to be an expression for the total energy provided that the integral is uniquely determined. Equation (22.3) is the reason why the system is called *conservative*. The singular points x_s of (22.1) lie on the x-axis and are defined by the equation

$$(22.4) \qquad\qquad\qquad f(x_s) = 0.$$

We may assume without loss of generality that $x = 0$ is one of the singular points. Equation (22.3) resp. the equation

$$(22.5) \qquad\qquad\qquad y^2 + F(x) = c$$

is the equation of the phase trajectories. This equation shows that the curves are arranged symmetrically about the x-axis which implies that the origin cannot be a focus. From (22.2) we conclude that x increases in the upper half-plane and decreases in the lower half-plane. Therefore, the curves encircle the origin in the clockwise sense.

If in a neighborhood of $x = 0$ we have

$$(22.6) \qquad\qquad\qquad x f(x) > 0 \qquad (0 < x < h)$$

then the left side of (22.5) is positive for small x and this equation defines a family of closed curves. In this case the origin is obviously a center. If the function $f(x)$ possesses a Taylor expansion in a neighborhood of the origin

$$(22.7) \qquad\qquad\qquad f(x) = f_1 x + f_2 x^2 + \cdots,$$

then (22.6) implies that the first non-zero term of the series is of odd degree and positive. This condition is therefore necessary and sufficient for the origin to be a center. If the origin is a center then the singular point adjacent to the left or right cannot also be a center; for the curve $y = f(x)$ which, because of (22.6), passes from the third quadrant through the origin into the first quadrant, cannot arrive at the next singular point in the same direction.

If the first non-zero term in (22.7) is of odd degree and negative then the family of curves (22.5) behaves like $y^2 - x^{2m} = c$ in the vicinity of the origin. In this case there exist exactly two trajectories through the origin which this time is a saddle point. In many concrete systems $f(x)$ is an odd function of x, in which case the origin can only be a saddle point or a center. If the first non-zero term is of even degree then the curve has a cusp at the origin.

The separatrices are found as follows: In (22.5) we replace x and y by the coordinates $(x_s, 0)$ of a singular point and obtain the value c_s of the constant c which belongs to $x = x_s$ (*i.e.* the potential energy belonging to x_s). The desired equation is

$$y^2 + F(x) = c_s.$$

We illustrate this procedure in a few **examples**.

a) $f(x) = x^{2k-1}$, resp. $\dot{x} = y$, $\dot{y} = -x^{2k-1}$, $k \geq 1$ an integer. The origin is singular and a center. All solutions are periodic. We obtain a 2-parameter representation of the general solution in the following manner. Obviously $x^{2k} + ky^2$ is a first integral. We set it equal to c^{2k}, introduce

$$\theta = c^{k-1}\, t + \gamma, \qquad\qquad \gamma \text{ constant,}$$

and define two functions Cs θ and Sn θ by

$$\frac{dCs\,\theta}{d\theta} = -\,Sn\,\theta, \quad \frac{dSn\,\theta}{d\theta} = (Cs\,\theta)^{2k-1},$$

$$Cs\,0 = 1, \quad Sn\,0 = 0.$$

Then we have

$$x(t) = c\,Cs\,\theta, \quad y(t) = -\,c^k\,Sn\,\theta.$$

We obviously have here generalizations of the trigonometric functions $\sin\theta$ and $\cos\theta$. They were developed by LIAPUNOV [2] in order to study certain critical cases (*cf.* sec. 79). In particular, the relation

$$Sn\,\theta = \frac{1}{\sqrt{k}}\cdot\sin\varphi\sqrt{1 + \cos^2\varphi + \cdots + \cos^{2k-2}\varphi}$$

holds in case Cs $\theta =: \cos\varphi$ and the half-period is given by

$$2\sqrt{k}\int_0^1 \frac{dx}{\sqrt{1 - x^{2k}}} = \frac{\sqrt{\pi}}{k}\,\frac{\Gamma(1/2\,k)}{\Gamma((k+1)/2\,k)}.$$

b) $f(x) = x - x^3$. The singular points are -1, 0, $+1$, and they are a saddle point, a center, a saddle point, resp. The equation of the separatrix is

$$y^2 - \frac{1}{4}(x^2 - 1)^2 = 0;$$

its graph is a pair of parabolas which intersect in the saddle points (fig. 22.1).

c) $f(x) = x - x^2$. The singular points are 0, $+1$ (center and saddle point). The separatrix

$$y^2 + \frac{x^2}{2} - \frac{x^3}{3} = \frac{1}{6}$$

has a crunode at $x = +1$.

d) $f(x) = x[1 - (1 + a)(1 + x^2)^{-1/2}]$ with $a > 0$. Singular points are 0 (saddle point), $\pm\sqrt{(1 + a)^2 - 1}$ (centers). The equation of the phase curves (fig. 22.3) is

$$y^2 + x^2 - 2(1 + a)\left(\sqrt{1 + x^2} - 1\right) = c.$$

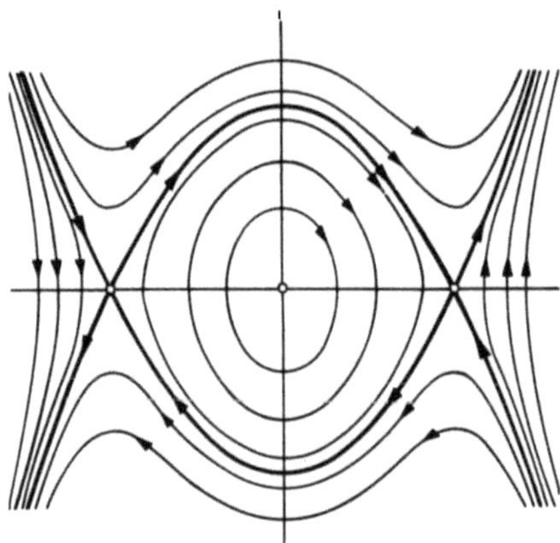

Fig. 22.1. (Example. 22b)

For the separatrix $c_s = 0$, it is a curve of the lemniscate type. For the centers we have

$$y^2 + \left(\sqrt{x^2 + 1} - a - 1\right)^2 = 0.$$

The differential equation describes the motion of the oscillating system sketched in fig. 22.2. The case of an unstable equilibrium in which both springs F_i are facing in the same direction corresponds to the saddle point[1].

e) $f(x) = x^2(1 - x^2)$. Singular points: $-1, 0, +1$, the first of which is a center, the third a saddle point. The equation of the trajectories is

$$\frac{y^2}{2} + \frac{x^3}{3} - \frac{x^5}{5} = c.$$

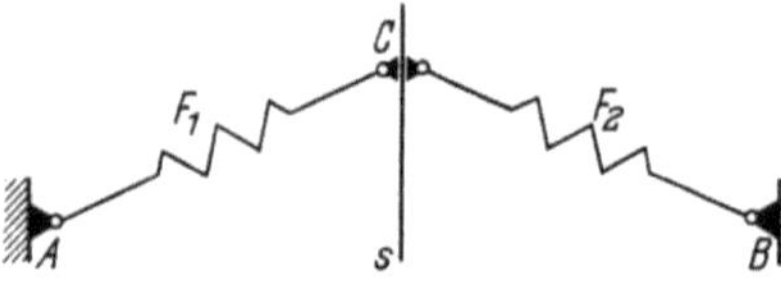

Fig. 22.2. (Example. 22d)

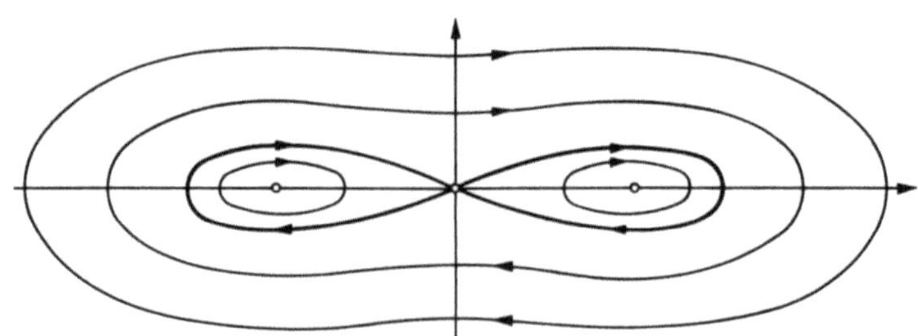

Fig. 22.3. (Example. 22d)

These curves have disconnected branches for values of c between $-2/15$ and $+2/15$. For $c = +2/15$ there is a separatrix through $(+1, 0)$ which has a crunode there and surrounds the other two singular points. But it also surrounds a branch of the phase curve belonging to $c = 0$, which forms a loop with a cusp at the singular point $(0, 0)$ (fig. 22.4).

---

[1] KAUDERER [1], sec. 44.

f) $f(x) = \sin x$. The equation describes the motion of an ideal simple pendulum. Singular points are the integral multiples of π. The even multiples correspond to centers, the odd multiples to saddle points. The equation of the two separatrices is

$$y^2 = 2\,(1 + \cos x) = 4 \cos^2 \frac{x}{2}\,.$$

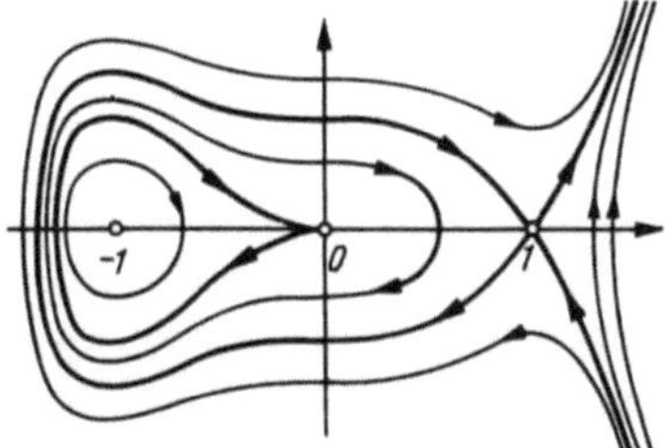
Fig. 22.4. (Example. 22 e)

The right side of this differential equation is periodic in x. Accordingly, the graph of the trajectories in the phase plane has a period and it suffices to discuss a strip one period wide. Imagining this strip bent into a cylinder we obtain in place of the phase plane a *phase cylinder*. In addition to the usual cycles of the differential equation there are other closed curves on the phase cylinder going completely around the surfaces of the zylinder. The corresponding motion is not periodic in t but rather, because of the nature of the equation, periodic in x.

g) The equation

$$(22.8) \qquad \ddot{x} + f_1(x) = 0 \ (x > 0); \quad \ddot{x} + f_2(x) = 0 \ (x < 0)$$

is of the type mentioned at the end of sec. 16; the right side has a discontinuity. We graph the curves for the individual equations in (22.8) separately, continue abutting curves across the y-axis, and interpret the resulting joined-together curves as phase trajectories of the solution of (22.8) (fig. 22.5). We must utilize a somewhat generalized concept of the solution here (*cf.* sec. 16). At the joints $x(t)$ is continuous but the derivative has a jump. The special case

$$\ddot{x} + x - a = 0 \ (x > 0);$$
$$(22.9)$$
$$\ddot{x} + x + a = 0 \ (x < 0)$$

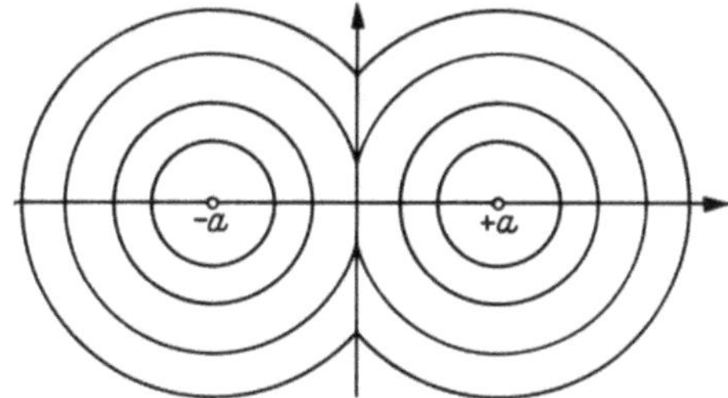
Fig. 22.5. Phase portrait of (22.9)

has the points $(+a,\, 0)$ and $(-a,\, 0)$ as singular points, both being centers. Each phase curve is made up of two circular arcs which have the singular points at their centers and which abut at the y-axis. If the starting points are sufficiently close to the singular points then the circles do not cut the y-axis and one of the two equations alone describes the whole motion. The phase curves show that we must complete (22.9) by requiring that $(0, 0)$ is also a solution; it is of course unstable (see also sec. 74).

As already remarked, the phase trajectory of (22.1) is only a projection of a motion into the phase plane and therefore does not give complete information on the motion. If we wish to know the velocity of the phase point, for instance, we must use the relation

$$\frac{ds}{dt} = \sqrt{\dot{x}^2 + \dot{y}^2} = \sqrt{y^2 + f(x)^2}$$

(s is arc length), and for this purpose x and y must be known explicitly as functions of t. On the other hand, the time needed to traverse the arc of the trajectory between two points of known abscissa can be computed using the first integral (22.3) resp. (22.5). For we have

$$\frac{dx}{dt} = \sqrt{c - F(x)}, \quad t = \int_{x_1}^{x_2} \frac{dn}{\sqrt{c - F(n)}} \ ;$$

c has the value which we found for the trajectory. If the trajectory approaches a singular point, then the denominator of the integrand tends to zero and the behavior of the integral depends on how quickly the denominator converges. For example, for $x_s = 0$ and $c = 0$ the denominator behaves like $\sqrt{F(x)}$. If $f(x)$ satisfies a Lipschitz condition then clearly $f(x) = O(x)$ and $F(x) = O(x^2)$ for small x. Then the integral diverges and this means the phase point cannot reach the singular point in finite time. This points out the fundamental difference between a cycle and a closed separatrix. The closer a cycle lies to a closed separatrix (a phase polygon) the larger is the time required to traverse the cycle, the *length of the period* because this quantity depends continuously on the initital values. In examples a), b), e) therefore, the length of the period grows from the inside out. In example d) there are two types of cycles. For the inner family the length of the period grows from the inside out and for the outer family from the outside in. If $f(x) = O(x^\alpha)$, $0 < \alpha < 1$, (the Lipschitz condition is not satisfied here) then $F(x) = O(x^{\alpha+1})$ and the integral converges. The singular point is reached in finite time.

Example.

$$\dot{x} = y, \quad \dot{y} = |x|^{1/2}; \quad F(x) = \frac{4}{3}|x|^{3/2}\operatorname{sgn} x,$$

for $x_1 < x_2 \leq 0$,

$$t = \sqrt{\frac{3}{4}} \int_{x_1}^{x_2} \frac{-dn}{|n|^{3/4}} = 2\sqrt{3}\left(|x_1|^{1/4} - |x_2|^{1/4}\right).$$

If $x_2 = 0$ then $t = 2\sqrt{6}\,|x_1|^{1/4}$, *i.e.* it is finite.

For concrete systems the function f in (22.1) depends in general on parameters and so does the position and the character of the singular points. In illustrating the pertinent problems we shall assume that only real parameters occur. The right side of (22.2) has the form $f(x, a)$ in this case. The equation for the singular points

$$(22.10) \qquad\qquad f(x_s, a) = 0,$$

which is formed according to (22.4) defines a curve in the (a, x_s)-plane. We assume that this curve has a continuous tangent except possibly at finitely many points. The curve (22.10) separates the region $f(x_s, a) < 0$ from the region $f(x_s, a) > 0$ (fig. 22.6). As a moves continuously, the corresponding differential equation has a singular point at the place where f changes sign. As we noted above the stability character depends here on the sign of the derivative $\frac{\partial f}{\partial x}(x_s, a)$: If the derivative is positive the equilibrium point is a center. A segment of the curve along which $\partial f/\partial x > 0$, lies above the region $f < 0$. Singular points of higher order belong to the crunodes of the curve as well as to points with a vertical tangent. The corresponding values of the parameter are called *branch values*. In their vicinity the phase curves change their entire topological structure: We see for instance

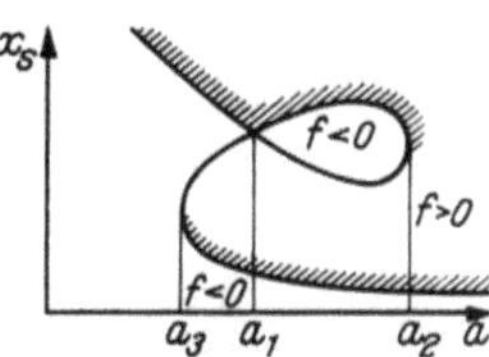

Fig. 22.6. System (22.10) depending on parameter; a_1, a_2, a_3 branch values. The hatched parts of the curve correspond to stable equilibria

in fig. 22.6, that the system has three singular points for $a_3 < a < a_2$ but only one for $a > a_2$. As a grows continuously and passes through a_2 from the left we must except the system to change discontinuously from one state of equilibrium to another in case it was in the state which corresponds to the upper part of the curve. A similar jump is to be expected as we pass through a_3 from right to left. Such jumps can actually be observed. On the other hand, the system remains in the state corresponding to the lower part of the curve, as a decreases through a_2. A closer inspection of the phase curves shows that not only the stable equilibrium on the upper part of the curve but also the surrounding cycles and the separatrix disappear as a passes through a_2 from the left.

An equation of the form

$$\ddot{x} + k\left(\frac{a}{c - x} - x\right) = 0$$

for example, describes the motion of a conductor through which an electric current flows; the conductor is elastically suspended (spring) and is being attracted by a parallel conductor. k and c are fixed constants, a depends on the magnitude of the current and is to be considered as a parameter. We see that the singular points are determined by the equation

$$x_s^2 - c x_s + a = 0 \quad (a \neq 0).$$

In case $a = 0$, only the origin is singular. If $a < 0$ then there are two centers. As a passes through the value 0 one of the centers becomes a saddle point. For $a = c^2/4$, the two singular points coincide and form a singular point of higher order. If $a > c^2/4$ then there exists no singular point[1]).

---

[1]) *cf.* ANDRONOV, WITT and KHAIKIN [1]; KAUDERER [1], sec. 44, where this example is discussed in more detail.

Chapter IV

The Direct Method of Liapunov [1]

23. Geometric Interpretation

Let the scalar system

$$(23.1) \qquad \begin{aligned} \dot{x} &= ax - y + kx(x^2 + y^2), \\ \dot{y} &= x - ay + ky(x^2 + y^2), \end{aligned} \qquad a^2 < 1,$$

be given and sketch the phase curves. Since the singular point of the linear part is a center the criteria of sec. 21 fail and nothing can be said immediately about the stability of the equilibrium of (23.1). In addition to the phase curves, graph the family of curves (fig. 23.1)

$$(23.2) \qquad v(x, y) := x^2 - 2axy + y^2 = \text{constant}.$$

This is a family of similar ellipses which share the origin as their common center. Introducing the vector

$$\text{grad } v = \text{col}\left(\frac{\partial v}{\partial x}, \frac{\partial v}{\partial y}\right)$$

we calculate the angle ψ between a trajectory and the outer normal line to one of the curves (23.2). We have

$$\cos \psi = \left(\frac{\partial v}{\partial x}\dot{x} + \frac{\partial v}{\partial y}\dot{y}\right)\Big/ \left(|\text{grad } v| \cdot \sqrt{\dot{x}^2 + \dot{y}^2}\right).$$

(23.3)

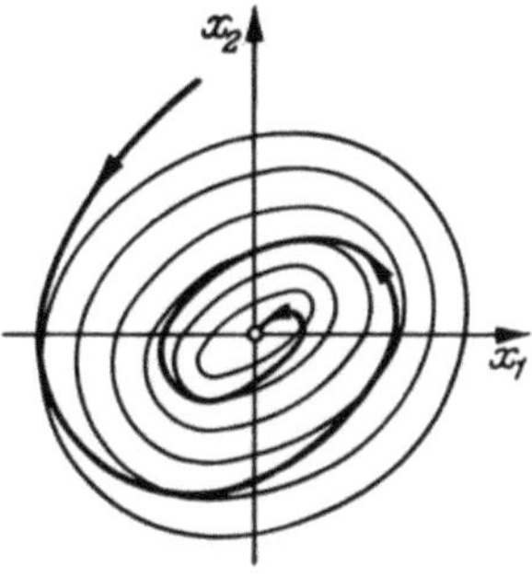

Fig. 23.1. Curves $v = \text{const.}$ and phase trajectory

If this cosine is constantly negative it means that the phase trajectories traverse the family of curves (23.2) from the outside in as t increases. If the angle of intersection is not too small, *i.e.* if (23.3) is not too close to zero, we can conclude that the trajectories come arbitrarily close to the origin: The origin is asymptotically stable.

The numerator of (23.3) is the derivative of (23.2) for the differential equation (23.1) (*cf.* sec. 16).

$$\dot{v} := 2\big((x - ay)\dot{x} + (y - ax)\dot{y}\big) = 2k(x^2 + y^2)(x^2 + y^2 - 2axy).$$

(23.4)

---

[1] References for this chapter: ANTOSIEWICZ [1], HAHN [4], KALMAN and BERTRAM [1], LaSALLE and LEFSCHETZ [1].

This numerator is formed by using as the argument of the function $v(x, y)$ the components of a solution and by taking the derivative of the resulting function of t. In the present case $\dot{v}$ is clearly negative if k is negative. Also (with $x^2 + y^2 = r^2$)

$$(1 - |a|)\, r^2 \leq x^2 - 2\,a\,x\,y + y^2 \leq (1 + |a|)\, r^2$$

hence

$$|\dot{v}| \geq 2\,|k|\,r^4 (1 - |a|)$$

and

$$|\operatorname{grad} v|^2 = 4\left((x - a\,y)^2 + (y - a\,x)^2\right) = 4\left(r^2(1 + a^2) - 4\,a\,x\,y\right)$$

$$< 4\,(1 + a^2)\left(1 + 2\,\frac{|a|}{1 + a^2}\right) r^2.$$

Finally we have

$$\dot{x}^2 + \dot{y}^2 = (a\,x - y)^2 + (x - a\,y)^2 + k^2\,r^6 + 2\,a\,k\,r^2(x^2 - y^2)$$

and for $r^2 < 1$ the right side becomes smaller than

$$(1 + a^2 + 4\,|a| + k^2 + 4\,|a\,k|)\, r^2.$$

Thus

$$\cos \psi < -\,b\,r,$$

where the constant b depends only on a and k. Along a fixed circle the angle ψ is therefore larger than a fixed angle different from zero. We conclude that for negative k the equilibrium is asymptotically stable. For $k > 0$ we have instability by the same reasoning. In the limiting case $k = 0$ the equilibrium is stable but not asymptotically stable.

Disregarding the peculiarities of the example, we recngnize the basic idea: The stability of the equilibrium is determined on the basis of the sign of a certain function namely the total derivative of the function $v(x, y)$ for the given differential equation. In this process equation (23.2) represents a family of closed curves including the origin which schlichtly cover a certain neighborhood of the origin. The stability is discussed without recourse to the explicit form of the solutions and using only the differential equations themselves. For this reason we speak of the *direct method*; it is named after LIAPUNOV who rigorously substantiated and systematically used it. Since Liapunov used still another method (see sec. 67) the direct method is often called Liapunov's *second method*.

To prove the main theorems of the direct method the ideas which we just discussed must be put in analytic form and freed of the limitations of second order systems. This is done in sec. 25.

24. Some Subsidiary Considerations

A. Comparison Functions. In studying the behavior of solutions of a differential equation for large arguments it is useful to work with comparison functions of a simple form, *i.e.* to estimate the solutions [see for instance (2.6) and (2.7)]. We shall preferably use the monotone functions introduced in Def. 2.5 and 2.6 and denote by $\varphi_1, \varphi_2, \ldots, \psi_1, \psi_2, \ldots$ functions of class K, by $\sigma_1, \sigma_2, \ldots$ functions of class L. Since these functions are always exclusively used in inequalities we can usually assume without loss of generality that they are differentiable. For a non-differentiable function in K, resp. L, can be replaced by a differentiable function of the same kind, maintaining the given inequality. The inverse function is denoted by the exponent I:

$$\varphi^I\left(\varphi(r)\right) \equiv r.$$

The following properties are more or less obvious.

a)
$$\varphi_1\left(\varphi_2(r)\right) \in K,$$

b)
$$\varphi\left(\sigma(s)\right) \in L$$

because for sufficiently large s, $r := \sigma(s)$ lies in the domain of definition of $\varphi(r)$, and $s_1 < s_2$ implies $\sigma(s_1) > \sigma(s_2)$, and hence $\varphi(\sigma(s_1)) > \varphi(\sigma(s_2))$.

c) Let $\varphi(r) \in K$, $0 \le r < r_0$, $\varphi(r_0) = t_0$. The inverse function $\varphi^I(t)$ is then defined at least for $0 \le t \le t_0$ and belongs to K. If $\varphi(r)$ is defined for all $r \ge 0$ and if $\lim_{r \to \infty} \varphi(r) = t_0$ is finite then $\varphi^I(t)$ is not defined for $t > t_0$.

d) $\varphi_1(r) > \varphi_2(r)$ implies $\varphi_1^I(t) < \varphi_2^I(t)$ for $0 \le r \le r_0$, resp. $0 \le t \le t_0$.

e) Let $\varphi(r) \in K$ for $0 \le r \le k^2$. The inequalities

$$\varphi(r_1 r_2) \le \varphi(k r_1), \quad \varphi(r_1 r_2) \le \varphi(k r_2), \quad 0 \le r_1 < k, \quad 0 \le r_2 < k,$$

imply the inequality

$$\varphi(r_1 r_2) \le \sqrt{\varphi(k r_1)} \cdot \sqrt{\varphi(k r_2)}$$

which can be written in the form

$$(24.1) \qquad \varphi(r_1 r_2) \le \varphi_1(r_1)\, \varphi_2(r_2)$$

if we define $\varphi_i(r_i) \in K$ appropriately.

If r_i takes on values in an infinite interval a similar inequality can clearly be derived if $\varphi(r)$ does not grow faster than a power of r. For if

$$\varphi(r) \le c r^\alpha, \quad \alpha > 0, \quad \text{resp. } \ln \varphi(r) = O(\ln r),$$

then

$$\varphi(r_1 r_2) \le c r_1^\alpha r_2^\alpha.$$

But now let

$$\varphi(r) \leq \varphi_3(r) \exp(\beta r^\gamma), \quad \beta > 0, \ \gamma > 0$$

and suppose that $\varphi_3(r)$ does not grow faster than a power of r. Then

$$\varphi(r_1 r_2) \leq \varphi_3(r_1 r_2) \exp(\beta r_1^\gamma r_2^\gamma)$$
$$\leq c r_1^\alpha r_2^\alpha \exp\left(\frac{\beta}{2}(r_1^{2\gamma} + r_2^{2\gamma})\right)$$

and this inequality can again be put into the form (24.1). This time the restriction on φ is $\ln \ln \varphi(r) = O(\ln r)$. This process can be continued. It is clearly always possible to estimate $\varphi(r_1 r_2)$ by a product $\varphi_1(r_1)\, \varphi_2(r_2)$ if the nth iterated logarithm of $\varphi(r)$ is of the order of magnitude of $\ln r$.

f) Let $\varphi(r) \in K$, $\psi(r) \in K$, $\sigma(s) \in L$, and either $\varphi'(r)$ or $\psi(r)$ bounded. Then there exists an estimate

$$\varphi\big(\psi(r)\,\sigma(s)\big) \leq \varphi_1(r)\,\sigma_1(s), \quad \varphi_1(r) \in K, \quad \sigma_1(s) \in L.$$

For if $\psi(r) \leq t_0$, then the left side is smaller than

$$\sqrt{\varphi\big(\sigma(s_0)\,\psi(r)\big)} \cdot \sqrt{\varphi\big(t_0\,\sigma(s)\big)}$$

and if $\varphi'(r) \leq \alpha$ then $\alpha\psi(r)\,\sigma(s)$ is a bound for $\varphi(\psi(r)\,\sigma(s))$.

g) If $s_1 \geq t_1$, $s_2 \geq t_2$, $\sigma(s) \in L$ for $s \geq t_1 + t_2$, then

$$\sigma(s_1 + s_2) \leq \sigma_1(s_1)\,\sigma_2(s_2), \quad \sigma_i \in L.$$

This follows in the same manner as the inequality in e).

h) If $0 < m_1 \leq r_1$, $0 < m_2 \leq r_2$, and $\varphi(r) \in K$, then

$$\varphi(r_1 r_2) \geq \varphi_1(r_1)\,\varphi_2(r_2), \quad \varphi_i \in K$$

again as in e). If φ is differentiable and if the derivative is bounded away from zero, $\varphi'(r) \geq \delta > 0$ then

$$\varphi(r_1 r_2) \geq \delta r_1 r_2$$

is an estimate of the same kind.

The estimates e) to h) are in a certain sense existence statements: It is essentially possible to estimate the function $\varphi(r_1 r_2)$ by a product of two functions belonging to K each of which depends on only one variable, etc. In case the comparison functions are explicitly given such estimates can often be made much easier and also much more advantageously; they become for instance trivial for power functions.

We shall also discuss comparison functions of two variables at this time although we will not need them immediately.

Def. 24.1. A real function $k(r, t)$ of the real variables r and t belongs to class KK, if a) it is defined for $0 \leq r \leq r_1$, resp. $0 < r < \infty$, and for $0 \leq t_0 \leq t < \infty$, b) for each fixed t it belongs to class K with respect to r, c) for each fixed r it is a continuous, positive, monotone increasing, and unbounded function of t.

Def. 24.2. A real function $l(r, s)$ of the real variables r and s belongs to the class KL, if a) it is defined for $0 \le r \le r_1$, resp. $0 \le r < \infty$, and for $0 \le s_0 \le s < \infty$, b) for each fixed s it belongs to class K with respect to r and for each fixed r it is monotone decreasing to zero as s increases (it need not be strictly monotone).

i) Let $l(r, s) \in KL$. If $l(r, s)$ is bounded with respect to r, $i.e.$ if $l(r, s) \le l_0(s)$ then there exists an estimate

$$l(r, s) \le \varphi(r)\,\sigma(s), \qquad \varphi \in K, \ \sigma \in L,$$

$e.g.$

$$\varphi(r) = \sqrt{l(r, s_0)}, \qquad \sigma(s) = \sqrt{l_0(s)}.$$

If r takes on values in an infinite interval and $l(r, s)$ is not uniformly bounded, such an estimate is certainly possible in case the function $l(r, s)/l(r_0, s)$ is monotone decreasing for all $r \ge r_0$, as s increases. For then we have

$$\frac{l(r, s)}{l(r_0, s)} < \frac{l(r, s_0)}{l(r_0, s_0)},$$

which implies

$$l(r, s) < \frac{l(r, s_0) \cdot l(r_0, s)}{l(r_0, s_0)}.$$

k) Let $k(r, t) \in KK$. Let $r = h(u, t)$ be obtained by solving the equation $k(r, t) = u$ for r, $i.e.$ $k(h(u, t), t) \equiv u$. This can also be written $r = k^I(u, t)$. If we choose $r_1 < r_2$, $t_1 > t_2$, so that

$$k(r_1, t_1) = k(r_2, t_2) =: u_0,$$

then

$$h(u_0, t_1) = r_1 < h(u_0, t_2).$$

Hence h is monotone decreasing as t increases. For finite t, $h(u, t)$ can vanish only for $u = 0$. If $u > 0$ then $h(u, t)$ tends to zero as $t \to \infty$. For if we assume $\lim_{t \to \infty} h(u, t) = \delta > 0$ then we have $\lim_{t \to \infty} k(\delta, t) = u$, a finite limit, which contradicts the hypothesis that $k \in KK$. The monotonicity of $k(r, t)$ with respect to r implies the monotonicity of $h(u, t)$ with respect to t, all of which implies that $h(u, t) \in KL$, resp. $k^I(u, t) \in KL$.

B. Definite Functions. We will next consider functions $v(x)$ of the variables $x_1, \ldots, x_n$, which are defined and continuous on a certain neighborhood K_h of the origin or on all of R_n; they vanish at the origin.

Def. 24.3. A function $v(x)$ is called *positive definite* if $v(0) = 0$ and if $v(x)$ is positive at every other point of a neighborhood U of the origin.

Because of the continuity of $v(x)$ we have for sufficiently small r, $0 < r_1 \leq r \leq r_2$,

$$v(x) \leq \max v(y), \quad |y| \leq r,$$
$$v(x) \geq \min v(y), \quad r \leq |y| \leq r_2.$$

on the sphere $|x| = r$. The functions of r on the right are monotone and may be estimated by means of functions of the class K, thereby maintaining the given inequalities. So there exist two functions $\varphi_1, \varphi_2 \in K$ such that

$$(24.2) \qquad \varphi_1(|x|) \leq v(x) \leq \varphi_2(|x|).$$

The left side of this inequality could also be used as a definition of the concept *positive definite*; the domain on which it is valid should then be indicated.

Def. 24.3 a. The function $v(x)$ is called *positive definite in the domain B* if there exists a function $\varphi(r) \in K$ such that

$$(24.3) \qquad \varphi(|x|) \leq v(x), \quad x \in B.$$

B must contain the origin at least as a boundary point.

A function $v(x)$ is called *negative definite* if $-v(x)$ is positive definite.

A function is called *positive*, resp. *negative semi-definite* in a neighborhood of the origin if it is never negative, resp. positive there.

Def. 24.4. A function $v(x)$ is called *indefinite* if $v(0) = 0$ and if $v(x)$ assumes both positive and negative values in each neighborhood of the origin.

Examples.

$$v(x) = x_1^2 + x_2^2 + x_3^2 \text{ is positive definite on } R_3,$$

$$v(x) = x_1^2 + (x_2 + x_3)^2 \text{ is semi-definite because the}$$

function vanishes on the plane $x_2 = -x_3$.

$$v(x) = x_1^2 + x_2^2 \text{ is positive definite in the plane},$$

semi-definite in R_3, since it vanishes on the x_3-axis.

$$v(x) = x_1^2 + x_2^2 - (x_1^4 + x_2^4)$$

is positive definite in the interior of the unit circle since clearly

$$v(x) \geq |x|^2 - |x|^4, \quad |x| < 1.$$

Outside the unit circle v can assume negative values.

$$v(x) = \frac{x_1^2}{1 + x_1^2} + x_2^2$$

admits an estimate (24.3) only for finite $|x|$ since for fixed x_2 and increasing x_1 v tendsto $x_2^2 + 1$.

Theorem 24.1. Let $v(x)$ be positive definite and, in the notation of (24.2), let $\varphi_1(h) =: c_1$. Then for $c < c_1$ the equation $v(x) = c$ defines a closed hypersurface in R_n.

Proof. The points y which satisfy the equation $v(y) = c$ lie within the sphere $|x| = h$; for $v(y) = c < c_1 = \varphi_1(h)$ and $\varphi_1(|x|) \leq v(x)$. Let $\bar{x}$ be a fixed point on the sphere $|x| = h$. Then the function $v(t\bar{x})$, considered as a function of the parameter t in $0 \leq t \leq 1$, varies between zero and a number $c_2 \geq c_1$. Thus there exists at least one real number t_0, $0 < t_0 < 1$ such that $v(t_0\bar{x}) = c$. If we choose on each ray $O\bar{x}$ the minimal number t_0, then the points $t_0\bar{x}$ determine a closed hypersurface.

The surface $v = c$ is closed for an arbitrary c only if (24.3) is valid for arbitrarily large r and if $\lim \varphi(r)$ is infinite. The surface

$$v(x) = \frac{x_1^2}{x_1^2 + 1} + x_2^2 = c > 1$$

is not closed. The surface

$$\frac{x_1^2 + x_2^2}{x_1^2 + x_2^2 + 1} = c$$

is not even defined in R_2 for $c > 1$.

Def. 24.5. A function $v(x)$ is called *radially unbounded* if there exists an estimate

$$v(x) \geq \varphi(|x|)$$

which holds for all x and if in addition $\lim \varphi(|x|) = \infty$.

C. Homogeneous Functions

Theorem 24.2. Let $v(x)$ be a function of the type considered in B, that is, either definite or indefinite, and which in addition is homogeneous of degree $k \geq 1$. Let $w(x)$ be continuous and assume that for $|x| \leq h$

$$(24.4) \qquad |w(x)| \leq a\,|x|^k, \quad a > 0.$$

Then in each case there exists a constant b such that the function

$$u(x) := v(x) + b\,w(x)$$

is positive definite, negative definite, or indefinite together with $v(x)$.

Proof. Because of the homogenity

$$v(x) = v\left(|x| \cdot \frac{x}{|x|}\right) = |x|^k v\left(\frac{x}{|x|}\right) \quad (x \neq 0).$$

If we define

$$a_1 := \max v(y), \quad a_2 := \min v(y), \quad |y| = 1.$$

then we have, accordingly,

$$a_2\,|x|^k \leq v(x) \leq a_1\,|x|^k.$$

7*

If $v(x)$ is positive definite then

$$u(x) \geq v(x) - |b|\,|w(x)| \geq (a_2 - |b|\,a)\,|x|^k$$

hence u is positive definite if $a_2 > |b|\,a$. If $v(x)$ is negative definite, we must have $|a_1| > b\,a$. Finally if $v(x)$ is indefinite, we take $a\,b = \min\,(|a_1|, |a_2|)$. For then the function $u(x)$ has the same sign as $v(x)$ near the points where $v(x)$ assumes its (positive) maximum, resp. its (negative) minimum; v is therefore also indefinite.

If $w(x)$ is a homogeneous function of degree $l > k$, then the function

$$u = v + w$$

has the same type of definiteness as $v(x)$. For then inequality (24.4) can be satisfied for an arbitrary a if $|x|$ is chosen sufficiently small; we have

$$|w(x)| \leq a_1\,|x|^l = a_1\,|x|^{k+r}, \quad r > 0$$

and so

$$|w(x)| \leq a\,|x^k| \quad \text{if} \quad |x| < \left(\frac{a}{a_1}\right)^{1/r}.$$

The definition implies immediately:

A positive definite homogeneous function is always radially unbounded.

D. Quadratic Forms. Let

$$v(x) = x^T B x = \sum_{i,k=1}^{n} b_{ik}\,x_i\,x_k$$

be a quadratic form with symmetric matrix

$$B = (b_{ik}), \quad b_{ik} = b_{ki}.$$

The following theorems are listed here without proof[1]).

Theorem 24.3. The quadratic form $v(x)$ is positive definite if and only if the main determinants

$$b_{11}, \quad b_{11}b_{22} - b_{21}b_{12},$$

$$\det(b_{ik}), \quad i,k = 1, 2, \ldots, m; \quad m = 3, 4, \ldots, n$$

are positive.

These inequalities are also known as the *Sylvester inequalities*.

An immediate consequence is: The form $v(x)$ is negative definite if the pth main determinant has the sign $(-1)^p$.

---

[1]) *cf.* for instance BELLMAN [2], SCHMEIDLER [1].

Theorem 24.4. Let $\beta_1, \ldots, \beta_n$ be the characteristic roots of the matrix B. Then

$$\max_{|x|=1} v(x) = \max_i \beta_i, \qquad \min_{|x|=1} v(x) = \min_i \beta_i.$$

Corollary: The form $v(x)$ is definite if all the characteristic roots have the same sign; it is indefinite if it possesses both positive and negative characteristic roots.

If all the characteristic roots of a matrix have the same sign and some of them are zero then $v(x)$ is semi-definite.

E. Scalar Equations of the First Order. Let

$$\dot{y} = -f(y), \quad f \in K, \quad 0 \le y \le h$$

and denote by $G(y)$ a primitive function for $-1/f$. Then

$$\int_{y_0}^{y} \left(-\frac{dy}{f(y)} \right) = G(y) - G(y_0) = t - t_0$$

and

$$(24.5) \qquad y(t) = G^I\big(t - t_0 + G(y_0)\big), \quad 0 \le y \le h.$$

If the function $1/f(y)$ is integrable on an interval containing zero then $G(y)$ can be normalized so that $G(0) = 0$ and so that $-G(y)$ belongs to class K. Then $-G^I(y)$ also belongs to class K. It follows from (24.5) that a finite argument of G^I corresponds to the value $y = 0$; the value $y = 0$ is attained after finite time. The integrability condition for f is incompatible with the Lipschitz condition; this explains the absence of uniqueness. If $1/f(y)$ is not integrable on an interval containing zero then $G(y)$ is monotone increasing and unbounded as y approaches zero. In this case the inverse function is a function of class L, and utilizating Ag) with $s_1 = t - t_0$, $t_1 = 0$, $s_2 = G(y_0)$, $t_2 = G(h)$, we obtain from (24.5) an estimate

$$y(t) \le \sigma_1(t - t_0)\,\sigma_2\big(G(y_0)\big), \quad \sigma_i \in L.$$

The second factor on the right belongs to class K with respect to y_0. Using an appropriate notation, we have

$$y(t) \le \varphi(y_0)\,\sigma(t - t_0), \quad \varphi \in K, \ \sigma \in L.$$

This type of an estimate exists a fortiori for solutions which already vanish on a finite time interval; it is hence valid in general. In case $f \in E$ this is equivalent to asymptotic stability of the equilibrium (see Def. 2.9). Simple examples are given by the powers of the variable. For $\dot{y} = -\sqrt{y}$ we have $G(y) = -2\sqrt{y}$ and the solution $y(t) = \left(\sqrt{y_0} - \frac{1}{2}(t - t_0) \right)^2$. For $\dot{y} = -y^2$ we have $G(y) = 1/y$ and the solution $y(t) = (t - t_0 + y_0^{-1})^{-1}$.

25. The Principal Theorems of the Direct Method for Autonomous Differential Equations

Consider the differential equation

$$(25.1) \qquad \dot{x} = f(x), \qquad 0 \leq |x| \leq h, \quad f \in E.$$

To justify rigorously and analytically the method sketched in sec. 23 we must first of all define precisely the properties of the Liapunov function v which was only characterized geometrically there. From the results of sec. 24B we recognize that definiteness is essential; in all other respects the Liapunov functions which we will consider next are of the type described in sec. 24B. The *derivative of v for the differential equation* is defined in accordance with sec. 16; it is here a function of x alone.

Theorem 25.1. If there exists a positive definite function $v(x)$ whose derivative $\dot{v}$ for (25.1) is negative semi-definite or identically zero then the equilibrium of (25.1) is stable.

Proof. By hypothesis there exist a function $v(x)$ and two functions $\varphi_1(r)$, $\varphi_2(r)$ in class K such that [see (24.2)]

$$(25.2) \qquad \varphi_1(|x|) \leq v(x) \leq \varphi_2(|x|).$$

If we replace the argument x of $v(x)$ by a solution $x(t)$ then $v(x)$ becomes a function of t which is non-increasing because $\dot{v} \leq 0$:

$$v(x(t)) \leq v(x(t_0)), \quad t \geq t_0.$$

If for $x(t)$ we use the solution $p(t, x_0, t_0)$ then

$$v(x) \leq v(x_0) \leq \varphi_2(|x_0|),$$

which implies, taking inverses, that

$$(25.3) \qquad |x| \leq \varphi_1^I(v(x)) \leq \varphi_1^I(\varphi_2(|x_0|)).$$

On the right we have a function of class K which will be denoted by φ_3. Then we have

$$|p(t, x_0, t_0)| \leq \varphi_3(|x_0|),$$

which implies the stability of the equilibrium by (2.6).

Theorem 25.2. If there exists a positive definite function $v(x)$ whose derivative for (25.1) is negative definite then the equilibrium of (25.1) is asymptotically stable.

Proof. In addition to (25.2) we now have the inequality

$$(25.4) \qquad \dot{v} \leq - \varphi_4(|x|), \quad \varphi_4 \in K.$$

We apply sec. 24A and conclude

$$(25.5) \qquad \varphi_2^I(v) \leq |x| \leq \varphi_1^I(v),$$

and hence

$$(25.6) \qquad \dot{v} \leq - \varphi_4\big(\varphi_2^I(v)\big) = : - \chi(v), \quad \chi \in K.$$

We now consider the auxiliary scalar differential equation

$$\dot{w} = - \chi(w), \quad w \geq 0.$$

From sec. 24E we have

$$w(t) \leq q(w_0)\, \varrho(t - t_0); \quad q \in K, \varrho \in L, w_0 = w(t_0).$$

If $v_0 = w_0$ then $v(t) \leq w(t)$ for all t. Hence

$$v(t) \leq q(v_0)\, \varrho(t - t_0)$$

and because of (25.5)

$$|\boldsymbol{x}| \leq \varphi_1^I\big(q(v_0)\, \varrho(t - t_0)\big) \leq \varphi_1^I\big(q(\varphi_2(|\boldsymbol{x}_0|))\big)\, \varrho(t - t_0))$$

and finally, utilizing sec. 24A f,

$$|\boldsymbol{x}| = |\boldsymbol{p}(t, \boldsymbol{x}_0, t_0)| \leq \varphi(|\boldsymbol{x}_0|)\, \sigma(t - t_0); \quad \varphi \in K, \sigma \in L.$$

This is the desired statement.

This argument makes the geometric considerations of sec. 23 rigorous.

Theorem 25.3. Let a function $v(\boldsymbol{x})$ be given, which has the following properties: a) there exist points $\boldsymbol{x}$ arbitrarily close to the origin such that $v(\boldsymbol{x}) < 0$; they form the "region $G: v < 0$", which is bounded by the surface $v = 0$ and the sphere $|\boldsymbol{x}| = h$. b) in the interior of G, v is bounded. c) in the interior of G, $\dot{v}$ is always negative.

Then the equilibrium is unstable.

Proof. First we realize that from a) and c) we can obtain an estimate

$$\dot{v} \leq - \varphi(|v|), \quad \varphi \in K, \boldsymbol{x} \in G.$$

This follows as in (25.6). Let $\boldsymbol{x}_0 \in G$ and $v(\boldsymbol{x}_0) = -a < 0$. Then

$$(25.7) \qquad v\big(\boldsymbol{x}(t)\big) = v_0 + \int_{t_0}^{t} \dot{v}\, dt \leq - a - (t - t_0)\, \varphi(a).$$

We conclude that v takes on arbitrarily large negative values and since v is bounded in G the phase curve along which we integrate must leave the region G. It can leave the region only through the sphere $|\boldsymbol{x}| = h$ since at the other boundary surfaces of G the equation $v = 0$ holds. Therefore the phase curve goes through a point for which $|\boldsymbol{x}| = h$, no matter how small $|\boldsymbol{x}_0|$ is. This implies instability (*cf.* sec. 2)●

This theorem is due to CHETAEV [1] and it contains as a special case the so-called First Instability Theorem of Liapunov:

Theorem 25.4. If there exists a function $v(\boldsymbol{x})$ with a negative definite derivative and if v is either negative definite or indefinite then the equilibrium is unstable.

If v and $\dot{v}$ are both negative then a proof for this theorem can also be made in analogy to that of Theorem 25.2. To do so we start with the inequality (25.4) using the opposite sign, arrive at a differential equation for v of the form $\dot{v} \geq \chi(v)$, $\chi \in K$, and derive from it an inequality of the form

$$(25.8) \qquad |\boldsymbol{x}(t)| \geq \varphi(|\boldsymbol{x}_0|)\,\varkappa(t - t_0),$$

where the function $\varkappa$ is monotone increasing at least for small arguments. In this case we also speak of *complete instability*: All the phase curves tend away from the origin. In the linear case (sec. 20) the origin is then an unstable focus or an unstable node but not a saddle point.

Theorem 25.5. Suppose the derivative of the function v has the form

$$\dot{v} = gv \quad \text{or} \quad \dot{v} = gv + w(\boldsymbol{x}),$$

where g is a positive constant. In the second case assume that v and w are not both semi-definite and of opposite sign. Then the equilibrium of (25.1) is unstable. (This is Liapunov's so-called Second Instability Theorem.)

Proof. Let us assume that $w(\boldsymbol{x})$ is positive semi-definite; the argument in the other case is analogous. By hypothesis v cannot be negative semi-definite and therefore admits positive values. Let $v(\boldsymbol{x}_0) =: v_0 > 0$. Then clearly

$$\dot{v} \geq gv.$$

Along the trajectory which begins at $\boldsymbol{x}_0$, $\dot{v} \geq gv_0$; it is thus positive, *i.e.* v is increasing. It follows by integration that

$$v = v_0 + \int_{t_0}^{t} \dot{v}\, dt \geq v_0 + gv_0(t - t_0),$$

and we see as in the proof of Theorem 25.4, that the trajectory leaves any given domain, irrespective of the magnitude of $\boldsymbol{x}_0$.

As shall be seen later (sec. 52), Theorems 25.3 and 25.5 are equivalent.

The theorems in this section may all be stated for functions v with the opposite sign, if in their formulations we interchange positive and negative everywhere.

Def. 25.1. A function $v(\boldsymbol{x})$, which satisfies the hypotheses of Theorems 25.1 to 25.5 and which can therefore be used for testing the stability behavior of a differential equation, shall be called a *Liapunov function for the differential equation*.

The principal theorems give only sufficient conditions, and say nothing about how a suitable Liapunov function is found in a particular case. In fact, the question whether such functions exist remains open. We shall show later that in some cases a Liapunov function can be constructed by using a fixed, practically feasible procedure; but generally we are reduced to a method of systemized trial and error. We shall also treat the problem of the existence of Liapunov functions but we will see that the results are not constructive. Eventually, the main significance of the principal theorems is not that they give a tool with which to attack the stability problem for a fixed differential equation but rather that they frequently give conditions for the stability of certain classes of differential equations, *e.g.* conditions on the parameter involved (*cf.* end of sec. 1). A very simple but typical condition resulted in sec. 23: From the inequality $\dot{v} < 0$ required in Theorem 25.2 we could read off the condition $k < 0$ for stability.

Examples for the principal theorems.

1) If for equation (22.1), resp. (22.2), we choose as the Liapunov function v the first integral (22.3) then $\dot{v} = 0$. Theorem 25.1 applies if v is positive definite. This is the case if the integral $F(x)$ is positive in a neighborhood of $x = 0$. This again leads to inequality (22.6) and to the results of sec. 22.

$$2) \qquad \dot{x}_1 = x_1(a^2 - x_1^2 - x_2^2) + x_2(x_1^2 + x_2^2 + a^2),$$

$$\dot{x}_2 = - x_1(a^2 + x_1^2 + x_2^2) + x_2(a^2 - x_1^2 - x_2^2),$$

$$v = x_1^2 + x_2^2; \quad \dot{v} = - 2(x_1^2 + x_2^2)(x_1^2 + x_2^2 - a^2).$$

In case $a = 0$, Theorem 25.2 implies asymptotic stability. If $a \neq 0$ the equilibrium is unstable by Theorem 25.4 since $\dot{v}$ is positive definite in the domain $x_1^2 + x_2^2 < a^2$.

$$3) \qquad \dot{x}_1 = (x_1 - \beta x_2)(a x_1^2 + b x_2^2 - 1),$$

$$\dot{x}_2 = (\alpha x_1 + x_2)(a x_1^2 + b x_2^2 - 1),$$

$$v = \alpha x_1^2 + \beta x_2^2; \quad \dot{v} = 2(\alpha x_1^2 + \beta x_2^2)(a x_1^2 + b x_2^2 - 1).$$

If the four constants a, b, α, β are all positive then v is positive definite and $\dot{v}$ is negative definite in the domain $a x_1^2 + b x_2^2 < 1$; the equilibrium is asymptotically stable.

4) Let

$$\dot{p}_i = \frac{\partial H}{\partial q_i}, \quad \dot{q}_i = - \frac{\partial H}{\partial p_i}, \quad i = 1, 2, \ldots n,$$

be the equations of a conservative system in canonical form. The Hamiltonian function $H(\boldsymbol{p}, \boldsymbol{q})$ is the sum of the kinetic energy $T(\boldsymbol{p})$ and the potential energy $W(\boldsymbol{q})$, where

$$\boldsymbol{p} = \mathrm{col}\,(p_1, \ldots, p_n); \quad \boldsymbol{q} = \mathrm{col}\,(q_1, \ldots, q_n)$$

and

$$T(\boldsymbol{p}) = T_2(\boldsymbol{p}) + T_3(\boldsymbol{p}) + \cdots$$

is assumed representable as a sum of homogeneous functions whose respective degrees are 2, 3, ... T_2 is positive definite by assumption. If $W(0) = 0$ is an isolated minimum for $q = 0$, then $W(q)$ is positive definite and hence $H(p, q)$ is positive definite with respect to p and q. If we choose the Hamiltonian function as our Liapunov function v then, since $\dot{H} = 0$, Theorem 25.1 applies; the equilibrium is stable if the potential energy has an isolated minimum. Assume now that $W(0) = 0$ is an isolated maximum and furthermore that W has an expansion

$$W(q) = W_k(q) + W_{k+1}(q) + \cdots, \quad k \geq 2,$$

into homogeneous functions whose degrees are $k, k+1, \ldots$. The assumption on $W(0)$ implies that $W_k(q)$ is negative definite. Choosing

$$v = p^T q = \sum_{i=1}^{n} p_i q_i$$

and using Euler's theorem on homogeneous functions we obtain

$$\dot{v} = \sum_{i=1}^{n} (\dot{p}_i q_i + \dot{q}_i p_i) = - \sum_{i=1}^{n} p_i \frac{\partial H}{\partial p_i} + \sum_{i=1}^{n} q_i \frac{\partial H}{\partial q_i}$$

$$= - \sum_{i=1}^{n} p_i \frac{\partial T_2}{\partial p_i} - \sum_{i=1}^{n} p_i \frac{\partial T_3}{\partial p_i} - \cdots + \sum_{i=1}^{n} q_i \frac{\partial W_k}{\partial q_i} + \cdots,$$

$$\dot{v} = - 2T_2 - 3T_3 - \cdots + kW_k + (k+1) W_{k+1} + \cdots.$$

By Theorem 24.2, the type of definiteness of $\dot{v}$ is determined by the term $-2T_2 + kW_k$. Thus $\dot{v} < 0$. On the other hand v is indefinite and therefore the equilibrium is unstable.

If the function W is of the above type but $W(0) = 0$ is not an extremum then we use the Liapunov function $v = Hp^T q$. Then there clearly exists a domain in which W, and hence H, is negative. The function $p^T q$ is indefinite. Therefore there exists a subdomain B of the domain $H < 0$ in which $p^T q < 0$ and $v > 0$. Since $\dot{H} = 0$, the derivative of v has the form

$$H(- 2T_2 - 3T_3 - \cdots + kW_k + \cdots)$$

and is positive in B. Thus the hypothesis of Theorem 25.3 is satisfied: the equilibrium is unstable.

$$\begin{aligned}
5) \quad \dot{x}_1 &= x_2 - 3x_3 - x_1(x_2 - 2x_3)^2, \\
\dot{x}_2 &= - 2x_1 + 3x_3 - x_2(x_1 + x_3)^2, \\
\dot{x}_3 &= 2x_1 - x_2 - x_3, \\
v &= 2x_1^2 + x_2^2 + 3x_3^2, \\
\dot{v} &= - 4x_1^2(x_2 - 2x_3)^2 - 2x_2^2(x_1 + x_3)^2 - 6x_3^2.
\end{aligned}$$

The derivative is negative semi-definite (it vanishes for $x_2 = x_3 = 0$); therefore the equilibrium is stable.

$$6) \quad \dot{x}_1 = -x_1 - x_2 + x_3 + x_1 r^2; \quad r^2 := x_1^2 + x_2^2 + x_3^2,$$
$$\dot{x}_2 = x_1 - 2x_2 + 2x_3 + x_2 r^2,$$
$$\dot{x}_3 = x_1 + 2x_2 + x_3 + x_3 r^2,$$
$$v = x_1^2 + x_2^2 - x_3^2,$$
$$\dot{v} = -2\left(x_1^2(1-r^2) + x_2^2(2-r^2) + x_3^2(1+r^2)\right).$$

$\dot{v}$ is negative definite in the domain $r^2 < 1$, v is indefinite. The equilibrium is unstable.

$$7) \qquad \dot{x}_1 = a x_1 + b x_2 + c_1 x_1(x_1^2 + x_2^2),$$
$$\dot{x}_2 = -b x_1 + a x_2 + c_2 x_2(x_1^2 + x_2^2).$$

And more generally

$$\dot{x} = (a E + B)\, x + \varphi(x)\, \mathrm{col}\,(c_1 x_1, \ldots, c_n x_n),$$

where B is skew-symmetric, $B^T = -B$, and $\varphi(x)$ is a positive definite scalar function. For $v = x^T x$ we find that

$$\dot{v} = \dot{x}^T x + x^T \dot{x} = 2 a x^T x + 2\varphi(x) \sum_{i=1}^{n} c_i x_i^2.$$

If $a > 0$ and $c_i > 0$, then the hypotheses of Theorem 25.5 are satisfied; the equilibrium is unstable.

8) The second order scalar equation

$$\ddot{x} = r(x, \dot{x}), \quad (r(0, 0) = 0)$$

can be written as a system by introducing $\dot{x} = y$ and $\dot{y} = r(x, y)$. For

$$v = y^2 - 2 \int_0^x r(\xi, 0)\, d\xi,$$

we have

$$\dot{v} = 2 y \dot{y} - 2 r(x, 0)\, \dot{x} = 2\left(r(x, y) - r(x, 0)\right) y = 2 y^2 r_y(x, \theta y),$$
$$0 < \theta < 1.$$

The mean value $r_y(x, \theta y)$ can be represented as an integral, yielding

$$\dot{v} = 2 y^2 \int_0^1 r_y(x, \tau y)\, d\tau.$$

If $r_y(0, 0) < 0$, then $\dot{v}$ is negative semidefinite. If $x r(x, 0) < 0$ $(x \neq 0)$, then v is positive definite. The two inequalities together therefore imply stability. If $r_y(0, 0) > 0$ or $x r(x, 0) > 0$, or if $r(x, 0)$ has the same sign for positive and negative values of x than the equilibrium is unstable[1].

---

[1] Leighton [1].

26. Supplements to the Principal Theorems

A variant of Theorem 25.2, useful for many practical applications, permits us to infer asymptotic stability from the inequality $\dot{v} \leqq 0$ alone, provided we know certain things about the point set defined by the condition $\dot{v} = 0$[1]).

Theorem 26.1. Let $v(x)$ be a function with continuous first order partial derivatives. Suppose there exists a region $G: 0 < v(x) < a$, in which $\dot{v} \leq 0$. Let M be the largest invariant subset of the set $\dot{v} = 0$. Then M is an attractive set for the set G, *i.e.* every motion which begins in G tends toward M (*cf.* sec. 16.).

Proof. Since as a function of t, v is non-increasing in G, every motion which begins in G remains entirely in G for $t \geq t_0$, and the $\lim\limits_{t\to\infty} v = a_\infty \in G$ exists. Let Γ_+ be the limit set (sec. 16) of the motion under consideration. On this limit set $v(x) = a_\infty$ because of continuity. Hence $\dot{v} = 0$ on Γ_+ and since Γ_+ is an invariant set, it follows that $\Gamma_+ \subset M$, which is the assertion of the theorem.

We point out that v need not be assumed to be definite. This theorem implies immediately the desired modification of the Stability Theorem.

Theorem 26.2. Let $v(x)$ be a positive definite function with a negative semi-definite derivative. Suppose it is known that apart from the equilibrium $x = 0$ no positive half-trajectory lies entirely in the region $\dot{v} = 0$. Then the equilibrium is asymptotically stable.

Under the condition of the theorem the set denoted by M in Theorem 26.1 consists exactly of the origin.

It follows from the proof of Theorem 26.1 that under the assumptions of Theorem 26.2 every motion tends toward the equilibrium provided its initial point lies in a domain in which v is positive definite (Def. 24.3a) and $\dot{v} \leq 0$. This conclusion goes beyond the assertion of Theorem 25.2 in the sense that it concerns the behavior of the motion "in the large" whereas Theorem 25.2 appears to characterize only the "local" behavior. The proof of Theorem 25.2 also contains of course an "in the large" statement: Motions with initial values satisfying the inequalities (25.2) and (25.4) tend to zero.

Def. 26.1. The *domain of attraction* of the origin (or *domain of stability*) for a differential equation (25.1) is the set of all points x_0 with the property

$$(26.1) \qquad \lim p(t, x_0) = 0 \quad (t \to \infty).$$

---

[1]) BARBASHIN and KRASOVSKII [1], LaSALLE [2, 3].

Thus the inequalities (25.2) and (25.4) provide estimates for the domain of attraction (see also sec. 33). In place of the term "domain of attraction" we often find the word "domain of stability"; this concept must however be carefully distinguished from the concept of a "domain of stability in the parameter space" (*cf.* sec. 11). If the domain of attraction is all of R_n we speak of *asymptotic stability in the whole,* (*cf.* sec. 2) or also of *global asymptotic stability.* We have

Theorem 26.3. In addition to the hypotheses of Theorem 25.2 assume further that $v(x)$ is radially unbounded. Then the equilibrium is globally asymptotically stable.

The proof of Theorem 25.2 can be repeated for each finite x. The additional condition is used as we go over to (25.3) (*cf.* sec. 24Ac). This proof does not in general furnish an estimate of the form $|p(t, x_0)| \leq \varphi(|x_0|)\,\sigma(t)$, which is valid for all x_0, *i.e.* globally, but only an estimate

$$(26.2) \qquad |p(t, x_0)| \leq l(|x_0|, t), \quad l \in KL; t \geq t_0 = 0$$

(*cf.* Def. 24.2 and the discussion following it). It is however possible for each fixed finite x_0-domain, to estimate the comparison function $l(|x_0|, t)$ in (26.2) by means of a product, as in sec. 24 A. The condition "v is radially unbounded" is equivalent to the statement "v is positive definite on all of R_n" in the sense of Def. 24.3a. It guarantees that the hyper-surfaces $v = c$ are closed for each arbitrary c. The following **counter-example** shows that these conditions are needed[1]).

$$\dot{x}_1 = -\frac{6\,x_1}{u^2} + 2x_2; \quad \dot{x}_2 = -\frac{2\,(x_1 + x_2)}{u^2}; \quad u := 1 + x_1^2,$$

$$v = \frac{x_1^2}{u} + x_2^2; \quad \dot{v} = -\frac{4}{u^2}\left(\frac{3\,x_1^2}{u^2} + x_2^2\right).$$

(the curves $v = c$ are not closed for $c > 1$).

The conditions for asymptotic stability of the region are satisfied. On the other hand the expression $\dot{x}_2/\dot{x}_1$ equals

$$-\frac{1}{1 + 2\,\sqrt{2}\,x_1 + 2\,x_1^2}$$

on the hyperbola $x_2 = 2/(x_1 - \sqrt{2})$ whereas the slope of the tangents to the hyperbola is

$$\frac{dx_2}{dx_1} = -\frac{1}{1 - \sqrt{2}\,x_1 + x_1^2/2}.$$

For $x_1 > \sqrt{2}$ the first expression is larger than the second. This implies that the trajectories cannot cut the branch of the hyperbola which lies in the first quadrant, in the direction toward the axes. (On the hyperbola we have $\dot{x}_1 > 0$, in case $x_1 > \sqrt{2}$.) So we obtain trajectories which do not tend toward the origin: The origin is not globally asymptotically stable.

---

[1]) Barbashin and Krasovskii [1].

In this example we focused on the angle with which the trajectories cut a fixed curve, the hyperbola. Similarly the angle between the trajectories and the curves $v = c$ played a part in the geometric interpretation of the principal theorems (see sec. 23). The same line of reasoning can also be applied somewhat differently, as the following simple *example* shows.

$$\dot{x}_1 = x_1; \quad \dot{x}_2 = - x_2; \quad v = x_1^2 + x_2^2; \quad \dot{v} = 2(x_1^2 - x_2^2).$$

The point set $\dot{v} > 0$ consists in this case of two sectors with their center at the origin, that is, of a plane cone bounded by the lines $x_1 = x_2$ and $x_1 = -x_2$. On these lines the second derivative of v for the differential equation, that is the expression $d\dot{v}/dt =: \ddot{v} = 4v$ is positive. This allows us to conclude that the phase trajectories cannot leave the region $\dot{v} > 0$ since because of the sign of the second derivative, $\dot{v}$ cannot pass from positive to negative values. Similar considerations apply to equations of higher order in case the region $v > 0$, $\dot{v} > 0$ is bounded by the surface of a cone on which $\dot{v} = 0$ and $\ddot{v} > 0$[1]).

We now mention a further variant of Theorem 25.2.

Theorem 26.4. Let the differential equation (25.1) be defined on all of R_n. Suppose the scalar function $h(x)$ is continuous and non-negative on all of R_n, and the set H defined by $h(x) = 0$ is an invariant set for the differential equation. Let the function $w(x)$ be continuous on all of R_n and have continuous first order partial derivatives. Suppose also that w is radially unbounded but bounded below, $w(x) \geq k > -\infty$, and that its derivative for (25.1) satisfies an estimate

$$\dot{w} \leq - \varphi(h(x)), \quad \varphi \in K, \quad \text{for all } x.$$

Then the set H is globally asymptotically stable.

Proof. The hypotheses imply that along the path of a motion $p(t, x_0)$ the function $w(t) := w(p(t, x_0))$ is monotonically non-increasing and bounded below. Hence the limit $\lim w(t) = w_\infty$ exists. We see that all the motions $p(t, x_0)$ are bounded and therefore possess a limit set Γ_+ (*cf.* the proof of Theorem 26.1). From the inequality

$$w(0) - w(t) \geq \int_0^t \varphi(h(p(\tau, x_0))) \, d\tau \geq 0,$$

which is valid for all $t > 0$, we conclude that the integral converges as $t \to \infty$ and hence that there exists a sequence $t_n \to \infty$, such that $h(p(t_n, x_0))$ tends to zero. Let $x := \lim p(t_n, x_0)$. Obviously x belongs to the limit set Γ_+ and also to H, and since H and Γ_+ are invariant we have $\Gamma_+ \subset H$. That is the assertion of the theorem.

---

[1]) KUDAEV [1].

Examples for Theorems 26.1 to 26.3.

a) The Liénard equation (18.3)

$$\ddot{x} + \dot{x} f(x) + g(x) = 0.$$

is equivalent (*cf.* sec. 18) to the system of equations

$$(26.3) \qquad \dot{x} = y - F(x), \quad \dot{y} = - g(x), \quad F(x) := \int_0^x f(u)\, du.$$

The derivative of

$$v(x, y) = \frac{1}{2} y^2 + \int_0^x g(u)\, du$$

for this system is

$$\dot{v} = - g(x)\, F(x).$$

We make the following assumptions on the non-linear functions: 1) There exists a $\beta > 0$ and an $\alpha > 0$ such that the inequality $\int_0^x g(u)\, du < \beta$ implies the inequality $|x| < \alpha$. 2) For $0 < |x| < \alpha$ we have $g(x)\, F(x) > 0$. Then $v(x, y)$ is positive definite and the equation $\dot{v} = 0$ can hold only for $x = 0$. Since $x = 0$ is not a trajectory, Theorem 26.2 applies: The origin is asymptotically stable.

b) A special case of the Liénard equation is the equation of van der Pol,

$$(26.4) \qquad \ddot{x} + a(x^2 - 1)\, \dot{x} + x = 0, \quad a > 0,$$

which plays a part in high frequency engineering. It is immediate that the origin is unstable (the damping term is negative for small x). Replacing t by $\tau = -t$ and denoting the derivative with respect to τ by means of a prime, we obtain the equation

$$x'' + a(1 - x^2)\, x' + x = 0,$$

resp. the equivalent system (26.3). The derivative of $v = (x^2 + y^2)/2$ for this system is

$$v' = + a x^2 (x^2/3 - 1).$$

Theorem 26.1 tells us that the disc $x^2 + y^2 < 3$ is contained in the domain of attraction of the origin. For the van der Pol equation we thus have that all the solutions which start in the interior of $x^2 + y^2 < 3$ leave this disc regardless of the size of a. We know from a different source[1] that equation (26.4) has exactly one limit cycle and we can conclude that its diameter must be greater than $2\sqrt{3}$[2] if it is interpreted in the (x, y) — plane.

---

[1] *cf.* for instance STOKER [1].   [2] LaSALLE [3].

c) Introducing into the scalar equation

$$(26.5) \qquad \ddot{u} + 2a\,|u|\,\dot{u} + bu = c, \qquad a > 0,$$

the variable $x = u - k$, $k := c/b$, we obtain

$$\ddot{x} + 2a\,|x + k|\,\dot{x} + bx = 0.$$

Obviously the point $(0, 0)$ of the $(x, \dot{x})$-plane is an equilibrium. The function

$$v = \frac{1}{2}\,(\dot{x}^2 + bx^2)$$

is positive definite in case $b > 0$. Its derivative

$$\dot{v} = \dot{x}\,\ddot{x} + bx\dot{x} = -\,2a\,|x + k|\,\dot{x}^2$$

is non-positive. It vanishes for $\dot{x} = 0$ and for $x = -k$. But the equation does not have a constant solution other than zero. The equilibrium is therefore asymptotically stable and the solutions of (26.5) tend toward the value c/b.

d) The scalar system

$$\dot{x} = y\left(a\,(x^2 + y^2) + b\right), \qquad a > 0,\, b > 0,$$
$$\dot{y} = -\,x\left(a\,(x^2 + y^2) - c\right) - dy, \qquad c > 0,\, d > 0,$$

has three singular points, namely $(0, 0)$ and $(\pm\,\sqrt{c/a},\, 0)$. The origin is unstable, the other two equilibria are asymptotically stable. If we set

$$v = \frac{a}{2}\,(x^2 + y^2)^2 + by^2 - cx^2$$

and take into account the relations

$$\dot{x} = \frac{1}{2}\,\frac{\partial v}{\partial y}, \qquad \dot{y} = -\,\frac{1}{2}\,\frac{\partial v}{\partial x} - dy,$$

then we see that

$$\dot{v} = -\,2dy^2\left(a\,(x^2 + y^2) + b\right).$$

The invariant subset of the region $\dot{v} = 0$ which lies on the line $y = 0$, consists of exactly the three singular points. The function v has a minimum at each of the points $(\pm\,\sqrt{c/a},\, 0)$, in fact, $v = -\,c^2/2a$ there. The origin is a saddle point for v. v is bounded in each disc $x^2 + y^2 < r^2$; definiteness was not required in the hypotheses of Theorem 26.1. We can conclude therefore that all the trajectories starting at a finite point will tend toward one of the three singular points. Since v is radially unbounded this statement is valid in the whole. The origin is a saddle point for the equation. Hence there exist exactly two trajectories leading into the origin; these form the separatrices. All the other trajectories tend toward one of the other two equilibria, depending on the position of their initial point.

e) The scalar system

$$\dot{x} = y, \quad \dot{y} = x$$

possesses the particular solutions $x = x_0 e^t$, $y = x_0 e^t$. The line $x = y$ corresponds to the totality of these solutions. Setting $v = x - y$, we have

$$\dot{v} = - v.$$

Theorem 26.1 is applicable in the domain $x - y > 0$ and asserts that all the trajectories which start in the half-plane below the line $x = y$ approach this line arbitrarily closely. The same is true for the trajectories in the upper half-plane as is seen by setting $v = y - x$.

f) A similar reasoning applies to the system

$$\dot{x} = y^3 - x, \quad \dot{y} = x - \frac{1}{2} y \quad ^{1)}.$$

The function $v = 2x^2 - y^4$ satisfies

$$\dot{v} = - 2v.$$

The real branch of the curve $2x^2 - y^4 = 0$, *i.e.* the curve

$$y^2 = \sqrt{2}\,|x|$$

is asymptotically stable in the whole. With the aid of the Liapunov function $x^2 + y^2$ we can furthermore establish that the origin is asymptotically stable. The domain of attraction is clearly larger than the disc $x^2 + y^2 < 0.5$. The other two singular points $\left(\dfrac{\sqrt{2}}{4}, \dfrac{\sqrt{2}}{2}\right)$ and $\left(-\dfrac{\sqrt{2}}{4}, -\dfrac{\sqrt{2}}{2}\right)$ are saddle points.

As is indicated in the proof of Theorem 25.2 there exists a connection between the comparison functions which describe the behavior of the solution and the functions $\varphi_i(r)$, which are used to estimate v and $\dot{v}$. In general the connection is complicated. But we have

Theorem 26.5. If the three comparison functions $\varphi_i(r)$, $i = 1, 2, 4$ of Theorem 25.2 are of the same order of magnitude, *i.e.* if there exist certain constants a_{ij}, a'_{ij} such that

$$a_{ij}\, \varphi_i(r) \leq \varphi_j(r) \leq a'_{ij}\, \varphi_i(r), \quad i \neq j,$$

then the equilibrium is exponentially stable, as defined in

Def. 26.2. The equilibrium of a differential equation is called *exponentially stable* if it is possible to find an estimate of the form

$$(26.6) \qquad |\boldsymbol{p}(t, \boldsymbol{x}_0, t_0)| \leq a\,|\boldsymbol{x}_0|\,e^{-b(t-t_0)}, \quad a > 0, \quad b > 0,$$

in a certain neighborhood B of the origin. The constants a and b may depend on B.

---

1) Szegö and Geiss [1].

Proof. If $v(x)$ is a Liapunov function satisfying the conditions of Theorem 25.2 and if $\psi(r) \in K$ is differentiable then $v_1(x) = \psi(v(x))$ is also a Liapunov function for (25.1). The comparison functions are, resp.,

$$\psi(\varphi_1(|x|)), \quad \psi(\varphi_2(|x|)), \quad \psi'(v(x))\,\varphi_4(|x|).$$

If need be the last function must be estimated by a member of K. By a suitable ψ, one of the two comparison functions of (25.2) can be put into a preselected form. So we might choose $\varphi_1(|x|) = |x|$, then because of the hypothesis the inequalities become

$$|x| \leq v(x) \leq a'|x|; \quad \dot{v} \leq -b'|x|$$

and this yields

$$\dot{v} \leq -\frac{b'}{a'}\,v, \quad v \leq v_0 e^{-\frac{b'}{a'}(t-t_0)}$$

and thus we obtain (26.6).

If in the proof of Theorem 25.2 we use in place of (25.4) an inequality

$$(26.7) \qquad\qquad \dot{v} \geq -\varphi_4(|x|), \quad \varphi_4 \in K,$$

then we arrive at a differential inequality

$$(26.8) \qquad\qquad \dot{v} \geq -\chi_1(v), \quad \chi_1 \in K,$$

and hence at an estimate which bounds v and $|x(t)|$ below. This says that $p(t, x_0)$ does not approach the equilibrium faster than a certain comparison function ($cf.$ sec. 36c). If (25.4) and (26.7) are both valid then both estimations are possible and we obtain

$$(26.9) \qquad \tilde{\varphi}(|x_0|)\,\tilde{\sigma}(t - t_0) \leq |p(t, x_0)| \leq \varphi(|x_0|)\,\sigma(t - t_0).$$

If the right side of (25.1) is bounded then $\dot{v}$ is bounded in a neighborhood of the origin and (26.7) can be derived from (25.5). In this case, that is for bounded right side, the hypotheses of Theorem 25.2 guarantee not only asymptotic stability of the equilibrium but also the existence of a two-sided estimate such as (26.9).

If in the hypothesis of Theorem 25.1 we replace the assumptions on $\dot{v}$ by "$\dot{v}$ is positive semi-definite" then an estimate

$$|p(t, x_0)| \geq \varphi(|x_0|), \quad \varphi \in K,$$

is possible. But this estimate does not imply the instability of the equilibrium; it only says: For each neighborhood U of the origin there exists a neighborhood U' such that each solution which originates outside of U remains entirely (for all $t \geq t_0$) outside the neighborhood U'.

If inequalities of the form (25.6) and (26.8) are available we can estimate the time which the phase point requires to traverse a certain segment of the trajectory. For then

$$-\chi_2(v) \leq \dot{v} \leq -\chi_1(v)$$

and it follows by integration that

$$(26.10) \qquad \int_{v_1}^{v_2} \left(-\frac{dv}{\chi_2(v)} \right) \le t_2 - t_1 \le \int_{v_1}^{v_2} \left(-\frac{dv}{\chi_1(v)} \right).$$

So far this estimates the time which the phase point needs to travel from the hypersurface $v(\pmb{x}) = v_1$ to the hypersurface $v(\pmb{x}) = v_2$. If we now set

$$m_i : = \max |\pmb{x}| \quad \text{for} \quad v(\pmb{x}) = v_i ,$$

$$\mu_i : = \min |\pmb{x}| \quad \text{for} \quad v(\pmb{x}) = v_i , \quad i = 1,\, 2,$$

we have the result that the phase point has come closer to the origin by at least $\mu_1 - m_2$ on the interval $(t_1,\, t_2)$, provided this difference is positive. The length of approach cannot exceed the distance $m_1 - \mu_2$.

Example. Let v and $\dot{v}$ be quadratic forms in $\pmb{x}$. Then there exist estimates

$$\alpha_1 |\pmb{x}|^2 \le v(\pmb{x}) \le \beta_1 |\pmb{x}|^2 ,$$

$$-\alpha_2 |\pmb{x}|^2 \le \dot{v} \le -\beta_2 |\pmb{x}|^2 , \quad \text{resp.} \quad -\frac{\alpha_2}{\alpha_1} v \le \dot{v} \le -\frac{\beta_2}{\beta_1} v ,$$

$$\frac{\alpha_1}{\alpha_2} \ln \frac{v_1}{v_2} \le t_2 - t_1 \le \frac{\beta_1}{\beta_2} \ln \frac{v_1}{v_2} .$$

It follows then that

$$\exp\left(\frac{\beta_2}{\beta_1} (t_2 - t_1) \right) \le \frac{v_1}{v_2} \le \exp\left(\frac{\alpha_2}{\alpha_1} (t_2 - t_1) \right),$$

$$\frac{\alpha_1}{\beta_1} \exp\left(\frac{\beta_2}{\beta_1} (t_2 - t_1) \right) |\pmb{x}(t_2)| \le \pmb{x}(t_1) \le \frac{\beta_1}{\alpha_1} \exp\left(\frac{\alpha_2}{\alpha_1} (t_2 - t_1) \right) |\pmb{x}(t_2)| .$$

This estimation, equivalent to (26.3), describes the fading of the solutions.

27. Construction of a Liapunov Function for a Linear Equation

Although the stability problem for the linear equation

$$(27.1) \qquad \dot{\pmb{x}} = A\,\pmb{x}$$

has been completely solved in sec. 4, further investigation is necessary in order to find a suitable Liapunov function for (27.1), because by modifying such a function we can find suitable functions for a large class of nonlinear equations.

If the Liapunov function is a quadratic form we can apply Theorem 24.3 to test its definiteness. In order to do so we set

$$(27.2) \qquad v(\pmb{x}) = \pmb{x}^T B \pmb{x}, \ B^T = B ,$$

and determine the so far unknown symmetric matrix B in such a way that $\dot{v} = \dot{\pmb{x}}^T B \pmb{x} + \pmb{x}^T B \dot{\pmb{x}} = \pmb{x}^T (A^T B + BA)\,\pmb{x}$ becomes equal to a preas-

8*

signed negative definite form $-\boldsymbol{x}^T C \boldsymbol{x}$. For this purpose we must solve
the matrix equation

(27.3) $A^T B + B A = - C$, B and C symmetric,

which can also be thought of as a system of determining equations for
the $n(n + 1)/2$ elements b_{ik} of B. We introduce a matrix H by execut-
ing a similarity transformation with a nonsingular matrix S

$$A = S^I H S.$$

Then (27.3) becomes

$$H^T S^{IT} B S^I + S^{IT} B S^I H = - S^{IT} C S^I.$$

In this process B and C are subjected to a congruence transformation
and retain the type of definiteness which they had. Let us now assume
that this transformation has made a triangular matrix out of A, $a_{ik} = 0$
for $i < k$. Then the characteristic roots $\lambda_1, \ldots, \lambda_n$ appear in the main
diagonal. The system of equations

$$2 \lambda_1 b_{11} = - c_{11}$$
$$a_{21} b_{11} + (\lambda_1 + \lambda_2) b_{12} = - c_{12}$$
$$\cdots\cdots\cdots\cdots\cdots\cdots\cdots$$

is triangular. Its determinant is

$$2^n \lambda_1 \lambda_2 \ldots \lambda_n \mathop{\Pi}_{i<j} (\lambda_i + \lambda_j)$$

(it is thus equal to the last Hurwitz determinant in sec. 7 except for its
sign). So the matrix B can be determined if and only if this determinant
is different from zero, $i.e.$ if all of the characteristic roots of A are non-
zero and if no two of them are equal and of opposite sign. This condition
is not affected by a similarity transformation and therefore is valid for
(27.3) also. The construction assures that the derivative of (27.2) is
negative definite. We must still check for the definiteness of B. This can
be done in a purely algebraic way[1]). But in the present case it is much
easier to apply the results of sec. 4 and to argue as follows.

a) If all the characteristic roots λ_i have negative real parts then the
equilibrium is asymptotically stable and B must be positive definite.
For otherwise we could apply Theorem 25.4 and obtain a contradiction.

b) If at least one of the real parts is positive and no real part is zero
then B cannot be positive definite. Otherwise we could apply Theorem
25.2 and again obtain a contradiction. If all the real parts are positive B
must be negative definite.

---

[1]) HAHN [2].

For the purpose of illustration we write the result in terms of a system fo two scalar equations. Let

$$\dot{x} = ax + by, \quad \dot{y} = cx + dy$$

be given. We seek

$$v = \alpha x^2 + 2\beta xy + \gamma y^2$$

such that

$$\dot{v} = -(\varrho x^2 + 2\sigma xy + \tau y^2).$$

Solving the equation for x, β, γ we find

$$\alpha = \frac{1}{2pq}\left(\varrho(q + d^2) - 2\sigma cd + \tau c^2\right),$$

(27.4)
$$\beta = -\frac{1}{2pq}\left(\varrho bd - 2\sigma ad + \tau ac\right),$$

$$\gamma = \frac{1}{2pq}\left(\varrho b^2 - 2\sigma ab + \tau(q + a^2)\right),$$

$$p := -(a + d), \quad q := ad - bc.$$

If A has characteristic roots with positive real parts and if we have the exceptional case in which one of the expressions $\lambda_i + \lambda_j$ vanishes then B cannot be constructed in the manner prescribed. In that case we form a matrix $A_1 = A - \delta E$ and choose δ so that A_1 has as many characteristic roots with positive real part as A but that the exceptional case does not occur. The equation

$$A_1^T B + B A_1 = -C \quad (C \text{ positive definite})$$

can then be solved for B and B is then clearly not positive definite. The derivative of the function $v(x) = x^T B x$ is a form whose matrix is

$$(A_1 + \delta E)^T B + B(A_1 + \delta E) = -C + 2\delta B$$

and which is negative definite for a sufficiently small δ. The function $v(x)$ now satisfies the hypothesis of Theorem 25.4. Summarizing we have

Theorem 27.1. If all the characteristic roots of the matrix A have negative real parts or if at least one characteristic root has a positive real part then there exists a Liapunov function of the form (27.2) whose derivative is definite.

Therefore in the case of the stable, resp. unstable matrix (sec. 4), the conditions of Theorems 25.2, resp. 25.4, are also necessary. If the matrix A is critical we cannot construct a Liapunov function whose derivative is definite. To see this we first put the matrix into its Jordan canonical form and then transform it in such a way that the elements of the blocks which belong to characteristic roots with vanishing real parts are real.

If zero is a simple characteristic root then there is a block with only one element. To it corresponds an equation of the form

$$\dot{z}_1 = 0. \tag{27.5}$$

If an elementary divisor of second order belongs to the double characteristic root zero then the corresponding equations are

$$\dot{z}_1 = z_2, \quad \dot{z}_2 = 0. \tag{27.6}$$

A pair of conjugate imaginary characteristic roots are associated with the equations

$$\dot{z}_1 = k z_2, \quad \dot{z}_2 = -k z_1 \tag{27.7}$$

and for an elementary divisor of second order we have a fourth order real system with matrix

$$\begin{pmatrix} 0 & k & 1 & 0 \\ -k & 0 & 0 & 1 \\ 0 & 0 & 0 & k \\ 0 & 0 & -k & 0 \end{pmatrix}. \tag{27.8}$$

Equation (27.5) has a family of constant solutions, $z_1 = c$. Equation (27.7) has a first integral of the form $z_1^2 + z_2^2 = c$. For sufficiently small values of the constant c, the corresponding trajectories lie arbitrarily close to the origin. Now let $v(z)$ be given. As the point z traverses a closed curve $z_1^2 + z_2^2 = c$, $v(z)$ is a periodic function of t. On the other hand

$$v = v_0 + \int_{t_0}^{t} \dot{v}\, dt. \tag{27.9}$$

If $\dot{v}$ were definite then $|\dot{v}|$ would be bounded away from zero on each fixed closed curve. The integral in (27.9) would not be bounded for increasing t and v could not be periodic on the closed trajectory. Considering a constant solution $z_1 = c$, we have $\dot{v}$ identically equal to zero and hence again not definite. In the case of equations (27.6), resp. (27.8), we must go through the same argument with the solutions $z_1 = c$, $z_2 = 0$, resp. $z_1^2 + z_2^2 = c$, $z_3 = z_4 = 0$.

Obviously the conclusion is quite independent of the special character of the differential equations and we have

Theorem 27.2. If there exist arbitrarily close to the origin of an autonomous differential equation constant solutions or closed phase trajectories, then there cannot exist a Liapunov function whose derivative is definite.

It is, however, possible to construct Liapunov functions with a semi-definite derivative for the above systems. For (27.5) and (27.7) the functions

$$v = z_1^2, \text{ resp. } v = z_1^2 + z_2^2$$

satisfy the hypotheses of Theorem 25.1, since the derivative is identically zero. For (27.6) we set

$$v = z_1 z_2, \quad \dot{v} = z_2^2$$

and for (27.8)

$$v = z_1 z_3 + z_2 z_4, \quad \dot{v} = z_3^2 + z_4^2.$$

Each time we can apply Theorem 25.3 and arrive at instability; the derivative is not definite, but semi-definite.

The same functions can also be utilized for elementary divisors of higher order. If A has m characteristic roots with negative and $n - m$ characteristic roots with positive real parts then we transform A to a diagonal form $\mathrm{diag}(A_1, A_2)$ in such a way that A_1 contains the m characteristic roots with negative real part. We then solve (27.3) once with $A = A_1$ and once with $A = A_2$ and with preassigned C_i, $i = 1, 2$, form the diagonal matrix $\mathrm{diag}(B_1, B_2)$ and take the inverse of the transformation. The matrix B can be utilized for finding a Liapunov function for the original system. It is indefinite by construction.

Theorem 27.3. If all the characteristic roots of A have negative real parts we can solve the matrix equation (27.3) in closed form.

Proof. The matrix

$$P(t) := \int_t^\infty e^{A^T(\tau - t)} C \, e^{A(\tau - t)} \, d\tau$$

satisfies the differential equation

$$\dot{P} = -C - A^T P - PA.$$

The integral converges because of the hypothesis on the real parts. Introducing $\tau' = \tau - t$ as the new variable of integration, we see that the integral only appears to depend on t. Thus $\dot{P} = 0$ and $A^T P + PA = -C$, and

$$(27.10) \qquad B = \int_0^\infty e^{A^T t} C \, e^{At} \, dt$$

is the solution of (27.3). The matrix C is arbitrary but, of course, the proof of the definiteness of B is valid only if C is definite.

The matrix

$$C = \boldsymbol{a}\boldsymbol{a}^T, \quad \boldsymbol{a} \text{ a vector,}$$

is positive semi-definite and our previous argument only implies that the corresponding matrix (27.10) is not negative. For a closer look we consider the quadratic form

$$y^T By = \int_0^\infty (y^T e^{A^T t} a)\,(y^T e^{A^T t} a)^T\, dt\,,$$

in which y is a parameter vector. This form clearly is non-negative and vanishes only for such vectors y for which

$$(27.11) \qquad\qquad y^T e^{A^T t} a \equiv 0 \quad (0 \leq t < \infty)\,.$$

We now apply a transformation R to the matrix A, to put it into its Jordan normal form, $R^I A R = J$. Then we have

$$e^A = Re^J R^I\,.$$

The condition (27.11) takes on the form

$$y^T R^{IT} e^{J^T t} R^T a \equiv 0\,.$$

But since J is a partitioned matrix, the left side can vanish for arbitrary $y \neq 0$ only if $R^T a$ has at least one zero component. Such a zero component appears if and only if a is orthogonal to at least one characteristic vector of A. This follows from the structure of the transformation matrix R whose columns are formed from the characteristic vectors of A. This implies: If the vector a is orthogonal to none of the characteristic vectors of A, then the matrix

$$B = \int_0^\infty (e^{A^T t} a)\,(e^{A^T t} a)^T dt$$

is positive definite.

If furthermore z is a characteristic vector of A, $A z = \lambda z$, then

$$z^T B = \int_0^\infty (z^T e^{A^T t} a\, a^T e^{At})\, dt = \int_0^\infty e^{\lambda t} z^T a\, a^T e^{At} dt\,.$$

If z is orthogonal to a then $z^T B = 0$.

We shall use these results in sec. 32. They also apply to complex vectors a if orthogonality is interpreted accordingly in the complex space C_n and if B is defined as a Hermitian matrix by

$$B = \int_0^\infty (e^{A^T t} a)\,(e^{A^T t} \bar{a})^T dt\,.$$

28. Liapunov Functions for Perturbed Linear Equations

The equations considered in sec. 21 are special cases of the more general equation

$$(28.1) \qquad\qquad \dot{x} = A x + f(x)\,.$$

At first we assume nothing about the added function $f(x)$ other than that it belongs to class E (sec. 16). We thus have, in particular, $f(0) = 0$. Also we shall assume that the stability behavior of the reduced equation

$$(28.2) \qquad\qquad \dot{x} = A\,x$$

is known. In fact, we shall assume the hypotheses of Theorem 27.1. That is, the critical cases have been excluded. [Example (21.4) shows that in the critical cases the terms of higher degree can be decisive.]

Equations of type (28.1) are often found in practice. They arise, for example, if we start with an arbitrary nonlinear equation

$$(28.3) \qquad\qquad \dot{x} = g(x), \quad g(0) = 0,$$

develop the components of the right side in Taylor expansions about the origin (if possible) and write the linear terms by themselves,

$$\dot{x} = \frac{\partial g}{\partial x}\bigg|_{x=0} x + f(x).$$

In this case the perturbation terms are at least of the second degree. But the functions $f_i(x)$ in (28.1) may also contain linear terms which could for example be caused by inaccuracy in measurements or other factors with which we cannot deal precisely. Frequently we know no more about them than that there exist estimates of the form

$$(28.4) \qquad\qquad |f(x)| \le a\,|x|, \quad a > 0$$

or (fig. 28.1 on p. 124)

$$(28.5) \qquad \sum_{j=1}^{n} \alpha_{ij} x_j \le f_i(x) \le \sum_{j=1}^{n} \beta_{ij} x_j, \quad i = 1, \ldots, m \le n.$$

We would like to emphasize that the use of the word "nonlinear" in the literature is not standardized. Many authors speak of nonlinear functions $f(x)$ only in case these contain at least second degree terms. We shall mean by nonlinear functions quite generally functions which are not linear.

Since the critical case has been excluded there exists for the reduced system (28.2) a Liapunov function $v(x) = x^T B x$ with a negative definite derivative $\dot{v} = -x^T C x$. If we choose the same Liapunov function for (28.1) and form the derivative for this equation we obtain (the subscripts indicate the number of the equation):

$$(28.6) \quad \dot{v}_{(28.1)} = \dot{v}_{(28.2)} + \sum_{i=1}^{n} \frac{\partial v}{\partial x_i} f_i(x) = -x^T C x + 2 \sum_{i,k} b_{ik} x_k f_i(x).$$

If the f_i are of second degree then the second term on the right is of third degree. On the basis of sec. 24C we conclude that this term does not influence the type of definiteness of the first term. But this means that for (28.1) and (28.2) the same principal theorem applies. Thus we have proved the important *Principle of the Stability in the First Approximation.*

Theorem 28.1. If the matrix A of the system in the first approximation either has only characteristic roots with negative real parts or at least one characteristic root with a positive real part, then the equilibrium of the complete system shows the same stability behavior as that of the linearized system.

In the technical literature this is often called the *Principle of the Small Oscillations.* We have now, belatedly, proved Theorem 21.1.

If for the nonlinear terms only estimates of the form (28.4) or (28.5) are available, we have to argue somewhat differently. Usually we are concerned here with the more special question whether the inequalities (28.5) for the additional terms are sufficient to guarantee asymptotic stability for (28.1) if it exists for (28.2). The problem can also be formulated in this way: We seek two systems of numbers $\alpha_{ij}, \beta_{ij}, i, j = 1, \ldots, n$ such that the asymptotic stability of (28.1) is assured if the $f_i(\boldsymbol{x})$ satisfy the inequalities (28.5).

To find appropriate bounds for the constants α_{ij}, β_{ij}, we work with an auxiliary linear equation

$$(28.7) \qquad\qquad \dot{\boldsymbol{x}} = A\,\boldsymbol{x} + G\boldsymbol{x},$$

in which the matrix G remains undetermined at first. The derivative of the Liapunov function v used above for (28.7) is

$$\dot{v}_{(28.7)} = \dot{v}_{(28.2)} + \boldsymbol{x}^T(G^T B + BG)\,\boldsymbol{x} = \boldsymbol{x}^T(-C + G^T B + BG)\,\boldsymbol{x}.$$

If G is so chosen that the matrix

$$M := -C + G^T B + BG$$

is negative definite then the equilibrium of the auxiliary system is asymptotically stable. The conditions for G can easily be given (sec. 24) and yield n inequalities for the n^2 elements of the matrix G. Their common solutions determine a domain in the space of the parameters g_{ik} which clearly is not void since it contains the origin $G = 0$. If this domain is known the problem is solved: The numbers α_{ik}, β_{ik} must be taken from the interval over which the g_{ik} vary. For then $\dot{v}_{(28.1)}$ is also negative definite.

As an example we consider a second order system

$$\dot{x} = ax + by + f(x, y),$$
$$\dot{y} = cx + dy$$

and we assume that

$$\gamma_1' x + \gamma_2' y \leq f(x, y) \leq \gamma_1 x + \gamma_2 y.$$

We must estimate the constants γ_i, γ_i'. To simplify this let $C = \mathrm{diag}(\varrho, \tau)$. G has only one nonzero row. The elements of B are calculated in (27.4). The elements of the matrix M are

$$m_{11} = -\varrho + 2 b_{11} g_{11}, \quad m_{12} = m_{21} = b_{11} g_{12} + b_{12} g_{11},$$

$$m_{22} = -\tau + 2 b_{12} g_{12}.$$

M is negative definite if the Sylvester inequalities

$$m_{11} < 0, \quad m_{11} m_{22} - m_{12}^2 > 0$$

are satisfied. In the (g_{11}, g_{12})-plane the first inequality describes a half-plane, the second a parabolic region. The constants of the estimate must lie in the intersection of these regions. Since in the present case the line $m_{11} = 0$ is tangent to the parabola we may simply choose the constants from the "interior" of the parabola. The parabola is further dependent on ϱ and τ. Each time we construct the parabola for an arbitrary pair of values (ϱ, τ) the interior of this parabola lies within the domain of stability. We recognize immediately that the estimate depends on the choice of the numbers ϱ and τ (see below).

The considerations in connection with (28.7) reduce the general problem to a finite algebraic problem. However the practical treatment of the inequalities is in general quite laborious. Even so, frequently not all the components of the vector f are non-zero, i.e. not all of the equations are "perturbed". In such a case it becomes considerably simpler to deal with the inequalities. For if exactly $m < n$ of the equations of (28.1) are perturbed then the highest degree of terms appearing in the inequalities is no larger than $2m$[1]) (in the example $n = 2$, $m = 1$, and the inequalities are quadratic). This is seen as follows. If only m equations are perturbed then G has only m rows and we may assume that these are the first m rows. Then $s := \mathrm{rank}\, BG \leq m$ and we can find two non-singular matrices U and V such that the matrix

$$S := UBGV$$

has in its upper left hand corner a unit matrix E_s and contains only zeros otherwise. We form the matrix

$$T := U(G^T B + BG)V = UG^T BV + S,$$

whose first summand is of rank s. Adding S only changes the first rows, resp. columns. The total number of independent rows and columns of T can thus be larger by at most s than the corresponding number for the matrix $UG^T BV$, i.e. we have

$$\mathrm{rank}\, T \leq 2s, \quad \mathrm{rank}\, (G^T B + BG) \leq 2s \leq 2m.$$

---

1) Hahn [1].

By row and column transformations on the matrix M we obtain that its rows from the $(2m + 1)$th row on, no longer depend on the parameters g_{ik}. The value of the principal minors remains unchanged in this process and we see that the degree of these minors with respect to the g_{ik} is at most $2m$.

The domain of stability in the parameter space of the g_{ik} depends very strongly on the choice of the matrix C. Choosing $C = E$, which is most convenient for the calculations, furnishes by no means always the most advantageous, *i.e.* the largest domain of stability. The optimalization problem which arises here, to determine C so that the resulting domain in the g_{ik}-space becomes maximal, has not yet been solved[1]).

A modification of the argument is necessary if the inequalities (28.5) are valid only in the domain

$$(28.8) \qquad |x_i| \leq a_i, \qquad i = 1, 2, \ldots, k \leq n,$$

or for

$$(28.9) \qquad |x_i| \geq b_i, \qquad i = 1, 2, \ldots, k \leq n.$$

In the first case we must forgo global asymptotic stability, but we can estimate the domain of attraction. In the second case nothing can be said about the stability of the equilibrium but a terminal domain can be given toward which the trajectories tend and which they will not leave again. In the first case we determine the largest of the hypersurfaces

$$v = x^T B x,$$

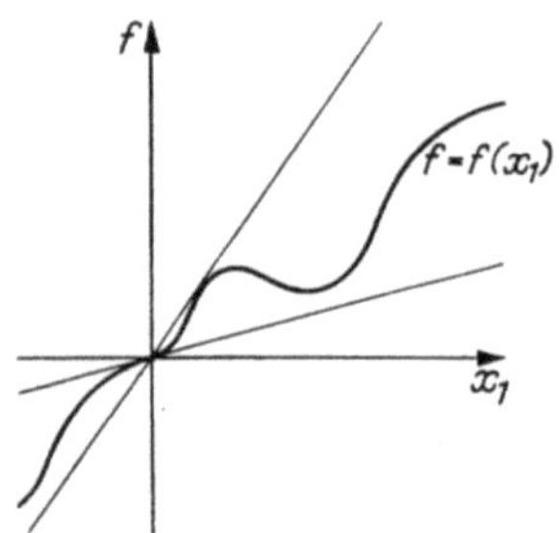

Fig. 28.1. Linear estimation [(28.5) for $m = 1$]

which still lies in the interior of the parallelepiped (28.8) (fig. 28.2). On this hypersurface and in its interior $\dot{v}$ is negative definite by construction; therefore the trajectories can pass through this surface only from outside in. Let $\bar{a}_j$ be one of the numbers $a_1, \ldots, a_n$ in (28.8) and consider the hypersurface with the hyperplane $x_j = \bar{a}_j$, which is perpendicular to the x_j-axis, as its tangent plane. The equation of the tangent plane to the hyperellipsoid at the point $\bar{x}$ is

$$\frac{\partial v}{\partial x_1} (x_1 - \bar{x}_1) + \cdots + \frac{\partial v}{\partial x_n} (x_n - \bar{x}_n) = 0.$$

The partial derivatives are to be evaluated at the place $x = \bar{x}$. For this equation to be valid for $x_j = \bar{a}_j$ we must have

$$\left. \frac{\partial v}{\partial x_i} \right|_{x_j = a_j} = 0, \qquad i = 1, 2, \ldots, n; \ i \neq j.$$

---

1) *cf.* LEHNIGK [2].

These are linear equations in the unknown coordinates $\bar{x}_i$, $i \neq j$, of the point of tangency. The equation of the hypersurface is then

$$\mathbf{x}^T B \mathbf{x} = \bar{\mathbf{x}}^T B \bar{\mathbf{x}}.$$

The number $\bar{\mathbf{x}}^T B \bar{\mathbf{x}}$ on the right side depends on j since we started with a fixed $\bar{a}_j$. We denote this number by c_j and choose $c := \min_j c_j$. Then

$$\mathbf{x}^T B \mathbf{x} = c$$

is the equation of the hypersurface witch lies completely within the domain of attraction. Of course, this domain may be much larger. It may even happen that we have asymptotic stability in the whole whereas the construction naturally furnishes a finite domain.

In the plane case (see sec. 27) we are dealing with the ellipses $v = \alpha x_1^2 + 2\beta x_1 x_2 + \gamma x_2^2$. And we have

$$\frac{\partial v}{\partial x_1} = 2(\alpha x_1 + \beta x_2), \quad \frac{\partial v}{\partial x_2} = 2(\beta x_1 + \gamma x_2).$$

If, say, a_1 is fixed, then the equation

$$\beta a_1 + \gamma \bar{x}_2 = 0$$

furnishes the missing second coordinate of the point of tangency. The equation of the ellipse through the point $\left(a_1, -\dfrac{\beta}{\gamma} a_1 \right)$ is

$$\alpha x_1^2 + 2\beta x_1 x_2 + \gamma x_2^2 = a_1^2 \left(\alpha - \frac{\beta^2}{\gamma} \right).$$

Similarly, we fix a_2 and get an ellipse

$$\alpha x_1^2 + 2\beta x_1 x_2 + \gamma x_2^2 = a_2^2 \left(\gamma - \frac{\beta^2}{\alpha} \right)$$

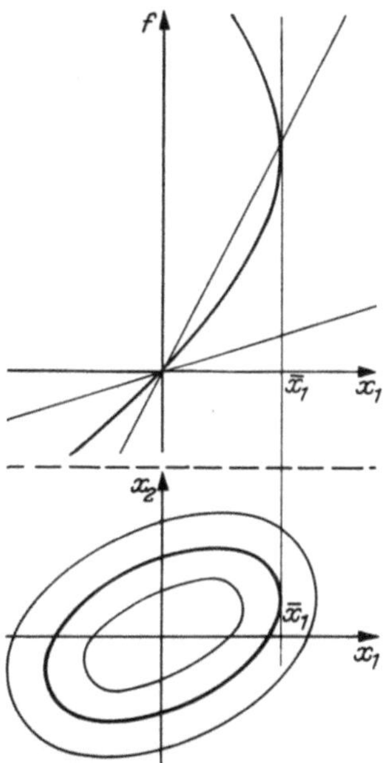

and the required ellipse has the equation with the right side $c := \min\,(a_1^2\,(\alpha - \beta^2/\gamma),$ $a_2^2\,(\gamma - \beta^2/\alpha))$. In fig. 28.2, the construction is carried through for $n = 2$, $m = k = 1$.

In the case (28.9), the conditions $v > 0$, $\dot{v} < 0$ are satisfied outside of the parallelepiped $|x_i| \leq b_i$. Therefore the trajectories tend toward this domain and fineally reach it. Consequently, the terminal domain is estimated by the inequalities

$$|x_i| \leq b_i, \quad i = 1, 2, \ldots, k.$$

If $k < n$, the domain is unbounded and the result is of no value.

Fig. 28.2. Construction of the domain of attraction.

If the terminal domain is sufficiently small the physical system described by the equations (28.1) can in most cases be considered as "practically stable" even if the equilibrium is not stable in the sense of Liapunov.

If the components $f_i(\boldsymbol{x})$ of the perturbation vector are such that (28.6) is negative definite for arbitrary $\boldsymbol{x}$, global asymptotic stability results. Sufficient conditions for this are obtained as follows: We have

$$\left| \sum_{i=1}^{n} \frac{\partial v}{\partial x_i} f_i(\boldsymbol{x}) \right|^2 \le \sum_{i=1}^{n} |f_i(\boldsymbol{x})|^2 \sum_{i=1}^{n} \left| \frac{\partial v}{\partial x_i} \right|^2 = \sum_{i=1}^{n} |f_i(\boldsymbol{x})|^2 \sum_{i=1}^{n} \left(2 \sum_{k=1}^{n} b_{ik} x_k \right)^2 .$$

We set $z_i := 2 \sum_{k=1}^{n} b_{ik} x_k$ and denote by $\boldsymbol{h}(\boldsymbol{x})$ a differentiable vector such that $|\boldsymbol{h}(\boldsymbol{x})| < \dfrac{1}{n}$. We further set

$$\gamma_i(\boldsymbol{x}) := h_i(\boldsymbol{x}) \, |\boldsymbol{x}^T C \boldsymbol{x}| \, \frac{z_i}{1 + z_i^2} .$$

Then if

(28.10) $$\qquad\qquad |f_i(\boldsymbol{x})| \le |\gamma_i(\boldsymbol{x})| ,$$

we have

$$\left| \sum_{i=1}^{n} \frac{\partial v}{\partial x_i} f_i(\boldsymbol{x}) \right|^2 \le \sum_{i=1}^{n} \gamma_i^2 \sum_{i=1}^{n} z_i^2 \le |\boldsymbol{x}^T C \boldsymbol{x}|^2 \sum_{i=1}^{n} h_i^2 \sum_{i=1}^{n} z_i^2 \sum_{i=1}^{n} \frac{z_i^2}{(1 + z_i^2)^2} .$$

Hence

$$\sum_{i=1}^{n} \frac{z_i^2}{(1 + z_i^2)^2} \sum_{i=1}^{n} z_i^2 \le \sum_{i=1}^{n} \frac{1}{1 + z_i^2} \sum_{i=1}^{n} z_i^2 < n^2$$

because the second term assumes its maximum when all the z_i are equal, and it is smaller than n^2 in this case. It follows that

$$\left| \sum_{i=1}^{n} \frac{\partial v}{\partial x_i} f_i(\boldsymbol{x}) \right| \le |\boldsymbol{h}(\boldsymbol{x})| \, |\boldsymbol{x}^T C \boldsymbol{x}| \, n < |\boldsymbol{x}^T C \boldsymbol{x}| ,$$

and if we substitute this in (28.6) we obtain $\dot{v} < 0$. The inequalities (28.10) therefore guarantee global asymptotic stability.

The considerations following (28.7) can be applied at least in theory to perturbed nonlinear equations

$$\dot{\boldsymbol{x}} = \boldsymbol{a}(\boldsymbol{x}) + \boldsymbol{f}(\boldsymbol{x})$$

provided that for the unperturbed equation $\dot{\boldsymbol{x}} = \boldsymbol{a}(\boldsymbol{x})$ a Liapunov function $v(\boldsymbol{x})$ is known. The auxiliary system has the form

$$\dot{\boldsymbol{x}} = \boldsymbol{a}(\boldsymbol{x}) + G \boldsymbol{x} ,$$

and the matrix G must be such that the expression

$$(\operatorname{grad} v)^T \boldsymbol{a} + (\operatorname{grad} v)^T G \boldsymbol{x}$$

is negative definite. As in the linear case this condition furnishes a system of inequalities for the coefficients g_{ik} and hence bounds for the α_{ij}, β_{ij} in (28.5). However it will only be possible in very special cases to evaluate this condition analytically.

29. The Problem of Aizerman

If we consider the linear system (28.7) by itself, without reference to (28.1), we can give the domain of stability in the space of the parameters g_{ik} immediately. We need only solve the Hurwitz inequalities (6.9) for the characteristic equation of the matrix $A + G$. This poses the question whether this domain is at the same time the optimum domain for the nonlinear problem, *i.e.* whether the nonlinear components in (28.5) are allowed to range over the same domain which is defined for the linear components by the Hurwitz inequalities. AIZERMAN [1] conjectured in 1949 that the answer was affirmative and this has stimulated a great deal of individual research. The outcome was that already in the general case of a system of order two the conjecture failed. The Hurwitz inequalities are not sufficient and additional conditions are needed. This is even more true for $n > 2$. On the other hand the conjecture holds for a number of systems of a special form. We shall presently treat a number of these cases which in part have practical significance. Since there is no general procedure these examples may at the same time serve to illustrate the various methods of investigation[1]).

We write the system of second order in scalar form

$$\dot{x} = x\,a(x) + y\,b(y),$$
$$\dot{y} = x\,c(x) + y\,d(y).$$

(29.1)

The functions on the right are assumed as usual to belong to class E. The Hurwitz inequalities for the corresponding linear system are given by

$$a + d < 0, \quad ad - bc > 0.$$

For the nonlinear system we would therefore have to require

(29.2) $a(x) + d(y) < 0, \quad a(x)\,d(y) - b(y)\,c(x) > 0 \quad (x,\,y \neq 0).$

a) Suppose $b,\,c,\,d$ are constant. Then the inequalities (29.2) become

$$a(x) + d < 0, \quad a(x)\,d - bc > 0.$$

If $b \neq 0$ we introduce (following MALKIN [2]) the Liapunov function

(29.3) $$v = \int\limits_0^x \left(u\,a(u)\,d - bc\,u\right) du + \frac{1}{2}(dx - by)^2.$$

---

[1]) See also KRASOVSKII [4], PLISS [2].

It is positive definite as the second term vanishes only for points on the line $dx = by$, and the first term is positive for $x \neq 0$ by (29.2). For the derivative we obtain

$$\dot{v} = \left(x\,a(x)\,d - b\,c\,x\right)\left(x\,a(x) + b\,y\right)$$
$$+ (d\,x - b\,y)\left(d\left(x\,a(x) + b\,y\right) - b(c\,x + d\,y)\right)$$
$$= \left(x\,a(x)\,d - b\,c\,x\right)\left(x\,a(x) + b\,y + d\,x - b\,y\right)$$
$$= x^2\left(a(x)\,d - b\,c\right)\left(a(x) + d\right).$$

Even though the derivative is not definite it is non-positive. And since the line $x = 0$ contains no trajectory other than the equilibrium we can apply Theorem 26.2. The equilibrium is asymptotically stable but we would further like to assure asymptotic stability in the whole. For that purpose we must add to the given proof the requirement that the integral in (29.3) diverges. For then v is positive definite in all of R_n and the equation $v = $ const. represents a closed curve even for arbitrarily large constants, as is seen immediately if we solve the equation for y. Thus v is radially unbounded. For the divergence of the integral, for example, the condition

$$d\,a(x) - b\,c \geq \delta > 0 \text{ for sufficiently large } |x| \,.$$

is sufficient. Hence the argument does not furnish the desired result since in addition to the Hurwitz inequalities a further condition is required. Now ERUGIN [1] showed by means of elementary but involved computations that we can forgo the condition on the integral if $d^2 + bc \neq 0$. We can even wave the uniqueness requirement and need only assume that $a(x)$ is continuous. On the other hand, the Hurwitz inequalities are certainly not sufficient in the exceptional case as the example

(29.4) $$\dot{x} = f(x) + y, \quad \dot{y} = -x - y$$

shows. Here $x\,a(x) = f(x)$, where

$$f(x) := x - \frac{e^{-2x}}{1 + e^{-x}} \ (x \geq 1), \quad f(x) := x\left(1 - \frac{e^{-2}}{1 + e^{-1}}\right)(x \leq 1).$$

We have $d^2 + bc = 0$, and

$$a(x) + d = -\frac{e^{-2x}}{x(1 + e^{-x})}, \quad \text{resp. } = -\frac{e^{-2}}{1 + e^{-1}},$$

$$a(x)\,d - b\,c = \frac{e^{-2x}}{x(1 + e^{-x})}, \quad \text{resp. } = \frac{e^{-2}}{1 + e^{-1}}.$$

The Hurwitz inequalities are satisfied and we see immediately that $f(x)$ satisfies a Lipschitz condition since the derivative is discontinuous only at $x = 1$ and there has a finite jump.

For $x \geq 1$ the system (29.4) is equivalent to the equation

$$y' = -\frac{x+y}{x+y-(e^{2x}+e^x)^{-1}} \cdot$$

A particular solution is

$$y = e^{-x} - x \quad (x \geq 1).$$

Along this curve we have

$$\dot{x} = x + y - \frac{e^{-2x}}{1+e^{-x}} = \frac{e^{-x}}{1+e^{-x}} > 0.$$

Hence x is increasing. We integrate and find

$$t = x + e^x - (1 + e).$$

The trajectory of (29.4) which starts at the point $x_0 = 1$, $y_0 = e^{-1} - 1$ tends therefore toward ∞.

This example already disproves the Aizerman conjecture in its most general form.

If $b = 0$ then (29.3) is no longer definite and the proof fails. But it follows from (29.2) that then $a(x) < 0$ and from the first equation (29.1) we see that x is monotone decreasing to zero. The equation

$$\dot{y} = cx + dy, \quad d < 0,$$

can be solved for y; it follows that y also tends to zero.

b) The somewhat more general system (fig. 29.1)

$$\dot{x} = xa(x) + by, \quad \dot{y} = xc(x) + dy$$
(29.5)

can be treated in the same manner as the special case a). If $b \neq 0$ the discussion of the Liapunov function

$$v = \int_0^x \big(ua(u)\,d - buc(u)\big)\,du$$
$$+ \frac{1}{2}(dx - by)^2$$

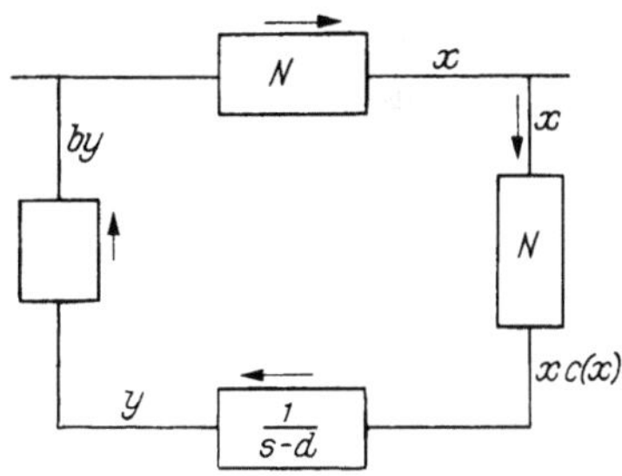

Fig. 29.1. Blockdiagram for (29.5)

leads to the result. The inequalities

$$a(x) + d < 0, \quad a(x)\,d - bc(x) > 0 \quad (x \neq 0)$$

guarantee asymptotic stability. The condition

$$\lim \int_0^x \big(ua(u)\,d - bc(u)\,u\big)\,du = \infty \quad \text{as } |x| \to \infty$$

assures global asymptotic stability. If $b = 0$ then x is monotone decreasing to zero. Then for any given $\varepsilon > 0$ we can choose a number T such that

$$|x(t)\, c(x(t))| < |d|\,\varepsilon \quad \text{for } t > T$$

(d is negative). From the equation for $\dot{y}$ we conclude

$$y(t) = y_0\, e^{dt} + \int_0^t x(\tau)\, c(x(\tau))\, e^{d(t-\tau)}\, d\tau,$$

$$|y(t)| \leq |y_0|\, e^{dt} + \varepsilon(1 - e^{dT}), \quad \text{for } t > T,$$

and since ε is arbitrarily small it follows that $\lim y(t) = 0 \;(t \to \infty)$.

This implies a stability condition for the Liénard equation (18.3), resp. for the system (18.4): The equilibrium of

$$\ddot{x} + f(x)\,\dot{x} + g(x) = 0$$

is globally asymptotically stable if the conditions

$$x\,F(x) > 0, \quad x\,g(x) > 0, \quad \lim_{|x| \to \infty} \int_0^x g(u)\, du = \infty,$$

$$F(x) := \int_0^x f(u)\, du$$

are satisfied.

c) The system

$$(29.6) \qquad \dot{x} = x\,a(x) + b\,y, \quad \dot{y} = c\,x - y\,d(v)$$

has two nonlinear terms with different arguments. The inequalities have the form

$$(29.7) \qquad a(x) + d(y) < 0, \quad a(x)\,d(v) - bc > 0.$$

The special case $bc = 0$ can be treated in the same manner as in a) and b) by direct integration. The equilibrium is globally asymptotically stable in this case. Assume therefore that $bc \neq 0$, and at first let $bc < 0$. Furthermore let

$$(29.8) \qquad a(x) \leq 0, \quad d(y) \leq 0.$$

Then because of (29.7) at least one of the two inequalities is strict. The function

$$v = (c\,x^2 - b\,y^2)\,\text{sgn}\, c$$

is positive definite because $bc < 0$ and is radially unbounded. Its derivative

$$\dot{v} = 2\,\text{sgn}\, c\,(c\,x^2\,a(x) - b\,y^2\,d(y))$$

is non-positive. It can vanish only on one of the two axes which contain no trajectory. Theorems 26.2 and 26.3 imply global asymptotic stability.

If the inequalities (29.8) are not both valid then at least one of the functions $a(x)$, $d(y)$ can assume positive values but because of the first inequality (29.7) both functions cannot be positive simultaneously. We may assume without loss of generality that $d(y)$ is positive. We abbreviate, setting $\beta := \max d(y)$ and $\alpha := bc/\beta$; β is a finite number because of (29.7). Then we have

$$(29.9) \qquad a(x) + \beta < 0, \quad a(x)\,\beta - bc > 0,$$

also $a(x) > bc/\beta = x$, which implies

$$\alpha + d(y) < a(x) + d(y) < 0.$$

Since by definition

$$d(y) \le \beta = \frac{bc}{\alpha},$$

that is $x\,d(y) \le bc$, we obtain in addition to (29.8) the inequalities

$$(29.10) \qquad d(y) + \alpha < 0, \quad \alpha\,d(y) - bc \ge 0.$$

If $bc > 0$ the same inequalities are obtained; but in this case the constants α and β are negative.

The Liapunov function

$$v = \frac{1}{2}\,(\beta^2 - bc)\,x^2 + \frac{1}{2}\,(b^2 - b^3 c\,\alpha^{-2})\,y^2$$

$$+ \beta \int\limits_0^x \xi\,a(\xi)\,d\xi + b^2\,\alpha^{-1} \int\limits_0^y \eta\,d(\eta)\,d\eta - b\,\beta\,x\,y$$

has as its derivative

$$\dot{v} = (a(x) + \beta)\,(a(x)\,\beta - bc)\,x^2 + b^2\,\alpha^{-2}\,(d(y) + \alpha)\,(d(y)\,\alpha - bc)\,y^2.$$

Because of inequalities (29.9) and (29.10) this function is non-positive. It can vanish only on the axes which contain no trajectory. We can again apply Theorems 26.2 and 26.3 after first showing that v is positive definite and radially unbounded. However,

$$v = \frac{1}{2}\,(by - \beta x)^2 + \int\limits_0^x (\beta\,\xi\,a(\xi) - bc\,\xi)\,d\xi$$

$$(29.11)$$

$$+ b^2\,\alpha^{-2} \int\limits_0^y (\alpha\,\eta\,d(\eta) - bc\,\eta)\,d\eta.$$

This implies the definiteness, and global asymptotic stability is assured if at least one of the two integrals in (29.11) diverges. KRASOVSKII [2] showed that this condition on the integrals is not needed. Thus the inequalities (29.7) are necessary and sufficient for global asymptotic stability.

9*

KRASOVSKII [2] has treated two further sub-cases of the system (29.1). His calculations are somewhat lengthy so that we will only give the results here.

d) $$\dot{x} = a x + y b(y), \quad \dot{y} = x c(x) + d y.$$

The Hurwitz conditions

$$a + d < 0, \quad a d - c(x) b(y) > 0$$

are necessary and sufficient for global asymptotic stability.

e) $$\dot{x} = x a(x) + y b(y), \quad \dot{y} = c x + d y.$$

The conditions are not sufficient here even in the strengthened form

$$a(x) + d \leq - \delta_1 < 0, \quad a(x) d - b(y) c \geq \delta_2 > 0.$$

However, the additional condition

$$x a(x) + d x \text{ is monotone decreasing}$$

guarantees global asymptotic stability.

30. Further Applications of the Direct Method

A. Systems with definite first integrals. We shall assume that the differential equation

$$(30.1) \qquad \dot{x} = f(x), \; f \in E, \; x \in K,$$

has a definite first integral $V(x)$. Using the notation of sec. 2 we have

$$V\big(p(t, x_0, t_0)\big) = \text{const.}$$

The derivative of such a function for (30.1) is of course identically zero. If the function $V(x)$ is definite we can apply the principal Theorem 25.1 and conclude the stability of the equilibrium of (30.1).

The equations for the motion of a rigid body (gyroscope) which rotates about a fixed point are

$$(30.2) \qquad \begin{aligned}
a_1 \dot{x}_1 + (a_3 - a_2) x_2 x_3 &= 0, \\
a_2 \dot{x}_2 + (a_1 - a_3) x_3 x_1 &= 0, \\
a_3 \dot{x}_3 + (a_2 - a_1) x_1 x_2 &= 0.
\end{aligned}$$

The a_i denote the main moments of inertia and x is the velocity vector with respect to the principal axes of inertia. In addition to the equilibrium, (30.2) has the constant solutions $(c_1, 0, 0)$, $(0, c_2, 0)$, $(0, 0, c_3)$. To determine the stability of $(c_1, 0, 0)$ we translate the origin and introduce

$z = \mathrm{col}\,(x_1 - c_1,\, x_2,\, x_3)$. Then

$$(30.3) \quad \begin{aligned}
a_1 \dot{z}_1 + (a_3 - a_2)\, z_2 z_3 &= 0,\\
a_2 \dot{z}_2 + (a_1 - a_3)\, z_1 z_3 + (a_1 - a_3)\, c_1 z_3 &= 0,\\
a_3 \dot{z}_3 + (a_2 - a_1)\, z_1 z_2 + (a_2 - a_1)\, c_1 z_2 &= 0.
\end{aligned}$$

Two first integrals of this system are

$$\frac{a_2 - a_1}{a_3}\, z_2^2 + \frac{a_3 - a_1}{a_2}\, z_3^2 \pm (a_1 z_1^2 + a_2 z_2^2 + a_3 z_3^2 + 2 a_1 c_1 z_1)^2.$$

One of these is definite if $a_1 < a_2 \leq a_3$ or $a_1 > a_2 \geq a_3$. Hence the rotations about the largest and about the smallest axis are stable.

The system of the first approximation for (30.3) is

$$\dot{z}_1 = 0,$$

$$\dot{z}_2 = -\,\frac{a_1 - a_3}{a_2}\, c_1 z_3; \quad \dot{z}_3 = -\,\frac{a_2 - a_1}{a_3}\, c_1 z_2.$$

The characteristic polynomial is

$$\lambda\!\left(\lambda^2 - c_1^2\, \frac{(a_1 - a_3)\,(a_2 - a_1)}{a_2 a_3}\right).$$

If a_1 lies between a_2 and a_3, then the polynomial has a positive root and the equilibrium is unstable. Otherwise the roots have zero real parts and the Principle of Stability in the First Approximation fails.

At times the first integrals $V_1, V_2, \ldots$ are known but not definite. We can try in that case to combine these integrals to obtain a definite integral and criteria for stability. (We note that for instance the form $V_1^2 + V_2^2$ is a first integral which is clearly semi-definite but not necessarily definite.)

This procedure is especially suitable for the theory of the stability of gyroscopes[1]).

With the help of appropriate first integrals it can be shown that the stability behavior in critical cases is not determined by the linear part. Let

$$\dot{x} = A x + f(x),$$

where $f(x)$ is at least of second degree. The matrix A is assumed to be "critical", *i.e.* we assume that it has no characteristic root with positive real part but at least one characteristic root with a zero real part. As LIAPUNOV [3] has shown, the equilibrium may be stable or unstable depending on the nature of the non-linear part $f(x)$. We conclude this as follows.

---

[1]) *cf.* CHETAEV [2].

Obviously it suffices to assume that A is a kth order block of a Jordan normal form. If the characteristic root equals zero then the corresponding system of equation is

$$
\text{(30.4)} \qquad
\begin{aligned}
\dot{x}_1 &= f_1(x), \\
\dot{x}_i &= x_{i-1} + f_i(x), \quad i = 2, \ldots, k.
\end{aligned}
$$

The functions h_i, $i = k, k-1, \ldots, 1$, are chosen so that

$$
h_i(x) = x_i^2 + \big(h_{i+1}(x)\big)^2, \quad h_{k+1}(x) = 0,
$$

and we set

$$
f_i(x) = -2x_{i+1}h_{i+1}(x), \quad i = 1, \ldots, k.
$$

Then the function $h_1(x)$ is a first integral of (30.4). It is positive for $x \neq 0$ and vanishes at the origin, hence it is definite. For this choice of f the equilibrium is therefore stable.

For the case $k = 3$ (30.4) has the form

$$
\begin{aligned}
\dot{x}_1 &= -2x_2(x_2^2 + x_3^4), \\
\dot{x}_2 &= x_1 - 2x_3^3, \\
\dot{x}_3 &= x_2.
\end{aligned}
$$

The first integral is

$$
h_1(x) = x_1^2 + (x_2^2 + x_3^4)^2.
$$

If we are concerned with a pair of conjugate imaginary characteristic roots then the order of the block is even, $k = 2m$, and the equations have the form

$$
\begin{aligned}
\dot{x}_1 &= -y_1 + f_1(x, y); \quad \dot{y}_1 = x_1 + g_1(x, y), \\
\dot{x}_i &= -y_i + x_{i-1} + f_i(x, y); \; \dot{y}_i = x_i + y_{i-1} + g_i(x, y), \\
&\qquad\qquad\qquad\qquad\qquad i = 2, 3, \ldots, m.
\end{aligned}
$$

For $i = m, m-1, \ldots, 1$, we again define

$$
h_i(x, y) = x_i^2 + y_i^2 + \big(h_{i+1}(x, y)\big)^2; \quad h_{m+1}(x, y) \equiv 0,
$$

and set

$$
f_i(x, y) = -2x_{i+1}h_{i+1}(x, y); \quad g_{i+1}(x, y) = -2y_{i+1}h_{i+1}(x, y).
$$

Then $h_1(x, y)$ is a first integral which is definite and the equilibrium is stable.

Instability of the equilibrium is more easily obtained. If the numbers k and m are greater than 1 then the linear part by itself is already unstable. If $k = 1$, resp. $m = 1$, then the equations

$$
\dot{x} = x^2, \text{ resp. } \dot{x} = -y + ax(x^2 + y^2), \; \dot{y} = x + ay(x^2 + y^2), \; a > 0
$$

each yield an example for an unstable equilibrium.

B. A criterion of Krasovskii[1]). Let us make the assumption on (30.1) that f has continuous first order partial derivatives at each $x \in K_r$ and let $J(x)$ be the functional matrix

$$(30.5) \qquad\qquad J(x) : = \frac{\partial f}{\partial x}.$$

Theorem 30.1. If there exists a positive definite matrix B with constant elements such that the characteristic roots of the matrix

$$M : = \frac{1}{2}\,(J^T B + B J)$$

are bounded above by a fixed negative bound $-c$ for all x in K_r, then the equilibrium of (30.1) is asymptotically stable. If the inequality holds for all x in R_n then the equilibrium is globally asymptotically stable.

Proof. The real parts of the characteristic roots of the matrix BJ lie between the largest and the smallest characteristic root of its symmetric part and are therefore smaller than $-c$. We therefore have in K_r

$$|\det BJ| \ge c^n$$

and we conclude that the function $w(x) := |\det J(x)|$ has a positive minimum α in the domain K_r. But this implies that the mapping of R_n into the space of components f_i, characterized by $f = f(x)$, is one to one in a neighborhood of the origin, and from this it follows that the origin is an isolated singularity for the equation (30.1). (In this exceptional case, therefore, we need not explicitly require that the origin is isolated — cf. the definition of the class E in sec. 16 — because this follows from the remaining hypotheses.) The Liapunov function

$$(30.6) \qquad\qquad v(x) = f^T B f$$

is positive definite in the f-space and also in the x-space. Its derivative for (30.1) is

$$\dot{v} = f^T (J^T B + B J) f$$

and this expression is negative definite because of the assumption on J. This assures asymptotic stability. If the inequality $w(x) \ge \alpha$ obtains in every finite domain then (30.1) cannot have any singularities in R_n other than the origin, and to prove global asymptotic stability we need only show that (30.6) is radially unbounded. For this purpose we integrate the volume element in the f-space and in the x-space:

$$\int df = \int w(x)\, dx \ge \alpha \int dx.$$

As $|x|$ increases the integral on the right becomes arbitrarily large. Therefore at least one component of $f(x)$ must grow without bound as x increases, and this furnishes the desired assertion.

---

[1]) Krasovskii [4].

Example.

$$(30.7) \qquad \dot{x}_1 = f_1(x_1) + f_2(x_2); \quad \dot{x}_2 = x_1 + a x_2.$$

Here we have

$$J(x) = \begin{pmatrix} f_1'(x_1) & f_2'(x_2) \\ 1 & a \end{pmatrix}.$$

We can choose $B = E$. As a sufficient condition for stability we obtain

$$2 f_1'(x_1) + 2a \leq -\delta_1 < 0; \quad 4 a f_1'(x_1) - (1 + f_2'(x_2))^2 \geq \delta_2 > 0$$

for all $(x_1, x_2) \in K_r$.

C. The method of the variable gradient[1]) depends on the fact that we
have

$$\dot{v} = (\operatorname{grad} v)^T \dot{x}, \quad \text{resp. } dv = (\operatorname{grad} v)^T dx.$$

We can thus write $v(x)$ formally as a line integral:

$$v(x) = \int_{P_0}^{P_1} (\operatorname{grad} v)_1 \, dx_1 + \int_{P_1}^{P_2} (\operatorname{grad} v)_2 \, dx_2 + \ldots + \int_{P_{n-1}}^{P_n} (\operatorname{grad} v)_n \, dx_n,$$

$$(30.8)$$

where

$$\operatorname{grad} v = : \operatorname{col}\left((\operatorname{grad} v)_i\right); \quad P_0 : = 0, \quad P_i : = \operatorname{col}(x_1, \ldots, x_i, 0, \ldots, 0).$$

The condition

$$(30.9) \qquad \frac{\partial (\operatorname{grad} v)_i}{\partial x_j} = \frac{\partial (\operatorname{grad} v)_j}{\partial x_i},$$

which assures integrability, must be satisfied. We set

$$\operatorname{grad} v = : A(x) \, x$$

and try to choose the elements of the matrix $A(x)$ so that (30.9) holds and
so that

$$\dot{v} = x^T A^T(x) f(x)$$

becomes negative definite. If the functional matrix $J(x)$ [see (30.5)] is
symmetric then $f(x)$ can be considered as a gradient and by (30.8) we
obtain the function

$$v(x) = -\int_{P_0}^{P_1} f_1(x) \, dx_1 - \ldots - \int_{P_{n-1}}^{P_n} f_n(x) \, dx_n.$$

Its derivative for (30.1) equals $-|f(x)|^2$ and hence is clearly negative
definite.

----

[1]) Schultz and Gibson [1].

Examples. a) We consider a transfer system consisting of two units in series

$$(30.10) \qquad y = x f(x), \quad \ddot{x} + \dot{x} = - (\dot{y} + \beta y)$$

with simple feedback (fig. 30.1).

$$\dot{x}_1 = x_2, \quad \dot{x}_2 = - x_2 - f(x_1) x_2 - \beta x_1 f(x_1) - x_1 x_2 f'(x_1)$$

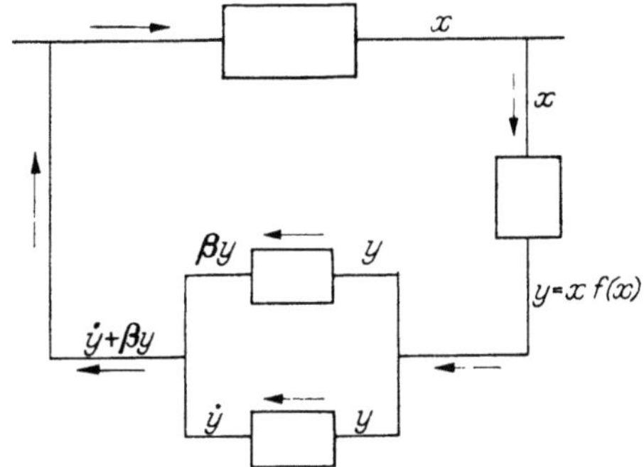

Fig. 30.1. Blockdiagram for (30.10)

is an equivalent system. We set

$$\operatorname{grad} v = \operatorname{col}\left(a_{11}(x_1) x_1 + 2 x_2, \quad a_{21} x_1 + 2 x_2\right),$$

so that only one element of the matrix A is not constant. (30.9) implies

$$a_{21} = 2,$$

and we obtain

$$v = x_1 x_2 \left(a_{11} - 2 - 2(1 + \beta) f(x_1)\right) - 2 f(x_1) \left(\beta x_1^2 + x_2^2\right)$$
$$- 2 x_1 x_2 (x_1 + x_2) f'(x_1).$$

For negative definiteness it is necessary that

$$f(x) > 0, \quad \beta > 0.$$

Again introducing the variable $y = x_1 f(x_1)$ we see that $f' = y'/x_1 - f(x_1)/x_1$ and

$$\dot{v} = x_1 x_2 \left(a_{11} - 2 - 2\beta f(x_1) - 2y'\right) - 2y' x_2^2 - 2\beta x_1^2 f(x_1).$$

We make the substitution

$$a_{11}(x_1) = 2 + 2\beta f_1(x_1) + 2y',$$

and obtain

$$\dot{v} = - 2 x_2^2 \frac{dy}{dx_1} - 2\beta x_1^2 f(x_1).$$

Since $\dot{v}$ is required to be negative definite we must have $\dfrac{dy}{dx_1} > 0$. Using (30.8) we compute the expression

$$v(\boldsymbol{x}) = x_1^2 + 2x_1 x_2 + x_2^2 + 2\beta \int^{x_1} x_1 f(x_1)\, dx_1 + 2 \int^{x_1} x_1 \frac{dy}{dx_1}\, dx_1.$$

It must now be tested for definiteness and radial unboundedness.
b) A third order equation with a strong nonlinearity[1]),

$$(30.11) \quad \dot{x}_1 = x_2, \quad \dot{x}_2 = x_3, \quad \dot{x}_3 = -\left(a x_1 + b x_2 + f(x_2)\, x_3\right).$$

We start with a function $\dot{v}$ of as simple a form as possible, assuming to begin with only that it is semi-definite,

$$\dot{v} = -x_3^2.$$

We obtain

$$v = -\int x_3^2\, dt = -\int x_3 \dot{x}_2\, dt = -x_2 x_3 + \int x_2 \dot{x}_3\, dt$$

and integrating again

$$-v(\boldsymbol{x}) = x_2 x_3 + \int_0^{x_2} \xi f(\xi)\, d\xi + \frac{1}{2} a x_1^2 + b \int x_2^2\, dt.$$

The limits of integration are chosen so that $v(0) = 0$. Now

$$\int x_2^2\, dt = \int x_2 \dot{x}_1\, dt = x_2 x_1 - \int x_1 \dot{x}_2\, dt$$

and the last integrand is equal to $-x_3(\dot{x}_3 + x_3 f(x_2) + b x_2)\,\dfrac{1}{a}$. Finally we obtain

$$-v(\boldsymbol{x}) = x_2 x_3 + b x_1 x_2 + \frac{a}{2} x_1^2 + \frac{b^2}{2a} x_2^2 + \frac{b}{2a} x_3^2$$

$$+ \int_0^{x_2} \xi f(\xi)\, d\xi + \frac{b}{a} \int f(x_2)\, x_3^2\, dt.$$

We choose the Liapunov function

$$v_1(\boldsymbol{x}) = -v(\boldsymbol{x}) - \frac{b}{a} \int f(x_2)\, x_3^2\, dt.$$

Its derivative is

$$\dot{v}_1 = -\frac{b}{a}\left(f(x_2) - \frac{a}{b}\right) x_3^2,$$

which vanishes only for $x_3 = 0$ (because from $x_2 = $ const. it follows that again $x_3 = 0$) and this is not a trajectory. On the other hand we can write

$$v_1(\boldsymbol{x}) = \frac{1}{2a}(b x_2 + a x_1)^2 + \frac{1}{2ab}(b x_3 + a x_2)^2 + \int_0^{x_2}\left(f(\xi) - \frac{a}{b}\right)\xi\, d\xi,$$

---

[1]) Reiss and Geiss [1].

and we see that the conditions

$$a > 0, \quad b > 0, \quad f(\xi) \geq a/b + \delta, \quad \delta > 0$$

are sufficient for global asymptotic stability by Theorems 26.2 and 26.3.

D. Occasionally this procedure can be modified along the following lines. We can consider the matrix A in (27.1) as the functional matrix (30.5) for the equation (27.1). The matrix B arises from the Liapunov function (27.2) if we form the second partial derivatives of v: $2 b_{ik} = \partial^2 v/\partial x_i \partial x_k$. Accordingly v can be obtained from B by integrating twice: The first step yields the vector grad v, the second is the integration in (30.8). We now formally solve a matrix equation

$$J^T B + B J = - C,$$

in which $- C$ is a suitably chosen preassigned matrix, which is definite or at least semi-definite. The elements of B are of course functions of the variables $x_1, \ldots, x_n$, and from them we wish to obtain, by integrating twice, a suitable Liapunov function. Throughout the construction we will have to manipulate the coefficients so that the functions will have the desired properties. For example, it is indicated to alter the matrix $B(\pmb{x})$ so that b_{ik} depends only on x_i and x_k[1]).

Examples. a)

$$(30.12) \qquad \dot{x}_1 = x_2 \quad \dot{x}_2 = - (a_1 x_1 + a_2 x_1^2 x_2),$$

$$J(\pmb{x}) = \begin{pmatrix} 0 & 1 \\ - (a_1 + 2 a_2 x_1 x_2) & - a_2 x_1^2 \end{pmatrix}.$$

If we choose the matrix C so that only $c_{22} \neq 0$ and, in particular, so that the coefficient γ in (27.4) is equal to 1, then $\beta = 0$ and $\alpha = a_1 + 2 a_2 x_1 x_2$. Working with the modified matrix

$$B = \begin{pmatrix} a_1 & 0 \\ 0 & 1 \end{pmatrix},$$

we integrate and obtain

$$\operatorname{grad} v = \operatorname{col}(a_1 x_1, x_2), \quad v = \frac{1}{2} (a_1 x_1^2 + x_2^2), \quad \dot{v} = - a_2 x_1^2 x_2^2.$$

By Theorems 26.2 and 26.3, we see that the system (30.12) has a globally asymptotically stable equilibrium in case $a_1 > 0$, $a_2 > 0$.

b) Writing the scalar equation

$$(30.13) \qquad \dddot{x} + a_1 \ddot{x} + a_2 \dot{x} + x f(x) = 0$$

---

[1]) Ingwerson [1].

as a system and choosing for C the matrix

$$\left|\begin{array}{ccc} 2\,a_2(x\,f' + f) & 0 & 2\,(x\,f' + f) \\ 0 & 0 & 0 \\ 2\,(x\,f' + f) & 0 & 2\,a_1 \end{array}\right|,$$

we obtain

$$B = \left|\begin{array}{ccc} a_2^2 + a_1(x_1 f' + f) & a_1 a_2 & a_2 \\ a_1 a_2 & a_1^2 + a_2 & a_1 \\ a_2 & a_1 & 2 \end{array}\right|,$$

$$\operatorname{grad} v = \operatorname{col}(a_2^2 x_1 + a_1 x_1 f(x_1) + a_1 a_2 x_2 + a_2 x_3,$$

$$a_1 a_2 x_1 + (a_1^2 + a_2) x_2 + a_1 x_3, \qquad a_2 x_1 + a_1 x_2 + 2 x_3),$$

$$v = \frac{1}{2} a_2^2 x_1^2 + a_1 \int_0^{x_1} u f(u)\, du + a_1 a_2 x_1 x_2 + \frac{1}{2} (a_1^2 + a_2) x_2^2$$

$$+ a_2 x_1 x_3 + a_1 x_2 x_3 + x_3^2,$$

$$\dot v = - \left(a_2 x_1^2 f(x_1) + 2 x_1 x_3 f(x_1) + a_1 x_3^2\right).$$

Again we must apply Theorem 26.3. The conditions for global asymptotic stability are

$$a_1 > 0, \quad a_2 > 0, \quad f(y) > 0, \quad a_1 a_2 - f(y) > 0,$$

and these are exactly the "generalized Hurwitz conditions" (*cf.* sec. 29). For the special equation (30.13) therefore the Aizerman conjecture is correct[1]).

As already indicated, the procedures which we sketched in this section form a type of systemized trial and error. We can certainly expect to be successful only if we are dealing with equations of a low order whose nonlinearities are of a simple analytical construction, especially of polynomial type[2]).

31. Absolute Stability

In many systems in applications there is only one nonlinear transfer unit. For control systems this is usually the motor. Its characteristic often deviates so much from a linear function that we cannot substitute a linearized system to study such a system. The direct method has proved to be a suitable tool to deal with a certain problem which is important in practice. To begin with we assume that the equations for the motion are

---

[1]) *cf.* also BERGEN and WILLIAMS [1], PLISS [2].

[2]) Further examples are found in KU and PURI [1], SZEGÖ [1, 2, 3].

given in the form

(31.1) $$\dot{z} = \tilde{A}\,z + b\,\xi, \quad \xi = f(\sigma), \quad \sigma = \tilde{g}^T z.$$

ξ and σ are scalars, z, b, $\tilde{g}$ n-dimensional vectors. The nonlinear function $f(\sigma)$ satisfies a so-called *sector condition*

$$0 \le \sigma f(\sigma) \le \varkappa_0 \sigma^2, \quad 0 < \varkappa_0 \le \infty.$$

The physical system described by (31.1) is called *absolutely stable* in the sector $[0, \varkappa_0]$, if the equilibrium is globally asymptotically stable and independent of the particular choice of the nonlinear function f, as long as this function satisfies the sector condition. If $\varkappa_0 = \infty$, then the graph of $f(\sigma)$ passes through the first and third quadrants of the (σ, f)-plane. Since the function $f \equiv 0$ is admissible, the matrix $\tilde{A}$ is of necessity stable in case we have absolute stability.

If $\tilde{A}$ is critical and in particular, if $\tilde{A}$ has one zero characteristic root and otherwise stable characteristic roots then we may assume without loss of generality that

$$\tilde{A} = \begin{pmatrix} A & 0 \\ 0 & 0 \end{pmatrix},$$

where A is stable; this can always be arranged by means of a linear transformation. The variable

$$\dot{\sigma} = \tilde{g}^T \dot{z} = \tilde{g}^T \tilde{A}\,z + \tilde{g}^T b\,\xi,$$

which is obtained by taking the derivative of the third equation in (31.1) does not contain the component z_n on the right. Since, because of the form of $\tilde{A}$, z_n is absent from the right side of the first equation also, we can ignore this component. Using the notation

$$y = \mathrm{col}(z_1, \ldots, z_{n-1}), \quad b = \mathrm{col}(b_1, \ldots, b_{n-1}), \quad g = \mathrm{col}(g_1, \ldots, g_{n-1}),$$

the equations of the motion can be written in the form

$$\dot{y} = A\,y + b\,\xi, \quad \xi = f(\sigma), \quad \dot{\sigma} = g^T A\,y + g^T b\,\xi,$$

resp.

(31.2) $$\dot{y} = A\,y + b\,f(\sigma), \quad \dot{\sigma} = g^T A\,y + g^T b\,f(\sigma).$$

The equations of the motion

(31.3) $$\dot{x} = A\,x + \eta\,b, \quad \dot{\eta} = f(\sigma), \quad \sigma = c^T x - \varrho\,\eta$$

describe a controlled circuit with a nonlinear servomotor and feedback (see fig. 31.1). The input which is represented by the components of the vector x is transformed into the scalar variable $c^T x$ by the measuring unit M. The nonlinear transfer unit N takes its input σ into $f(\sigma)$. The output is integrated by the unit I, and after being multiplied by a nega-

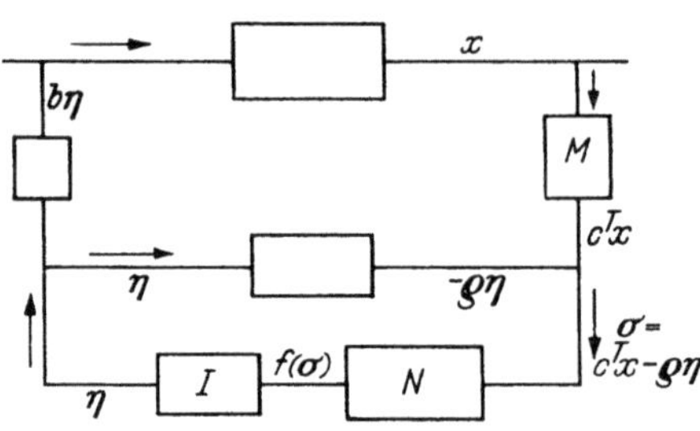

Fig. 31.1. Block diagram for (31.3)

tive amplification factor $-\varrho$ it is added to the output of M by the feedback, so that the scalar σ becomes the input of N. Since the function f, the characteristic of the unit N (the servomotor), is usually only approximately known, it is desirable to make the stability of the system independent of the special properties of the characteristic. One would therefore try, for example, to choose the parameters of the controller which are contained in the components of c in (31.3), in such a way that the total system is absolutely stable in a sector $[\varepsilon, \varkappa_0]$ with an arbitrarily small positive ε, i.e. for

$$\varepsilon \sigma^2 \le \sigma f(\sigma) \le \varkappa_0 \sigma^2.$$

We shall now see that we have to exclude the case $f \equiv 0$, by changing the sector condition. For if $f \equiv 0$ then $y = 0$, $\sigma = \text{const.}$ is a solution of (31.2) which then is clearly not absolutely stable. An appropriate choice of notation, however, shows the systems (31.2) and (31.3) equivalent as far as absolute stability is concerned. By differentiating the first and third equations in (31.3), eliminating $\dot\eta$, and setting $y = \dot x$ we obtain

$$(31.4) \qquad \dot y = A y + f(\sigma) b, \quad \dot\sigma = c^T y - \varrho f(\sigma).$$

This is (31.2) with $g^T A = c^T$, $g^T b = -\varrho$. Therefore, if $x \to 0$ and $\eta \to 0$, then $\sigma \to 0$ and $y \to 0$. Conversely, as y and σ approach zero, x and η tend toward constant values x^0, η^0, where

$$A x^0 + b \eta^0 = 0, \quad c^T x^0 - \varrho \eta^0 = 0.$$

This in turn implies $x^0 = 0$, $\eta^0 = 0$; for the determinant of the matrix

$$G := \begin{pmatrix} A & b \\ c^T & -\varrho \end{pmatrix}$$

is different from zero. This follows from the fact that in the case of a linear characteristic $f(\sigma) = \varkappa\sigma$, (31.2) must have an asymptotically stable equilibrium; for the linear characteristic satisfies the sector condition for $0 < \varkappa < \varkappa_0$. Hence the linear system

$$\dot y = A y + b \varkappa\sigma, \quad \dot\sigma = c^T y - \varrho\varkappa\sigma$$

has an asymptotically stable equilibrium and accordingly a non-zero determinant, which is equal to $\varkappa \det G$. Therefore the two equations (31.2) and (31.3) are completely equivalent. There is a further way to write this. Since $y = A x + b \eta$, we have

$$\sigma = c^T x - \varrho \eta = c^T (A^I y - A^I b \eta) - \varrho \eta,$$

and the system of equations assumes the form

$$\dot{y} = A y + b f(\sigma), \quad \dot{\eta} = f(\sigma), \quad \sigma = c^T A^I y - (c^T A^I b + \varrho) \eta.$$
(31.5)

Historically the problem of absolute stability arose first in connection with a system of equations of the form (31.3), in conjunction with the control of the course of an airplane. The sector was $[\varepsilon, \infty)$; the parameter vector c needed to be found. Since then the problem has been treated in a great number of papers. They either consider the system (31.3) with a stable $\tilde{A}$ or else one critical characteristic root is admitted [leading to equations (31.2) through (31.5)], or again, several critical characteristic roots or even unstable characteristic roots are admitted. Each of these cases requires a special procedure even though the basic idea of the method is always the same[1]. For the present we limit our attention to the special case of the system of equations (31.4). As already noted in part, we assume for this purpose that the matrix A is n-dimensional and stable, that G is non-singular, and $\varrho > 0$.

If both A and G are non-singular the same is true for the matrices

$$H : = \operatorname{diag}(A^I, 1), \quad HG = \begin{pmatrix} E & A^I b \\ c^T & -\varrho \end{pmatrix}.$$

The condition $\det HG \neq 0$ leads to the inequality

(31.6) $$\varrho + c^T A^I b \neq 0.$$

For (31.4) we construct a Liapunov function which consists of "a quadratic form + an indefinite integral of the nonlinear function". Functions of this type were partly used in sec. 29 also. We choose an arbitrary positive definite matrix C, define the matrix B by the equation $A^T B + B A = -C$ [see (27.3)], and set

(31.7) $$v = y^T B y + \int_0^\sigma f(s) \, ds.$$

The derivative of this function for (31.4) is

$$\dot{v} = - y^T C y - \varrho (f(\sigma))^2 + 2 f(\sigma) \left(b^T B + \frac{1}{2} c^T \right) y.$$

---

[1] *cf.* AIZERMAN and GANTMACHER [2].

This expression is considered as a quadratic form in the $n + 1$ variables $y_1, \ldots, y_n, f(\sigma)$. By construction, $\dot{v}$ is negative definite with respect to the first n variables $y_1, \ldots, y_n$ (*i.e.* for $f \equiv 0$). The last of the Sylvester inequalities is

$$\det \begin{pmatrix} C & -(B\boldsymbol{b} + \boldsymbol{c}/2) \\ -(B\boldsymbol{b} + \boldsymbol{c}/2)^T & \varrho \end{pmatrix} > 0$$

or

$$(31.8) \qquad \varrho > (B\boldsymbol{b} + \boldsymbol{c}/2)^T C^I (B\boldsymbol{b} + \boldsymbol{c}/2).$$

It guarantees that $\dot{v}$ is negative definite with respect to all $n+1$ variables and that therefore the equilibrium is asymptotically stable. If we assume in addition that the integral in (31.7) grows without bound as σ increases, then v becomes radially unbounded and we can apply Theorem 26.3. The stability has been proved quite independently of the choice of the function $f(\sigma)$. Hence we have:

Theorem 31.1. The inequalities (31.6) and (31.8), in conjunction with the divergence of the integral $\int_0^\infty f(s)\, ds$ are sufficient for the absolute stability of the system of equations (31.4) in the sector $[\varepsilon, \infty)$[1]).

The inequalities determine in the space of the parameters c_i a *domain of absolute stability*. It depends on the matrix C. We can construct this domain for all possible positive definite matrices C and form the union of all such domains. It is not yet known whether in this manner a maximal domain of absolute stability is obtained, because here the stability has been dealt with by means of a Liapunov function of a *special* kind.

If $A = \mathrm{diag}(\alpha_1, \ldots, \alpha_n)$, α_i real, then by (27.3) the choice $C = \mathrm{diag}(\gamma_1^2, \ldots, \gamma_n^2)$, $\gamma_i \neq 0$, leads to

$$B = \mathrm{diag}(-\gamma_1^2/2\alpha_1, \ldots, -\gamma_n^2/2\alpha_n),$$

and inequality (31.8) becomes simply

$$\varrho > \frac{1}{4} \sum_{i=1}^n \left(-\frac{b_i \gamma_i}{\alpha_i} + \frac{c_i}{\gamma_i}\right)^2.$$

Lur'e [1] originally attacked the problem in a different way. He first of all used a linear transformation

$$(31.9) \qquad \boldsymbol{y} = P\boldsymbol{z}$$

on (31.4). Such a transformation leaves the stability condition (31.6), unchanged, as we can easily check, since B and C are subjected to a congruence transformation (*cf.* sec. 27) and $\boldsymbol{b}$ and $\boldsymbol{c}$ go over into $P^I\boldsymbol{b}$,

---

[1]) *cf.* LaSalle and Lefschetz [1], Lefschetz [2].

resp. $P^T c$. Lur'e chose P so that A was changed into a Jordan canonical form and used his freedom in the choice of P to make the vector $P^I b$ as simple as possible. If all the characteristic roots of A are simple and if b is not orthogonal to any left-characteristic vector of A then none of the components of the transformed vector $P^I b$ vanish and by a suitable normalization it can be arranged that $P^I b = \mathrm{col}(1, 1, \ldots, 1)$. If b is orthogonal to $m \geq 1$ left-characteristic vectors of A then we can still have $P^I b = \mathrm{col}(1, \ldots, 1, 0, \ldots, 0)$ with m zero components. In the first case the system of equations becomes

$$(31.10) \qquad \dot{z}_i = \alpha_i z_i + f(\sigma), \quad \dot{\sigma} = \sum_{k=1}^{n} \beta_k z_k - \varrho f(\sigma),$$

$$i = 1, 2, \ldots, n,$$

where $c^T P z =: \sum_{k=1}^{n} \beta_k z_k$; in the second case the last m equations which have the form

$$\dot{z}_i = \alpha_i z_i \quad (i = m + 1, \ldots, n)$$

are completely independent of the others, and can therefore be treated separately. In investigating the absolute stability we can limit our attention to the remaining nonlinear system of $n - m$ equations together with the equation for $\dot{\sigma}$; we need not pay attention to the terms involving $z_{m+1}, \ldots, z_n$. From the point of view of control engineering the canonical transformation (31.9), *i.e.* the transition to (31.10), amounts to replacing the linear part of (31.3), resp. (31.1), by an equivalent system of first order transfer units connected in parallel and connecting the integrating unit somewhere into the circuit (fig. 31.2).

Before pursuing the ideas of Lur'e and the *canonical system* (31.10) any further we note that the

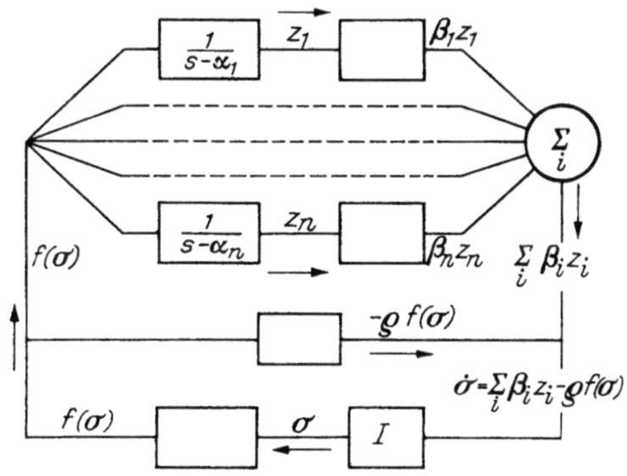

Fig. 31.2. Block diagram for (31.10) (canonical transform)

condition "b is orthogonal to no left-characteristic vector of A" can be replaced by the equivalent condition "the vectors $b, A b, \ldots, A^{n-1} b$ are linearly independent". This is seen as follows: If there is a left-characteristic vector y of A which is orthogonal to b then $y^T b = 0$ and $y^T A = \alpha y^T$. This implies $y^T A^k b = 0$, $k = 0, \ldots, n - 1$, and hence the linear dependence of $b, Ab, \ldots, A^{n-1}b$, since $y \neq 0$ cannot be orthogonal to n linearly independent vectors. To prove conversely the existence of a left-characteristic vector y which is orthogonal to b if we are given that $b, Ab, \ldots, A^{n-1}b$ are linearly dependent, we transform A into its Jordan normal form and write down the condition for linear dependence. For this it suffices to consider only one of the "blocks" of the normal form. If it is one-dimensional then the vector b (which has been subjected to the same transformation as A) has at least one zero component. Then y is the vector which has a 1 in the corresponding place and zeros otherwise. A similar argument is used for a block of higher dimension,

$$J := \begin{vmatrix} \alpha & 0 & 0 & \cdots & 0 & 0 \\ 1 & \alpha & 0 & \cdots & 0 & 0 \\ 0 & 1 & \alpha & \cdots & 0 & 0 \\ \hdotsfor{6} \\ 0 & 0 & 0 & \cdots & 1 & \alpha \end{vmatrix}_k .$$

Because of the linear dependence, we have a relation

$$c_1 b + c_2 J b + \cdots + c_k J^{k-1} b = 0$$

whence either $b_1 = 0$ or $c_1 + \alpha c_2 + \cdots + \alpha^{k-1} c_k = 0$. Considering the second row of the relation in the second case we yield either $b_2 = 0$ or $c_2 + 2\alpha c_3 + \cdots + (k-1)\alpha^{k-2} c_k = 0$ etc. At any rate, one component of b must be zero.

If A has elementary divisors of higher order then the vector $P^I b$ can have zero components without b being orthogonal to a left-characteristic vector of A; the original system does not decompose into systems of lower order. Example:

$$A = \begin{vmatrix} 1 & 0 & 0 \\ 1 & 1 & 0 \\ 0 & 1 & 1 \end{vmatrix}, \quad b = \mathrm{col}(1, 0, 1), \quad y^T = (1, 0, 0).$$

Here $P = E$; $P^I b = \mathrm{col}(1, 0, 1)$ has a zero component. But $y^T b = 1$ and the system $\dot{x} = Ax + bf$ does not decompose. KALMAN [2] calls *completely controllable*[1]) a control system which cannot be decomposed into systems of lower order, resp. a matrix-vector pair (A, b) which satisfies the above condition "$b, Ab, A^2 b, \ldots, A^{n-1}b$ are linearly independent", or an equivalent condition.

---

[1]) See also sec. 4.

For the system (31.10) Lur'e [1] chooses a Liapunov function

$$v = \sum_{i=1}^{n} k_i z_i \bar{z}_i + \sum_{i,k} \frac{g_i g_k z_i \bar{z}_k}{-\alpha_i - \alpha_k} + \int_{0}^{\sigma} f(\eta)\, d\eta.$$

The constants k_i, to be discussed later, are positive to begin with. Assuming that A has m real characteristic roots and that all the characteristic roots of A are enumerated by

(31.11)
$$\alpha_1, \ldots, \alpha_m \text{ real,}$$
$$\alpha_{m+1} = \bar{\alpha}_{m+2}, \ldots, \alpha_{n-1} = \bar{\alpha}_n \text{ conjugate complex,}$$

we consider v as a quadratic form in the n variables $z_1, \ldots, z_m$, $\mathrm{Re}\, z_{m+1}$, $\mathrm{Im}\, z_{m+1}, \ldots, \mathrm{Im}\, z_n$. The constants g_i are such that analogously to (31.11)

(31.12)
$$g_1, \ldots, g_m \text{ real,}$$
$$g_{m+1} = \bar{g}_{m+2}, \ldots, g_{n-1} = \bar{g}_n \text{ conjugate complex.}$$

We further set $k_{m+1} = k_{m+2}, \ldots, k_{n-1} = k_n$,
Since by hypothesis $\mathrm{Re}\,\alpha_i < 0$, v is a positive definite function of the variables just named. To find the derivative of v for (31.10) we perform a short calculation and obtain

$$\dot{v} = 2 \sum_{i=1}^{n} k_i \,\mathrm{Re}\,\alpha_i |z_i|^2 - \left(f(\sigma)\sqrt{\varrho} + \sum_{i=1}^{n} g_i z_i \right)^2$$
$$+ f(\sigma)\left(\sum_{i=1}^{n} \tilde{c}_i z_i + 2\sqrt{\varrho}\sum_{i=1}^{n} g_i z_i - 2\sum_{i,k}\frac{g_i g_k}{\alpha_i + \alpha_k} z_i + 2\sum_{i=1}^{n} k_i z_i \right),$$

where $\tilde{c} := P^I c$; the term $2 f(\sigma)\sqrt{\varrho}\sum_{i=1}^{n} g_i z_i$ has been added and subtracted. The derivative is clearly negative definite if the coefficient of $f(\sigma)$, which is real by construction, vanishes, *i.e.* if

(31.13) $\quad \tilde{c}_i + 2k_i + 2g_i\sqrt{\varrho} - 2g_i\sum_{k=1}^{m}\frac{g_k}{\alpha_i + \alpha_k} = 0 \qquad (i = 1, \ldots, n).$

It is therefore sufficient for asymptotic stability that for the given values of the k_i the n equations (31.13) can be solved for the g_i in such a way that (31.12) holds. This is possible for arbitrarily small k_i and in the limit we arrive at the system of equations

(31.14) $\quad \tilde{c}_i + 2g_i\sqrt{\varrho} - 2g_i\sum_{k=1}^{m}\frac{g_k}{\alpha_i + \alpha_k} = 0 \qquad (i = 1, \ldots, n),$

which must be solvable for the g_i in such a way that (31.12) holds. In the same manner as (31.8), this condition determines a domain in the space of the parameters $c_1, \ldots, c_n$. To assure global asymptotic stability we further need v to be radially unbounded, for which the divergence of the integral $\int^{\pm\infty} f(\eta)\, d\eta$ is sufficient.

10*

The evaluation of these conditions is rather involved even in the simplest cases and must be expected to be quite impractical for systems of a degree higher than four. It is, however, of fundamental interest that the direct method allows us to reduce the transcendental stability problem to a purely algebraic problem, namely the solution of a system of quadratic inequalities with side conditions: Theoretically the treatment of (31.8) as well as that of (31.14) is essentially simpler than that of the original problem.

If the linear part of (31.1) is unstable, *i.e.* if the matrix A has characteristic roots with positive real part, then the problem is unsolvable in the given form. This is seen as follows: If we set $f(\sigma) = \gamma\sigma$, $0 < \gamma < \varkappa_0$, then we obtain a linear equation whose equilibrium is asymptotically stable since the particular function $f(\sigma)$ satisfies the sector condition. However, if A is unstable and the α_i are sufficiently large then this linear equation cannot be stable, as appears immediately from (31.10). The conditions for $f(\sigma)$ must therefore be changed for the case of an unstable matrix A; for instance, it could be required that for a fixed γ

$$(31.15) \qquad f(\sigma) = \gamma\sigma + \sigma\varphi(\sigma), \quad 0 < \sigma\varphi(\sigma).$$

This effects a rotation of the sector over which the nonlinearity is allowed to range, through the angle $\arctan\gamma$[1]) which depends on the unstable roots of A. We are not going to discuss these calculations in detail here but only mention that this idea can also be reversed. For sometimes the nonlinearity is fixed [for instance in the form (31.15)] and the linear part can be modified. As can be seen from the preceding discussion, under certain circumstances a stable total system can be obtained if γ is sufficiently large even though A has unstable characteristic roots, *i.e.* at times unstable linear parts can also be considered[2]).

32. Popov's Criterion

V. M. Popov[3]) has found a criterion for absolute stability, which is based on the frequency response diagram of the linear part. We shall illustrate the mathematical idea involved, using the system of equations (31.1) with stable $\tilde{A}$. If $f(\sigma) = \varkappa\sigma$ is linear and satisfies the sector condition $0 < \varkappa < \varkappa_0$, we eliminate σ and obtain the linear equation

$$(32.1) \qquad \dot{z} = \tilde{A}z + \varkappa\,\tilde{b}\,\tilde{g}^T z.$$

---

[1]) REKASIUS and GIBSON [1].   [2]) *cf.* LETOV [1, 2].

[3]) V. M. POPOV [1]; *cf.* also the discussion in AIZERMAN and GANTMACHER [2] or HALANAY [1].

Its characteristic equation is

$$\det\left(sE - \tilde{A} - \varkappa\,\boldsymbol{b}\,\tilde{\boldsymbol{g}}^T\right) = 0$$

and can be written in the form

$$(32.2) \qquad\qquad \varphi_2(s) + \varkappa\,\varphi_1(s) = 0.$$

The left side consists of polynomials in s, and $\varphi_2(s) = \det(sE - \tilde{A})$ is the characteristic polynomial of the matrix $\tilde{A}$. The last equation has an interpretation in control engineering: If we consider the quotient

$$(32.3) \qquad\qquad \varkappa\,\varphi_1(s)/\varphi_2(s)$$

as a transfer function of an open control loop ($cf.$ sec. 8) then (32.2) is the denominator of the transfer function of the corresponding closed loop. According to the Nyquist criterion (sec. 10), the closed loop is stable if 1) the open loop is stable and 2) the frequency response diagram belonging to (32.3) does not encircle the point $(-1, 0)$. The first condition is satisfied here: $\varphi_2(s)$ is a Hurwitz polynomial by hypothesis. The second condition is equivalent to the condition that the response diagram

$$(32.4) \qquad\qquad \varkappa\,\frac{\varphi_1(i\,\omega)}{\varphi_2(i\,\omega)} = \varkappa\left(u(\omega) + i\,v(\omega)\right)$$

either does not meet the interval $(-\infty, -1)$ of the real axis at all or cuts it downward as often as upward. This holds for a fixed $\varkappa$. If the closed loop is to be stable for all values of the parameter $\varkappa$, which may be interpreted as an amplification factor, in a given interval $0 \le \varkappa \le \varkappa_0$ then the response diagram

$$(32.5) \qquad\qquad \frac{\varphi_1(i\,\omega)}{\varphi_2(i\,\omega)} = u(\omega) + i\,v(\omega), \qquad 0 \le \omega < \infty,$$

must not cut the interval $\left(-\infty, -\dfrac{1}{\varkappa_0}\right)$. For if such a cut point exists then there exists an ω_1, such that $u(\omega_1) < -\dfrac{1}{\varkappa_0}$, and $v(\omega_1) = 0$, and such that a response diagram (32.4) involving a parameter value $\varkappa$ from the interval $\varkappa_1 = \dfrac{-1}{u(\omega_1)} < \varkappa < \varkappa_0$ cuts the interval $(-\infty, -1)$. In place of the response diagram (32.5) we can also consider the response diagram

$$(32.6) \qquad\qquad u(\omega) + i\,\omega\,v(\omega), \qquad 0 \le \omega < \infty.$$

It also must not cut the interval $\left(-\infty, -\dfrac{1}{\varkappa_0}\right)$. This condition is satisfied if there exists a line through the point $\left(-\dfrac{1}{\varkappa_0}, 0\right)$ which does not intersect the curve (32.6) or, in other words, if there exists a positive number β such that the inequality

$$(32.7) \qquad \operatorname{Re}(1 + i\,\omega\,\beta)\,\frac{\varphi_1(i\,\omega)}{\varphi_2(i\,\omega)} + \frac{1}{\varkappa_0} > 0, \qquad 0 \le \omega < \infty,$$

is satisfied. For (32.7) implies

$$\operatorname{Re}\frac{\varphi_1(i\,\omega)}{\varphi_2(i\,\omega)} - \beta\,\omega\,\operatorname{Im}\frac{\varphi_1(i\,\omega)}{\varphi_2(i\,\omega)} + \frac{1}{\varkappa_0} = u(\omega) - \beta\,\omega\,v(\omega) + \frac{1}{\varkappa_0} > 0\,.$$

In the distorted coordinate system $u^* = u$, $v^* = \omega v$ the equation $u^* - \beta v^* + \dfrac{1}{\varkappa_0} = 0$ represents a line through $\left(-\dfrac{1}{\varkappa_0}, 0\right)$. Inequality (32.7) characterizes a half-plane and the condition says that the response diagram lies entirely within this half-plane. Hence, if there exists a β satisfying (32.7) then the equilibrium of (32.1) is asymptotically stable for all parameter values of the interval $0 \le \varkappa \le \varkappa_0$.

V. M. Popov [1] recognized that inequality (32.7) can also be used in the nonlinear case. It is sufficient for the system of equations (31.1) to be absolutely stable in the sector $[0, \varkappa_0]$.

We now consider the system of equations (31.5) which we write in the form

$$(32.8) \qquad \dot{\boldsymbol{x}} = A\,\boldsymbol{x} - \boldsymbol{g}\,f(\sigma)\,, \qquad \dot{\xi} = -\,f(\sigma)\,, \qquad \sigma = \boldsymbol{k}^T\boldsymbol{x} + \gamma\,\xi$$

with a slight change of notation. As above we set $f(\sigma) = \varkappa\sigma$ and apply the Laplace transformation in order to determine the transfer function of the linear part. It follows (the initial values are zero) that

$$s\,\bar{\boldsymbol{x}} = A\,\bar{\boldsymbol{x}} - \varkappa\,\boldsymbol{g}\,\bar{\sigma}\,, \qquad s\,\bar{\xi} = -\,\varkappa\,\bar{\sigma}\,, \qquad \bar{\sigma} = \boldsymbol{k}^T\bar{\boldsymbol{x}} + \gamma\,\bar{\xi}$$

and hence

$$(32.9) \qquad \bar{\sigma}\left(1 + \varkappa\left(\boldsymbol{k}^T(s\,E - A)^{-1}\boldsymbol{g} + \frac{\gamma}{s}\right)\right) = 0\,.$$

The transfer function is the multiplier of $\bar{\sigma}$ in (32.9) and the term

$$(32.10) \qquad \varkappa\left(\boldsymbol{k}^T(s\,E - A)^{-1}\boldsymbol{g} + \frac{\gamma}{s}\right)$$

corresponds exactly to the transfer function (32.3). It has an additional pole $s = 0$.

We next prove the criterion for the equation (32.8), giving Kalman's proof[1] which goes further: As Popov [1] already recognized, his criterion is equivalent in a certain sense to the criteria of sec. 31. For if we can prove absolute stability from the Popov criterion then we can also do so using a Liapunov function of the type (31.7), and conversely. Of course, the Popov criterion does not immediately furnish inequalities for the parameter variables. To use it for determining the domain of stability we must draw the response diagrams for various constellations of the parameters and check the Popov condition.

---

[1] Kalman [1].

We first prove a *lemma*. Let α be a non-negative number, Q a stable matrix, and $\boldsymbol{p}$ and $\boldsymbol{q}$ vectors such that the pairs $(Q, \boldsymbol{p})$ and $(Q^T, \boldsymbol{q})$ are completely controllable in the sense of the definition of sec. 31. Let the real vector $\boldsymbol{u}$ and the real symmetric non-negative matrix P be connected by the equations

$$(32.11) \qquad Q^T P + PQ = -\,\boldsymbol{u}\,\boldsymbol{u}^T, \qquad P\boldsymbol{p} - \boldsymbol{q} = \alpha\,\boldsymbol{u}.$$

Then the lemma says that such a pair $(\boldsymbol{u}, P)$ exists if and only if the inequality

$$(32.12) \qquad \tfrac{1}{2}\alpha^2 + \mathrm{Re}\left(\boldsymbol{q}^T (i\,\omega E - Q)^I \boldsymbol{p}\right) \geq 0$$

is satisfied for all ω, $0 \leq \omega \leq \infty$. Inequality (32.12) follows also if the first equation (32.11) is replaced by

$$(32.13) \qquad Q^T P + PQ = -(\boldsymbol{u}\,\boldsymbol{u}^T + R), \qquad R = R^T > 0$$

for arbitrary R. In place of $(i\,\omega E - Q)^I$ we can also write $-Q\,(\omega^2 E + Q^2)^I - i\,\omega\,(\omega^2 E + Q^2)^I$. The condition then takes the form

$$\frac{1}{2}\alpha^2 - \boldsymbol{q}^T Q\,(\omega^2 E + Q^2)^I \boldsymbol{p} \geq 0.$$

Proof. A) Suppose that P and $\boldsymbol{u}$ satisfy the equations (32.11). The first of these equations yields successively

$$P(Q - i\,\omega E) + (Q^T + i\,\omega E)\,P = -\,\boldsymbol{u}\boldsymbol{u}^T,$$

$$(i\,\omega E + Q^T)^I P - P(i\,\omega E - Q)^I = (i\,\omega E + Q^T)^I \boldsymbol{u}\,\boldsymbol{u}^T (i\,\omega E - Q)^I$$

$$= (i\,\omega E + Q^T)^I \boldsymbol{u}\,[(i\,\omega E - Q^T)^I \boldsymbol{u}]^T.$$

We scalar multiply by $\boldsymbol{p}$ and $\boldsymbol{p}^T$:

$$\boldsymbol{p}^T (i\,\omega E + Q^T)^I P\boldsymbol{p} - \boldsymbol{p}^T P(i\,\omega E - Q)^I \boldsymbol{p}$$

$$= \boldsymbol{p}^T (i\,\omega E + Q^T)^I \boldsymbol{u}\,[(i\,\omega E - Q^T)^I \boldsymbol{u}]^T \boldsymbol{p}.$$

We use the second equation to replace $P\boldsymbol{p}$ and $\boldsymbol{p}^T P$ on the left and re-arrange:

$$\boldsymbol{p}^T (i\,\omega E + Q^T)^I \boldsymbol{q} - \boldsymbol{q}^T (i\,\omega E - Q)^I \boldsymbol{p} + \alpha\,\boldsymbol{p}^T (i\,\omega E + Q^T)^I \boldsymbol{u}$$

$$- \alpha\,\boldsymbol{u}^T (i\,\omega E - Q)^I \boldsymbol{p}$$

$$= \boldsymbol{p}^T (i\,\omega E + Q^T)^I (\boldsymbol{q} + \alpha\,\boldsymbol{u}) - (\boldsymbol{q}^T + \alpha\,\boldsymbol{u}^T)\,(i\,\omega E - Q)^I \boldsymbol{p}.$$

The right side of the next to last equation can be written as

$$[\boldsymbol{p}^T (i\,\omega E + Q^T)^I \boldsymbol{u}]\,[\boldsymbol{p}^T (i\,\omega E - Q^T)^I \boldsymbol{u}]^T.$$

On both sides of the resulting equation we have scalar quantities which may be replaced formally by their transposes. This results in

$$(32.14) \quad 2\,\mathrm{Re}\,[\boldsymbol{q}^T (i\,\omega\,E - Q)^I\,\boldsymbol{p}] = |\,\boldsymbol{u}^T (i\,\omega\,E - Q)^I\,\boldsymbol{p}\,|^2$$
$$- 2\alpha\,\mathrm{Re}\,\big(\boldsymbol{u}^T (i\,\omega\,E - Q)^I\,\boldsymbol{p}\big)$$
$$= |\,\boldsymbol{u}^T (i\,\omega\,E - Q)^I\,\boldsymbol{p} - \alpha\,|^2 - \alpha^2,$$

which implies (32.12).

If in addition we add on the right side of the first equation (32.11) the non-negative matrix R, then the right side of the resulting equation (32.14) is not diminished. This proves the statement concerning (32.13).

B) Assume (32.12) is satisfied. We must show that a transformation P and a vector $\boldsymbol{u}$ can be found which are connected by (32.11). Let

$$\psi(s) : = \det(s\,E - Q) = : s^n + a_1\,s^{n-1} + \cdots + a_n,$$

and

$$Q^* : = \begin{pmatrix} 0 & 1 & 0 \cdots & 0 \\ 0 & 0 & 1 \cdots & 0 \\ \multicolumn{4}{c}{\dotfill} \\ -a_n & -a_{n-1} & \cdots & -a_1 \end{pmatrix}$$

and set

$$\boldsymbol{p}^* : = \mathrm{col}\,(0,0, \ldots, 0\,,1).$$

The matrix Q^* occurs if in accordance with sec. 4, the equation $\dot{\boldsymbol{x}} = Q\boldsymbol{x}$ is written as a scalar equation of nth order. The matrices Q and Q^* are similar and can be transformed into the same Jordan canonical form by means of non-singular matrices R, resp. R^*. This transformation takes $\boldsymbol{p}$ and $\boldsymbol{p}^*$ into $R\boldsymbol{p}$, resp. $R^*\boldsymbol{p}^*$. By hypothesis $(Q, \boldsymbol{p})$ is completely controllable and it is immediately seen that the same is true for the pair $(Q^*, \boldsymbol{p}^*)$. Noting that the transformation matrices R and R^* have been formed using the respective characteristic vectors, we see that the transformed vectors $R\boldsymbol{p}$ and $R^*\boldsymbol{p}^*$ have no zero components. Hence there exists a diagonal matrix D such that $DR\boldsymbol{p} = R^*\boldsymbol{p}^*$. The matrix

$$S : = R^{*I}\,DR$$

takes Q into Q^* and $\boldsymbol{p}$ into $\boldsymbol{p}^*$: $SQS^I = Q^*$, $S\boldsymbol{p} = \boldsymbol{p}^*$. For an arbitrary vector $\boldsymbol{h}$ let

$$\boldsymbol{h}^* = \mathrm{col}\,(h_1^*, \ldots, h_n^*) : = S^{IT}\,\boldsymbol{h},$$

then

$$\boldsymbol{h}^T (s\,E - Q)^I\,\boldsymbol{p} = \boldsymbol{h}^{*T} (s\,E - Q^*)^I\,\boldsymbol{p}^*$$

holds identically in s. The denominator of the rational function on the right is $\psi(s)$. The numerator is the polynomial $h_1^* - h_2^*\,s + h_3^*\,s^2 - \cdots + (-1)^{n-1} h_n^*\,s^{n-1}$. This follows from the special form of Q^* and $\boldsymbol{p}^*$.

Hence

$$(32.15) \qquad \boldsymbol{h}^T (sE - Q)^I \boldsymbol{p} = \frac{h_1^* - h_2^* s + \cdots + (-1)^{n-1} h_n^* s^{n-1}}{\psi(s)}.$$

This expression gives us the components of the transformed vector $\boldsymbol{h}^*$; they are simply the coefficients of the powers of s in the numerator of $\boldsymbol{h}^T (sE - Q)^I \boldsymbol{p}$.

We now return to (32.12) and define a polynomial $\chi(s)$ by the formula

$$(32.16) \qquad \alpha^2 + 2\,\mathrm{Re}\left(\boldsymbol{q}^T (i\omega E - Q)^I \boldsymbol{p}\right) =: \left|\frac{\chi(i\omega)}{\psi(i\omega)}\right|^2.$$

$\chi(s)$ is a polynomial of degree n and has real coefficients which are determined as follows. The numerator on the left side of (32.16) is a real polynomial in ω^2, namely

$$\left(\alpha^2 - 2\,\boldsymbol{q}^T Q (\omega^2 E + Q^2)^I \boldsymbol{p}\right) \det(\omega^2 E + Q^2).$$

Since it assumes only positive values it can be decomposed into factors of the form $\omega^2 + a^2$ and $\omega^4 + 2\omega^2(b^2 - c^2) + (b^2 + c^2)^2$. However,

$$\omega^2 + a^2 = |i\omega + a|^2,$$

$$\omega^4 + 2\omega^2(b^2 - c^2) + (b^2 + c^2)^2 = |(i\omega)^2 + 2i\omega b + (b^2 + c^2)|^2,$$

which implies

$$\chi(s) = \alpha\, \Pi(s + a)\, \Pi\left(s^2 + 2sb + (b^2 + c^2)\right).$$

The polynomial

$$\varphi(s) := -\chi(s) + \alpha\,\psi(s)$$

is of degree $n - 1$. Considering its coefficients as the components of a vector $\boldsymbol{u}^*$ and observing (32.15) and (32.16), we define a vector $\boldsymbol{u}$ by

$$\frac{\varphi(i\omega)}{\psi(i\omega)} = \boldsymbol{u}^T (i\omega E - Q)^I \boldsymbol{p},$$

and we have

$$\alpha - \frac{\chi(i\omega)}{\psi(i\omega)} = \boldsymbol{u}^T (i\omega E - Q)^I \boldsymbol{p},$$

$$\left|\frac{\chi(i\omega)}{\psi(i\omega)}\right|^2 = |\boldsymbol{u}^T (i\omega E - Q)^I \boldsymbol{p} - \alpha|^2 = \alpha^2 + 2\,\mathrm{Re}\left(\boldsymbol{q}^T (i\omega E - Q)^I \boldsymbol{p}\right).$$

This relation is identical with (32.14) and we can start from it and return to (32.11) and thereby prove that the vector $\boldsymbol{u}$ which we constructed has the desired properties. This proves the lemma.

We now return to (32.8) and make the assumption that the pairs $(A, \boldsymbol{g})$ and $(A^T, \boldsymbol{k})$ are completely controllable. For the nonlinearity the sector condition

$$(32.17) \qquad\qquad 0 < \sigma f(\sigma)$$

holds. Suppose that for (32.8) there exists a Liapunov function of the form "quadratic form in $\boldsymbol{x}$, σ + integral of $f(\sigma)$",

$$(32.18) \qquad v(\boldsymbol{x}, \sigma) = \boldsymbol{x}^T P \boldsymbol{x} + \beta_1 (\sigma - \boldsymbol{k}^T \boldsymbol{x})^2 + \beta \int_0^\sigma f(\eta)\, d\eta.$$

Its derivative for (32.8) is

$$(32.19) \qquad \dot{v} = \boldsymbol{x}^T (PA + A^T P)\, \boldsymbol{x} - 2f(\sigma)\, \boldsymbol{x}^T \left(P\boldsymbol{g} - \beta_1 \gamma \boldsymbol{k} - \frac{\beta}{2} A^T \boldsymbol{k} \right)$$
$$- \beta(\gamma + \boldsymbol{k}^T \boldsymbol{g}) f(\sigma)^2 - 2\beta_1 \gamma f(\sigma)\, \sigma.$$

For this expression to be non-positive we must have

$$(32.20) \qquad \alpha^2 := \beta(\gamma + \boldsymbol{k}^T \boldsymbol{g}) \geq 0.$$

This is inequality (31.6) in the notation of (32.8). If we define a vector $\boldsymbol{u}$ by

$$(32.21) \qquad \alpha \boldsymbol{u} := P\boldsymbol{g} - \beta_1 \gamma \boldsymbol{k} - \frac{\beta}{2} A^T \boldsymbol{k}$$

and set

$$(32.22) \qquad U := -(A^T P + PA),$$

then (32.19) becomes

$$(32.23) \qquad \dot{v} = -\boldsymbol{x}^T (U - \boldsymbol{u}\boldsymbol{u}^T)\, \boldsymbol{x} - (\alpha f(\sigma) + \boldsymbol{u}^T \boldsymbol{x})^2 - 2\beta_1 \gamma \sigma f(\sigma).$$

This expression is negative semi-definite for an arbitrary function $f(\sigma)$ which satisfies the sector condition (32.17), if and only if $U - \boldsymbol{u}\boldsymbol{u}^T \geq 0$, $\alpha^2 \geq 0$. If $\alpha^2 > 0$ then $\boldsymbol{u}$ is well defined by (32.21). In case $\alpha = 0$ we must have $P\boldsymbol{g} - \beta_1 \gamma \boldsymbol{k} = \frac{\beta}{2} A^T \boldsymbol{k}$. Then $\boldsymbol{u}$ is not well defined but for $U \geq 0$ it can always be chosen so that $U - \boldsymbol{u}\boldsymbol{u}^T \geq 0$.

The theorem of V. M. Popov runs as follows [*cf.* (32.7) and (32.10)]:

Theorem 32.1. The absolute stability of the equation (32.8) is assured if the inequality

$$(32.24) \qquad \operatorname{Re}(1 + i\omega\beta) \left(\boldsymbol{k}^T (i\omega E - A)^I \boldsymbol{g} + \frac{\gamma}{i\omega} \right) \geq 0$$

is satisfied for all real ω and a fixed positive β.

Popov's theorem is contained in the three statements of the following theorem due to KALMAN [1]:

Theorem 32.2. a) Let the inequality

$$(32.25) \qquad \operatorname{Re}(2\beta_1 \gamma + i\omega\beta) \left(\boldsymbol{k}^T (i\omega E - A)^I \boldsymbol{g} + \frac{\gamma}{i\omega} \right) \geq 0$$

be satisfied for all real ω; β and β_1 are two non-negative numbers which do not vanish together. Then a positive definite function of the form

(32.18) can be constructed such that its derivative (32.19) for (32.8) is negative semi-definite.

b) Let a positive definite Liapunov function of the form (32.18) with a negative semi-definite derivative (32.19) be given. Then (32.25) holds.

c) The equilibrium of (32.8) is absolutely stable if the function (32.18) is positive definite and has a negative semi-definite derivative and if $\beta_1 > 0$.

Before proving the Kalman theorem we mention that the assertions can be strengthened somewhat. The most general function of the type "quadratic form + integral" formally involves besides (32.18) a term of the form $\sigma c^T x$. Popov has given an example which shows that c must equal 0 if $v(x, \sigma)$ is to be positive definite and have a negative definite derivative for arbitrary $f(\sigma)$ satisfying (32.17). Kalman extended the assertion in c) to the case $\beta_1 = 0$ under an additional hypothesis.

Proof of a). Assume that the two numbers β_1, β are given in accordance with the hypothesis and that (32.25) is satisfied. We first show that the number α^2 defined in (32.20) is non-negative. In case $\beta = 0$, $\alpha^2 = 0$. If $\beta > 0$, consider inequality (32.25) and allow the parameter ω in it to grow without bounds. Then the left side tends toward $\beta(\gamma + k^T g)$; this quantity is of necessity non-negative, which implies $\alpha^2 \geq 0$.

We now introduce the vector

$$q := \beta_1 \gamma k + \tfrac{1}{2} \beta A^T k,$$

rearrange (32.25) somewhat,

$$\mathrm{Re}\left(2\beta_1 \gamma k^T (i\omega E - A)^I g + i\omega\beta k^T (i\omega E - A)^I g\right) + \beta\gamma \geq 0,$$

and replace the term $\beta_1 \gamma k^T$ on the left by q. It follows that

$$\mathrm{Re}\left(2 q^T (i\omega E - A)^I g + \beta k^T (i\omega E - A)(i\omega E - A)^I g\right) + \beta\gamma \geq 0,$$

$$\mathrm{Re}\left(q^T (i\omega E - A)^I g\right) + \tfrac{1}{2}(\beta k^T g + \beta\gamma) \geq 0.$$

If we introduce α^2 and identify A with Q and g with p, we have inequality (32.12). By the lemma there exists a vector u satisfying (32.11), resp. the equations

$$A^T P + PA = -uu^T, \qquad Pg - q = \alpha u.$$

Then $U = uu^T$. Hence a function v of the desired type exists; it is non-negative and its derivative

$$(32.26) \qquad \dot{v} = -\left(\alpha f(\sigma) + u^T x\right)^2 - 2\beta_1 \gamma \sigma f(\sigma)$$

is non-positive. Actually, we conclude that P and hence v is positive definite. For if P were only semi-definite there would exist (*cf.* the end of sec. 27) a left characteristic vector of A orthogonal to $\boldsymbol{u}$. This vector would also be orthogonal to $\boldsymbol{q}$ and therefore to $\boldsymbol{k}$, which contradicts one of the hypotheses.

Proof of b). The inequalities $v > 0$, $\dot{v} \leq 0$ imply that the numbers β, β_1 must be non-negative and cannot vanish together. Further, P, U, $\boldsymbol{u}$ exist according to (32.22), (32.21) and we must have $\alpha^2 \geq 0$ and $U - \boldsymbol{u}\boldsymbol{u}^T =: R \geq 0$. Thus (32.12) is valid by the lemma and (32.25) follows. The condition is therefore necessary. If $\beta_1 \neq 0$, then (32.25) becomes (32.24); for we can normalize the parameters so that $2\beta_1 \gamma = 1$.

Proof of c). We begin by proving that the solutions exist in the future and are bounded. Since v is non-increasing along each trajectory, v is bounded. This implies that $\boldsymbol{x}$ is bounded for each solution; for $|\boldsymbol{x}| \to \infty$ implies $v \to \infty$. If $\beta_1 = 0$ then, for the same reason, $\sigma(t)$ must be bounded. Furthermore we have

$$\frac{d}{dt}\,|\sigma| = \dot{\sigma}\,\operatorname{sgn}\sigma = \boldsymbol{k}^T A\,\boldsymbol{x}\,\operatorname{sgn}\sigma - (\boldsymbol{k}^T\boldsymbol{g} + \gamma)\,|f(\sigma)|$$

$$\leq \boldsymbol{k}^T A\,\boldsymbol{x}\,\operatorname{sgn}\sigma.$$

Since the right side is bounded for finite t, $\sigma(t)$ is also bounded for finite t. This proves the existence of the solutions in the future. We now apply Theorem 26.2. If $\boldsymbol{x}(t)$ is a trajectory lying in the manifold $\dot{v} = 0$, then (32.26) implies

$$\alpha f(\sigma) = -\boldsymbol{u}^T\boldsymbol{x}, \quad \beta_1 \gamma \sigma f(\sigma) = 0,$$

and therefore $\sigma(t) \equiv 0$. Consequently $\xi(t) \equiv \xi(0)$ and

$$\boldsymbol{k}^T\boldsymbol{x}(t) = -\gamma\,\xi(0), \quad \boldsymbol{x}(t) = e^{At}\boldsymbol{x}(0), \quad \boldsymbol{k}^T e^{At}\boldsymbol{x}(0)\ \text{constant},$$

follows because of (32.8). The last relation is possible only for $\boldsymbol{k}^T A^k \boldsymbol{x}(0) = 0$, $k = 1, 2, \ldots$, and if $\boldsymbol{x}(0) \neq 0$ we have a contradiction to one of the hypotheses on $\boldsymbol{k}$. This completes the proof.

33. The Domain of Attraction

The domain of attraction (or the domain of stability) of the differential equation

$$(33.1) \qquad\qquad \dot{\boldsymbol{x}} = f(\boldsymbol{x}), \quad \boldsymbol{f} \in E$$

was defined in Def. 26.1 as the set of all points $\boldsymbol{x}_0$ which are initial points of motions which eventually approach the origin. In the case of a plane system it follows from the Bendixson theory (sec. 18) that the boundary of the domain of attraction is formed by whole trajectories. Let P be a regular point of the boundary Γ and γ the trajectory which goes through P. Obviously Γ and γ cannot intersect; for otherwise there would exist points $Q \in \gamma$ such that Q and the origin lie on opposite sides of Γ, and we would arrive at a contradiction to the definition of the domain of attraction, regardless whether the direction of the motion is from P to Q or from Q to P. Again it is impossible that γ is tangent to the boundary Γ from the inside at the point P. For then the trajectory passing through a neighboring point $P_1 \in \Gamma$ would likewise have to be tangent to Γ from the inside; P and P_1 could be arbitrarily close to each other. Since the two trajectories cannot intersect the tangential direction would have to change by 180° over an arbitrarily small distance, contradicting continuity. Hence the trajectory γ is part of the boundary Γ. If γ is a cycle then it is the boundary; otherwise the boundary is a phase polygon. It can also happen that the boundary is formed by a closed curve consisting only of singular points. For instance, for the system

$$(33.2) \qquad \dot{x} = -x(1 - x^2 - y^2), \quad \dot{y} = -y(1 - x^2 - y^2)$$

every point of the unit circle is singular. The trajectories are the radii of the unit circle.

As already mentioned in sec. 26, the domain of attraction can on occasion be estimated by means of the direct method. Let $v(x)$ be positive definite in a domain B (sec. 24). Let there exist a closed hypersurface F entirely contained in the interior of B with the following properties: a) The origin is on the inside of F, b) $\dot{v} = 0$ for $x \in F$, c) for $x \neq 0$, $\dot{v} < 0$ if x lies on the inside of F, $\dot{v} > 0$ if x lies outside of F. d) The hypersurface $v = c_1$ (closed by hypothesis) lies entirely inside of F. Then the domain $v \leq c_1$ is a subset of the domain of attraction, because for each point of this domain the proof of Theorem 25.1 can be carried through; *cf.* also the end of sec. 28.

If the surface F consists entirely of trajectories we have exactly the boundary of the domain of attraction. For example, the function $v = x^2 + y^2$ is a Liapunov function for (33.2). The derivative $\dot{v} = -2v(1-v)$ is negative definite exactly in the interior of the unit circle and vanishes on the circle. The domain of attraction can be the entire phase space; then we have global asymptotic stability. The domain can be bounded, but it can also be unbounded without being the whole phase space. This is shown in a simple

Example. The system of equations

$$\dot{x} = -x + y^2, \quad \dot{y} = -y + x^2$$

has the two singular points $P = (0, 0)$ and $Q = (1, 1)$, the first of which is a node, the second a saddle point (*cf.* sec. 21). By Theorem 18.4 a periodic solution cannot exist. The trajectory $x - y = 0$ passes through the singular points. It consists of the three segments $(-\infty, P)$, (P, Q), $(Q, +\infty)$; the first two are traversed in the direction toward P. Since Q

is a saddle point there must exist two further segments beginning at infinity and ending at Q. These form the boundary of the domain of attraction (see fig. 21.1).

Another **example** with only one singular point is due to PLISS [1]. Let

$$(33.3) \qquad \dot{x} = y - g(x), \quad \dot{y} = - g(x).$$

The function $g(x)$ has the following properties: a) $g(x) \in C_1$, b) $xg(x) > 0$ for $x \neq 0$, $g(0) = 0$, c) $\int_0^\infty g(x)\,dx =: d$ is finite, d) $\lim\sup\limits_{x\to\infty} g(x) =: a$ is finite.

From (33.3) and the properties of $g(x)$ the solution

$$(33.4) \qquad \boldsymbol{p}(t; x_0, y_0; t_0) = \mathrm{col}\,\big(x(t),\, y(t)\big)$$

in which we can set $t_0 = 0$ without loss of generality, satisfies the following: 1) If $x(t) > 0$, then $y(t)$ decreases. 2) If $y_0 < a$ then $\dot{x}$ is negative or is positive at most in a finite interval $0 \leq t \leq T_1$. 3) If $y_0 > a$ then there exists $x_0 > 0$ such that the x-component of (33.4) tends to ∞. This can be seen in the following way. Let $\gamma := \frac{1}{3}(y_0 - a)$. Because of d) there exists a number A such that for $x > A$ the inequality $g(x) < a + \gamma$ holds. By continuity a number $T > 0$ can be so chosen that the y-component of (33.4) satisfies the inequality

$$(33.5) \qquad y_0 - y(t) < \gamma, \quad \text{resp. } a + 2\gamma < y(t)$$

on the interval $0 \leq t \leq T$. If $x > A$ then we have because of (33.3) and (33.5),

$$(33.6) \qquad \dot{x} = y(t) - g(x) > \gamma$$

for $0 \leq t \leq T$, and also

$$y_0 - y(t) = \int_0^t g(x)\,dt = \int_{x_0}^{x(t)} g(x)\,\frac{dt}{dx}\,dx \leq \frac{1}{\gamma}\int_{x_0}^{x} g(x)\,dx.$$

Since the integral converges for $x \to \infty$, the number A can be so chosen that in addition the inequality

$$(33.7) \qquad \int_{x_0}^{x} g(x)\,dx < \gamma^2 \text{ for } x \geq x_0 > A$$

holds. Now we can conclude that (33.5) and (33.6) are valid for all $t > 0$. For if there existed a finite t_1 such that $\dot{x}(t) > \gamma$ for $0 \leq t < t_1$ and $\dot{x}(t_1) = \gamma$ then we would have $y_0 - y(t_1) < \gamma$, as above, which again implies $\dot{x}(t_1) > \gamma$. Since therefore $\dot{x}(t) > \gamma$ throughout, it follows that $\lim x(t) = \infty$.

4) Let $y_0 > a$ be fixed. The greatest lower bound of all those x_0 for which (33.4) has an unbounded x-component is a function of y_0 which we will denote by $\psi(y_0)$. Let Γ be the trajectory passing through the point $(\psi(y_0), y_0)$; let $\bar{x}(t)$ and $\bar{y}(t)$ be the components of the corresponding solution. Then we have

$$(33.8) \qquad \lim \bar{x}(t) = \infty, \quad \lim \bar{y}(t) = a, \quad \text{as } t \to \infty.$$

The proof is indirect. $\lim \bar{y}(t) \geq a$ follows from 2). Let us therefore assume that $\lim \bar{y}(t) = b > a$. If x_0' is chosen sufficiently large, then the point (x_0', y_0'), with $y_0' = (b + a)/2$, does not lie on Γ. Let $x_0'' := \psi(y_0')$. Obviously $x_0'' \leq x_0'$. Therefore the trajectory Γ' passing through the point (x_0'', y_0') lies between the x-axis and Γ. The point on Γ' whose ordinate is y_0 must have an abscissa which is smaller than $\psi(y_0)$. But this contradicts the definition of $\psi(y_0)$. Hence (33.8) is correct. Since $g(x) \in C_1$, the trajectory Γ is uniquely determined.

Let us assume in addition that

$$\int_0^x g(\xi)\,d\xi \to +\infty, \quad x \to -\infty.$$

The Liapunov function

$$(33.9) \qquad v = 2 \int_0^x g(\xi)\,d\xi + y^2$$

is positive definite. The equation $v = c^2$ defines a closed curve for $c^2 < 2d$; in fact, the abscissa of the point of intersection with the negative x-axis is bounded whereas the other point of intersection, for c^2 sufficiently close to d, has an arbitrarily large positive abscissa. The derivative

$$\dot{v} = -2\,(g(x))^2$$

is non-positive. Since $g(x) = 0$ cannot be a trajectory, the curves $v = c^2$ are traversed from outside in; they therefore lie in the domain of attraction. From the reasoning leading up to (33.8) we see that for a motion which trails off to infinity we have of necessity $x(t) \to \infty$ and $\lim y(t) = a$. For such a motion y decreases in the fourth quadrant (because g has property b)) and in the second and third quadrants y increases. Finally we see that Γ intersects the y-axis, and in such a way that $\bar{y}(0)$ is a maximum. For otherwise $\bar{y}(\bar{x})$ would be unbounded as $\bar{x}$ approaches zero, which is not compatible with (33.3). All this implies that the trajectory Γ which approaches the line $y = a$ asymptotically in the first quadrant, while in the third quadrant $\bar{x}$ and $\bar{y}$ are unbounded, is exactly the boundary of the domain of attraction: No motion which starts above Γ can approach the origin and each motion which starts below Γ does approach the origin.

The theorem which was explained above for $n = 2$, is valid in general:

Theorem 33.1. The boundary of the domain of attraction is formed by whole trajectories.

To prove this it suffices to show that the domain of attraction is an open invariant set. Then the boundary of the domain of attraction is its topological boundary and Theorem 33.1 follows from Theorem 16.3.

The invariance of the domain of attraction A follows from the definition. To see that A is open, choose $P \in A$ and consider the half-trajectory $\boldsymbol{p}_+(P, t)$. We choose T so large that $|\boldsymbol{p}(P, T)| < \frac{a}{2}$, where a is an arbitrary positive number chosen so small that the domain $|\boldsymbol{x}| < a$ is contained in A. Let P_1 be a second point and assume that the distance PP_1 is so small that the distance from $\boldsymbol{p}(P, T)$ to $\boldsymbol{p}(P_1, T)$ is less than $a/2$. Then $\boldsymbol{p}(P_1, T) \in A$; the half-trajectory $\boldsymbol{p}_+(\boldsymbol{p}(P_1, T), t)$ tends toward the origin and P_1 also belongs to A: A is open.

We mention only one further property of the boundary of the domain of attraction.

Theorem 33.2. Let $\boldsymbol{x}_0$ be a point of the boundary Γ of the domain of attraction and let $\boldsymbol{x}_n$ be a sequence converging to $\boldsymbol{x}_0$, $\boldsymbol{x}_n \in A$. Assume the sphere K_r is contained in A, $K_r \subset A$, and let t_n be the time at which the motion $\boldsymbol{p}(\boldsymbol{x}_n, t)$ reaches the fixed sphere $K_{r'}$, where $r' < r$. Then the sequence t_n is unbounded.

Proof. Assume the sequence t_n is bounded, $t_n \leq T < \infty$. By continuity there exists for each $\varepsilon > 0$ an $\eta > 0$ such that $|\boldsymbol{p}(\boldsymbol{x}', t) - \boldsymbol{p}(\boldsymbol{x}'', t)| < \varepsilon$ for $0 \leq t \leq T$, provided only that $|\boldsymbol{x}'' - \boldsymbol{x}'| < \eta$. Choosing $\varepsilon < r - r'$, we can conclude that all the trajectories which start in an η neighborhood of $\boldsymbol{x}_n$ arrive at the sphere $K_{r'}$ at a time no later than $t = T$, and therefore clearly lie in the domain of attraction. Since this is true for all n this assertion also holds for the trajectory $\boldsymbol{p}(\boldsymbol{x}_0, t)$, contradicting the assumption $\boldsymbol{x}_0 \in \Gamma$.

Consider example (33.2) in polar coordinates; we have

$$t_n = \ln \frac{r'}{r_n} - \frac{1}{2} \ln \frac{(1 + r')(1 - r')}{(1 + r_n)(1 - r_n)} \to \infty \text{ for } r_n \to 1.$$

Occasionally we have need for the concept of the domain of attraction of an invariant set. It is given in analogy to Def. 26.1.

Def. 33.1. Let M be an asymptotically stable closed invariant set of the differential equation (33.1). The domain of attraction of M is the set of all points $\boldsymbol{x}_0$ with the property that $\boldsymbol{x}_0 \notin M$ and

$$\varrho(M, \boldsymbol{p}(t, \boldsymbol{x}_0)) \to 0 \quad (t \to \infty).$$

Theorem 33.3. The domain of attraction of an asymptotically stable closed set M is an open invariant set; its boundary is formed by trajectories.

The proof is very much like that of Theorem 33.1; we need only replace $|\boldsymbol{p}(P, T)|$ and $|\boldsymbol{x}|$ by $\varrho(M, \boldsymbol{p}(P, T))$, resp. $\varrho(M, \boldsymbol{x})$.

Sometimes the set M is considered a subset of its domain of attraction. Theorem 33.3 must then be formulated accordingly.

34. Zubov's Theorem

The principle involved in estimating the domain of attraction, which we described in the last section can be refined to become a constructive procedure. ZUBOV [1, 2, 4] has found that the boundary of the domain of attraction can be exactly determined with the help of an appropriately chosen Liapunov function. This function is obtained by solving a partial differential equation. Even though this theorem has the character of an existence theorem it is practically applicable in many cases, either because the differential equation can be solved explicitly or because an approximation procedure is available to determine the solution.

We again start with the differential equation (33.1). Let A be a simply connected domain containing a neighborhood of the origin. The following theorem gives a sufficient condition for A to be exactly the domain of attraction of the origin.

Theorem 34.1. Let two scalar functions $v(\boldsymbol{x})$, $h(\boldsymbol{x})$ exist with the following properties: a) v is defined, continuous, and positive definite in A and satisfies in A the inequality $0 < v(\boldsymbol{x}) < 1$ $(\boldsymbol{x} \neq 0)$. b) $h(\boldsymbol{x})$ is defined for all finite $\boldsymbol{x}$, $h(0) = 0$, h is continuous and positive for $\boldsymbol{x} \neq 0$.

c) For $\boldsymbol{x} \in A$ we have

$$(34.1) \qquad \dot{v} = -h(\boldsymbol{x})\,(1 - v(\boldsymbol{x}))\,\sqrt{1+|f(\boldsymbol{x})|^2}\,.$$

d) As $\boldsymbol{x} \in A$ approaches a point of the boundary of A or in case of an unbounded region A, as $|\boldsymbol{x}| \to \infty$, $\lim v(\boldsymbol{x}) = 1$. Then A is exactly the domain of attraction of the equilibrium.

Proof. Hypotheses a) through c) guarantee the asymptotic stability. If we introduce a new independent variable [cf. (16.9)] by means of the substitution

$$ds^2 = (1 + |f|^2)\,dt^2$$

then (34.1) becomes

$$(34.2) \qquad dv/ds = -h(\boldsymbol{x})\,(1 - v(\boldsymbol{x})),$$

while the stability properties of the origin remain unchanged.

Going over again to the independent variable t, writing equation (34.2) in the form

$$\frac{d}{dt} \ln\,(1 - v(\boldsymbol{x})) = h(\boldsymbol{x})$$

and integrating along a trajectory $\boldsymbol{p}(t, \boldsymbol{x}_0)$ we obtain

$$(34.3) \quad 1 - v\big(\boldsymbol{p}(t, \boldsymbol{x}_0)\big) =: 1 - v(t) = (1 - v_0)\exp\left(-\int_0^t h\big(\boldsymbol{p}(r, \boldsymbol{x}_0)\big)\,dr\right).$$

Let $\boldsymbol{x}_0 \in A$. If $\boldsymbol{x}_0$ belongs to the domain of attraction, $\boldsymbol{p}(t, \boldsymbol{x}_0)$ tends to zero. Otherwise $\limsup |\boldsymbol{p}(t, \boldsymbol{x}_0)| =: \delta > 0$ and also $\liminf |\boldsymbol{p}(t, \boldsymbol{x}_0)|$ $=: \delta' > 0$. The possibility that the lim sup is different from zero and the lim inf is equal to zero is out of the question because of

Theorem 34.2. If the differential equation is autonomous and the equilibrium stable, and if $\liminf |\boldsymbol{p}(t, \boldsymbol{x}_0)| = 0$, then $\boldsymbol{p}(t, \boldsymbol{x}_0)$ tends to zero.

Proof. By hypothesis there exists for each $\varepsilon > 0$ a sequence t_n such that

$$|\boldsymbol{p}(t_n, x_0)| < \varepsilon.$$

Assume there exists also a sequence t'_n such that $|\boldsymbol{p}(t'_n, \boldsymbol{x}_0)| \geq \delta > 0$. Since the equilibrium is stable there exists an estimate

$$|\boldsymbol{p}(t, \boldsymbol{x}_0)| < \varphi(|\boldsymbol{x}_0|), \quad \varphi \in K.$$

Now let ε be so small that $\varphi(\varepsilon) < \delta/2$, and N so large that $t'_N > t_1$. Then

$$|\boldsymbol{p}(t'_N, \boldsymbol{x}_0)| = |\boldsymbol{p}(t'_N - t_1, \boldsymbol{p}(t_1, \boldsymbol{x}_0))| \leq \varphi(|\boldsymbol{p}(t_1, \boldsymbol{x}_0)|) \leq \varphi(\varepsilon) < \frac{\delta}{2}$$

contradicting the definition of the sequence t'_n.

So if $\boldsymbol{x}_0$ does not lie in the domain of attraction then $h(\boldsymbol{p}(t, \boldsymbol{x}_0))$ is certainly larger than a fixed positive number throughout. The exponential function in (34.3) tends to zero since the integral diverges. It follows that $1 - v(t) \to 0$, contradicting hypothesis a). Hence $\boldsymbol{x}_0$ lies in the domain of attraction.●

An immediate consequence of Theorem 34.1 is

Theorem 34.3. Let the function $h(\boldsymbol{x})$ satisfy the hypotheses of Theorem 34.1. Assume that the function $v(\boldsymbol{x})$ is positive definite in A and satisfies the inequality $0 \leq v(\boldsymbol{x}) \leq 1$, $\boldsymbol{x} \in A$, as well as the differential equation

$$\sum_{i=1}^n \frac{\partial v}{\partial x_i} f_i(\boldsymbol{x}) = - h(\boldsymbol{x})(1 - v(\boldsymbol{x}))\sqrt{1 + |\boldsymbol{f}(\boldsymbol{x})|^2}.$$

Then the boundary of the domain of attraction is defined by the equation

$$(34.4) \qquad\qquad v(\boldsymbol{x}) = 1.$$

If the domain of attraction A is all of R_n, we have global asymptotic stability. The condition on $v(\boldsymbol{x})$ in this case is

$$v(\boldsymbol{x}) \to 1 \quad (|\boldsymbol{x}| \to \infty).$$

We can also work with a different Liapunov function. If we set, for instance,

$$w(\boldsymbol{x}) = - \ln(1 - v(\boldsymbol{x}))$$

then (34.1) becomes

$$\dot{w} = - h(x) \sqrt{1 + |f|^2}$$

and the condition (34.4) defining the boundary becomes $w(x) \to \infty$.

The function $h(x)$ is largely arbitrary. We will as far as possible try to choose it so that the partial differential equation becomes easy to solve. From the proof we obtain the additional result that the relation

$$v = c, \quad 0 < c < 1,$$

defines a family of closed hypersurfaces which schlichtly cover the domain A. The origin corresponds to $c = 0$, the boundary of A to $c = 1$.

Examples. a)

$$(34.5) \quad \begin{aligned} \dot{x} &= 2x \frac{1 - x^2 + y^2}{(x + 1)^2 + y^2} + xy =: f_1; \\ \dot{y} &= \frac{1 - x^2 + y^2}{2} - \frac{4 x^2 y}{(x + 1)^2 + y^2} =: f_2. \end{aligned}$$

Here the singular point is $(1, 0)$. The partial differential equation is

$$v_x f_1 + v_y f_2 = - 2 \frac{(x - 1)^2 + y^2}{(x + 1)^2 + y^2} (1 - v),$$

if we set

$$h(x) = 2 \frac{(x - 1)^2 + y^2}{(x + 1)^2 + y^2} (1 + f_1^2 + f_2^2)^{-1/2}.$$

From it we obtain

$$v = \frac{(x - 1)^2 + y^2}{(x + 1)^2 + y^2}$$

and we have $v = 1$ if $x = 0$. The boundary of the domain of attraction is therefore the y-axis. The family of curves $v = 1 - c$ $(0 < c < 1)$ consists of the circles

$$(x + 1 - 2/c)^2 + y^2 = (1 - 2/c)^2 - 1.$$

b)

$$(34.6) \quad \dot{x} = - a(x) + b(y), \quad \dot{y} = - c(x)$$

where

$$z a(z) > 0, \quad z b(z) > 0, \quad z c(z) > 0 \text{ for } z \neq 0.$$

The differential equation for v for suitably chosen $h(x) = a(x) c(x)$ is

$$v_x (b(y) - a(x)) - v_y c(x) = - c(x) a(x) (1 - v).$$

Its solution is

$$v(x, y) = 1 - \exp\left(- \int_0^x c(\xi) \, d\xi - \int_0^y b(\eta) \, d\eta \right).$$

The condition for global asymptotic stability is the divergence of the integrals. Of course the reasoning must be somewhat modified since the functions $h(x)$ and $\dot{v}$ are not positive definite but only semi-definite.

11*

Therefore we must complete our argument by introducing Theorem 26.2. If the integrals do not diverge, the condition $v = 1$ cannot be satisfied. Then the theorem does not apply [*cf.* Example (33.3)].

Zubov's theorem can also be used to start with a given domain A (which of course must contain the origin in its interior) and to construct a differential equation whose domain of attraction is exactly A. For this purpose we choose a positive definite function $v(x)$ defined in A which satisfies the inequality $0 \leq v < 1$ in A and which tends to the value 1 as the boundary of A is approached. Also let $h(x)$ be a function which is positive in R_n except at the origin. We begin by formally writing a system of equations

$$(34.7) \qquad \sum_{k=1}^{n} p_{ik} f_k(x) = q_i(x), \quad i = 1, 2, \ldots, n,$$

and setting

$$p_{1k} := \frac{\partial v}{\partial x_k}, \qquad q_1(x) := - h(x) \left(1 - v(x)\right).$$

Then the first equation in (34.7) takes on the form

$$\sum_{k=1}^{n} \frac{\partial v}{\partial x_k} f_k(x) = - h(x) \left(1 - v(x)\right).$$

On the left we have the derivative of v for the differential equation $\dot{x} = f(x)$, whose right side is formed from the formal solution $f(x) = \mathrm{col}(f_1, \ldots, f_n)$. Next the remaining coefficients p_{ik} and the right sides q_i, $i \geq 2$, are chosen so that the solution $f(x)$ has the desired properties, as for instance membership in class E. Theorem 34.1 implies that A is the domain of attraction. It is easy to see that conversely we can set up a system of the form (34.7) if we are given the functions $f_i(x)$; for instance, using the coefficients

$$p_{ii} = 1, \quad p_{ik} = 0 \quad (i \neq k),$$

$$q_1(x) = - h(x) \sqrt{1 + |f(x)|^2} \left(1 - v(x)\right); \; q_i(x) = f_i(x) \; (i = 2, 3, \ldots, n).$$

Under the more special hypothesis that the right side of the differential equation (33.1) is analytic, Zubov has developed a procedure for approximating the boundary of the domain of attraction which we shall sketch briefly. Accordingly let

$$(34.8) \qquad \dot{x} = A x + g(x);$$

$g(x)$ has a Taylor expansion starting with second degree terms. We assume that the real parts of the characteristic roots of A are negative. For $h(x)$ we choose a positive definite quadratic form. The solution of the corresponding partial differential equation in Theorem 34.3 is a series of the form

$$(34.9) \qquad v = v_2 + v_3 + \cdots$$

in which the terms $v_i, i = 2, 3, \ldots$, are homogeneous functions of degree i. It can be shown that a convergent series of this kind actually exists. For the purposes of this proof the existence is, however, not required since at each stage we only need finitely many terms. The function v_2 is obviously a quadratic form and can be found by the procedure of sec. 27; v_3 is obtained from the differential equation if only terms of the third degree are taken into account, etc. Let $\dot{v}$ be the derivative of (34.9) for (34.8). This function is either negative in all of R_n $(\boldsymbol{x} \neq 0)$ or there exists a manifold F in which $\dot{v} = 0$. Let

$$\alpha: = \min v_2(\boldsymbol{x}), \quad \beta: = \max v_2(\boldsymbol{x}), \quad \boldsymbol{x} \in F.$$

Since v is positive definite, $\alpha > 0$. We have

Theorem 34.4. The point set defined by

$$(34.10) \qquad\qquad v_2(\boldsymbol{x}) = \alpha$$

lies entirely inside the domain of attraction A of the differential equation (34.8).

Proof. Assuming that part of the set (34.10) lies outside of A, there exist certain phase trajectories on the boundary of A ($cf.$ Theorem 33.1), which meet this set in at least two points. Let $\boldsymbol{x}(\tau_1)$, $\boldsymbol{x}(\tau_2)$ be two such points and let τ be the parameter of the trajectory. By hypothesis

$$v\big(\boldsymbol{x}(\tau_1)\big) = v\big(\boldsymbol{x}(\tau_2)\big) = \alpha.$$

There exists therefore a value τ_3, $\tau_1 < \tau_3 < \tau_2$, such that $\dot{v}(\boldsymbol{x}(\tau_3)) = 0$. Since the point $\boldsymbol{x}(\tau_3)$ of F lies within the surface (resp. curve) (34.10), $v_2(\boldsymbol{x}(\tau_3)) < \alpha$; this contradicts the definition of α. The reasoning remains correct if there are singular points on the trajectory under consideration because for those $\dot{v}_2 = 0$, anyway.

Analogously we obtain

Theorem 34.5. The point set defined by

$$(34.11) \qquad\qquad v_2(\boldsymbol{x}) = \beta$$

lies entirely outside of the domain of attraction A of the differential equation (34.8). If $\beta = \infty$ then global asymptotic stability obtains.

The two theorems together imply that the boundary of the domain of attraction lies in the domain

$$\alpha < v_2(\boldsymbol{x}) < \beta.$$

In the proof we did not use the fact that $v_2(\boldsymbol{x})$ was of second degree. The argument applies therefore equally well to the function $v_2 + v_3$, etc.; possibly this function furnishes better bounds.

Chapter V

The Direct Method for General Motions

35. The General Stability Concept

In secs. 1 and 2 we gave a tentative definition of the concepts *stable, attractive,* etc., which were adequate for the applications considered so far. Subsequently we shall examine these concepts more closely and introduce a number of extensions and refinements. First, however, we must give a more exact definition of the concept of a motion defined in sec. 1. It will be made so general that the various applications obtain a unified foundation. In this manner it is possible to approach the very different parts of the theory in a uniform way.

Let X be a normed linear space. If x is an element of X, let $\|x\|$ denote its norm. We have, of course, $\|x\| = 0$ only for the zero element which is denoted by 0 — there is no danger of confusing it with the number 0 — and we have $\|x + y\| \leq \|x\| + \|y\|$ ($x \in X$, $y \in X$). We consider elements $p \in X$ which depend on three parameters t, t_0, a. t_0 and $t \geq t_0$ range continuously or discretely over a half-line R of real numbers

$$t_0 \leq t \leq \infty \text{ or } t = t_0,\ t_0 + \tau,\ t_0 + 2\tau,\ \cdots.$$

The parameter a ranges over a sub-domain A^* of a second linear space A with norm $\|a\|_1$. A and X may be the same space but this is by no means necessary[1]. The dependence of the elements on the parameters is expressed by the notation

$$(35.1) \qquad\qquad p = p(t,\, a,\, t_0),$$

p is assumed to be continuous with respect to a, t and t_0 being fixed. In case t and t_0 vary continuously p is assumed to be continuous with respect to t and t_0, a being fixed.

Def. 35.1. For a fixed a and a fixed t_0, the set of elements $p(t, a, t_0)$, $t_0 \leq t \leq t_1$, is called the *motion* determined by a and t_0 on the interval (t_0, t_1). If $t_1 = \infty$, it is simply called the *motion* determined by a and t_0.

Examples. 1) Differential equations in Euclidean space. We have $X = A = R_n$, $a = x_0$ the initial value, $t = t_0$ the initial time.

---

[1] As for relations between the two norms, *cf.* MOVCHAN [2].

2) Difference equations in R_n. $\boldsymbol{a}$ is the initial value $\boldsymbol{x}_0$; the variable t is discrete.

3) Differential difference equations, *cf.* secs. 12, 13. Here the parameter $\boldsymbol{a}$ is the initial function vector $\boldsymbol{a}(\theta)$. It belongs to the space of continuous functions on $[0, h]$, and the same is true for the solution $\boldsymbol{p}(t, \boldsymbol{a}, t_0)$. For a more complete discussion see sec. 43.

As the parameter $\boldsymbol{a}$ ranges over the domain A^* of A, (35.1) defines a *family of motions.* Let $\boldsymbol{p}(t, \boldsymbol{a}, t_0)$ and $\boldsymbol{p}(t, \boldsymbol{a}', t_0)$ be two different motions of the family belonging to the same t_0 but to different parameter values $\boldsymbol{a}$. For a fixed t the norm

$$(35.2) \qquad \|\boldsymbol{p}(t, \boldsymbol{a}, t_0) - \boldsymbol{p}(t, \boldsymbol{a}', t_0)\| =: d(t, t_0; \boldsymbol{a}, \boldsymbol{a}')$$

can be interpreted as the distance of the points on the two motions which correspond to this t-value, or more briefly as the *distance of the two motions at time t.* This distance is a scalar function of four variables $t, t_0, \boldsymbol{a}, \boldsymbol{a}'$. (We shall not employ the term "functional" which is frequently used to denote a mapping of a general space into the set of real numbers.) In stability problems we are concerned with estimating the distance (35.2) by means of suitable comparison functions. Primarily comparison functions of the classes K and L introduced in sec. 24 will be used. The main arguments of these functions will be either the distance $\|\boldsymbol{a} - \boldsymbol{a}'\|_1$ of the parameters $\boldsymbol{a}$ and $\boldsymbol{a}'$ in A or the length $t - t_0$ of the interval of the motion. The comparison function often depends on other parameters as well, which are written as second, third, ... arguments. The notation $\varphi \in K$, $\sigma \in L$, etc., always refers to the principal argument which appears in the first place.

We next hold $\boldsymbol{a}$ fixed, and call the well-defined motion $\boldsymbol{p}(t, \boldsymbol{a}, t_0)$ of the family *the unperturbed motion,* in the language of the theory of differential equations. The parameter is then allowed to range over a certain neighborhood of $\boldsymbol{a}$, for instance a spherical neighborhood

$$(35.3) \qquad \|\boldsymbol{b} - \boldsymbol{a}\|_1 < r.$$

A motion $\boldsymbol{p}(t, \boldsymbol{b}, t_0)$ is then called a *perturbed motion.*

Def. 35.2. The unperturbed motion $\boldsymbol{p}(t, \boldsymbol{a}, t_0)$ is called *stable* if there exists a function $\varphi \in K$ and a number t_0' such that

$$(35.4) \qquad d(t, t_0; \boldsymbol{a}, \boldsymbol{b}) \leq \varphi(\|\boldsymbol{a} - \boldsymbol{b}\|_1; t_0), \quad t_0' \leq t_0 \leq \ .$$

If (35.4) is valid only in a finite time interval $t_0 \leq t \leq t_1$ then the unperturbed motion is called *stable on the finite time interval* (t_0, t_1). This can also be expressed by writing

$$(35.5) \qquad d(t, t_0; \boldsymbol{a}, \boldsymbol{b}) \leq \varphi(\|\boldsymbol{a} - \boldsymbol{b}\|_1; t_0, t_1).$$

Def. 35.3. The unperturbed motion is called *attractive* if there exists a function $\sigma \in L$, a number t_0'. and a domain $B \subseteq A$, which may depend on t_0, such that

$$(35.6) \quad d(t, t_0; a, b) \leq \sigma(t - t_0; b, t_0), \quad b \in B, \quad t_0' \leq t_0 \leq t.$$

Def. 35.4. If the unperturbed motion is stable and attractive then it is called *asymptotically stable*. In this case there exists an estimate of the form

$$(35.7) \qquad d(t, t_0; a, b) \leq \varphi(\|a - b\|_1; t_0)\, \sigma(t - t_0; b, t_0).$$

Defs. 35.2 through 35.4 are obviously generalizations of the definitions given in sec. 2. In them the equilibrium is the unperturbed motion. If the unperturbed motion is defined as the solution $p(t, x_0, t_0)$ of the differential equation

$$(35.8) \qquad \dot{x} = f(x, t)$$

in R_n, then its stability can always be reduced to the stability of the equilibrium. For this purpose we set

$$x = y + p(t, x_0, t_0)$$

in (35.8) and obtain the differential equation

$$(35.9) \qquad \dot{y} = f\big(y + p(t, x_0\, t_0), t\big) - f\big(p(t, x_0, t_0), t\big).$$

The right side of this equation is a function of y and t, which vanishes for $y = 0$. The stability behavior of the solution $p(t, x_0, t_0)$ is equivalent to the stability behavior of the equilibrium of (35.9). (35.9) is also called *the differential equation of the perturbed motion*. For example (30.3) is the equation of the perturbed motion for (30.2). The goal of this transformation is a simplification of the stability problem; however, the new differential equation is often more complicated then the original one. So the differential equation of the perturbed motion for the solution of an autonomous differential equation is almost always nonautonomous.

In sec. 2 we gave two different definitions for the concepts *stable* and *attractive* and we now furnish the proof for their equivalence. The $\varepsilon - \delta$-formulation of Def. 35.2 is:

Def. 35.5. The unperturbed motion $p(t, a, t_0)$ is called stable, if for each $\varepsilon > 0$ there exists a $\delta > 0$ such that

$$(35.10) \qquad d(t, t_0; a, b) < \varepsilon$$

provided only that

$$\|a - b\|_1 < \delta.$$

The concept *stable* has thus a definition very similar to that of the concept *continuous* and we can actually think of stability as a type of continuity if we look at the motion as a whole and consider its dependence on a, resp. b.

Theorem 35.1. Defs. 35.2 and 35.5 are equivalent.

Proof. a) If (35.4) is satisfied then $\varepsilon = \varphi(\delta; t_0)$ is a number which satisfies (35.10); therefore $\delta = \varphi^I(\varepsilon; t_0)$. b) Assume (35.10) is satisfied in accordance with Def. 35.5. For a fixed ε we denote by $\bar{\delta}(\varepsilon)$ the supremum of all applicable δ. Then $\|a - b\|_1 < \bar{\delta}(\varepsilon)$ implies $d(t, t_0; a, b) < \varepsilon$, and if $\delta_1 \geq \bar{\delta}(\varepsilon)$ then there exists at least one parameter value b' such that

$$\|a - b'\|_1 < \delta_1, \quad \sup_{t \geq t_0} d(t, t_0; a, b') \geq \varepsilon.$$

The function $\bar{\delta}(\varepsilon)$ is positive and strictly monotone, but not necessarily continuous. We choose a function $\zeta(r) \in K$ so that $\zeta(r) \leq \bar{\delta}(r)$, and denote its inverse function by φ:

$$\varphi(r; t_0) := \zeta^I(r; t_0).$$

This function can be used as a comparison function φ in (35.4). In general it depends on t_0 since the number δ of Def. 35.5 also depends on t_0 (see also sec. 36).

The $\varepsilon - \delta$-formulation of Def. 35.3 is:

Def. 35.6. The unperturbed motion $p(t, a, t_0)$ is called attractive if for each $\eta > 0$ there exists a number T such that

$$(35.11) \qquad\qquad d(t, t_0; a, b) < \eta, \quad t - t_0 > T,$$

provided that b belongs to a fixed ball (35.3)

Theorem 35.2. Defs. 35.3 and 35.6 are equivalent.

Proof. a) If (35.6) holds then $T = \sigma^I(\eta; b, t_0)$ is a number of the type required in Def. 35.6. b) Assume (35.11) is satisfied in accordance with Def. 35.6. For a fixed η we denote by $\bar{T}(\eta)$ the infimum of all numbers T applicable. Then we have

$$d(t, t_0; a, b) < \eta, \text{ in case } t - t_0 > \bar{T}(\eta)$$

and

$$\sup d(t, t_0; a, b) \geq \eta \text{ for } t_0 \leq t \leq t_0 + \bar{T}(\eta).$$

The function $\bar{T}(\eta)$ is positive and monotone increasing as η approaches zero. If $\bar{T}(\eta)$ is bounded then $d(t, t_0; a, b)$ becomes zero after only a finite time has elapsed. Then the estimate (35.6) is possible a fortiori. If $\bar{T}(\eta)$ is unbounded then we choose a continuous, positive, increasing function $U(\eta)$ such that $U(\eta) \geq \bar{T}(\eta)$. Its inverse function

$$\sigma(s) := U^I(s)$$

belongs to class L and can be used as a comparison function in (35.6). This function in general depends on b and t_0.

As we already pointed out, the properties *stable* and *attractive* are by definition properties of a special motion, the unperturbed motion. We obtain a characterization of the family of motions if in (35.4) we allow the parameters a and b to vary.

Def. 35.7. A family of motions is called a *stable system* if there exists a function $\varphi \in K$ such that (35.4) holds for all a, b belonging to a certain sub-domain of the space A.

The concept *attractive* has a comparable extension to families.

Def. 35.8. A family of motions is called an *attractive system* if for each pair a, b in a certain sub-domain of the space A, there exists a function $\sigma \in L$ such that

$$d(t, t_0; a, b) \le \sigma(t - t_0; a, b; t_0).$$

The definition says that all the motions of the family come arbitrarily close to each other in time, respectively that all the motions converge toward a single one. This, for instance, is the case for the linear differential equation

$$\dot{x} = A x + f(t)$$

with asymptotically stable equilibrium. Various other terms are in use for the properties described by Defs. 35.7 and 35.8[1]).

36. Extensions and Modifications of the Basic Definitions

We henceforth shall assume that the spaces X and A of the previous section are identical and that

$$(36.1) \qquad p(t_0, a, t_0) = a.$$

This introduces a as an *initial point*. We further assume that the motions $p(t, b, t_0)$ exist for all $b \in K_r(a)$ and for all $t \ge t_0$, and that the relation

$$(36.2) \qquad p(t, p(t_1, b, t_0), t_1) = p(t, b, t_0), \quad t_0 \le t_1 \le t,$$

holds [*cf.* (16.3)]. This relation makes it possible to interpret the motion as a two-parameter family of mappings of the space A into itself. Incidentally, some of the considerations which follow are still valid if A and X are not identical or if the motion is not uniquely determined by a.

a) Stability of invariant sets. A set $M \subset A$ is called *invariant* if $p(t, a, t_0) \in M$ for a fixed $t = t_1$ implies $p(t, a, t_0) \in M$ for all $t \ge t_1$ (*cf.* Def. 16.3). A special case of an invariant set is the set $M(a, t_0)$ of all points of the motion $p(t, a, t_0), t \ge t_0$. The *distance* of a point x from a set M is defined as usual by

$$(36.3) \qquad \varrho(x, M) := \inf \|x - y\|, \quad y \in M.$$

---

[1]) *cf.* Yoshizawa [3, 4], Lakshmikantham [2, 3].

If $x \in M$ then, of course, $\varrho(x, M) = 0$.

Def. 36.1. An invariant set M is called *stable* if an estimate

$$(36.4) \qquad \varrho\big(p(t, b, t_0), M\big) \leq \varphi\big(\varrho(b, M); t_0\big), \qquad \varphi \in K,$$

is possible.

Def. 36.2. An invariant set is called *attractive* if an estimate

$$(36.5) \qquad \varrho\big(p(t, b, t_0), M\big) \leq \sigma(t - t_0; b, t_0), \qquad \sigma \in L,$$

is possible whenever the distance $\varrho(b, M)$ is smaller then a fixed number.

Def. 36.3. An invariant set is called *asymptotically stable* if it is stable and attractive.

In place of Defs. 36.1 and 36.2 we can of course also give $\varepsilon - \delta$-definitions: M is stable if for each ε there exists a δ such that $\varrho(b, M) < \delta$ implies the inequality $\varrho(p(t, b, t_0), M) < \varepsilon$; M is attractive if $\lim \varrho(p(t, b, t_0), M) = 0$, etc.

For the invariant set $M(a, t_0)$ (see above) we define especially

Def. 36.4. The motion $p(t, a, t_0)$ is called *orbitally stable*, resp. *orbitally attractive*, if the invariant set $M(a, t_0)$ is stable, resp. attractive. If both of these obtain simultaneously the motion is *orbitally asymptotically stable*. In this case there exists an estimate

$$(36.6) \qquad \varrho\big(p(t, b, t_0), M(a, t_0)\big) \leq \varphi\big(\varrho(b, M(a, t_0)); t_0\big)\, \sigma(t - t_0; b, t_0).$$

The concepts given in the last definition were introduced because the concept of stability in the sense of Liapunov is rather too restrictive for certain purposes. This is seen, for instance, in the example of an autonomous second order system in the neighborhood of a center. The phase trajectories form closed curves (see fig. 22.1). In the linear case they are ellipses. If we consider the motion given by one of these closed curves as the "unperturbed motion", it is in general not stable in the sense of Liapunov. The reason for this is that the period, *i.e.* the time required for the phase point to traverse the closed trajectory varies in the nonlinear case from curve to curve (*cf.* sec. 22). Therefore there arises between points on neighboring trajectories a phase difference, possibly as large as 180°, this maximum value actually being attained now and then. If we interpret the motion as a curve in a three-dimensional (x, y, t)-space then the closed trajectory corresponds to a helix about the t-axis. The pitch of the turns of this helix depends on the period and is therefore not uniform. The spatial distance of two points which are close together at time t_0, but which belong to different trajectories, becomes accordingly arbitrarily large as t increases.

None the less, the closed trajectories have a certain stability-like property: Curves which at one time are close to each other remain close. To arrive at the proper concept we have to look at the motion as a whole, *i.e.* we must introduce the set $M(a, t_0)$, and thus we arrive at orbital stability. For the equilibrium the distinction between Liapunov and orbital stability is empty since in this case the set $M(0, t_0)$ contains only the origin.

Incidentally, the concept of orbital stability is interesting only in the case of a closed trajectory (*i.e.* of a periodic motion). It makes, for instance, little sense to talk about orbital stability of an almost periodic motion whose trajectory, for $n = 2$, covers an annulus.

We give an example of a cycle which is orbitally stable but not stable in the sense of Liapunov.

$$\dot{x} = \frac{x(1 - r^2)^3}{r^2} - y\left(1 + (1 - r^2)^2\right),$$

$$r^2 = x^2 + y^2,$$

$$\dot{y} = x\left(1 + (1 - r^2)^2\right) + \frac{y(1 - r^2)^3}{r^2}.$$

In polar coordinates these equations become

$$\frac{d}{dt} r^2 = 2(1 - r^2)^3; \qquad \frac{d\varphi}{dt} = 1 + (1 - r^2)^2.$$

Setting $1 - r^2 = z$, we get the system

$$\dot{z} = -2z^3, \quad \dot{\varphi} = 1 + z^2$$

whose solution is

$$z(t) = z_0(1 + 4z_0^2 t)^{-1/2}, \qquad \varphi = \varphi_0 + t + \frac{1}{4} \ln(1 + 4z_0^2 t).$$

The cycle $r = 1$ of the original system corresponds to the values $z_0 = 0$, $\varphi = \varphi_0 + t$. The trajectory of each neighboring solution is also a spiral, these spirals coming arbitrarily close to the unit circle. On the other hand the phase difference $\varphi(t, z_0) - \varphi(t, 0) = \ln(1 + 4z_0^2 t)$ increases without bound.

b) Uniformity. The dependence of the comparison functions on the secondary arguments t_0 and x_0 or b respectively is of considerable importance. Consider, for instance, the equilibrium of the differential equation $\dot{x} = -x$. The general solution, *i.e.* the perturbed motion is $x_0 \exp(t_0 - t)$, and we have

$$|p(t, x_0, t_0)| \le \varphi(|x_0|) \, \sigma(t - t_0)$$

where $\varphi(r) = r$, $\sigma(s) = e^{-s}$. Both comparison functions can be chosen independently of t_0. For

$$(36.7) \qquad \dot{x} = -\frac{x}{1 + t}, \quad p(t, x_0, t_0) = x_0 \frac{1 + t_0}{1 + t}$$

such an estimate is not possible: The σ-function must be written in the form $\sigma(t - t_0; t_0)$. Using Def. 35.6 we see that the number

$$T = \frac{|x_0| - \eta}{\eta} (1 + t_0)$$

depends on the given η, and increases arbitrarily with t_0.

As a consequence, it seems necessary to refine the definitions of sec. 35 and to emphasize the dependence of the initial values of the perturbed motion. We write $q(t)$ for the unperturbed motion $p(t, a, t_0)$ and introduce

$$D(t, t_0; b): = \|q(t) - p(t, b, t_0)\|.$$

Of course, we have

$$a = q(t_0), \quad D(t, t_0; b) = d(t, t_0; q(t_0), b).$$

Def. 35.2 was formulated for a fixed time t_0. A motion which is stable lor $t = t_0$ is not necessarily stable for all $t_1 > t_0$, cf., for instance, sec. 44. But stability for $t = t_0$ implies stability for $t = t_1$ if the mapping defined by the motions is continuous at $q(t_0) = a$, *i.e.* if a neighborhood of a is mapped onto a neighborhood of $q(t_1)$. The trajectories of an ordinary differential equation in R_n define a mapping of this type.

Def. 36.5. The unperturbed motion is called *uniformly stable* if an estimate

$$D(t, t_0; b) \leq \varphi(\|b - q(t_0)\|)$$

exists where the comparision function $\varphi(r)$ can be chosen independently of t_0, or equivalently, if for each ε there exists a $\delta = \delta(\varepsilon)$ independent of t_0 such that $D(t_0, t_0; b) = \|q(t_0) - b\| < \delta$ implies $D(t, t_0; b) < \varepsilon$ for all $t \geq t_0$.

The equivalence of the two definitions can be proved as for Def. 35.2.

Def. 36.6. The unperturbed motion is called *uniformly attractive with respect to* b if there exists an estimate

$$D(t, t_0; b) \leq \sigma(t - t_0; t_0)$$

uniformly for all b with $\|q(t_0) - b\| \leq \gamma, \gamma > 0$.

The unperturbed motion is called *uniformly attractive with respect to* t_0 if an estimate

$$D(t, t_0; b) \leq \sigma(t - t_0; b)$$

holds uniformly for all $t_0, t_0' \leq t_0 < \infty$.

If the motions are defined by differential equations the unperturbed motion and the equilibrium of the differential equation of the perturbed motion have the same stability behavior regarding uniformity. Some special terms are frequently used. We define them in the case of the equilibrium. Here $q(t) \equiv 0$, $D(t, t_0; b) = \|p(t, b, t_0)\|$.

Def. 36.7. The equilibrium is called *equiasymptotically stable* if it is stable and in addition uniformly attractive with respect to b. This means

$$(36.8) \qquad \|p(t, b, t_0)\| \leq \varphi(\|b\|; t_0)\, \sigma(t - t_0; t_0)$$

Def. 36.8. The equilibrium is called *uniformly asymptotically stable* if it is uniformly stable and uniformly attractive with respect to $\boldsymbol{b}$ as well as with respect to t_0. This means

$$(36.9) \qquad \|\boldsymbol{p}(t, \boldsymbol{b}, t_0)\| \leq \psi(\|\boldsymbol{b}\|)\, \sigma(t - t_0).$$

According to Def. 35.3, the domain of attraction of the motion $\boldsymbol{p}(t, \boldsymbol{a}, t_0)$ for $t = t_0$ consists of all points $\boldsymbol{b}$ which satisfy (35.6). Generally, the domain of attraction depends on t_0. Let $\varrho(t_0)$ be the radius of the largest ball $\|\boldsymbol{q}(t_0) - \boldsymbol{x}\| < \varrho(t_0)$ which lies within the domain of attraction. It may happen that $\varrho(t_0)$ tends to zero with increasing t_0. Consider, for instance, the solution $x = 0$ of the differential equation

$$\dot{x} = -x - \frac{x}{t}(1 - x^2 t^3).$$

For fixed t_0, its domain of attraction is given by

$$|x_0| \leq t_0^{-1},$$

for the equation is derived from $\dot{z} = z(z^2 - 1)$ by the substitution $z = t x$. Here, the undisturbed motion $x = 0$ is non-uniformly stable. The equilibrium of the equation

$$\dot{x} = \begin{cases} -t^2 \,(x > t^{-2}) \\ -x^2 \,(0 \leq x \leq t^{-1}) \\ -x \,(x \leq 0) \end{cases}$$

is asymptotically stable and uniformly stable since $|p(t, x_0, t_0)| \leq |x_0|$. But the domain of attraction is bordered by the upper branch of the hyperbola $x t = 1$. Therefore, $\varrho(t_0)$ tends to zero.

If the t_0-dependent domains of attraction have a non-empty intersection it is called the *domain of attraction* of $\boldsymbol{q}(t)$.

The asymptotic stability of the equilibrium of an autonomous differential equation is uniform, *cf.* the proof of Theorem 25.2.

For differential equations, respectively for motions defined in R_n, a further concept is needed.

Def. 36.9. Let $\boldsymbol{p}(t, \boldsymbol{b}, t_0) \in R_n$ be defined for all $\boldsymbol{b} \in R_n$. The equilibrium is called *uniformly asymptotically stable in the whole* if 1) it is uniformly stable, 2) it is attractive for arbitrary $\boldsymbol{b}$, and 3) for each pair ζ and η of numbers there exists a number $T = T(\zeta, \eta)$ such that

$$|\boldsymbol{p}(t, \boldsymbol{b}; t_0)| < \eta \text{ whenever } |\boldsymbol{b}| < \zeta \text{ and } t - t_0 \geq T.$$

As in the proof of Theorem 35.2 and 35.3 we recognize that this definition could also be formulated as follows: There exists a function $l(r, s) \in KL$ defined for all $r > 0$, (*cf.* Def. 24.2) such that

$$(36.10) \qquad |\boldsymbol{p}(t, \boldsymbol{b}, t_0)| \leq l(|\boldsymbol{b}|, t - t_0), \text{ for all } \boldsymbol{b}.$$

Occasionally the right side may be estimated by means of a product (*cf*. sec. 24 A i).

The definitions for *uniformly orbitally stable* resp. *a uniformly stable system*, etc., are obtained from Defs. 36.1 through 36.3, resp. Defs. 35.7 and 35.8, by specialization of the comparison functions.

c) Increasing comparison functions and estimates from below. Inequalities of the form

$$(36.11) \qquad d(t, t_0; \boldsymbol{a}, \boldsymbol{b}) \geq \varphi(\|\boldsymbol{a} - \boldsymbol{b}\|; t_0),$$

$$(36.12) \qquad d(t, t_0; \boldsymbol{a}, \boldsymbol{b}) \geq \sigma(t - t_0; \boldsymbol{b}, t_0)$$

can also be used to characterize the unperturbed motion. The first one says that the perturbed motion cannot come arbitrarily close to the unperturbed motion, the second means that the approximation if possible at all, cannot progress arbitrarily fast. Again another type of inequality is

$$(36.13) \qquad d(t, t_0; \boldsymbol{a}, \boldsymbol{b}) \geq \varkappa(t - t_0), \text{ resp. } \varkappa(t - t_0; \boldsymbol{b}, t_0),$$

where $\varkappa$ denotes a function which is monotone increasing in its first argument. Inequalities (36.12) and (36.13) correspond in a certain sense to attractivity, whereas (36.11) describes a property related to stability, a type of "antistability". Incidentally (36.11) does not imply that the unperturbed motion is unstable.

d) Classification of stability types. Since there exist several types of comparison functions which in addition may depend on the secondary variables, we can quite formally discern a large number of stability types and name them as the situation demands. We obtain a synopsis of the various possibilities in the following manner: 1) There are three types of comparison functions, functions φ of class K, functions σ of class L, and the functions $\varkappa$ introduced in (36.13). 2) The functions $d(t, t_0; \boldsymbol{a}, \boldsymbol{b})$, resp. $D(t, t_0; \boldsymbol{b})$, can be estimated from below or from above. 3) One of the two motions can be held fixed (thus singling out an unperturbed motion) and the other can be allowed to vary, or both motions can be allowed to vary. 4) The comparison function may in addition depend on t_0 and on $\boldsymbol{b}$, and in case of stable systems on $t_0, \boldsymbol{a}$, and $\boldsymbol{b}$. 5) In place of $d(t, t_0; \boldsymbol{a}, \boldsymbol{b})$ we can estimate the expression

$$(36.14) \qquad \delta(t, t_0; \boldsymbol{a}, \boldsymbol{b}) := \varrho(M(\boldsymbol{a}, t_0), \boldsymbol{p}(t, \boldsymbol{b}, t_0)).$$

On the right the argument $\|\boldsymbol{a} - \boldsymbol{b}\|$ must then be replaced by $\varrho(\boldsymbol{b}, M(\boldsymbol{a}, t_0))$. Taking into account all of these possibilities we obtain a total of eight groupings of eleven inequalities each.

In the first group, A, $\boldsymbol{p}(t, \boldsymbol{a}, t_0) = \boldsymbol{q}(t)$ is a fixed motion, the unperturbed motion. The distance $d(t, t_0; \boldsymbol{a}, \boldsymbol{b})$ or $D(t, t_0; \boldsymbol{b})$, if the uniformity with respect to t_0 is emphasized, is estimated from above. Group A contains the following inequalities:

$$1)\ d(t, t_0; a, b) \leq \varphi(||a - b||; t_0, t_1)$$

$$2)\ d(t, t_0; a, b) \leq \varphi(||a - b||; t_0)$$

$$3)\ D(t, t_0; b) \quad \leq \varphi(||q(t_0) - b||)$$

$$4)\ d(t, t_0; a, b) \leq \sigma(t - t_0; b, t_0)$$

$$5)\ D(t, t_0; b) \quad \leq \sigma(t - t_0; b)$$

$$6)\ d(t, t_0; a, b) \leq \sigma(t - t_0; t_0)$$

$$7)\ D(t, t_0; b) \quad \leq \sigma(t - t_0)$$

$$8)\ d(t, t_0; a, b) \leq \varkappa(t - t_0; b, t_0)$$

$$9)\ D(t, t_0; b) \quad \leq \varkappa(t - t_0, b)$$

$$10)\ d(t, t_0; a, b) \leq \varkappa(t - t_0; t_0)$$

$$11)\ D(t, t_0, b) \quad \leq \varkappa(t - t_0)$$

In all these inequalities, b belongs to a fixed sub-domain of X which contains the point $q(t_0) = a$ as an interior point. The variable t ranges over a finite interval, $t_0 \leq t \leq t_1$ in the first inequality. In the following two inequalities, we have $t_0 \leq t < \infty$. Inequalities 4) through 11) must hold for $t_0 + \bar{t} < t < \infty$; $\bar{t}$ must be chosen so large that the comparison functions are defined for $t - t_0 > \bar{t}$. $\varkappa$ is a monotone increasing function of its first argument. In 1), 2), 4), 6), 8), 10), t_0 may be fixed, in 3), 5), 7), 9), 11) t_0 is a parameter.

In accordance with the definitions given above, A 3 characterizes uniform stability (Def. 36.5), A 6 characterizes uniform attractivity with respect to b.

If we consider the equilibrium, we set $q(t) \equiv 0$, $a = 0$, and the left side of all inequalities is $||p(t, b, t_0)||$.

Group A' is obtained from group A by reversing the inequality signs. Thus (36.11) corresponds to $A' 2$, (36.13) to $A' 11$, resp. $A' 8$. For the inequalities $A' 4$ through $A' 11$, the last four of which describe the divergence of the neighboring motions from the unperturbed motion, b must be taken from an annular neighborhood of a, resp. $q(t_0)$, $0 < \alpha_1 \leq ||q(t_0) - b|| \leq \alpha_2$.

The stability types of group A' can of course also be defined by $\varepsilon - \delta$-relations. For instance, the definition for $A' 2$ is: For each ε there exists a δ, such that $d(t, t_0; a, b) > \varepsilon$ if $||a - b|| > \delta$. For $A' 6$ we have: For each η there exists a T such that $d(t, t_0; a, b) > \eta$ whenever $t - t_0 < T$, provided that $\|\boldsymbol{q}(t_0) - \boldsymbol{b}\|$ is larger than a fixed number $\alpha > 0$. T depends on t_0 and is independent of $\boldsymbol{b}$. For A' 10 we have the same estimate for $d(t, t_0; \boldsymbol{a}, \boldsymbol{b})$ but with $t - t_0 > T$.

Groups B and B' contain the same inequalities as A and A'; but both motions are viewed as variable, the initial time being t_0 for both of them. Therefore, the left side of all inequalities is $d(t, t_0; \boldsymbol{a}, \boldsymbol{b})$. Inequality B 3 is

$$d(t, t_0; \boldsymbol{a}, \boldsymbol{b}) \leq \varphi(\|\boldsymbol{a} - \boldsymbol{b}\|).$$

Groups C, C' and D, D' arise from A through B' if instead of d, resp. D, we estimate the expression δ defined in (36.14). So, for instance, C 3 characterizes uniform orbital stability of the unperturbed motion,

$$\delta(t, t_0; \boldsymbol{a}, \boldsymbol{b}) \leq \varphi\big(\varrho(M(\boldsymbol{a}, t_0), \boldsymbol{b})\big).$$

We do not intend to introduce further special nomenclature; we shall simply use the number of the inequality as given in the list above, if the need arises. The formal classification does not, of course, tell us which of the individual types are realizable. We shall deal with that question later. Example (36.7) shows that A 6 and A 7 are different types. Another important question is in which cases two different estimates are both valid simultaneously — e.g. in the case of asymptotic stability, A 2 and A 4 — and which combinations are possible. Some combinations are trivially impossible, for instance A 4 and A' 8; at times, however, the relations between the classes are not easily seen (see also sec. 38). We mention a further property of a certain class of motions which is characterized by inequalities.

Theorem 36.1. The set of all motions $\boldsymbol{p}(t, \boldsymbol{b}, t_0)$, which for a fixed $\boldsymbol{a} = \boldsymbol{q}(t_0)$ satisfy a strict inequality of the type A 6 or A 7 is open.

To prove this we hawe to show the following: If there exists a domain $B \subset A$ such that for $\boldsymbol{b} \in B$ we have

$$(36.15) \qquad d(t, t_0; \boldsymbol{a}, \boldsymbol{b}) < \sigma(t - t_0; t_0), \qquad \sigma \in L,$$

then there exists an open $B^* \supset B$ and a function $\sigma^* \in L$ such that

$$(36.16) \qquad d(t, t_0; \boldsymbol{a}, \boldsymbol{b}) < \sigma^*(t - t_0; t_0), \qquad \boldsymbol{b} \in B^*.$$

If this were not the case we could find a closed domain $\bar{B} \supset B$ such that for each point $\boldsymbol{b}$ not belonging to $\bar{B}$ an estimate (36.16) is impossible. Let $\boldsymbol{b}'$ be a boundary point of $\bar{B}$. Then for the motion $\boldsymbol{p}(t, \boldsymbol{b}', t_0)$ an estimate (36.16) is possible and therefore for a given η a number $T(\eta)$ can be found such that

$$(36.17) \qquad d(t, t_0; \boldsymbol{a}, \boldsymbol{b}) < \eta \text{ for } t - t_0 \geq T,$$

for all b in $\bar{B}$, hence, in particular, for the boundary point b'. On the other hand, there exist in every neighborhood of b' points $\bar{b}$ such that for an arbitrarily long time and for arbitrarily small ε

$$(36.18) \qquad d(t, t_0; a, \bar{b}) > \eta, \quad ||\bar{b} - b'|| < \varepsilon.$$

This is shown in the same way as the assertion of Theorem 33.2. But now the expression d is a continuous function of b and t, hence in conjunction with (36.16) we have the inequality

$$(36.19) \qquad |d(t, t_0; a, b_1) - d(t, t_0; a, b_2)| < \psi(||b_1 - b_2||)\, g(t - t_0),$$
$$\psi \in K, \;\; g \in K.$$

We now choose $\hat{T}$ so large that

$$d(t_0 + \hat{T}, t_0; a, b') < \frac{\eta}{2}$$

and determine $\bar{b}$ so that on the one hand (36.17) holds with $t = t_0 + T_0$, $T_0 > \hat{T}$, and on the other hand $\psi(||b' - \bar{b}||)\, g(T_0) < \frac{\eta}{2}$. Then (31.19) implies for $b_1 = b'$, $b_2 = \bar{b}$

$$d(t_0 + T_0, t_0; a, \bar{b}) \le \frac{\eta}{2} + d(t_0 + \hat{T}, t_0; a, b') < \eta,$$

a contradiction to (36.18). Therefore the set B under consideration is open.

e) Types of boundedness. Let $p(t, a, t_0)$ be a family of motions which contains the equilibrium $p(t, a, t_0) = 0$. (36.1) implies that the parameter value belonging to the equilibrium is $a = 0$.

Def. 36.10. An individual motion $p(t, b, t_0)$ of the family is called *bounded* if

$$\sup ||p(t, b, t_0)|| < \infty \;\; \text{for} \;\; t \ge t_0.$$

If this is valid for all motions for which $b \in B$ then

$$||p(t, b, t_0)|| \le \beta(b; t_0),$$

where β is a positive scalar function, bounded for $b \in B$.

Def. 36.11. The family is called *uniformly bounded* if it is possible to find an estimate

$$(36.20) \qquad ||p(t, b, t_0)|| \le \gamma(||b||; t_0), \quad b \in B,$$

$\gamma(r, t_0)$ positive and bounded for $r > 0$.

This estimate calls to mind the definition of stability by means of (35.4); but it is not necessary here that $\lim\limits_{r \to 0} \gamma(r, t_0) = 0$. A property corresponding to attractivity of type $A\,6$ is given by the estimate

$$(36.21) \qquad ||p(t, b_0, t_0)|| \le g(t - t_0; ||b||), \quad b \in B,$$

where the comparison function $g(s;r)$ is defined for all r and sufficiently large s, is a monotone increasing function of r and a monotone decreasing function of s, but does not necessarily tend to zero.

Example. The general solutions of the scalar differential equations

$$\dot{x} = -x^3/t^3 \text{ and } \dot{x} = +x^3/t^3$$

are

$$p_1(t, x_0, t_0) = \left(\frac{1}{t_0^2} - \frac{1}{t^2} + \frac{1}{x_0^2}\right)^{-1/2},$$

$$p_2(t, x_0, t_0) = \left(\frac{1}{t^2} - \frac{1}{t_0^2} + \frac{1}{x_0^2}\right)^{-1/2}.$$

For the first solution we have

$$|x_0| \, t_0 (t_0^2 + x_0^2)^{-1/2} \leq |p_1(t, x_0, t_0)| \leq |x_0|,$$

for the second

$$|x_0| \leq |p_2(t, x_0, t_0)| < |x_0| \cdot t_0 (t_0^2 - x_0^2)^{-1/2}.$$

In the first case there exists an estimate of the form (36.21) and a fortiori one of the form (36.20), in fact for all x_0. In the second case we must require $|x_0| < t_0$ and obtain then an estimate (36.20) but not one of the form (36.21). To obtain it we have to restrict x_0 to a suitably chosen fixed domain, e.g. $|x_0| \leq a$, $t_0 > a + \delta$, $\delta > 0$.

Boundedness types can be associated with the other types of stability and attractivity as well[1]).

f) Further modifications. MASSERA [6] has pointed out a certain lack of symmetry in the concept of a perturbed motion introduced by Liapunov: The disturbance concerns only the initial values, it is spatial; the time scale, *i.e.* the "clock" with which time is measured along the motion is not affected. If we wish to take into account disturbances of the time scale as well, we have to replace the distance (35.2) in the definitions of sec. 35 by

$$\|p(s(t, a), a, t_0) - p(s(t, b), b, t_0)\|,$$

which expresses the fact that time is measured by a specific function $s(t, a)$ along each motion. We will assume here that the error $|t - s(t, a)|$ of each of the time scales is sufficiently small. The error $|t - s|$ will then also occur in the estimates. Massera showed several possibilities for modifying Liapunov's definition in this manner. We shall not deal with them further nor shall we concern ourselves with further modifications which have very recently been introduced. They deal with the behavior of "motions" $p(t, a, t_0)$, whose course is determined by a chance process,

---

[1]) *cf.* YOSHIZAWA [1, 2].

12*

as for instance the solutions of differential equations with stochastic coefficients. Statements about stable behavior, etc., are then probability statements. The basic definitions are of the type: The equilibrium is stable with probability one if for each two numbers $1 > \alpha > 0$, $\varepsilon > 0$, there exists a $\delta = \delta(\alpha, \varepsilon) > 0$ with the property that the probability for the inequality $\|\boldsymbol{p}(t, \boldsymbol{x}_0, t_0)\| \geq \varepsilon$ is smaller than α provided that $\|\boldsymbol{x}_0\| < \delta$. If in addition the probability that $\|\boldsymbol{p}(t, \boldsymbol{x}_0, t_0)\|$ is larger than an arbitrarily small η approaches zero with increasing t then we speak of asymptotic stability with probability one, etc. A theory of stability based on such definitions requires the tools of probability theory and is of considerable practical significance, for example, for investigating control systems which are subject to random disturbances[1].

LaSalle [1] defines the points

$$\boldsymbol{b}_1 = \boldsymbol{p}(t_0 + \tau, \boldsymbol{b}_0, t_0),$$

$$\boldsymbol{b}_i = \boldsymbol{p}(t_0 + \tau, \boldsymbol{b}_{i-1}, t_0), \quad i = 2, 3, \ldots, \tau \text{ fixed},$$

calls the sequence $\{\boldsymbol{b}_i\}$ a *motion* and develops a theory of stability. We shall not go into its details but refer to LaSalle's paper and the discussion in Reissig, Sansone and Conti [1], sec. 2.10.

37. Instability and Non-Uniform Stability

Def. 2.4 defines the concept *unstable* to mean "not stable". In general we have

Def. 37.1. The unperturbed motion is called *unstable* (for a fixed t_0), if it is not stable (for a fixed t_0) (sec. 2).

In this case there exists a number $\varepsilon > 0$, a sequence $\boldsymbol{b}_n \to \boldsymbol{a}$, and a sequence t_n such that

$$(37.1) \qquad\qquad d(t_0 + t_n, t_0; \boldsymbol{a}, \boldsymbol{b}_n) \geq \varepsilon.$$

Instability defined in this way cannot be described by one of the inequalities of the previous sections. The reason is that the properties *stable*, *attractive*, etc., are defined in terms of the behavior of a *family* of motions $\boldsymbol{p}(t, \boldsymbol{b}, t_0)$, where $\boldsymbol{b}$ ranges over a neighborhood of $\boldsymbol{a}$. On the other hand, the exceptional values $\boldsymbol{b}_n$ in (37.1) need by no means constitute a neighborhood.

Instability types can also be defined by inequalities. An inequality

$$d(t, t_0; \boldsymbol{a}, \boldsymbol{b}) \geq \varphi(\|\boldsymbol{a} - \boldsymbol{b}\|) \, \varkappa(t - t_0; t_0)$$

(that is $A'\,3$ together with $A'\,10$ in the table of sec. 36) implies of course instability; but this is a much stronger type of instability than (37.1).

---

[1] *cf.* for instance Kushner [1] and Caughey and Gray [1], where also further references can be found.

Occasionally instability is not defined as above but instead so that (37.1) holds for all b_n in a neighborhood of a[1]). In this case we can again work with comparison functions.

To define instability we may not require, as we did in the case of stability, that the motion $p(t, b, t_0)$ remains finite for finite t. For if arbitrarily close to the point a there exist points b such that $p(t, b, t_0)$ becomes infinite on a finite time interval, then the unperturbed motion is unstable by definition. A similar reasoning applies to the concept *not uniformly stable*. The equilibrium is not uniformly stable if the hypothesis of Def. 36.5 is not satisfied. Stated positively, this says: There exist an $\varepsilon > 0$, a sequence $b_n \to 0$, a sequence t_{0n}, and a sequence t_n such that

$$(37.2) \qquad |p(t_{0n} + t_n, b_n, t_{0n})| \geq \varepsilon.$$

Again no claim is made that the b_n form a neighborhood of 0. Of course (37.1) implies (37.2). Therefore we must at times explicitly state whether the statement "the equilibrium is not uniformly stable" is to include the possibility of instability or not.

The property *attractive* is negated similarly. The equilibrium is not attractive if in each neighborhood of the origin at least one motion originates which does not eventually approach the origin, or positively: The unperturbed motion is not attractive if there exists a sequence $b_n \to a$ such that

$$\limsup_{t \geq t_0} d(t, t_0; a, b_n) > 0.$$

38. Relationships between the Stability Types

Theorem 38.1. If the family of motions $p(t, a, t_0)$ depends continuously on a and t_0 then the stability of the unperturbed motion follows from its A 6-attractivity. Analogously, B 6-attractivity implies B 2-stability.
Proof. Let $\varepsilon > 0$ be given. By A 6-attractivity there exist numbers $T = T(\varepsilon, t_0)$ and $r > 0$ such that

$$(38.1) \qquad d(t, t_0; a, b) < \varepsilon \quad \text{for} \quad \|b - a\| < r, \quad t \geq t_0 + T.$$

The assumption that the unperturbed motion is unstable would imply that (37.1) is satisfied. For sufficiently large n, $\|b_n - a\| < r$, and thus (38.1) implies that almost all t_n are smaller than T. But the inequality

$$d(t_0 + t_n, t_0; a, b_n) \geq \varepsilon \quad (b_n \to a, t_n \geq T)$$

contradicts the assumed continuity. In investigating stability of systems a must also be approximated by a sequence a_n.

We assume below that the unperturbed motion is the equilibrium.

---

[1]) As for instance in LEFSCHETZ [1].

Theorem 38.2. Let a function $\varphi \in K$ and a continuous function $\psi(s)$ exist such that

$$\| \boldsymbol{p}(t, \boldsymbol{b}, t_0) \| \leq \varphi(\| \boldsymbol{b} \|) \, \psi(t - t_0).$$

Then uniform stability (A 3) follows from A 7-attractivity.

The hypothesis is given, for instance, if

$$(38.2) \qquad \frac{d}{dt} \| \boldsymbol{p}(t, \boldsymbol{b}, t_0) \| < c \, \| \boldsymbol{p}(t, \boldsymbol{b}, t_0) \|, \qquad c \text{ const}.$$

Condition (38.2) is satisfied if the family is defined by a differential equation $\dot{\boldsymbol{x}} = \boldsymbol{f}(\boldsymbol{x}, t)$ in R_n whose right side satisfies a uniform Lipschitz condition with respect to t, $\boldsymbol{f} \in \bar{C}_1$.

Proof. If A 7 is valid we have (for $\boldsymbol{a} = 0$)

$$\| \boldsymbol{p}(t, \boldsymbol{b}, t_0) \| \leq \sigma(t - t_0), \qquad \boldsymbol{b} \in K_r.$$

For a given T choose the number $\varkappa$ so that

$$\varphi(\varkappa) \, \psi(T) = \sigma(T).$$

If now $\| \boldsymbol{b} \| \leq \varkappa$, then the hypothesis implies that

$$\| \boldsymbol{p}(t, \boldsymbol{b}, t_0) \| \leq \varphi(\| \boldsymbol{b} \|) \, \psi(T) \quad \text{for} \quad t \geq t_0 + T$$

and this is equivalent to A 3 because the right side is independent of t_0.

From the somewhat more general estimate

$$\| \boldsymbol{p}(t, \boldsymbol{b}, t_0) \| \leq \varphi(\| \boldsymbol{b} \|) \, \psi(t - t_0, t_0),$$

we can conclude similarly that simple stability (A 2) follows from A 6 or A 7.

Theorem 38.3. Let the family $\boldsymbol{p}(t, \boldsymbol{a}, t_0)$ be defined in R_n by the differential equation $\dot{\boldsymbol{x}} = \boldsymbol{f}(\boldsymbol{x}, t)$, where $\boldsymbol{f} \in E$, then A 3 and together A 4 imply A 6.

Proof. By hypothesis (writing $|\boldsymbol{x}|$ for the norm of $\boldsymbol{x}$)

$$(38.3) \qquad |\boldsymbol{p}(t, \boldsymbol{x}_0, t_0)| \leq \varphi(|\boldsymbol{x}_0|), \qquad t \geq t_0,$$

and at the same time

$$|\boldsymbol{p}(t, \boldsymbol{x}_0, t_0)| \leq \sigma(t - t_0; \boldsymbol{x}_0, t_0), \qquad \boldsymbol{x}_0 \in B = B(t_0).$$

For a fixed initial value $\bar{\boldsymbol{x}}_0$ we choose the number $T(\bar{\boldsymbol{x}}_0)$ which depends on $\bar{\boldsymbol{x}}_0$, so large that for $t > t_0 + T(\bar{\boldsymbol{x}}_0)$ we have

$$(38.4) \qquad \sigma(t - t_0; \bar{\boldsymbol{x}}_0, t_0) \leq \frac{1}{2} \, \varphi(|\bar{\boldsymbol{x}}_0|).$$

If $\bar{\boldsymbol{x}}_0$ is restricted to a closed subdomain $H \subset B$ containing the origin the numbers $T(\bar{\boldsymbol{x}}_0)$ are bounded, $T(\bar{\boldsymbol{x}}_0) \leq T$. If this were not true we could choose a sequence $\bar{\boldsymbol{x}}_{0n} \to \tilde{\boldsymbol{x}}_0 \in H$ so that $|\boldsymbol{p}(t, \tilde{\boldsymbol{x}}_0, t_0)|$ could not be estimated by a function of class L.

Since the solutions depend continuously on the initial values there exists a spherical neighborhood $K(\bar{\pmb{x}}_0)$ of $\bar{x}_0$ such that

(38.5) $|\pmb{p}(t,\pmb{x}'_0\,t_0)| \leq \varphi(|\bar{\pmb{x}}_0|)$ for $\pmb{x}'_0 \in K(\bar{\pmb{x}}_0)$, $t_0 \leq t \leq t_0 + T$.

For each $\bar{\pmb{x}}_0 \in H$ we select the ball $K(\bar{\pmb{x}}_0)$ and the number $T(\bar{\pmb{x}}_0)$. By the Heine-Borel Theorem, finitely many of these balls cover the domain H. Let their centers be denoted by $\bar{\pmb{x}}_1, \ldots, \bar{\pmb{x}}_N$. Then, because of (38.5), the inequality

$$|\pmb{p}(t,\pmb{x}_0,t_0)| \leq \varphi\left(\max_i |\bar{\pmb{x}}_i|\right) =: \beta, \quad t_0 \leq t < t_0 + T,$$

holds for all $\pmb{x}_0 \in H$, and for $t > t_0 + T$ we have, by construction [cf. (38.4)],

$$|\pmb{p}(t,\pmb{x}_0,t_0)| \leq \sigma(t - t_0;\ \hat{\pmb{x}}, t_0).$$

The vector $\hat{\pmb{x}}$ is defined by the equation $\varphi(|\hat{\pmb{x}}|) = 2\beta$. The comparison function on the right does not depend on $\pmb{x}_0$ but only on a domain and on t_0; the estimate is therefore uniform in $\pmb{x}_0$.

Further relationships can be given between the properties defined by inequalities of type A'. For example: If the family of motions $\pmb{p}(t,\pmb{a},t_0)$ depends continuously on $\pmb{a}$, then the existence of an inequality A' 10 implies an inequality A' 2. This theorem which is analogous to Theorem 38.1 can be proved as follows. By hypothesis there exists for each ε a number $T(\varepsilon,t_0)$, depending only on ε and t_0, such that $d(t,t_0;\pmb{a},\pmb{b}) > \varepsilon$ if $t - t_0 > T$ and if $\|\pmb{b} - \pmb{a}\|$ is greater than a fixed number α. If we could not realize A' 2 then there would exist a sequence t_n and a sequence $\pmb{b}_n$ not converging to $\pmb{a}$, such that $d(t_n,t_0;\pmb{a},\pmb{b}_n)$ becomes arbitrarily small, for instance smaller than ε. Since α can be chosen smaller than almost all the numbers of the sequence $\|\pmb{b}_n - \pmb{a}\|$, the inequality $t_n < T$ must hold for almost all t_n. This implies that certain motions come arbitrarily close to the unperturbed motion in finite time although their initial points lie at a definite distance from $\pmb{a}$. And this is impossible.

A further group of relationships exists for motions defined by differential equations with periodic coefficients. If the right side of the differential equation has period ω, i.e. if $\pmb{f}(\pmb{x},t + \omega) = \pmb{f}(\pmb{x},t)$ for all t, then the solution remains unchanged if both the initial time t_0 and the time t are shifted uniformly by one period. Then

(38.6) $\pmb{p}(t + \omega, \pmb{x}_0, t_0 + \omega) = \pmb{p}(t, \pmb{x}_0, t_0).$

We shall now ignore where the relation (38.6) originated and make it, by definition, the property of a family of motions.

Def. 38.1. A family of motions is called *periodic with respect to t_0*, if the relation (38.6) holds for all t, t_0.

A special case is given by the motions in Chapter III which were defined by autonomous differential equations depending only on $t - t_0$. For then every number ω is a possible period.

Theorem 38.4. Consider a family of motions which is periodic with respect to t_0 and which contains a motion $q(t)$ periodic with the same period with respect to t. Then A 2-stability of $q(t)$ implies A 3-stability and A 4-attractivity implies A 5-attractivity.

Proof. By hypothesis, we have an estimate

$$(38.7) \qquad \sup_{t \geq t_0} \| q(t) - p(t, b, t_0) \| \leq \varphi(\| q(t_0) - b \|; t_0).$$

The left side is periodic with respect to t_0 since, by (38.6),

$$q(t) - p(t, b, t_0 + \omega) = q(t) - p(t - \omega, b, t_0)$$
$$= q(t - \omega) - p(t - \omega, b, t_0).$$

Therefore, the comparison function $\varphi(r, t_0)$ of (36.7) can be chosen periodic with respect to t_0 whence

$$D(t, t_0, b) = \| q(t) - p(t, b, t_0) \| \leq \sup_{0 \leq u \leq \omega} \varphi(\| q(t_0) - b \|; u).$$

The right side is independent of t_0, and that means A 3-stability. In the same manner, we can start with

$$\| q(t) - p(t, b, t_0) \| \leq \sigma(t - t_0; b_0, t_0)$$

and obtain an estimate

$$D(t, t_0; b) = \| q(t) - p(t, b, t_0) \| \leq \sup_{0 \leq u \leq \omega} \sigma(t - t_0; b, u)$$

which is of type A 5.

Theorem 38.5. If the equilibrium of a differential equation $\dot{x} = f(x, t)$ ($f \in E$) with constant or periodic coefficients is stable then it is uniformly stable. If the equilibrium is asymptotically stable then it is uniformly asymptotically stable.

Proof. The first part of the theorem is contained in Theorem 38.4. To prove the second part we first use Theorem 38.3 to prove that A 6 holds and then apply Theorem 38.4 again to show that A 7 holds.

The natural surmise that a corresponding statement is valid for equations with almost periodic[1]) coefficients is not even true, in general, for scalar equations of the first order. This is shown by an equation given by Coppel [2], whose equilibrium is not uniformly attractive,[2])

$$(38.8) \qquad \dot{x} = - a(t) x, \ a(t) := \sum_{n=1}^{\infty} n^{-3/2} \sin(\pi t/n).$$

---

[1]) *cf.* sec. 73.

[2]) Conley and Miller [1] gave an example of a non-uniformly stable equilibrium.

For we have

$$\pi \int_0^t a(s)\, ds = \sum_{n=1}^\infty n^{-1/2} \left(1 - \cos (\pi t/n)\right)$$

$$= 2 \sum_{n=1}^\infty n^{-1/2} \sin^2 (\pi t/2 n) \geq 2 t^2 \sum_{n \geq t} n^{-5/2},$$

since for $0 \leq \varphi \leq \pi/2$, $\sin \varphi \geq 2\varphi/\pi$. Furthermore, for $t \geq 1$,

$$\pi \int_0^t a(s)\, ds > 2 t^2 \int_{t+1}^\infty u^{-5/2}\, du = (4/3)\, t^2 (t+1)^{-3/2} > \sqrt{2t}/3 .$$

On the other hand

$$\pi \int_0^t a(s)\, ds = 2 \left(\sum_{n \leq t} + \sum_{n > t} \right) n^{-1/2} \sin^2 (\pi t/2 n)$$

$$\leq 2 \sum_{n \leq t} n^{-1/2} + (\pi^2/2)\, t^2 \sum_{n > t} n^{-5/2} < 2 \int_0^t u^{-1/2}\, du + \frac{\pi^2 t^2}{2} \int_{[t]}^\infty u^{-5/2}\, du$$

$$= 4 \sqrt{t} + (\pi^2/3)\, t^2 [t]^{-3/2} < (4 + 2^{3/2}\, \pi^2/3) \sqrt{t}, \qquad \text{for} \quad t \geq 1 .$$

($[t]$ denotes the greatest integer $\leq t$.) Thus there exist two positive constants a_1 and a_2 such that

$$a_1 t^{1/2} \leq \int_0^t a(s)\, ds \leq a_2 t^{1/2},$$

so that we have

$$a_1 \sqrt{t_0} - a_2 \sqrt{t} \leq - \int_{t_0}^t a(s)\, ds \leq a_2 \sqrt{t_0} - a_1 \sqrt{t} .$$

Hence the solution of (38.7) satisfies an estimate

$$|x_0| \exp (a_1 \sqrt{t_0} - a_2 \sqrt{t}) < |p(t, x_0, t_0)| < |x_0| \exp (a_2 \sqrt{t_0} - a_1 \sqrt{t}) .$$

If the equilibrium were uniformly asymptotically stable, it would have to be exponentially stable by Theorem 58.5 (*cf.* sec. 58), *i.e.* an estimate

$$|p(t, x_0, t_0)| < a_3 |x_0| \exp (- a(t - t_0))$$

would be possible. This would imply the inequality

$$a_1 \sqrt{t_0} - a_2 \sqrt{t_0 + T} < - a_4 T , \quad \text{for arbitrary } t_0, T,$$

leading to a contradiction since the two sides are not of the same order of magnitude with respect to T.

Between the types of groups A and B the following connections exist.

Theorem 38.6. If the equilibrium is attractive of one of the types A 4 through A 7, then within the domain of attraction the system is attractive of type B 4 through B 7, respectively.

Proof. The hypothesis implies that for any two motions $p(t, a_1, t_0)$, $p(t, a_2, t_0)$ whose initial points lie in the domain of attraction, the inequalities

$$|p(t, a_1, t_0)| < \sigma(t - t_0; a_1, t_0),$$

$$|p(t, a_2, t_0)| < \sigma(t - t_0; a_2, t_0)$$

hold, where $\sigma(t - t_0; a, t_0) \in L$. Hence

$$d(t, t_0; a_1, a_2) < \sigma(t - t_0; a_1, t_0) + \sigma(t - t_0; a_2, t_0).$$

On the right we have a function of class L. According as we are concerned with type $A\,5$, $A\,6$, or $A\,7$, the appropriate secondary variables can be omitted on the right.

Theorem 38.7. If the right side of the differential equation of Theorem 38.3 satisfies a Lipschitz condition with a uniform Lipschitz constant and if the equilibrium is uniformly asymptotically stable ($A\,3 + A\,7$), then the system is stable of type $B\,3 + B\,7$.

Proof. By Theorem 38.6, $B\,7$ clearly holds. Assuming that the system is not $B\,3$-stable, there exists a $\bar{x}_0$, a sequence $x_{0n} \to \bar{x}_0$, a sequence t_{0n}, a sequence t_n, and an $\varepsilon > 0$ such that

$$d(t_{0n} + t_n, t_{0n}; x_{0n}, \bar{x}_0) \geq \varepsilon.$$

But because of the hypothesis (*cf.* p. XI)

$$d(t_{0n} + t_n, t_{0n}; x_{0n}, \bar{x}_0) < |x_{0n} - \bar{x}_0| \exp(K t_n),$$

where K depends upon the Lipschitz constant. The two inequalities are compatible only if the sequence t_n is unbounded. On the other hand, because of the $B\,7$-attractivity we have

$$d(t_{0n} + t, t_{0n}; a_1, a_2) < \varepsilon,$$

if t is chosen larger than a number T which is independent of t_0 and the initial points a_i. Thus the assumption leads to a contradiction: We have $B\,3$-stability.

39. Realizing Some Stability Types[1])

In this section the numerals I, II, III will denote the properties "unstable", $A\,2$, and $A\,3$, and the letters a through e will denote the five properties "not attractive", $A\,4$ through $A\,7$. Thus the combination $I\,e$ means unstable and at the same uniformly attractive, and Theorem 38.1 says that this combination cannot occur for families of motions which depend continuously on the initial values. We are interested in seeing which of the 15 combinations can actually be realized. This is clarified by means of examples. It will appear that apart from the combinations $I\,d$, $I\,e$, $III\,b$, $III\,c$, which are excluded by the theorems of sec. 38, all other

---

[1]) HAHN [6].

eleven combinations can be realized and indeed by differential equations of the first and second order. A few lemmas are necessary.

1) Let $f(t)$ be continuously differentiable and positive for $t > 0$. Let $\lim_{t \to \infty} f(t) = 0$, and let an estimate

$$(39.1) \qquad \frac{f(t+T)}{f(t)} \leq \sigma(T), \quad 0 < t, \ 0 \leq T_1 \leq T < \infty, \ \sigma \in L,$$

be given. Then there exists an estimate of the form

$$\frac{f(t+T)}{f(t)} \leq a e^{-bt}, \quad b > 0,$$

i.e. the comparison function is of exponential type.

Proof. Taking logarithms in (39.1) and utilizing the fact that $\lim f(t) = 0$, we obtain the inquality

$$\left| \ln f(t+T) \right| - \left| \ln f(t) \right| \geq \left| \ln \sigma(T) \right|$$

and hence applying the mean value theorem,

$$\frac{d}{dt} \left| \ln f(t) \right|_{t+\delta T} \geq \frac{\ln \sigma(T)}{T}, \quad 0 < \delta < 1.$$

The right side of the inequality is independent of t, the inequality can therefore be valid only if the left side remains larger then a fixed positive number for arbitrarily large values of t:

$$\frac{d}{dt} \left| \ln f(t) \right| \geq c > 0.$$

This implies

$$f(t) \leq e^{-ct},$$

which is our assertion.

If the quotient $f(t+T)/f(t)$ tends toward a positive value which is independent of T, as t increases, then an estimate (39.1) is not possible. Then the differential equation formed with $f(t)$

$$(39.2) \qquad \dot{x} = \frac{\dot{f}(t)}{f(t)} x$$

whose solution is

$$p(t, x_0, t_0) = \frac{f(t)}{f(t_0)} x_0$$

has an attractive equilibrium; but the attractivity is not uniform with respect to t_0. This is, for instance, the case for $f(t) = (1 + t)^{-1}$.

2) Let two continuous, positive, and continuously differentiable functions $h(t)$ and $k(t)$ be defined for $t > 0$. Let the function $k(t)$ have infinitely many maxima and minima at the points $t = t_i'$, resp. $t = t_i''$, $(t_i' < t_i'' < t_{i+1}' < t_{i+1}'')$, with the property that the sequence $k(t_i')$ of maxima is monotone increasing and unbounded whereas the sequence $k(t_i'')$ of minima tends to zero. Let the function $g(t) := h(t) k(t)$ be bound-

ed and tend to zero; and let the sequence $h(t'_{i+1})/h(t''_i)$ be bounded below by a fixed positive number $c > 0$. Then

$$\lim \frac{g(t)}{g(t_0)} = 0 \quad \text{for fixed } t_0, \text{ as } t \to \infty,$$

and

$$\lim_{i \to \infty} \frac{g(t'_{i+1})}{g(t''_i)} = \lim_{i \to \infty} \frac{h(t'_{i+1})\,k(t'_{i+1})}{h(t''_i)\,k(t''_i)} \geq c \lim \frac{k(t'_{i+1})}{k(t''_i)} = \infty.$$

We recognize that the equilibrium of the differential equation

$$\dot{x} = \frac{\dot{g}(t)}{g(t)}\, x$$

is stable and attractive but not uniformly stable: If we choose $t_{0n} = t''_n$, $t_n = t'_{n+1}$, then $p(t_{0n} + t_n, x_0, t_{0n})$ becomes arbitrarily large. (See also sec. 37. Since the differential equation is linear we can neglect the initial points x_0.) Examples are

$$h_1(t) = t^{-2}, \quad k_1(t) = t^{\cos t}$$

or

$$h_2(t) = e^{-t^2}, \quad k_2(t) = e^{2\pi t \cos t}.$$

(The function $g_2(t)$ is not exactly of the type characterized in the lemma.) The extrema occur at places approximated by

$$\tilde{t}'_i = 2i\pi, \quad \tilde{t}''_i = (2i+1)\pi$$

where

$$\frac{g_1(\tilde{t}'_{i+1})}{g_1(\tilde{t}''_i)} = \frac{(2i+1)^3}{2i+2}\,\pi^2; \qquad \frac{g_2(\tilde{t}'_{i+1})}{g_2(\tilde{t}''_i)} = e^{(4i+3)\pi^2}.$$

The function $g(t) = t^{-2+\cos t}$ belongs to the type considered under 1); the equilibrium of the differential equation formed with g is therefore neither uniformly stable nor uniformly attractive with respect to t_0. On the other hand, the equation formed with $g(t) = e^{-t^2+2\pi t \cos t}$ has an equilibrium which is uniformly attractive with respect to t_0. For we have

$$\exp(2\pi t \cos t - 2\pi t_0 \cos t_0 + t_0^2 - t^2) \leq \exp\big((t+t_0)(t_0-t+2\pi)\big) < \eta$$

in case

$$t - t_0 > |\ln \eta| + 2\pi.$$

Theorem 38.2 does not apply here since the right side of the differential equation formed with $g(t)$ does not satisfy a Lipschitz condition with uniform constant.

3) The function

$$(39.3) \qquad k(t, a, \alpha) := \frac{1+2at^2}{1+t+a^\alpha t^3}, \quad 0 \leq t, \quad 0 \leq a \leq 1, \quad \alpha \geq 1,$$

has the following properties:

a) $k(t, 0, \alpha) = 1/(1 + t)$ is monotone decreasing.

b) For $a \neq 0$ and for large t, k behaves asymptotically like $2a^{1-\alpha}t^{-1}$. Therefore $\lim\limits_{t\to\infty} k(t, a, \alpha) = 0$; but this limit is not uniform with respect to a.

The equation

$$k(t, a, \alpha) = 1$$

has in addition to the root $t_1 = 0$ two further roots t_2, t_3, which satisfy the equation

$$a^\alpha t^2 - 2at + 1 = 0.$$

They are not real for $\alpha < 2$. In case $\alpha = 2$ we have $t_2 = t_3 = a^{-1}$ and hence

$$k(t, a, \alpha) \leq 1 \quad \text{for} \quad 1 \leq \alpha \leq 2, \quad t \geq 0.$$

c) For $\alpha > 2$ we have

$$t_{2,3} = a^{1-\alpha} \pm a^{1-\alpha}\sqrt{1 - a^{\alpha-2}} \sim a^{1-\alpha}\left(1 \pm \left(1 - \tfrac{1}{2}a^{\alpha-2}\right)\right).$$

If a is very small then the roots behave like

$$2a^{1-\alpha} - \frac{1}{2a}, \quad \text{resp.} \ \frac{1}{2a},$$

and the value $t_4 = a^{-\alpha/2}$ lies between t_2 and t_3. Since for small values of a the functional value $k(t_4, a, \alpha)$ behaves like $a^{1-\alpha/2}$, it is clear that

$$\sup k(t, a, \alpha) \geq \frac{1}{2} a^{1-\alpha/2}, \quad \alpha > 2, \quad t \geq 0.$$

d) The function $k(t, a, 2)$ has an extremum which lies between $t = 0$ and $t = a^{-1}$. Since the derivative

$$\frac{\partial k(t, a, 2)}{\partial t} = (1 - at)\frac{-1 + 3at + 2at^2 + 2a^2t^3}{(1 + t + a^2 t^3)^2}$$

is negative for $t = 0$ the extremum is a minimum. Near $t_5 = a^{-1/2}$, $k(t, a, 2)$ behaves like $a^{1/2}$ for small values of a. The minimum clearly is even smaller.

We now form the system of equations

$$\text{(39.4)} \qquad \dot{r} = \frac{\dfrac{\partial}{\partial t}k(t,a,\alpha)}{k(t,a,\alpha)}\, r, \quad \dot{\varphi} = 0,$$

resp. the equivalent system

$$\text{(39.5)} \qquad \dot{x} = \frac{x}{k(t,a,\alpha)}\frac{\partial k(t,a,\alpha)}{\partial t}; \ \dot{y} = \frac{y}{k(t,a,\alpha)}\frac{\partial k(t,a,\alpha)}{\partial t},$$

and we set

$$a = a(\varphi) = \sin^2 \varphi \quad \text{or} \quad a = a(x, y) = \frac{y^2}{x^2 + y^2}.$$

For $x = y = 0$, $a(x, y)$ is not defined. But the right sides of (39.5) are continuous at $x = y = 0$ and the first order partial derivatives are bounded except at the origin. (39.5) therefore satisfies a Lipschitz condition. The solution of (39.4) is

$$r = r_0 \frac{k(t, a, \alpha)}{k(t_0, a, \alpha)}; \quad \varphi = \varphi_0; \quad a = \sin^2 \varphi_0.$$

It follows from properties a) and b) that the equilibrium is attractive but not uniformly attractive with respect to a and not uniformly attractive with respect to t_0.

If $\alpha > 2$ then the equilibrium is unstable. This is seen by choosing a sequence $\varphi_n \to 0$ (there corresponds to it a sequence $a_n \to 0$) and setting

$$r_{0n} = a_n^{\frac{\alpha}{2}-1}, \quad t_0 = 0, \quad t_n = a_n^{-\frac{\alpha}{2}}.$$

Then $r(t_n)$ is nearly equal to one and $\sup r(t)$ is certainly even larger, although the sequence r_{0n} tends to zero. If $\alpha = 2, t_{0n} = t_5, t_{0n} + t_n = a_n^{-1}$, then

$$r(t_{0n} + t_n) = r_0 \frac{k(a_n^{-1}, a_n, 2)}{k(t_5, a_n, 2)} = r_0 \frac{1}{k(t_5, a_n, 2)} = O\left(a_n^{-\frac{1}{2}}\right).$$

The expression is unbounded; the stability is not uniform.

If we form the differential equation for the function $k(e^t, a, \alpha)$, then because of what was said in 1), $r \to 0$ uniformly with respect to t_0; the remaining properties are unchanged.

The following table shows that actually eleven of the fifteen combinations mentioned in the beginning of this section can be realized.

	I	II	III
a)	$f = e^t$	$f = (1 + t)^{-\cos^2 t}$	$f = e^{\sin t}$
b)	$k(t, a, 3)$	$k(t, a, 2)$	—
c)	$k(e^t, a, 3)$	$k(e^t, a, 2)$	—
d)	—	$f = (1 + t)^{-2+\cos t}$	$f = (1 + t)^{-1}$
e)	—	$f = e^{-t^2+2\pi t\cos t}$	$f = e^{-t}$

The f's are to be used to form the differential equation (39.2) and the k's for (39.4).

If we replace the functions f, resp. k, of the table by their negatives then the time factors in the solutions x and r of the corresponding equations are replaced by their reciprocals. In this manner we obtain examples for the properties A_2', A_3', and A_8' through A_{11}', as well as for the negation of A_2' and A_8', and we can then read from the corresponding table, to what extent the properties of the types in this group are independent of each other. We further mention the solution of the scalar equation

$$\dot{x} = - \left((1 + t)^{-1} + 2\,t\,\cos^2 t - t^2 \sin 2\,t\right) x,$$

which is

$$p(t, x_0, t_0) = x_0 \,\frac{1 + t_0}{1 + t}\, \exp\left(t_0^2 \cos^2 t_0 - t^2 \cos^2 t\right).$$

It satisfies a two-sided estimate

$$\sigma_1(t - t_0; x_0, t_0) < |p(t, x_0, t_0)| < \sigma_2(t - t_0; x_0, t_0)$$

and hence represents the types A_4' and A_4. That the limits are not uniform with respect to t_0 is seen if we set $x_0 = 1$ and

$$t_{0k} = k\,\pi, \quad t = t_{0k} + m\,\pi,$$

resp.

$$t_{0k}' = (2\,k + 1)\,\frac{\pi}{2}, \quad t' = t_{0k}' + m\,\pi.$$

The functional value

$$p(t, 1, t_{0k}) = \frac{1 + t_{0k}}{1 + t}\, e^{-(2km + m^2)\pi^2}$$

cannot be estimated from below uniformly with respect to t_0, *i.e.* to k. It decreases too fast while $p(t', 1, t_{0k}') = \dfrac{1 + t_{0k}'}{1 + t'}$ decreases too slowly and therefore it cannot be estimated from above uniformly with respect to k.

40. An Example for Instability

Example I b of the table in sec. 39 shows that the properties *unstable* and *attractive* are not mutually exclusive. The differential equation is nonautonomous of order two. In sec. 13 an analogous example was given. The equation is autonomous but must be interpreted (as a differential difference equation) in a general space. The following example constructed by Vinograd [4] shows that the combination "unstable and attractive" can be realized even in an autonomous system of equations of second order; to be sure, this system is not linear. Let

$$(40.1) \quad \dot{x} = \frac{x^2(y - x) + y^5}{(x^2 + y^2)(1 + (x^2 + y^2)^2)}; \quad \dot{y} = \frac{y^2(y - 2x)}{(x^2 + y^2)(1 + (x^2 + y^2)^2)}.$$

The right sides are defined to be zero for $x = y = 0$. Then the Lipschitz condition is satisfied. If we set $r^2 = x^2 + y^2$ and $u = \tan \varphi$ then (40.1) can be written in the form

$$(40.2) \quad \frac{dr}{dt} = \frac{r}{(1 + r^4)(1 + u^2)^2}\,(u^4 - 2\,u^3 + u - 1 + u^3 r^2 \sin^2 \varphi).$$

The origin is the only singular point of the differential equation and it suffices to consider only the upper half-plane in the discussion of the trajectories, since the system is invariant under the transformation $x' = -x,\, y' = -y$.

Along the curves $y = 0$, $y = 2x$, and $y^5 + x^2 y - x^3 = 0$ the derivatives $\dot{x}$ and $\dot{y}$ change sign. The third curve is the isoclinal line Γ defined by $\dot{x} = 0$. It has a parametric representation

$$x = \left(\frac{1-\lambda}{\lambda^5}\right)^{1/2}, \qquad y = \lambda x = \left(\frac{1-\lambda}{\lambda^3}\right)^{1/2}, \qquad 1 \geq \lambda > 0,$$

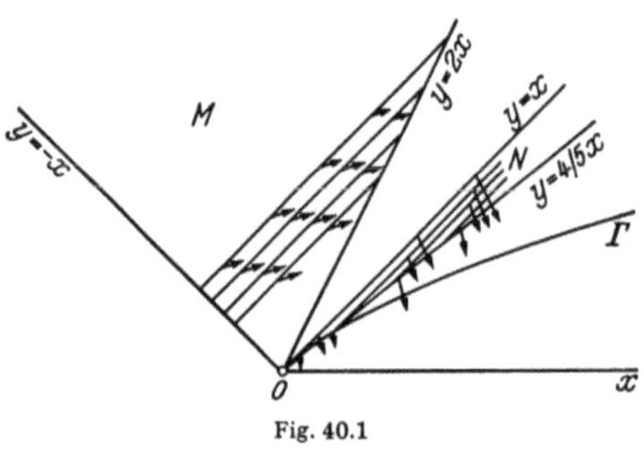

Fig. 40.1

which implies that this curve is monotone increasing with x but remains constantly below the line $y = x$.

Let M and N be the sectors of the plane formed by the rays $y = -x$ and $y = 2x$, resp. $y = x$ and $y = 4x/5$ (fig. 40.1). Also let $v := x - y$. In the numerator of $\dot{v} = \dot{x} - \dot{y}$ we have the polynomial

$$Z := x^2(y - x) + y^5 - y^2(y - 2x).$$

In the sector M we have $-y \leq x \leq y/2$ and hence

$$Z \geq 0 + y^5 - y^2(y + 2y) = y^5 - 3y^3 > 0 \quad \text{for} \quad y > \sqrt{3}.$$

In N we have $y \leq x \leq 5y/4$, and therefore

$$Z \geq y^2(2y - y) - \frac{25}{16} y^2 \left(\frac{5}{4} y - y\right) + y^5 = \frac{39}{64} y^3 + y^5 > 0.$$

The denominator of $\dot{v}$ is positive. This allows us to conclude as we did in the geometric interpretation of the direct method, that the trajectories in M and N intersect the family of curves $v = c$ from left to right.

Each trajectory which begins to the left of the isoclinal line Γ intersects that line after a finite time has elapsed. For if the initial point lies between the lines $y = 0$ and $y = -x$, then $\dot{x}$ and $\dot{y}$ are positive and the trajectory enters the sector M after a finite time. In M the trajectories for $y > \sqrt{3}$ move to the right toward the line $y = 2x$. Also, the trajectories remain for only a finite time span in the triangle formed by the lines $y = -x$, $y = 2x$, and $y = x + 2\sqrt{3}$ [of the lines $v = c$ this is the one which passes through the point $(-\sqrt{3}, +\sqrt{3})$], and similarly for the trajectories whose initial points are in the region between Γ and the line $y = 2x$.

The isoclinal line Γ meets the line $y = 4x/5$ in the point $(25/32, 5/8)$ $=: P$. Let S denote the sector between the ray OP and the x-axis. Each trajectory which begins on Γ enters S after a finite time since for points between Γ and the x-axis we have $\dot{x} < 0$ and $\dot{y} < 0$; thus the trajectories pass to the left.

Each trajectory which begins in the sector S tends toward the origin. This follows from (40.2), for on S we have

$$0 \leq u \leq \frac{4}{5}, \quad 0 \leq r \leq \frac{5\sqrt{41}}{32}, \quad 0 \leq y \leq \frac{5}{8},$$

$$u^3 r^2 \sin^2 q = u^3 y^2 \leq \left(\frac{4}{5}\right)^3 \left(\frac{5}{8}\right)^2 = \frac{1}{5}.$$

Let

$$f(u) := u^4 - 2u^3 + u - 1$$

be the polynomial appearing in (40.2). Its derivative $f'(u) = (2u - 1)(2u^2 - 2u - 1)$ vanishes only for

$$u_1 = \frac{1}{2}, \quad u_2 = \frac{1 + \sqrt{3}}{2}, \quad u_3 = \frac{1 - \sqrt{3}}{2}.$$

The interval $(0, 4/5)$ of the sector S contains only the value u_1, a maximum occurs there. In S, therefore, $f(u) \leq f(1/2) = -11/16$, *i.e.* the numerator of dr/dt is negative. In fact

$$\frac{dr}{dt} < -\frac{r}{15};$$

r tends exponentially to zero.

The arguments so far show that each trajectory will eventually tend toward the origin. In spite of this, the equilibrium is unstable. To see this we consider the triangle formed by the y-axis, the line $y = 3x$, and a line $y = a$ such that $a < 1/\sqrt{27}$ (fig. 40.3). Inside this triangle and on its edges $\dot{y} > 0$, so that the trajectories rise. The trajectories entering through the

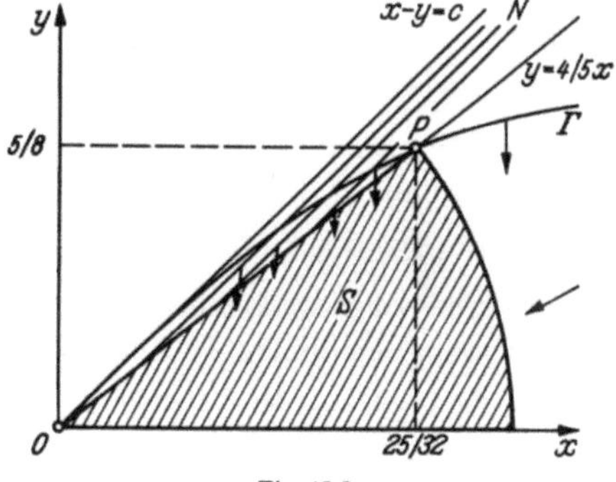

Fig. 40.2

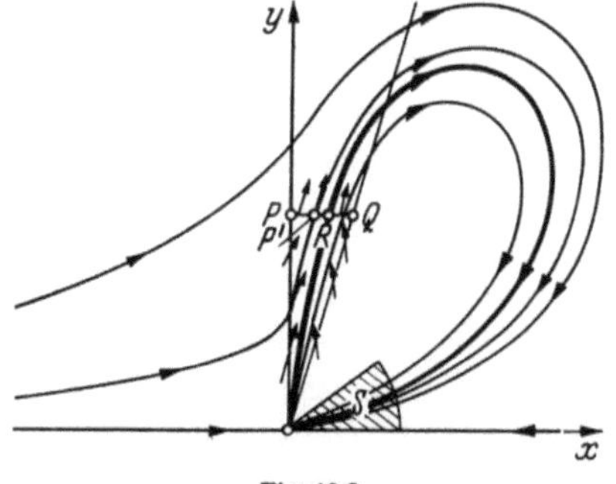

Fig. 40.3

y-axis go from left to right whereas the ray $y = 3x$ is cut upward, from below. Trajectories leaving the triangle pass through the horizontal side. On this side therefore there must be a point R with the following property: The trajectories which leave the triangle and pass through the horizontal line on the left of R have entered through the y-axis. Those leaving on the right of R have entered through the ray $y = 3x$ and not through the y-axis. The trajectory passing through R itself has come from the origin and returns to it as $t \to \infty$. It forms a loop. The tangents at the origin determine an elliptic sector (sec. 19) and the origin is unstable.

41. Liapunov Functions

In applying the direct method of Liapunov to general motions, we work, as we did in the case of differential equations in Euclidean space (*cf.* sec. 25), using scalar functions $v(x, t)$, which are, however, defined for arguments x in a normed linear space. We need functions with the following properties.

a) $v(x, t)$ is defined in a "half cylinder" K_{h,t_0}, *i.e.* for all x in a neighborhood of the origin, or even for all x in the space, as well as for all $t \geq t_0 \geq 0$.

b) $v(0, t) = 0$ for all $t \geq t_0$.

c) For a fixed t, $v(x, t)$ is continous with respect to $\|x\|$ and bounded in the domain of definition.

d) For fixed x, $v(x, t)$ is continuous with respect to t.

In place of a) we sometimes only need to have $v(x, t)$ defined for all x in a certain neighborhood of the origin and for the discrete values $t = t_0$, $t_0 + \tau$, $t_0 + 2\tau, \ldots$. In this case we must of course omit hypothesis d). Frequently we will also have to make certain differentiability conditions.

As in the definition of the general motion $p(t, a, t_0)$, we shall not use the term "functional" although frequently the distinction is made in the literature between Liapunov functions and Liapunov functionals, according as the argument x belongs to a Euclidean or a general space.

Def. 41.1. A function $v(x, t)$ is called *positive definite* if there exists a function $\varphi(r)$ of the class K such that

$$(41.1) \qquad\qquad v(x, t) \geq \varphi(\|x\|)$$

holds for all $t \geq t_0$ and all x belonging to a certain ball K_h.

In analogy to Def. 24.3a we call the function *positive definite in the domain B* if the inequality (41.1) holds for $x \in B$, $t \geq t_0$. The domain B must contain the origin at least as a boundary point. The function $v(x, t)$ is called *negative definite* if $-v(x, t)$ is positive definite.

Def. 41.2. The function $v(x, t)$ is called *positive (negative) semi-definite* in the domain B if it is non-negative (non-positive) there.

Def. 41.3. The function $v(x, t)$ is called *radially unbounded* if it is defined for all $x \in X$ and all $t \geq t_0$, if (41.1) holds for all x, and if the comparison function $\varphi(r)$ grows without bounds for $r \to \infty$.

If $v(x)$ is independent of t and $x \in R_n$, then according to sec. 24 B we can give an estimate from above. In the general case the existence of such an estimate must be especially required.

Def. 41.4. The function $v(x, t)$ is called *decrescent*, if there exists a function $\varphi(r)$ of class K such that in a neighborhood of the origin and for all $t \geq t_0$ we have

$$v(x, t) \leq \varphi(\|x\|).$$

Other expressions used to characterize the property of Def. 41.4 are "v admits an infinitely small upper bound" or "v becomes uniformly small".

Examples. Consider the following functions defined in R_2:

$$v_1 = x_1^2(1 + \sin^2 t) + x_2^2(1 + \cos^2 t),$$
$$v_2 = (x_1^2 + x_2^2) \sin^2 t,$$
$$v_3 = x_1^2 + (1 + t) x_2^2,$$
$$v_4 = x_1^2 + x_2^2/(1 + t),$$
$$v_5 = (x_1 - x_2)^2 (1 + t).$$

v_1 is positive definite and decrescent, v_2 is positive semidefinite and decrescent, v_3 is positive definite but not decrescent (the function can become arbitrarily large for arbitrarily small $(x_1, x_2) \neq (0, 0)$); v_4 is decrescent and always positive but not positive definite (the function can become arbitrarily small for fixed $x_2 \neq 0$); v_5 is positive semi-definite and not decrescent.

Theorem 41.1. A function $v(x, t)$ which has a bounded derivative with respect to x in K_{h,t_0} is decrescent.

Proof. We have $v(x, t) = v(x, t) - v(0, t)$. The right side can be estimated by means of the mean value theorem. Hence

$$|v(x, t)| \leq m \|x\|,$$

where m is a bound for the derivative with respect to x. The examples above show that the concept *semi-definite* is not very specific. It is therefore necessary to introduce some further concepts.

Def. 41.5. (*cf.* sec. 24 A). The function $v(x, t)$ is called *strongly positive definite*, respectively *weakly positive definite*, if there exist estimates

$$v(x, t) \geq k(\|x\|, t), \quad k \in KK,$$

13*

resp.

$$v(\boldsymbol{x}, t) \geq l(\|\boldsymbol{x}\|, t), \quad l \in KL.$$

Def. 41.6. The function $v(\boldsymbol{x}, t)$ is called *strongly decrescent*, respectively *weakly decrescent*, if there exist estimates

$$v(\boldsymbol{x}, t) \leq l(\|\boldsymbol{x}\|, t), \quad l \in KL,$$

resp.

$$v(\boldsymbol{x}, t) \leq k(\|\boldsymbol{x}\|, t), \quad k \in KK.$$

The function v_4 is strongly decrescent and weakly positive definite.

Remark. The properties *strongly positive definite* and *weakly decrescent* defined in Defs. 41.5 and 41.6 essentially correspond to properties which MASSERA [1] considered without naming them. The first of the two concepts is also found in a paper of S. K. PERSIDSKII [1].

Let a family of motions $\boldsymbol{p}(t, \boldsymbol{a}, t_0)$ in the sense of sec. 35 be given, which satisfies in addition conditions (36.1) and (36.2); also let a function $v(\boldsymbol{x}, t)$ be given. The expression

$$(41.2) \qquad\qquad v\big(\boldsymbol{p}(t, \boldsymbol{a}, t_0), t\big)$$

depends on t and the parameters $\boldsymbol{a}$ and t_0. Since in general we are interested only in its dependence on t it is customary to write $v(t)$ for (41.2) if there is no danger of confusion. By definition we then have $v(t_0) = v(\boldsymbol{a}, t_0)$.

Def. 41.7. The total derivative $\dot{v}(t) = dv(t)/dt$ of the expression (41.2) is called the *derivative of $v(\boldsymbol{x}, t)$ along the motion $\boldsymbol{p}(t, \boldsymbol{a}, t_0)$.* The analytic expression for this derivative is

$$Dv := \limsup_{k \to 0} \frac{1}{k} \left[v\big(\boldsymbol{p}(t + k, \boldsymbol{a}, t_0), t + k\big) - v\big(\boldsymbol{p}(t, \boldsymbol{a}, t_0), t\big) \right].$$

It is a function of t and $\boldsymbol{a}$. But (36.2) implies the relation $\boldsymbol{b} = \boldsymbol{p}(t_0, \boldsymbol{p}(t_1, \boldsymbol{b}, t_0), t_1)$, or $\boldsymbol{a} = \boldsymbol{p}(t_0, \boldsymbol{x}, t)$, (this says that if t_0 is sufficiently large the initial value $\boldsymbol{a}$ can be expressed in terms of the value $\boldsymbol{x}$ at the instant t). Dv can therefore be written as a function of $\boldsymbol{x}$. Because of (36.2) we have

$$\boldsymbol{p}(t + k, \boldsymbol{a}, t_0) = \boldsymbol{p}\big(t + k, \boldsymbol{p}(t_0, \boldsymbol{x}, t), t_0\big) = \boldsymbol{p}(t + k, \boldsymbol{x}, t)$$

and it follows that

$$Dv = \limsup_{k \to 0} \frac{1}{k} \left[v\big(\boldsymbol{p}(t + k, \boldsymbol{x}, t), t + k\big) - v(\boldsymbol{x}, t) \right].$$

If the family of motions is defined by means of the differential equation

$$(41.3) \qquad\qquad \dot{\boldsymbol{x}} = \boldsymbol{f}(\boldsymbol{x}, t)$$

then Dv can be given without first computing $\boldsymbol{p}(t, \boldsymbol{x}, t_0)$:

$$(41.4) \qquad Dv = \limsup_{k \to 0} \frac{1}{k} \left[v\big(\boldsymbol{x} + k\boldsymbol{f}(\boldsymbol{x}, t), t + k\big) - v(\boldsymbol{x}, t) \right].$$

If (41.3) is defined in R_n and if $v(x, t)$ has continuous first order partial derivatives with respect to $x_1, \ldots, x_n, t$, then

$$(41.5) \qquad Dv = \frac{\partial v}{\partial x_1} f_1(x, t) + \cdots + \frac{\partial v}{\partial x_n} f_n(x, t) + \frac{\partial v}{\partial t}.$$

This expression will also be called the *derivative of $v(x, t)$ for the equation* (41.3); in the autonomous case it is identical with the expression defined in sec. 25.

If the parameter t of the motion is discrete, $t = t_0 + \tau, t_0 + 2\tau, \ldots$ then Def. 41.7 does not make sense. In this case we define

$$(41.6) \quad Dv := \frac{1}{\tau} \left[v \big(p(t + \tau, a, t_0), t + \tau \big) - v \big(p(t, a, t_0), t \big) \right],$$

resp.

$$Dv := \frac{1}{\tau} \Big(v \big(p(t + \tau, x, t), t + \tau \big) - v(x, t) \Big),$$

and if the motion is given by a difference equation (*cf.* sec. 14)

$$(41.7) \qquad \theta \, x = f(x, t), \quad x \in R_n,$$

where $\theta \, x(t) = x(t + 1)$,

$$(41.8) \qquad Dv := v(\theta \, x, t + 1) - v(x, t).$$

Many of the theorems which we shall now formulate for Dv are valid for (41.3) and (41.5), resp. for (41.7) and (41.8).

Occasionally we need functions $v(x, y, t)$, depending on three arguments, where x and y are elements of the same normed linear space. The continuity properties are the same as defined for $v(x, t)$ above. In place of b) we put

$$v(x, x, t) = 0, \quad t \ge t_0 \ge 0.$$

In the definition of the concepts *definite* and *decrescent* $\|x - y\|$ is the argument of the comparison functions. The function is thus called positive definite if it is possible to find an estimate

$$v(x, y, t) \ge \varphi(\|x - y\|), \quad \varphi \in K,$$

etc.

If we substitute for x and y the expressions $p(t, a, t_0)$ and $p(t, b, t_0)$ then an expression analogous to (41.2) arises whose total derivative with respect to t, *the derivative of $v(x, y, t)$ along the two motions* has a definition similar to the above.

42. Tests for Stability

In the stability theorems of sec. 25 the Liapunov function $v(x)$ plays the rôle of a generalized distance. In a certain sense it measures the distance from the phase point to the origin. The derivative $\dot v$ describes the behavior of the distance as a function of time. The direct method is used in the same way to study the stability behavior of general motions. Again the function $v(x, t)$ is a type of distance and the derivative along the motion, *i.e.* the expression Dv, describes how this distance varies with t. Thus stability is discussed, as in the autonomous case, on the basis of the signs of certain scalar functions.

Henceforth we shall assume that the motions have the properties of secs. 35 and 36; the functions satisfy the hypotheses formulated in sec. 41. Let it further be assumed that the motions contain the equilibrium

$$(42.1) \qquad\qquad \boldsymbol{p}(t, 0, t_0) \equiv 0.$$

At first the stability theorems are formulated for the equilibrium $\boldsymbol{a} = 0$ so that $d(t, t_0; \boldsymbol{a}, \boldsymbol{b}) = \|\boldsymbol{p}(t, \boldsymbol{b}, t_0)\|$. If the motions are defined by means of differential equations, then the stability of an arbitrary motion can immediately be reduced to the stability of the origin by going over to the differential equation of the perturbed motion (sec. 35). By the *derivative of the function* we shall always mean the derivative along the motion in the sense of Def. 41.7. The various types of stability are those of sec. 36.

Theorem 42.1. If there exists a positive definite function $v(x, t)$ with a negative semi-definite derivative Dv, then the equilibrium is stable $(A\ 2)$; in fact this is the case for all initial times $t_0' \geq t_0$.
Proof. By hypothesis

$$v(x, t) \geq \varphi(\|x\|).$$

Since $Dv \leq 0$,

$$v(t_1) \geq v(t_2), \quad \text{for } t_1 \leq t_2,$$

and hence

$$(42.2) \qquad v(t_0) = v(\boldsymbol{a}, t_0) \geq v(\boldsymbol{x}, t) \geq \varphi(\|\boldsymbol{x}\|)$$

for $\boldsymbol{x} = \boldsymbol{p}(t, \boldsymbol{a}, t_0)$ and $t > t_0$. This implies

$$\|\boldsymbol{x}\| \leq \varphi^I(v(\boldsymbol{a}, t_0)).$$

We define the function

$$\sup v(\boldsymbol{b}, t_0), \quad \|\boldsymbol{b}\| \leq r,$$

which either belongs to class K or can be estimated above by a function $\chi(r; t_0) \in K$, and it follows that

$$\|\boldsymbol{x}\| \leq \varphi^I(\chi(\|\boldsymbol{a}\|, t_0)),$$

and this is an inequality of type $A\ 2$.

In the proof the condition that Dv is negative semi-definite is used only in the derivation of inequality (42.2). Therefore this condition can be replaced by the condition "$v(t)$ is (not strictly) monotone decreasing along each motion". However for the application of the direct method this replacement has no advantage; for it is the virtue of this method to make stability assertions without first gaining information on the motions $\boldsymbol{p}(t, \boldsymbol{a}, t_0)$.

Theorem 42.2. If there exists a positive definite, decrescent function $v(\boldsymbol{x}, t)$ with a negative semi-definite derivative then the equilibrium is uniformly stable (.4 3).

Proof. We now have the additional hypothesis

$$v(\boldsymbol{x}, t) \leq \psi(\|\boldsymbol{x}\|),$$

which implies

$$\ddot{v}(\boldsymbol{a}, t_0) \leq \psi(\|\boldsymbol{a}\|), \quad \varphi^I(v(\boldsymbol{a}, t_0)) \leq \varphi^I(\psi(\|\boldsymbol{a}\|)).$$

The previous proof therefore leads to the inequality

$$\|\boldsymbol{x}\| \leq \varphi^I(\psi(\|\boldsymbol{a}\|)),$$

whose right side is independent of t_0.

Theorem 42.3. If there exists a strongly positive definite function with negative semi-definite derivative then the equilibrium is equiasymptotically stable ($A\ 2 + A\ 6$).

Proof. As in the proof of Theorem 42.1 we see that

$$v(\boldsymbol{x}, t_0) \leq v(\boldsymbol{a}, t_0).$$

Furthermore

$$v(\boldsymbol{x}, t) \geq k(\|\boldsymbol{x}\|, t), \quad k \in KK,$$

which implies

$$\|\boldsymbol{x}\| = \|\boldsymbol{p}(t, \boldsymbol{a}, t_0)\| \leq k^I(v(\boldsymbol{a}, t_0), t) \leq k^I(\chi(\|\boldsymbol{a}\|; t_0), t).$$

The function on the right belongs to class KL (sec. 24, A k). For bounded $\|\boldsymbol{a}\|$ (possibly even for all $\boldsymbol{a}$) it can be estimated by a product in accordance with sec. 24 A i. It follows that

$$\|\boldsymbol{x}\| = \|\boldsymbol{p}(t, \boldsymbol{a}, t_0)\| \leq \varphi_1(\|\boldsymbol{a}\|; t_0)\, \sigma(t - t_0; t_0), \quad \varphi_1 \in K, \quad \sigma \in L,$$

for all $\boldsymbol{a}$ in a fixed domain B which may depend on t_0 and which may become arbitrarily small as t_0 increases.

Theorem 42.4. If there exists a positive definite decrescent function with negative definite derivative then the equilibrium is uniformly asymptotically stable ($A\ 3 + A\ 7$).

Proof. Theorem 42.2 assures uniform stability ($A\ 3$). As in the proof of Theorem 25.2 we derive an inequality

$$Dv \leq -\chi(v), \quad \chi \in K,$$

from the inequalities

$$\varphi_1(\|\pmb{x}\|) \leq v(\pmb{x}, t) \leq \varphi_2(\|\pmb{x}\|),$$
$$Dv \leq -\psi(\|\pmb{x}\|).$$

The only difference consists in using general norms $\|\pmb{x}\|$ instead of the vector norms. If t varies continuously the proof of Theorem 25.2 applies as it is.

If the auxiliary equation

$$Dw = -\chi(w)$$

which occurs in the proof is a difference equation then the inequality

$$\theta v = v(t+1) - v(t) \leq -\chi(v) < 0$$

implies that $v(t)$ is monotone decreasing and because of $v(t) \geq 0$, $\lim v(t) = c \geq 0$ exists. From the auxiliary equation $\chi(c) = 0$ follows and since $\chi \in K$ we have $c = 0$. The argument also applies in case zero is reached in a finite number of steps. The estimate is uniform with respect to t_0.

The following theorem supplements Theorem 42.4 in a similar way as Theorem 26.3 supplements Theorem 25.2.

Theorem 42.5. Let the motion be defined in R_n by a differential equation

$$\dot{\pmb{x}} = f(\pmb{x}, t), \quad f \in E, \quad \pmb{x} \in R_n.$$

Let there exist a Liapunov function satisfying the hypothesis of Theorem 42.4 in all of R_n, which is also radially unbounded. Then the equilibrium is uniformly asymptotically stable in the whole.

The similarity of the proofs of Theorems 25.2 and 42.4 is of course no accident. By Theorem 38.5 the asymptotic stability of autonomous motions is always uniform. Liapunov himself was not aware of the concept of uniform stability and had stated Theorem 42.4 more weakly, asserting asymptotic stability only. The concept *uniformly asymptotically stable* has been formulated clearly for the first time by MALKIN [4].

The underlying principle of the proof of Theorem 42.4 still applies if instead of insisting on the hypothesis "Dv negative definite" we replace it by an inequality of the form

$$Dv \leq h(t, v(t)), \quad h \text{ continuous}, \quad h(t, 0) \equiv 0.$$

In that case we must consider the auxiliary equation

$$Dw = h(t, w(t)).$$

If for instance the equilibrium of this equation is asymptotically stable then

$$w(t) \leq \varphi(w_0; t_0)\, \sigma(t - t_0; w_0, t_0)$$

and we obtain an estimate for v: If $v_0 = w_0$ then

$$v(\pmb{x},\, t) \leq \varphi(v_0;\, t_0)\, \sigma(t - t_0;\, v_0,\, t_0)$$

and since

$$\|\pmb{x}\| \leq \varphi_1^I\big(v(\pmb{x},\, t)\big),\quad v_0 \leq \varphi_2(\|\pmb{x}_0\|),$$

we obtain an estimate for $\|\pmb{x}\|$ which implies the asymptotic stability of the equilibrium[1]).

Of course we do not obtain in this manner conditions for asymptotic stability without simultaneously assuring their uniformity with respect to the "spatial" coordinates, since the initial estimates for $v(\pmb{x},\, t)$ are a priori uniformly valid in a certain $\pmb{x}$-domain and accordingly the final estimate is always made in terms of the norm $\|\pmb{x}_0\|$, resp. $\|\pmb{a}\|$ (cf. also sec. 54).

The rôle played by the estimate for Dv in proving asymptotic stability becomes even clearer if we look at LIAPUNOV's original proof for Theorem 42.4[2]). He first infers stability of the equilibrium from Theorem 42.1 and then concludes that a δ can be chosen such that the motions remain in a preassigned domain K_{h,t_0} for $\|\pmb{a}\| < \delta$. The hypothesis further implies that $v(t)$ decreases throughout and therefore has a non-negative limit v_∞. We now argue indirectly: If $v_\infty > 0$ then we would have $\varphi_2(\|\pmb{p}(t,\, \pmb{a},\, t_0)\|) \geq v_\infty > 0$ for all t, and furthermore $\|\pmb{p}(t,\, \pmb{a},\, t_0)\| \geq \varphi_2^I(v_\infty) =: p_0$. We integrate (resp. sum) the inequality for Dv and obtain

$$v(t) = v(t_0) + \int_{t_0}^{t} Dv\, dt \leq v(t_0) - \psi(p_0)\,(t - t_0).$$

This leads to a contradiction since $v(t)$ is always positive. This line of reasoning shows that the inequality

$$Dv \leq -\,\psi(\|\pmb{x}\|)$$

can also be replaced by

$$Dv \leq -\,g(t)\,\psi(\|\pmb{x}\|),\quad \psi \in K,$$

provided the integral $\int_{t_0}^{\infty} g(u)\, du$ diverges.

We now prove the two main theorems on the instability of the equilibrium.

Theorem 42.6. Let $v(\pmb{x},\, t)$ be a function with the following properties:

a) For each $\varepsilon > 0$ and for each $t \geq t_0$ there exist points $\bar{\pmb{x}}$ such that $v(\bar{\pmb{x}},\, t) < 0$ and $\|\bar{\pmb{x}}\| < \varepsilon$. The set of all points $(\pmb{x},\, t)$ such that $\|\pmb{x}\| < h$ and $v(\pmb{x},\, t) < 0$ shall be called the "domain $v < 0$". It is bounded by

---

[1]) CORDUNEANU [2].

[2]) LIAPUNOV [1]; cf. also for instance MALKIN [3].

the hypersurfaces $||x|| = h$ and $v = 0$ and may consist of several component domains.

b) In at least one of the component domains G of the domain $v < 0$, v is bounded from below.

c) In this domain G, $Dv \leq - \varphi(v)$, $\varphi \in K$.

Then the equilibrium is unstable.

Proof. Let $(a, t_0') \in G$. Then $v(a, t_0') = v(t_0') = : -\beta$ is negative, and likewise Dv is negative at the point (a, t_0'). Because of c), v is decreasing along the motion $p(t, a, t_0')$. On the other hand we obtain by integration, respectively summation,

$$v(t) = v(t_0') + \int_{t_0'}^{t} Dv \, dt$$

or

$$v(t) = v(t_0') + \sum_{n=1}^{N} Dv \big|_{t=t_0'+n\tau}$$

and further

(42.3) $\qquad v(t) \leq - \beta - \varphi(\beta) \, (t - t_0)$, resp. $\leq - \beta - N \varphi(\beta)$,

and because of b), the motion must leave the domain G. This can only happen at a place where $||x|| = h$ since at the other boundary points $v = 0$. Hence there exists a t' such that $||p(t', a, t_0')|| = h$, and since a may lie arbitrarily close to zero this implies the instability.

If we replace a) by the weaker condition "for each $\varepsilon > 0$ there exist points $\bar{x}$ such that $||\bar{x}|| < \varepsilon$ and such that $v(\bar{x}, t_0) < 0$" then instability can be inferred only for $t = t_0$. In the autonomous case Theorem 42.6 is identical with Theorem 25.3.

The assumption on Dv is only used to prove that the integral, respectively the series, diverges or at least becomes sufficiently large. It can therefore be replaced by any other condition which implies the divergence of the integral, e.g. by

$$Dv \leq - g(t) \varphi(v), \quad \int^{\infty} g(u) \, du \text{ divergent.}$$

A small modification of the hypotheses furnishes sufficient conditions for non-uniform stability. If we add to the first sentence in a) the statement "provided t_0 is larger then a number $T(\varepsilon)$ depending on ε" then we again obtain inequality (42.3) but must expect that t_0' becomes arbitrarily large as a approaches the origin. If $T(\varepsilon)$ is bounded we have instability. If $T(\varepsilon)$ is unbounded then for a fixed t_0' exceptional values a cannot be found arbitrarily close to the origin. If, in addition, G is a neighborhood of the origin, then the equilibrium is stable but not uniformly stable.

Theorem 42.7. Let there exist a bounded function $v(x, t)$ in the domain K_{h,t_0} with the following properties:

a)
$$Dv = gv + w(x, t),$$

where g is a positive constant and $w(x, t)$ is either identically zero or semi-definite; b) if w is not identically zero then in each domain K_{h_1,t_1} with arbitrarily large t_1 and arbitrarily small h_1 there exist x-values such that $v(x, t)$ and $w(x, t)$ have the same sign for $t \geq t_1$.

Then the equilibrium is unstable.

The proof is similar to that of Theorem 25.5, to be modified for discretely varying t in the manner of the preceding proof. We also note that in Theorems 42.6 and 42.7 it need not be assumed that the functions $v(x, t)$ vanish at the origin.

As was the case for the stability theorems, the condition on Dv can be replaced by an inequality $Dv \geq h(t, v(t))$ if it is known that the equilibrium of the auxiliary scalar equation $Dv = h(t, v(t))$ is unstable.

A sufficient condition for stability for sufficiently large t_0 is given by

Theorem 42.8. Let $v(x, t)$ be positive definite and decrescent, $Dv \leq - \chi(t)$, and let $\chi(t)$ be such that

$$\limsup_{t_0 \to \infty} \left(\sup_{a \geq 0} \int_{t_0}^{t_0+a} \chi(t)\, dt \right)$$

is zero. Then there exists a t_0' such that the equilibrium is stable for $t_0 \geq t_0'$[1]).

Proof. Assuming the statement false, we would have three sequences $t_{0n} \to \infty$, $t_n \geq t_{0n}$, $x_{0n} \to 0$, such that $\|p(t_n, x_{0n}, t_{0n})\| \geq \varepsilon$, and so

$$v(t_n) := v\big(p(t_n, x_{0n}, t_{0n}), t_n\big) \geq \varphi(\varepsilon).$$

On the other hand we have (see abvoe)

$$v(t_n) = v(t_{0n}) + \int_{t_{0n}}^{t_n} Dv\, dt \leq v(t_{0n}) - \int_{t_{0n}}^{t_n} \chi(t)\, dt.$$

However, since $x_{0n} \to 0$, $v(t_{0n})$ becomes arbitrarily small, and the integral becomes arbitrarily small as t_{0n} increases: We obtain a contradiction.

Let $\dot{x} = f(x, t)$ be a differential equation in which x belongs to a Hilbert space. Assume that an estimate of the inner product $(x, f(x, t))$ is given,

$$\big(x, f(x, t)\big) \leq \|x\|\, \varphi(\|x\|)\, g(t),$$

such that $\varphi(r) \in K$ and $g(t)$ is continuous for $t \geq 0$. Suppose also that the integral

$$\int_{0+} \frac{dr}{\varphi(r)}$$

---

[1]) LaSalle and Rath [1].

is divergent. The solution of the differential equation $\chi'(r)\,\varphi(r) = \chi(r)$ is

$$\chi(r) = \exp\left(\int_c^r \frac{ds}{\varphi(s)}\right), \quad 0 < c,$$

for $r > 0$. If we set $\chi(0) = 0$ then χ belongs to class K. We select the function $\chi(\|\boldsymbol{x}\|) =: v(\boldsymbol{x})$ as the Liapunov function. Then

$$Dv = \chi'(\|\boldsymbol{x}\|)\,\frac{d}{dt}\,\|\boldsymbol{x}\| = \chi'(\|\boldsymbol{x}\|)\,\frac{(\boldsymbol{x},\dot{\boldsymbol{x}})}{\|\boldsymbol{x}\|} = \frac{\chi(\|\boldsymbol{x}\|)\,(\boldsymbol{x},\dot{\boldsymbol{x}})}{\varphi(\|\boldsymbol{x}\|)\,\|\boldsymbol{x}\|}.$$

Applying the estimate for the inner product we obtain

$$Dv \leq \chi(\|\boldsymbol{x}\|)\,g(t) = v(t)\,g(t),$$

$$v \leq v_0 \exp\left(\int_{t_0}^t g(u)\,du\right),$$

$$\|\boldsymbol{x}\| \leq \chi^I\left(\chi(\|\boldsymbol{x}_0\|)\exp\left(\int_{t_0}^t g(u)\,du\right)\right).$$

The stability behavior depends therefore only on the function $g(t)$[1]. Further and in part stronger instability theorems are mentioned in sec. 45.

Without a detailed discussion we also remark that a stability theory for the motions determined by random processes which we mentioned in sec. 36f, can be constructed with the help of Liapunov functions; it is formally very similar to the theory for the usual motions. The derivative defined in Def. 41.7 becomes a stochastic variable and must be replaced in the statements of the criteria by its expectation or a similar expression [2]).

43. Applications and Examples

I. Differential and Difference Equations

A. For the scalar system

$$\dot{x} = a(t)\,y + b(t)\,x\,(x^2 + y^2), \quad \dot{y} = -a(t)\,x + b(t)\,y\,(x^2 + y^2)$$

a Liapunov function is given by

$$v(x, y) = x^2 + y^2, \quad \dot{v} = 2\,b(t)\,(x^2 + y^2)^2.$$

The equilibrium is stable if $b(t) \leq 0$ and unstable if $b(t) > 0$. The condition $b(t) \leq q < 0$ is sufficient for (uniform) asymptotic stability. See also sec. 45.

---

[1]) Corduneanu [3].   [2]) Kushner [1].

B. For the matrix differential equation

$$\dot{X} = A X, \quad X = (x_{ij}),$$

a Liapunov function is given by

$$v = \operatorname{Tr} (X^T B X).$$

B is the matrix defined by (27.3). Obviously v is a sum of quadratic forms involving the matrix B. Therefore

$$c_1 \sum_{i,j} x_{ij}^2 \le v \le c_2 \sum_{i,j} x_{ij}^2,$$

the two constants c_1 and c_2 depending on the matrix B. Also

$$\dot{v} = \operatorname{Tr} (- X^T C X) \le - \|C\| \sum_{i,j} x_{ij}^2.$$

These are the inequalities in the hypothesis of Theorem 42.4 if we introduce the norm $\|X\|^2 := \sum_{i,j} x_{ij}^2$. The equilibrium is stable or unstable depending on the type of definiteness of B, *i.e.* it depends on the sign of the real parts of the characteristic roots of A.

C. Difference equations in R_n. Once more we consider the difference equation

$$(43.1) \qquad \theta x = A x, \quad x \in R_n,$$

of sec. 14, and as in sec. 27 we seek a Liapunov function of the form

$$(43.2) \qquad v(x) = x^T B x.$$

According to (41.8), the derivative of this function for (43.1) is

$$Dv = \theta v - v = x^T (A^T B A - B) x.$$

To obtain definiteness we set

$$(43.3) \qquad A^T B A - B = - C, \quad C \text{ positive definite,}$$

and compute the matrix B. As in sec. 27 we first put the matrix A into its Jordan form by means of a similarity transformation; this subjects B and C to a congruence transformation. On the diagonal of the transformed matrix A we have the characteristic roots $\alpha_1, \alpha_2, ..., \alpha_n$. The coefficients of the transformed system of equations involve the factors $(\alpha_i \alpha_j - 1)$ (if the Jordan form is not a diagonal matrix additional terms appear) and we can be sure that the system has a solution only if $\alpha_i \alpha_j \neq 1$ for all i, j. This is in particular the case if all the characteristic roots of A have absolute value less than 1. The type of definiteness of B can be examined in a purely algebraic way. However, we shall base our discussion, as we did in sec. 27, on the results of sec. 14. If all the characteristic roots have absolute values less than 1, then B must be positive definite; if at least one of the characteristic roots has absolute value greater than one, then B is indefinite or negative definite. A critical case occurs if no characteristic root has absolute value greater than one but some of the characteristic roots actually have absolute value equal to one. In that case we can carry out the stability discussion with appropriate Liapunov functions, as was done in sec. 27; we shall not do so here.

The equilibrium of the *perturbed* difference equation

$$(43.4) \qquad \theta x = A x + f(x, t), \quad f(x, t) = o(\|x\|),$$

has the same stability behavior as that of the unperturbed equation, if either all the characteristic roots of A have absolute value less than one, or if at least one characteristic root occurs whose absolute value is greater than one.

The principle of the Stability in the First Approximation is thus also valid for difference equations with constant linear part. This is immediately seen from the proof of Theorem 28.1 which, after all, was based on the mere fact that for the unperturbed and the perturbed system the same Liapunov function could be utilized. In critical cases the principle fails, as it does for differential equations.

For the more general autonomous equation

$$(43.5) \qquad \theta x = f(x),$$

we have

$$Dv = \theta v - v = v\big(f(x)\big) - v(x).$$

We recognize that the condition

$$\theta v \leq \alpha v, \quad 0 < \alpha \leq \alpha_1 < 1, \quad x \in R_n,$$

is sufficient for global asymptotic stability, for it guarantees the exponential fading of $v(x)$ and hence of $p(t, x_0, t_0)$ for all initial values.

The formal similarity between differential equations and difference equations in R_n permits us to apply the methods developed for differential equations for the estimation of the domain of stability and of the nonlinearities, to difference equations as well. If for instance an equation

$$\theta x = A x + g(x)$$

is given, whose nonlinearity admits a linear estimate (28.5) then the admissible bounds α, β can be estimated as in sec. 28: We work with the subsidiary equation $\theta x = (A + G) x$ and with the function (43.2) as determined by (43.3). The estimates are deduced from the condition

$$A^T B G + G^T B A + G^T B G - C \text{ negative definite}$$

(*cf.* an example explicitly worked out in KODAMA [1]). The condition can be weakened somewhat: It suffices to require that

$$\theta^{r_i+1} v \leq \alpha \, \theta^{r_i} v, \qquad 0 < \alpha \leq \alpha_1 < 1,$$

holds for a sequence

$$r_1, r_2, \ldots$$

such that the differences $r_{i+1} - r_i$ are bounded.

Theorems 34.1 and 34.3 have analogues formulated for equation (43.5):

Theorem 43.1. Let two functions $v(x)$, $h(x)$ with the following properties be given: 1) $v(x)$ is continuous and positive definite and satisfies the inequality $0 < v(x) < 1$ $(x \neq 0)$ in a simply connected domain A which contains a neighborhood of the origin. 2) $h(x)$ is defined for all finite x, is continuous and positive for $x_0 \neq 0$. 3) The condition

$$(43.6) \qquad \theta v - v = - h(x)(1 - v)$$

holds for $x \in A$. 4) As x approaches a boundary point of A, respectively as $x \to \infty$, $\lim v(x) = 1$.

Then A is exactly the domain of attraction of the origin. Furthermore, if two functions with the properties 2) and 3) are given and if

$$v\big(f(x)\big) - v(x) = - h(x)\big(1 - v(x)\big),$$

then $v(x) = 1$ is exactly the boundary of the domain of attraction.

The proof proceeds as in sec. 34. From (43.6) we obtain the relation

$$(1 - \theta v)/(1 - v) = 1 + h$$

and from it by summation of logarithms

$$\ln\big(1 - v(x(N))\big) - \ln\big(1 - v(x(n))\big) = \sum_{k=n}^{N-1} \ln\big(1 + h(x(k))\big).$$

If $x(n)$ lies in the domain of attraction then $\lim x(N) = 0$, $\lim v(x(N)) = 0$; the right side must converge and $x(n) \in A$ follows. If $x(n)$ does not lie in the domain of attraction then the right side is clearly not convergent, nor is the left side, and the hypothesis $1 - v(x(N)) < 1$ is not satisfied.

Example:

$$\theta x_1 = x_1^2 - x_2^2,$$

$$\theta x_2 = 2 x_1 x_2.$$

Let $h(x) = x_1^2 + x_2^2$, so that $\theta(1 + h) = 1 + (x_1^2 - x_2^2)^2 + 4 x_1^2 x_2^2 = 1 + (x_1^2 + x_2^2)^2$. In case of convergence the above formula yields

$$\frac{1}{1 - v(x(n))} = \prod_{r=1}^{\infty}\big(1 + \theta^{r-1} h(x(n))\big)$$

$$= \prod_{r=1}^{\infty}\big(1 + (x_1^2(n) + x_2^2(n))^{2^{r-1}}\big) = \frac{1}{1 - x_1^2(n) - x_2^2(n)}$$

and therefore in this example

$$v\big(\pmb{x}(n)\big) = x_1^2(n) + x_2^2(n),$$
$$v(\pmb{x}) = x_1^2 + x_2^2.$$

The domain of attraction is the unit circle[1].

Unfortunately the difference equation can be solved explicitly only in exceptional cases.

44. Applications and Examples

II. Functional and Partial Differential Equations

A. Functional differential equations. In sec. 12 we briefly dealt with a special class of functional differential equations, namely the differential difference equations with constant coefficients and constant delay terms of the form (12.6), as well as the somewhat more general equations (12.7). For these equations the general solution depends on an arbitrary function defined on a finite interval, respectively on an n-tuple of such functions, the initial functions or initial function vectors respectively. This is easily seen by integrating the original equation, progressing from interval to interval (see sec. 12). Because of the initial functions, the general solution cannot be represented like that of a differential equation, in a finite dimensional phase space; the geometric interpretation must take place in a suitable, more general space. The same is true for more general equations on whose right side the time t appears explicitly in the coefficients or in the delay terms, or in which the linear expression in (12.7) is replaced by a nonlinear expression.

In sec. 12 we briefly sketched a concrete transfer system which must be described by means of a differential difference equation because it has a time lag. The general characterization (3.1) of the transfer unit, the relation

$$(44.1) \qquad\qquad x_O = \Re\, x_I,$$

is a functional differential equation if the value of the output variable x_O at the instant t depends not only on the value of the input (resp. its derivatives) at the instant t but also on the values at earlier times. These may be discrete or a continuous variable on an interval. Strictly speaking, the description of physical processes by a differential equation always constitutes an idealization; for since all processes have only a finite velocity the response on the left of (44.1) will always be observed somewhat later then its cause which appears on the right. Usually, of course, this time difference is so small that it may be neglected. If this is not the case

---

[1] O'SHEA [1].

we simply have to introduce more general equations, such as functional differential equations. There also exist transfer units which are adequately described by partial differential equations. A very simple transfer unit of this kind is a long pipe filled with a liquid which is closed at both ends by an elastic membrane. The input consists of a displacement of one of the membranes and the output of the displacement of the other membrane caused by it. The equation for the motion is the wave equation with boundary conditions. If a transfer system contains units of this kind we speak of a system with *distributed parameters*[1]).

It stands to reason that from a practical point of view one would examine the same questions for transfer units with a delay or for systems with distributed parameters, as for the simple systems described by autonomous linear differential equations. So, especially questions concerning stability or asymptotic stability of the equilibrium, the domain of attraction, the influence of small disturbances, etc., are of interest and very recently the direct method has been successfully applied here.

For general functional differential equations the relation (36.2), which must be required if we wish to apply the direct method, is not necessarily guaranteed, and it can happen that the stability for t_0 does not necessarily imply the stability for $t_0' > t_0$.

Example (KRASOVSKII [3]). For $t_0 = 0$, the general solution of the scalar equation

$$(44.2) \qquad \dot{x} = x(t) - 2\,e^{-h(t)}\,x\big(t - h(t)\big),$$

$$h(t) = \begin{cases} \dfrac{1}{2}\,t, & 0 \le t \le 2, \\ 1, & t \ge 2, \end{cases}$$

is

$$x(t) = x_0 e^{-t}.$$

The trivial solution is therefore stable and attractive. For $t_0 \ge 2$ the equation assumes the form

$$\dot{x} = x(t) - 2e^{-1}\,x(t - 1).$$

Its trivial solution is unstable for all later initial times (see sec. 12). It is easy to see the reason for this behavior: The space of initial values for $t_0 = 0$ is the real axis, that for $t_0 \ge 2$ is the space of continuous functions on an interval of length one.

We can define sufficiently general functional differential equations which admit a reasonable stability theory, as follows[2]). Let $C([\alpha, \beta], R_n)$ denote the space of all continuous mappings of the real interval $[\alpha, \beta]$

---

[1]) WANG and TUNG [1]. [2]) HALE [1 to 3].

into the linear space of n-vectors. Let the norm of $\varphi \in C\left([\alpha, \beta], R_n\right)$ be defined by

$$(44.3) \qquad \|\varphi\| = \sup |\varphi(\theta)|, \quad \alpha \le \theta \le \beta.$$

If furthermore $x(u)$ is a continuous n-vector defined for $-r \le u \; (r > 0)$ then $x_t(.)$ shall denote the mapping of the interval $[-r, 0]$ into R_n defined by means of the function $x(t + \theta), -r \le \theta \le 0$. Then $x_t(.) \in C\left([-r, 0], R_n\right)$; t is to be considered as a parameter. Finally, let $C_h\left([-r, 0], R_n\right)$ be the subset of $C\left([-r, 0], R_n\right)$ the norm of whose elements is bounded by h, and $f(\varphi, t)$ a function defined for all $\varphi \in C_h\left([-r, 0], R_n\right)$ and for all $0 \le t < \infty$. The equations under consideration can then be written in the form

$$(44.4) \qquad \dot{x}(t) = f(x_t, t).$$

A function $p(t, \varphi, t_0)$ is called a *solution* with initial function vector $\varphi \in C_h\left([-r, 0], R_n\right)$, and initial point t_0, $t_0 \ge 0$ if 1) for each $t > t_0$, $p_t(., \varphi, t_0)$ belongs to $C_h\left([-r, 0], R_n\right)$, 2) $p_{t_0}(., \varphi, t_0) = \varphi$, and 3) (44.4) is valid for all $t > t_0$. In the linear autonomous case f is independent of t and is of the form

$$f(\varphi) = \int\limits_{-\tau}^{0} dH(\theta)\, \varphi(\theta), \quad H(\theta) \text{ a matrix.}$$

We will not take up the question here, what conditions on f guarantee the uniqueness and existence of the solutions[1]), but will instead assume suitable hypotheses, as for instance a Lipschitz condition for f. The general results of secs. 35, 36, 41, and 42 apply immediately. By means of the direct method sufficient stability criteria (and in part also necessary criteria, *cf.* Chapter VI) can be given. It is, however, in general quite difficult to construct suitable Liapunov functions explicitly.

In many cases the criteria of the direct method are too restrictive. It has been tried to extend their applicability by weakening the hypotheses. So, for example, it is not necessary to require that the functions v and $\dot{v}$ be definite in the whole set $C\left([-r, 0], R_n\right)$ if the $p_t(., \varphi, t_0)$, which are being examined, are already known to lie in a subset of C. The next two theorems are based on this idea. For the first of them we need the additional norms[2]) defined for $\varphi \in C\left([-r, 0], R_n\right)$

$$(44.5) \quad \|\varphi\|_1 := \sup |\varphi(\theta)|, \quad -r \le r_1 \le \theta \le 0, \; r_1 \text{ fixed,}$$

$$\|\varphi\|_2 := \sup |\varphi(\theta)|, \quad -r \le -r_2 \le \theta \le -r_2' \le 0, \; r_2, r_2' \text{ fixed.}$$

---

[1]) *cf.* for instance DRIVER [1], HALANAY [1], PINNEY [1].
[2]) KRASOVSKII [4].

Theorem 44.1. Let

$$\psi_1(\|x\|_1) \leq v(x(\cdot), t) \leq \psi_2(\|x\|),$$

$$\dot{v} \leq -\psi_3(\|x\|_2),$$

where $\psi_i \in K$, $i = 1, 2, 3$. If the right side of (44.4) is bounded then the equilibrium of (44.4). is asymptotically stable.

Proof. Let ε be given and choose the number $\delta \leq \varepsilon$ so small that $\psi_2(\delta) < \psi_1(\varepsilon)$. Then for $\|\varphi\| < \delta$ we have

$$v(\varphi, t_0) < \psi_1(\varepsilon),$$

and since $\dot{v} \leq 0$,

$$v(p_t(\cdot, \varphi, t_0), t) \leq \psi_1(\varepsilon),$$

and hence

$$\|p_t\|_1 \leq \varepsilon.$$

This proves the stability of the equilibrium. If the equilibrium were not asymptotically stable there would exist for each number $\eta > 0$ a φ and an unbounded sequence t_n, $n = 1, 2, \ldots$, such that

$$(44.6) \qquad |p(t_n, \varphi, t_0)| \geq \eta.$$

If we choose a sequence t'_n so that

$$-r_2 \leq t_n - t'_n \leq -r'_2$$

$[r_2$ and r'_2 are introduced in (44.5)], then (44.6) is equivalent to

$$(44.7) \qquad |p_{t'_n}(\cdot, \varphi, t_0)|_2 \geq \eta.$$

Now $\dot{x}$ is bounded because of the hypothesis on f. Therefore there exists a number $\gamma > 0$ such that we can derive from (44.7) the inequality

$$\|p_t(\cdot, \varphi, t_0)\|_2 > \frac{1}{2}\eta, \quad t'_n - \frac{1}{2}\gamma \leq t \leq t'_n + \frac{1}{2}\gamma.$$

It follows that

$$\dot{v} \leq -\psi_3\left(\frac{\eta}{2}\right), \quad \text{for } t'_n - \frac{1}{2}\gamma \leq t \leq t'_n + \frac{1}{2}\gamma,$$

and further

$$v = v_0 + \int_{t_0}^{t'_n} \dot{v}\, dt \leq v_0 - \psi_3\left(\frac{\eta}{2}\right) n\gamma.$$

Letting n increase, we obtain a contradiction to the definiteness of v.

Theorem 44.2. Let

$$\psi_1(\|x\|) \leq v(x(\cdot), t) \leq \psi_2(\|x\|), \quad \psi_i(r) \in K,$$

14*

be given. Let H be the subset of $C([-r, 0], R_n)$ of all elements $y(t)$ for which the form $v(y(t), t)$, considered as a function of t, is monotone decreasing. Let

$$\dot{v}(y(\cdot), t) \leq - \psi_3(\|y\|), \quad \text{for} \quad y \in H, \ \psi_3 \in K.$$

Then the equilibrium of (44.4) is asymptotically stable.

The proof follows from the remark preceding (44.5). To be able to show the asymptotic stability we need only know that $\dot{v}$ is negative definite with respect to the solutions, $i.e.$ to the elements p_t of $C([-r, 0], R_n)$. These elements however are contained in H by hypothesis and for $y \in H$, $\dot{v}$ is negative definite.

If $r_1 = 0$ or if $r_2 = r_2'$ then the estimates for v, resp. $\dot{v}$, are estimates for scalar functions defined on R_n since in that case the interval over which θ ranges degenerates to a point.

Examples (KRASOVSKII [3]). a) The scalar equation

$$(44.8) \qquad \ddot{x} = g(t, x, \dot{x}) + f(x(t - h(t))),$$

where $f(x)$ is continuously differentiable and g and h are periodic in t, and where

$$0 \leq h(t) \leq h; \quad g(t, x, y) < - ay, \quad a > 0,$$

$$g(t, x, 0) = 0; \quad f(x) < - bx, \quad b > 0; \quad f'(x) < k, \quad k > 0,$$

is equivalent to the system

$$\dot{x} = y, \ \dot{y} = g(t, x, y) + f(x) - \int_{-h(t)}^{0} f'(x(t + \tau)) y(t + \tau) \, d\tau.$$

The function

$$v(x, y(\cdot)) = - 2 \int_0^x f(\xi) \, d\xi + y^2 + \frac{a}{h} \int_{-h}^{0} d\tau \int_{t+\tau}^{t} (y(\theta))^2 \, d\theta$$

is positive definite and its derivative is the function

$$\dot{v} = \frac{dv}{dt} = 2y \, g(t, x, y) - 2y \int_{-h}^{0} f'(x(t + \tau)) y(t + \tau) \, d\tau$$

$$+ \frac{a}{h} \int_{-h}^{0} ((y(t))^2 - (y(t + \tau))^2) \, d\tau,$$

which under the given hypotheses admits an estimate

$$\dot{v} \leq - \int_{-h}^{0} \left(\frac{a}{h} (y(t))^2 - 2k |y(t) \, y(t + \tau)| + \frac{a}{h} (y(t + \tau))^2 \right) d\tau.$$

The derivative is negative semidefinite in the sense of the Euclidean norm if $h < a/k$; Then the equilibrium is stable. Using Theorem 26.2 we even infer asymptotic stability.

b) Let

$$\dot{x}_i = \sum_{j=1}^{n} a_{ij} x_j(t) + \sum_{j=1}^{n} c_{ij} x_j(t - h_j), \tag{44.9}$$

$$i = 1, 2, \ldots, n, \quad h_j \text{ constant}.$$

The matrix A is assumed to be stable and we set $v_0 = x^T B x$, where $A^T B + B A = -E$. We further set, with certain numbers μ_i,

$$v = v_0 + \sum_{i=1}^{n} \mu_i \int_{t-h_i}^{t} (x_i(\theta))^2 \, d\theta.$$

The derivative of this expression is

$$\dot{v} = -\sum_{i=1}^{n} (x_i(t))^2 + 2 \sum_{i,j,k} b_{ik} c_{kj} x_i(t) \, x_j(t - h_j)$$

$$+ \sum_{i=1}^{n} \mu_i \left((x_i(t))^2 - (x_i(t - h_i))^2 \right).$$

Accordingly, stability is assured if the quadratic form on the right in the $2n$ variables $x_i(t)$, $x_i(t - h_i)$, is negative definite. In the scalar case

$$\dot{x} = - a x(t) - b x(t - h);$$

this leads to the inequality

$$|b| < a - \varepsilon, \quad \varepsilon > 0.$$

Using $v = x^2$ in the last scalar equation we obtain

$$\dot{v} = - 2 \left(a (x(t))^2 + b x(t) \, x(t - h) \right).$$

The expression

$$v(y(t)) = (y(t))^2$$

is monotone decreasing if and only if $(y(t))^2$ is decreasing. For such functions $y(t)$, $\dot{v}$ is negative definite in case $a > 0$, $|b| < a$. By Theorem 44.2 we arrive at the same sufficient condition for stability as above.

B. Partial differential equations. The solutions of the system of partial differential equations

$$\frac{\partial u_s}{\partial t} = f_s \left(x_1, \ldots, x_k; \, u_1, \ldots, u_n; \, \ldots, \frac{\partial u_i}{\partial x_j}, \ldots; t \right), \tag{44.10}$$

$$s = 1, \ldots, n; \, i = 1, \ldots, n; \, j = 1, \ldots, k,$$

which depend on t, t_0, and the preassigned initial values, respectively initial functions, define a mapping of the space of all initial values into itself. We are making suitable assumptions on the right sides, to guarantee existence and uniqueness of the solutions. This mapping satisfies the group relations (36.2). We can therefore apply the stability theory developed above and the theorems of the direct method, provided we are able to construct functions $v(\boldsymbol{u})$, resp. $v(\boldsymbol{u}, t)$. This is possible in the following example[1]). Consider a special case of (44.10):

$$(44.11) \qquad \frac{\partial u_s}{\partial t} = f_s(u_1, \ldots, u_n) + \sum_{i=1}^{k} b_i \frac{\partial u_s}{\partial x_i}, \quad s = 1, \ldots, n,$$

$$f_s(0) = 0, \quad b_i \text{ constant};$$

f_s is continuous and bounded, and has continuous first order partial derivatives. In addition, we consider the ordinary differential equation

$$(44.12) \qquad \frac{du}{dt} = f(u_1, \ldots, u_n), \quad t_0 = 0.$$

Let $\boldsymbol{p}(t, \boldsymbol{u}_0)$ be the general solution of (44.12)(cf. sec. 16). We use it and form the vector

$$(44.13) \qquad \boldsymbol{u} = \boldsymbol{p}\big(t, \boldsymbol{\varphi}(x_1 + b_1 t, \ldots, x_k + b_k t)\big),$$

which depends on $\boldsymbol{x}$ and t and in which $\boldsymbol{\varphi} = \boldsymbol{\varphi}(\boldsymbol{x})$ denotes a continuously differentiable n-vector, defined for $\boldsymbol{x} \in R_k$ and bounded there together with its derivative. For finite values of t the components of (44.13) and all their first partial derivatives are bounded with respect to $\boldsymbol{x}$. For we have

$$\frac{\partial \boldsymbol{u}}{\partial \boldsymbol{x}} = \frac{\partial \boldsymbol{u}}{\partial \boldsymbol{\varphi}} \frac{\partial \boldsymbol{\varphi}}{\partial \boldsymbol{x}}.$$

Here, the elements of $\partial \boldsymbol{\varphi}/\partial \boldsymbol{x}$ are bounded. The argument of $\boldsymbol{\varphi}$ is $\boldsymbol{x} + t\boldsymbol{b}$. Therefore, $\boldsymbol{\varphi}$ varies in a bounded domain of R_n. For t fixed, $\boldsymbol{u}$ can be identified with the solution $\boldsymbol{p}(t, \boldsymbol{u}_0)$, $\boldsymbol{u}_0 = \boldsymbol{\varphi}(\boldsymbol{x} + t\boldsymbol{b})$, of (44.12), and by virtue of the continuability of the solutions of (44.12) to all values of t, the boundedness of $\partial \boldsymbol{u}/\partial \boldsymbol{\varphi}$ follows.

For $t = 0$, the right side of (44.13) assumes the value $\boldsymbol{\varphi}$. Let

$$(44.14) \qquad \|\boldsymbol{\varphi}\| = \sup |\boldsymbol{\varphi}(\boldsymbol{x})|, \quad \boldsymbol{x} \in R_h,$$

be the norm of $\boldsymbol{\varphi}$. The right side of (44.13) is continuous with respect to t as well as with respect to $\boldsymbol{\varphi}$ in the sense of the norm (44.14). This can be seen as follows. Since $\boldsymbol{u}$ is defined by means of the differential equation (44.12), the expression

$$\boldsymbol{p}\big(t, \boldsymbol{\varphi}(\boldsymbol{x} + t\boldsymbol{b})\big) - \boldsymbol{p}\big(\bar{t}, \boldsymbol{\varphi}(\boldsymbol{x} + \bar{t}\boldsymbol{b})\big)$$

approaches zero as $t - \bar{t} \to 0$ and $|\boldsymbol{\varphi}(\boldsymbol{x} + t\boldsymbol{b}) - \boldsymbol{\varphi}(\boldsymbol{x} + \bar{t}\boldsymbol{b})| \to 0$. Also

$$|\boldsymbol{\varphi}(\boldsymbol{x} + t\boldsymbol{b}) - \bar{\boldsymbol{\varphi}}(\boldsymbol{x} + \bar{t}\boldsymbol{b})| \leq |\boldsymbol{\varphi}(\boldsymbol{x} + t\boldsymbol{b}) - \boldsymbol{\varphi}(\boldsymbol{x} + \bar{t}\boldsymbol{b})|$$
$$+ |\boldsymbol{\varphi}(\boldsymbol{x} + \bar{t}\boldsymbol{b}) - \bar{\boldsymbol{\varphi}}(\boldsymbol{x} + \bar{t}\boldsymbol{b})|$$
$$\leq \|\boldsymbol{\varphi} - \bar{\boldsymbol{\varphi}}\| + M|t - \bar{t}|.$$

---

[1]) Zubov [2].

M is a bound for the partial derivatives $\partial \varphi_i / \partial x_j$. Hence we have

$$\| \boldsymbol{p}\left(t,\, \boldsymbol{\varphi}\left(\boldsymbol{x} + t\boldsymbol{b}\right)\right) - \boldsymbol{p}\left(\bar{t},\, \overline{\boldsymbol{\varphi}}\left(\boldsymbol{x} + \bar{t}\boldsymbol{b}\right)\right) \| \to 0$$

as $t - \bar{t} \to 0$, $\|\boldsymbol{\varphi} - \overline{\boldsymbol{\varphi}}\| \to 0$.

Finally we set $\boldsymbol{\xi} = \boldsymbol{x} + t\boldsymbol{b}$, and we see by differentiating the components of (44.13) that

$$\frac{\partial u_s}{\partial t} = \dot{p}_s(t,\, \boldsymbol{\varphi}) + \sum_{j=1}^{n} \frac{\partial u_s}{\partial \xi_j} \frac{\partial \xi_j}{\partial t}.$$

In forming the derivative denoted by the dot, $\boldsymbol{\varphi}$ is to be considered as constant. The last equation thus becomes

$$\frac{\partial \boldsymbol{u}}{\partial t} = \boldsymbol{f}(\boldsymbol{u}) + \frac{\partial \boldsymbol{u}}{\partial \boldsymbol{x}}\, \boldsymbol{b}.$$

All of this implies that (44.13) is a solution of (44.11). Now assume that the trivial solution of (44.12) is asymptotically stable and let A be its domain of attraction. Suppose we have two functions $v(\boldsymbol{u})$, $h(\boldsymbol{u})$ which satisfy the hypotheses of Theorem 34.1; in particular suppose that the derivative of $v(\boldsymbol{u})$ for (44.12) satisfies

$$(44.15) \qquad \dot{v}_{(44.12)} = - h(\boldsymbol{u})\left(1 - v(\boldsymbol{u})\right).$$

(It follows from the converse theorems in Chapter VI that in the presence of asymptotic stability two such functions always exist.) Let M_λ denote the set of points for which $v(\boldsymbol{u}) = 1 - \lambda$, $0 < \lambda \leq 1$; it is a closed hypersurface (sec. 24). Let $\boldsymbol{\Phi}_\lambda$ be the class of all continuous vector-valued functions $\boldsymbol{\varphi}(\boldsymbol{x})$ which map the space R_k onto the domain $v(\boldsymbol{u}) \leq 1 - \lambda$, and let $\boldsymbol{\xi}$ be a fixed point in R_k. We obtain a well-defined but not necessarily invertible mapping $\boldsymbol{\varphi} \to \boldsymbol{u}$ of the functions $\boldsymbol{\varphi}$ in $\boldsymbol{\Phi}_\lambda$ to the points $\boldsymbol{u} \in M_\lambda$ by assigning to the point $\boldsymbol{u}_0 \in M_\lambda$ the function $\boldsymbol{\varphi}_0(\boldsymbol{\xi}) = \boldsymbol{u}_0$. This mapping defines a function $\tilde{v}$ in $\boldsymbol{\Phi}_\lambda$. It is defined for $\boldsymbol{\varphi}$ in $\boldsymbol{\Phi}_\lambda$, $\lambda \leq 1$, and maps the space of all such functions into the real axis. We have $\tilde{v}(\boldsymbol{\varphi}_0) = v(\boldsymbol{u}_0)$, if $\boldsymbol{\varphi}_0$ corresponds to the point $\boldsymbol{u}_0$. If we apply v to a solution (44.13) of (44.11) and set

$$v\left(\boldsymbol{p}(t,\, \boldsymbol{u}_0)\right) = 1 - \lambda(t)$$

then it follows that

$$\boldsymbol{p}\left(t,\, \boldsymbol{\varphi}\left(\boldsymbol{x} + \boldsymbol{b}\,t\right)\right) \in \boldsymbol{\Phi}_{\lambda(t)}$$

provided $\boldsymbol{\varphi} \in \boldsymbol{\Phi}_\lambda$. For if we hold t fixed and set $\boldsymbol{x} = \boldsymbol{\xi} - t\boldsymbol{b}$, then $\boldsymbol{p}(t,\, \boldsymbol{\varphi}(\boldsymbol{\xi})) = \boldsymbol{p}(t,\, \boldsymbol{u}_0)$, where $\boldsymbol{\varphi}$ corresponds to $\boldsymbol{u}_0$. This says that the points $\boldsymbol{p}(t,\, \boldsymbol{\varphi}(\boldsymbol{x} + t\boldsymbol{b}))$ lie on $M_{\lambda(t)}$ or in the interior of the domain bounded by $M_{\lambda(t)}$. Hence

$$v\left(\boldsymbol{p}(t,\, \boldsymbol{u}_0)\right) = \tilde{v}\left(\boldsymbol{p}(t,\, \boldsymbol{\varphi}(\boldsymbol{x} + t\boldsymbol{b}))\right).$$

By taking the derivative with respect to t we conclude that

$$\dot{v}_{(44.12)} = \dot{\tilde{v}}_{(44.11)}$$

and hence, because of (44.15),

$$\dot{\tilde{v}} = - h\left(\boldsymbol{p}\left(t,\, \boldsymbol{u}_0\right)\right)\left(1 - \tilde{v}\left(\boldsymbol{\varphi}\right)\right).$$

If we set

$$h\left(\boldsymbol{p}\left(t,\, \boldsymbol{u}_0\right)\right) =: \tilde{h}\left(\boldsymbol{\varphi}\right)$$

we finally obtain

$$\dot{\tilde{v}} = \tilde{h}\left(\boldsymbol{\varphi}\right)\left(1 - \tilde{v}\left(\boldsymbol{\varphi}\right)\right)$$

and conclude on the basis of the theorems of sec. 42 that the equilibrium of (44.11) is asymptotically stable (in fact uniformly so). We also find by employing the argument of the proof of Theorem 34.1 that the domain of attraction of the equilibrium is equal to the union of the sets M_λ, $0 < \lambda < 1$.

Further examples for the general theorems are furnished by some boundary value problems[1]). Let

$$(44.16) \qquad \frac{\partial^2 u}{\partial t^2} = a^2\, \frac{\partial^2 u}{\partial x^2}, \quad u\left(0,\, t\right) = u\left(l,\, t\right) = 0.$$

The solution depends in part on the values $u\left(x,\, t\right)$, $u_t\left(x,\, t\right)$ along a certain curve; the underlying function space is therefore the space of all continuous, once differentiable functions defined on $0 \le x \le l$, $t \ge t_0 \ge 0$, and their first partial derivatives with respect to t. It is indicated to adapt the definition of the norm to the boundary conditions:

$$(44.17) \qquad \|u\|^2 := \int_0^l \left[\left(u\left(x,\, t\right)\right)^2 + \left(u_t\left(x,\, t\right)\right)^2\right] dx,$$

where t is fixed. For the Liapunov function we choose

$$v\left(u\right) = \int_0^l \left(u_t^2 + a^2\, u_x^2\right) dx.$$

The derivative of v for (44.16) with respect to t is

$$\frac{dv}{dt} = 2 \int_0^l \left(u_t u_{tt} + a^2\, u_x u_{xt}\right) dx.$$

We integrate by parts and obtain

$$\frac{dv}{dt} = 2 \int_0^l u_t u_{tt}\, dx + 2 a^2\, \left[u_x u_t\right]_0^l - 2 a^2 \int_0^l u_t u_{xx}\, dx$$

$$= 2 \int_0^l u_t \left(u_{tt} - a^2\, u_{xx}\right) dx + 2 a^2\, \left[u_x u_t\right]_0^l.$$

---

[1]) MOVCHAN [1].

Because of (44.16) this expression is identically zero. Since the function $v(u)$ is obviously definite with respect to the norm (44.17), the stability of the equilibrium of (44.16) with respect to the norm (44.17) follows.

If instead of (44.16) we consider the damped wave equation

$$(44.18) \qquad \frac{\partial^2 u}{\partial t^2} + \alpha \frac{\partial u}{\partial t} = a^2 \frac{\partial^2 u}{\partial x^2}, \qquad \alpha > 0,$$

then the derivative of $v(u)$ becomes $-2\alpha \int u_t^2 \, dx$. We cannot immediately conclude asymptotic stability because $\dot{v}$ is not negative definite, but we can apply Theorem 26.1 which is valid in general spaces. The set of points $\dot{v} = 0$ contains only those solutions which are constant with respect to t and which are therefore only functions of x. Because of (44.18) they must be linear in x and it follows from the boundary conditions that they are identically zero.

Research on the applications of the direct method to partial differential equations is still in its initial stages. To illustrate the many possibilities we cite another example[1]. In this case we are dealing with partial differential equations which generalize an ordinary differential equation of the type $J(x)\, \dot{x} = \operatorname{grad} h(x)$, $h(x)$ a scalar, with boundary conditions. The problem is to find an n-vector $u = u(y, t)$ depending on the two scalar variables y and t, where $t > 0, 0 < y < 1$. To involve the boundary conditions valid for $y = 0$ and $y = 1$ we introduce the $3n$-vector

$$(44.19) \qquad w(y, t) := \operatorname{col}\big(u(y, t),\, u(0, t),\, u(1, t)\big).$$

The generalization of the gradient is made as follows. Let $g(u, u_y), \varphi(u)$, $\psi(u)$ be continuous, once differentiable scalar functions of their vector variables. The derivative $\partial g/\partial u$ is as usual the vector $\operatorname{col}\left(\dfrac{\partial g}{\partial u_1}, \ldots, \dfrac{\partial g}{\partial u_n}\right)$. The derivative with respect to w of the scalar function

$$(44.20) \qquad f(w) := \int_0^1 g(u, u_y)\, dy + \varphi(u)\big|_{y=0} + \psi(u)\big|_{y=1},$$

respectively the "gradient" of $f(w)$ with respect to w, is the $3n$-vector

$$(44.21) \qquad f_w = \operatorname{col}\left(g_u - \frac{d}{dy} g_{u_y},\, (\varphi_u - g_{u_y})\big|_{y=0},\, (\psi_u - g_{u_y})\big|_{y=1}\right).$$

This fact is derived in the same manner as the Euler differential equation in the calculus of variations. The boundary value problem is then set up in terms of the matrices J_0, J_1, J_2, A_0 and the vectors b_0, b_1, b_2, all of which depend on u and y, in the form

$$
\begin{aligned}
J_0 u_t &= A_0 u_y + b_0, & 0 < y < 1,\ & t > 0, \\
(44.22)\qquad J_1 u_t &= b_1, & y = 0, \quad & t > 0, \\
J_2 u_t &= b_2, & y = 1, \quad & t > 0.
\end{aligned}
$$

---

[1] BRAYTON and MIRANKER [1].

Introducing, in addition, the notation

$$J := \operatorname{diag}(J_0, J_1, J_2), \quad A := \operatorname{diag}(A_0, 0, 0), \quad \boldsymbol{b} := \operatorname{col}(\boldsymbol{b}_0, \boldsymbol{b}_1, \boldsymbol{b}_2),$$

and applying (44.19) we can write the problem in the form

$$J\boldsymbol{w}_t = A\,\boldsymbol{w}_y + \boldsymbol{b}.$$

We make the following additional assumptions on the right side: There exists a scalar function $h(\boldsymbol{w})$ of type (44.20) such that $A\,\boldsymbol{w}_y + \boldsymbol{b} = h_{\boldsymbol{w}}$, where $h_{\boldsymbol{w}}$ is defined by (44.21). We then have

$$(44.23) \qquad\qquad J\,\boldsymbol{w}_t = h_{\boldsymbol{w}},$$

and this is the desired form of the problem.

The solution vectors $\boldsymbol{u}(y, t)$, respectively the associated vectors $\boldsymbol{w}(y, t)$, are interpreted in two Hilbert spaces H_0 and H_1 with the respective norms

$$\|\boldsymbol{w}\|_0^2 := \boldsymbol{u}^T(0)\,\boldsymbol{u}(0) + \boldsymbol{u}^T(1)\,\boldsymbol{u}(1) + \int_0^1 \boldsymbol{u}^T(y)\,\boldsymbol{u}(y)\,dy,$$

$$\|\boldsymbol{w}\|_1^2 := \|\boldsymbol{w}\|_0^2 + \int_0^1 \boldsymbol{u}_y^T(y)\,\boldsymbol{u}_y(y)\,dy,$$

the parameter t being dropped.

The set of points $\|h_{\boldsymbol{w}}\|_0 = 0$, which we denote by M, contains the "equilibrium points" in H_0. By means of a "Liapunov function" $f(\boldsymbol{w})$ defined on H_1 we can give a criterion for M to be asymptotically stable for all initial values in H_1, *i.e.* that all solutions approach M as t increases. The following conditions are sufficient:

1) $f(\boldsymbol{w})$ is bounded below, $f(\boldsymbol{w}) \geq \alpha > -\infty$.
2) $f(\boldsymbol{w})$ grows without bound if and only if $\|w\|_1$ grows without bound.
3) The total derivative of $f(\boldsymbol{w})$ in the sense of equation (44.23) satis-
fies an estimate $\dfrac{df}{dt} \leq -c\,\|h_{\boldsymbol{w}}\|_0^2,\ c > 0$.

This is proved like Theorem 26.4. However, additional arguments are necessary to guarantee the existence of the limit set. Incidentally, the derivative df/dt is equal to the expression $f_{\boldsymbol{w}}^T \boldsymbol{w}_t$.

BRAYTON and MIRANKER gave another example: To solve the telegraphic equation

$$L i_t = -v_y - R i, \quad -C v_t = i_y + G v$$

with the boundary conditions

$$0 = E - v_0 - R_0 i_0, \, y = 0, \, t > 0,$$

$$-C_1 \frac{\partial}{\partial t} v_1 = -i_1 + q(v_1),$$

where $v_0 = v(0, t)$, $v_1 = v(1, t)$, $i_0 = i(0, t)$, $i_1 = i(1, t)$, and q is a non-linear function. For h we can choose the expression:

$$h := \int_0^1 \left(-\frac{1}{2} R\, i^2 + \frac{1}{2} G\, v^2 - i\, v_y \right) dy - \frac{1}{2} R_0\, i_0^2$$

$$+ (E - v_0)\, i_0 + \int_0^v q(v)\, dv.$$

The function:

$$f := \lambda h + \frac{1}{2} \int_0^1 (L\, i_t^2 + C\, v_t^2)\, dy + \frac{1}{2} C_1 \left(\frac{\partial}{\partial t} v_1 \right)^2$$

is suitable as a Liapunov function if we choose λ so that

$$-\frac{q'(v)}{C_1} < \lambda < \frac{R}{L}.$$

The condition

$$- q'(v) < \frac{R C_1}{L}$$

proves to be sufficient for the stability of the equilibria defined by $h = 0$.

In the paper cited procedures are given for the construction of suitable functions $f(\boldsymbol{w})$ in simple cases.

45. System Stability and Stability of Invariant Sets

In the proofs of the stability theorems of sec. 42 we used only the estimates for v and Dv. They served to derive estimates for the norm of the motion $\boldsymbol{p}(t, \boldsymbol{a}, t_0)$. The argument can be carried through in the same way if instead of $v(\boldsymbol{x}, t)$ we consider a function $v(\boldsymbol{x}, \boldsymbol{y}, t)$ which depends on two elements $\boldsymbol{x}, \boldsymbol{y}$ of the space X, and if in the inequalities we replace the argument $\|\boldsymbol{x}\|$ of the comparison functions by $\|\boldsymbol{x} - \boldsymbol{y}\|$. We then obtain an estimate for the norm

$$\|\boldsymbol{x} - \boldsymbol{y}\| = \|\boldsymbol{p}(t, \boldsymbol{a}, t_0) - \boldsymbol{p}(t, \boldsymbol{b}, t_0)\|$$

by a comparison function, *i.e.* we obtain criteria for system stability (Def. 35.7 and 35.8, group B of the table in sec. 36). We shall not explicitly formulate the corresponding theorems but only cite a simple example here.

Let

$$\dot{\boldsymbol{x}} = A\boldsymbol{x} + \boldsymbol{f}(t), \quad \dot{\boldsymbol{y}} = A\boldsymbol{y} + \boldsymbol{f}(t); \quad \boldsymbol{x} \in R_n, \ \boldsymbol{y} \in R_n.$$

Assume that A has only characteristic roots with negative real parts so that a matrix B can be constructed as in (27.3). The derivative of the function

$$v(\boldsymbol{x}, \boldsymbol{y}) = (\boldsymbol{x} - \boldsymbol{y})^T B (\boldsymbol{x} - \boldsymbol{y})$$

along the two motions defined by the system was given at the end of sec. 41 and has the form

$$Dv = (\dot{x} - \dot{y})^T B (x - y) + (x - y)^T B (\dot{x} - \dot{y}) = -(x - y)^T C (x - y).$$

Obviously $v(x, y)$ and Dv can be estimated in terms of the Euclidean norm $|x - y|$. A theorem parallel to Theorem 42.4 shows that $\lim_{t \to \infty} (x - y) = 0$: The system of solutions of $\dot{x} = A x + f(t)$ is asymptotically stable. All the solutions tend toward a distinguished solution.

The same principle can be applied to gain statements of the following form. Let two differential equations in R_n be given:

$$(45.1) \qquad \dot{x} = f(x, t), \quad \dot{y} = g(y, t).$$

Also let $v(x, y, t)$ be a function of known definiteness with respect to the Euclidean norm $|x - y|$. Further let a scalar differential equation

$$(45.2) \qquad h(v, \dot{v}, t) = 0$$

be given which is satisfied by the derivative

$$\dot{v} = \sum_{i=1}^{n} \frac{\partial v}{\partial x_i} f_i + \sum_{i=1}^{n} \frac{\partial v}{\partial y_i} g_i + \frac{\partial v}{\partial t}$$

of v along the motions of (45.1), and assume that the stability behavior of the trivial solution of (45.2) is known. Then we can make assertions about the expression $|p(t, x_0, t_0) - p(t, y_0, t_0)|$. Again we shall not formulate individual theorems here but refer the reader to LAKSHMIKANTHAM [1—5], who proves a number of theorems of this kind about differential equations and functional differential equations.

If in an inequality for $\|p(t, b, t_0)\|$ we replace the argument $\|b\|$ of the comparison function by the distance $\varrho(b, M)$ of the point b from an invariant set M, then we obtain an estimate for the distance $\varrho(p(t, b, t_0), M)$. If $M = M(a, t_0)$ is the totality of points of a motion then we are dealing with orbital stability, etc., that is with assertions of the groups C through D'. It is therefore possible in principle to discuss orbital stability with the help of the direct method. Of course this application differs in a very essential aspect from the discussion of the stability of the equilibrium. For the latter we test the function $v(x, t)$, respectively its derivative, for definiteness and decrescence — properties of the functions themselves. In the case of orbital stability on the other hand we are concerned with estimates of the form $v(x, t) \leq \varphi(\varrho(x, M))$ etc., which in general cannot be tested without exactly knowing the set M, $i.e.$ without knowing the motion under consideration. The application of the direct method to orbital stability (especially to the stability behavior of periodic motions, $cf.$ sec. 81) is therefore primarily of theoretical value.

46. Boundedness Criteria. The Parallel Theorems

The essence of the proofs in sec. 42, as we have frequently emphasized, consists in using certain inequalities for v and $\dot{v}$ to gain an estimate for the argument of the comparison functions. The same idea applies if some of these inequalities are reversed. The inequalities in the statements must then be changed accordingly. We consider a few examples.

Theorem 46.1. If there exists a positive definite function $v(\boldsymbol{x}, t)$ with a non-negative derivative then an estimate of the form A' 2

$$(46.1) \qquad \|\boldsymbol{p}(t, \boldsymbol{a}, t_0)\| \geq \varphi(\|\boldsymbol{a}\|; t_0), \quad \varphi \in K,$$

is valid.

Theorem 46.2. If there exists a positive definite decrescent function $v(\boldsymbol{x}, t)$ whose derivative satisfies an inequality

$$Dv \geq -\psi(\|\boldsymbol{x}\|), \quad \psi \in K,$$

then an estimate of type A' 3 + A' 7 is valid:

$$(46.2) \qquad \|\boldsymbol{p}(t, \boldsymbol{a}, t_0)\| \geq \varphi(\|\boldsymbol{a}\|)\,\sigma(t - t_0), \quad \varphi \in K, \sigma \in L.$$

Theorem 46.3. If there exists a positive definite decrescent function $v(\boldsymbol{x}, t)$ whose derivative satisfies an estimate

$$Dv \geq \psi(\|\boldsymbol{x})\|), \quad \psi \in K,$$

then an estimate of type A' 3 + A' 11 is valid:

$$(46.3) \qquad \|\boldsymbol{p}(t, \boldsymbol{a}, t_0)\| \geq \varphi(\|\boldsymbol{a}\|)\,\varkappa(t - t_0).$$

The proofs of these theorems are obtained by making the appropriate sign changes and changes in the inequalities in the proofs of Theorems 42.2 and 42.4. The assertion (46.1) characterizes a behavior of the motion which is in a certain sense opposite to attractivity but which is entirely compatible with the stability of the equilibrium. The estimate (46.3) represents a strong instability: Every motion tends away from the equilibrium. The estimate (46.2) is of no interest by itself, it gains importance only if it is observed to supplement an estimate A 2 + A 7 (uniform asymptotic stability of the equilibrium). In that case the motion admits a two-sided estimate (*cf.* the example at the end of sec. 39).

The theorems which we have just discussed belong to the numerous *parallel theorems* which can be formulated and proved quite schematically from the principal classical theorems of the direct method. We will not go beyond these few examples, especially since most of these theorems have a rather limited field of application. The most interesting of them are instability statements of type (46.3). A special case of Theorem 46.3 is contained in Theorem 27.1: If all the characteristic roots of a matrix

A have positive real parts then the matrix C used in the proof of Theorem 27.1 is positive definite. In that case all solutions grow exponentially.

A much more important group of parallel theorems gives us criteria for the *boundedness* of solutions. In Definitions 36.10 and 36.11 some types of boundedness were characterized by means of comparison functions. The essential distinction from the "parallel" stability properties is that the estimates are required only for sufficiently large arguments. We thus obtain theorems on bounded solutions which run parallel to the stability and attractivity theorems if we require the validity of the inequalities for v, respectively Dv, only outside of a certain neighborhood of the origin or of the t-axis. Two theorems illustrate this idea.

Theorem 46.4. Let a function $v(\boldsymbol{x}, t)$ be given with the following properties:

$$(46.4)\quad \psi_1(\|\boldsymbol{x}\|) \leq v(\boldsymbol{x}, t) \leq \psi_2(\|\boldsymbol{x}\|) \text{ for } \|\boldsymbol{x}\| \geq h, \quad \psi_i \in K,$$

$$Dv < 0 \text{ for } \|\boldsymbol{x}\| \geq h_1 \geq h.$$

Then the motions are uniformly bounded (Def. 36.11).

Proof. As in the proof of Theorem 42.1 we have

$$\psi_2^I(v) \leq \|\boldsymbol{x}\| \leq \psi_1^I(v)$$

but this inequality is assured only for $\|\boldsymbol{x}\| \geq h$, $v \geq \psi(h)$. If $\|\boldsymbol{a}\| \geq h_1$ then $v(t) = v(\boldsymbol{p}(t, \boldsymbol{a}, t_0), t)$ decreases as t increases, at least as long as $\|\boldsymbol{p}(t, \boldsymbol{a}, t_0))\| > h_1$. It follows that

$$\|\boldsymbol{x}\| \leq \psi_1^I(v(\boldsymbol{a}, t_0)) \text{ for } \|\boldsymbol{a}\| \geq h_1$$

and this is an estimate of type (36.20).

Example. Let the scalar equation

$$(46.5)\qquad\qquad \ddot{x} + f(x, \dot{x})\dot{x} + h(x) = e(t)$$

be given. Suppose $f(x, y)$ is continuous and non-negative for all points (x, y) and $h(x)$ is continuous. Suppose also that the function $H(x) := \int_0^x h(u)\, du$ admits an estimate

$$H(x) \geq \chi(\|x\|), \quad \chi \in K, \quad \chi(r) \to \infty \text{ as } r \to \infty.$$

Let $e(t)$ be continuous for $t \geq 0$ and suppose that $\int_0^\infty |e(t)|\, dt$ is finite. If we replace the equation (46.5) by the equivalent system

$$\dot{x}_1 = x_2, \quad \dot{x}_2 = -f(x_1, x_2)x_2 - h(x_1) + e(t)$$

and utilize the function

$$v(x_1, x_2, t) = + \sqrt{x_2^2 + 2H(x_1)} - \int_0^t |e(\tau)|\, d\tau$$

then we obtain

$$Dv = \frac{1}{\sqrt{x_2^2 + 2H(x_1)}} \left(- x_2^2 f(x_1, x_2) + x_2\, e(t) - |e(t)| \sqrt{x_2^2 + 2H(x_1)} \right).$$

Obviously $Dv \leq 0$ and $v(x_1, x_2, t)$ satisfies an inequality (46.4) for $\sqrt{x_2^2 + 2H(x_1)} > \int_0^\infty |e(t)|\, dt$. The uniform boundedness of $x(t)$ and $\dot{x}(t)$ follows.

Theorem 46.5. If the hypothesis on Dv in Theorem 46.4 can be strengthened to an inequality

$$Dv \leq - \psi_3(\|x\|), \quad \|x\| \geq h_1 \geq h,$$

then the motions are uniformly bounded in the sense of an estimate (36.21).

Proof. As in the proof of Theorem 42.4 we arrive at a differential inequality $Dv \leq -\chi(v)$ which now, however, need only be valid for sufficiently large v. The estimate (*cf.* proof of Theorem 25.2)

$$v \leq q(v_0)\gamma(t - t_0)$$

is therefore also valid only for sufficiently large v and the comparison function γ need not belong to class L. It is monotone decreasing but need not tend to zero. Calculating from the estimate for v an estimate for $\|p\|$, we obtain an inequality of type (36.21)●

The principle on which Theorem 46.4 is based has found frequent application without the authors' adopting the language of the direct method. If we can prove with the help of a suitable function $v(x, y)$, $(x, y) \in R_2$, that the solutions of a second order scalar differential equation are bounded (the uniformity follows from the autonomy of the equation, *cf.* Theorem 38.5) and if we know that no stable equilibrium exists, then we can conclude by Theorem 18.1 that there exists at least one closed phase curve.[1]

The boundedness theorems can be extended to the situations discussed in secs. 44 und 45. For example, we can construct a theory of systems with bounded distance, the distance between two arbitrary motions being estimated by a function of the initial distance.[2]

We mention another possibility for obtaining boundedness statements.

---

[1] As a primary reference see REISSIG, SANSONE and CONTI and also their bibliography; further MÜLLER [1, 2], REISSIG [1, 2, 3].

[2] YOSHIZAWA [3, 4].

Theorem 46.6. Let a function $v(x, t)$ be given which satisfies the following estimates:

$$\sigma_1(\|x\|) \leq v(x, t) \leq \sigma_2(\|x\|), \tag{46.6}$$

$$Dv \geq \sigma_3(\|x\|),$$

$$\|x\| \geq h, \quad \sigma_i \in L, \quad i = 1, 2, 3.$$

Then the motions are uniformly bounded and, in fact, there exists an estimate (36.21).[1]

Proof. We again follow in outline the proof of Theorem 27.2, respectively Theorem 42.2. We start with (46.6), introduce inverse functions, and obtain

$$\sigma_1^I(v) \leq \|x\| \leq \sigma_2^I(v).$$

The solution of the auxiliary differential equation

$$Dw = \sigma_3\left(\sigma_2^I(w)\right)$$

whose right side is monotone increasing, admits an estimate

$$w \geq \varkappa(w_0)\,\lambda(t - t_0).$$

The functions on the right are increasing. It follows that $v \geq w$ for $v_0 = w_0$, and

$$v \geq \varkappa(v_0)\,\lambda(t - t_0) \geq \varkappa\left(\sigma_1(\|a\|)\right)\lambda(t - t_0),$$

$$\|x\| \leq \sigma_2^I\left(\varkappa\left(\sigma_1(\|a\|)\right)\lambda(t - t_0)\right).$$

This is an estimate of the desired type (36.21).

An application of this theorem is found in SKOWRONSKI and ZIEMBA [1]. This example deals with a mechanical system whose total energy $H(x)$ is a non-decreasing function of $|x|$. $v == e^{-H}$ can be used.

---

[1] BHATIA [1].

Chapter VI

The Converse of the Stability Theorems

47. Formulation of the Problem

The principal theorems of the direct method (secs. 25 and 42) furnish sufficient conditions for stability and instability. They do not say whether the given conditions are also necessary, *i.e.* whether from known stability properties we can infer the existence of suitable Liapunov functions. Only the case of a linear autonomous system has been completely settled by the results of sec. 27: It is known that in case of asymptotic stability a suitable Liapunov function always exists, in fact as a quadratic form. In the general case the answer was found relatively late. The first result is due to K. P. PERSIDSKII [1] and refers to simple, non-uniform stability. Persidskii was able to show that a Liapunov function can always be found which satisfies the conditions of Theorem 42.1. However, he only considered differential equations in R_n. The corresponding theorems, usually called *converse theorems*, on asymptotic stability could not be discovered until the concept of uniform stability had been recognized and clearly defined. After all, the so-called Second Stability Theorem of Liapunov (25.2, resp. 42.4) yields not merely asymptotic stability but considerably more, namely uniform asymptotic stability. It is thus impossible to prove the converse of the theorem in its original weaker form. Once the basic concepts had been clarified the conditions for the existence of Liapunov functions were carefully studied and the converse problem can now be considered as essentially settled.

In the converse theorems the stability behavior of a family of motions $p(t, a, t_0)$ is assumed to be known. For example, it might be assumed that the expression $\|p(t, a, t_0)\|$ is estimated by known comparison functions (secs. 35 and 36). Then one attempts to construct by means of a finite or transfinite procedure, a Liapunov function which satisfies the conditions of the stability theorem under consideration. Furthermore, one tries to construct a Liapunov function which is as "nice" as possible, that is one which is not only continuous but has partial derivatives of maximal order. The next few sections will show that it is relatively easy to construct a continuous Liapunov function. On the other hand it is in general quite difficult to guarantee the existence of continuous partial derivatives of a given order or even of arbitrary order. Some of the proofs are very long. The results also depend strongly on the assumptions made on the motions, whether they are defined in a general way or by differential equations under more or less special assumptions on the right sides (Lipschitz conditions, etc.).

The converse theorems which we have chosen here do not always yield the "nicest" Liapunov functions or utilize the weakest hypotheses on the motions. They have been selected so as to bring out the essential ideas and to acquaint the reader with the different methods of proof. We give preference to the comparison functions used in the definition of the concepts of stability, etc. Our proofs of the converse theorems will therefore occasionally deviate from the original proofs cited although the basic idea of the proof has been retained. A compilation comparing the results of the various authors is found in MASSERA [5]. See also HAHN [4].

We also mention that the converse theorems are primarily of theoretical significance. They find no application in the treatment of stability problems in practice, for the converse theorem assumes as known the solution of the practical problem, the stability behavior. Even the constructive procedures of the converse theorems give in general no hint (except in the linear case) as to how a Liapunov function can practically be obtained. It is obvious that knowledge about the existence of a Liapunov function can still occasionally be important (*cf.* sec. 56, total stability).

48. The Converse of the Theorems on Non-Asymptotic Stability

Theorem 48.1. Let $p(t, a, t_0)$ be a family of motions which satisfies the hypotheses of secs. 35 and 36 and which contains the equilibrium. Let the equilibrium be stable (A 2). Then there exists a positive definite function $v(x, t)$ with a non-positive derivative.[1]

Here and in the other converse theorems we always mean by the derivative of the function $v(x, t)$ the derivative along the motion under consideration.

Proof. Let $x = p(t_0 + t, a, t_0)$. By (36.2), $a = p(t_0, x, t_0 + t)$ and we have identically

$$p(t_0, p(t_0 + t, a, t_0), t_0 + t) = p(t_0, p(t_0, a, t_0), t_0) = a.$$

We now set

$$(48.1) \qquad v(x, t) := \|a\| = \|p(t_0, x, t_0 + t)\|;$$

---

[1] K. P. PERSIDSKII [1].

t_0 is to be considered as a parameter. Because of the above identity the function $v(t) := v(\boldsymbol{p}(t_0 + t, \boldsymbol{a}, t_0), t)$ does not depend on t at all, so that

$$Dv \equiv 0.$$

Also, because of the assumed stability

$$\|\boldsymbol{p}(t_0 + t, \boldsymbol{a}, t_0)\| = \|\boldsymbol{x}\| \leq \varphi(\|\boldsymbol{a}\|; t_0)$$

for a suitable $\varphi \in K$, that is

$$\|\boldsymbol{a}\| = v(\boldsymbol{x}, t) \geq \varphi^I(\|\boldsymbol{x}\|, t_0).$$

The right side does not depend on t. Hence $v(\boldsymbol{x}, t)$ is positive definite.

The principle involved in this proof can be utilized to extend the theorem. Let

$$\dot{y} = g(y, t), \quad g \in E,$$

be a scalar differential equation whose general solution satisfies an estimate

$$q(t, y_0, t_0) \geq \psi(y_0), \quad \psi \in K,$$

for sufficiently small $y_0 > 0$. The function

$$v(\boldsymbol{x}, t) := q\big(t, \|\boldsymbol{p}(0, \boldsymbol{x}, t)\|, 0\big)$$

formed with the motion $\boldsymbol{p}(t, \boldsymbol{a}, t_0)$ of Theorem 48.1 is positive definite; for we have

$$v(\boldsymbol{x}, t) \geq \psi\big(\|\boldsymbol{p}(0, \boldsymbol{x}, t)\|\big) \geq \psi\big(\varphi^I(\|\boldsymbol{x}\|; 0)\big).$$

Further $v(\boldsymbol{x}, t)$ satisfies the differential equation

$$(48.2) \qquad\qquad Dv = g\big(v(\boldsymbol{x}(t), t)\big).$$

This is so because

$$Dv = \frac{\partial}{\partial t} q(t, \|\boldsymbol{p}\|, 0) + \frac{\partial q}{\partial y} \frac{d}{dt} \|\boldsymbol{p}(0, \boldsymbol{p}(t, \boldsymbol{a}, 0), t)\|.$$

The last summand vanishes because it involves the derivative with respect to t of a term which is independent of t. (48.2) follows.[1])

If g is chosen to be identically zero then the solution is $q = y_0$ and we obtain Theorem 48.1.

The proof of Theorem 48.1 defines the Liapunov function $v(\boldsymbol{x}, t)$ only for those points $\boldsymbol{x}$ which are images of a certain ball K_{r, t_0} under the mapping induced by the motion. KURZWEIL [1] showed that by a slight modification of the construction we can directly show the existence of $v(\boldsymbol{x}, t)$ in a cylindrical domain.

---

[1]) CORDUNEANU [4].

The next theorem goes a little further.

Theorem 48.2. Under the conditions of Theorem 48.1 let the equilibrium be assumed uniformly stable (A 3). Then there exists a positive definite decrescent function with a non-positive derivative.

Proof. By hypothesis

$$(48.3) \qquad \| \boldsymbol{p}(t, \boldsymbol{a}, t_0) \| \leq \varphi(\|\boldsymbol{a}\|), \quad t \geq t_0 \geq 0.$$

Set

$$(48.4) \qquad w(\boldsymbol{x}, t) := \min_{t_0 \leq \tau \leq t} \| \boldsymbol{p}(\tau, \boldsymbol{x}, t) \|.$$

Obviously $w(0, t) = 0$. This expression is continuous in t and in $\boldsymbol{x}$. Further

$$w(\boldsymbol{x}, t) \leq \| \boldsymbol{p}(t, \boldsymbol{x}, t_0) \| = \|\boldsymbol{x}\|$$

i.e. w is decrescent. It also follows from (48.3) that

$$\|\boldsymbol{a}\| \geq \varphi^I(\| \boldsymbol{p}(t, \boldsymbol{a}, t_0) \|),$$

$$\| \boldsymbol{p}(\tau, \boldsymbol{x}, t) \| \geq \varphi^I(\| \boldsymbol{p}(t, \boldsymbol{p}(\tau, \boldsymbol{x}, t), t_0) \|).$$

The last inequality is valid for all t_0, thus also for $t_0 = \tau$, for which the argument on the right is equal to $\|\boldsymbol{x}\|$. Hence we have

$$\| \boldsymbol{p}(\tau, \boldsymbol{x}, t) \| \geq \varphi^I(\|\boldsymbol{x}\|)$$

so that $w(\boldsymbol{x}, t)$ is positive definite. Along any motion

$$w(t) := w\big(\boldsymbol{p}(t, \boldsymbol{x}, t_0), t\big)$$

$$= \min_{t_0 \leq \tau \leq t} \| \boldsymbol{p}(\tau, \boldsymbol{p}(t, \boldsymbol{x}, t_0), t) \| = \min_{t_0 \leq \tau \leq t} \| \boldsymbol{p}(\tau, \boldsymbol{x}, t_0) \|$$

is clearly non-increasing and hence $Dw \leq 0$.

Incidentally, it is not possible to give for each autonomous equation a function v independent of t which satisfies the hypotheses of the stability theorem. An example is given by the singular point of the character of a center-focus (example (21.14)). Let us assume here that a function $v(x, y)$ exists whose derivative along each motion is non-positive. It would follow that $v(x, y)$ cannot assume a smaller value at the origin than on any of the infinitely many circles among which the spirals run. For in every other annulus the spirals proceed from the inside out. But such a function is not positive definite.

KURZWEIL [2] has also shown that under somewhat stronger assumptions on the motions more can be said about the Liapunov function. For this purpose the function $w(\boldsymbol{x}, t)$ in Theorem 48.2 must be "smoothed out". We illustrate this procedure under the assumption that the motion is defined by a differential equation in R_n

$$\dot{\boldsymbol{x}} = \boldsymbol{f}(\boldsymbol{x}, t), \quad \boldsymbol{f} \in E, \quad \boldsymbol{f} \in C_1,$$

$$|\boldsymbol{x}| \leq h, \quad t \geq t_0 \geq 0,$$

which satisfies a Lipschitz condition. We first prove two lemmas.

Lemma 48.3. Let $P(x, t)$ be a scalar function defined for $x \in R_n$, $t \geq 0$, and continuous with respect to x and t. Suppose $P(0, t) = 0$, $P(x, t) > 0$ for $x \neq 0$, and $P(x, t_1) \geq P(x, t_2)$ for $t_1 \leq t_2$. Then there exists a continuous scalar function $g(x, t)$ defined for $x \in R_n$, $t \geq 0$, with the following properties:

   a) $g(0, t) = 0$, $g(x, t) > 0$ for $x \neq 0$, $g(x, t_1) \geq g(x, t_2)$ for $t_1 \leq t_2$.

   b) $2P(x, t) \geq g(x, t) \geq \frac{1}{2} P(x, t)$.

   c) $g(x, t)$ has continuous first order partial derivatives with respect to $x_1, \ldots, x_n, t$, for $|x| > 0$, $t \geq 0$.

Proof. Let $r \geq 0$ and s be integers, s fixed. Let I_s denote the spherical shell

$$2^{s-3} \leq |x|^2 \leq 2^{s+4}.$$

Define the sequence $\{\eta_r\}$ by

$$\eta_r := \min P(x, t); \quad x \in I_s, \quad 0 \leq t \leq r.$$

(η_r is obviously monotone decreasing) and define the sequence $\{t_i\}$ by

$$t_0 := 0, \quad t_1 := \frac{1}{u_1}, \quad t_2 := \frac{2}{u_1}, \ldots, \quad t_{u_1} := 1$$

where the positive integer u_1 is chosen so large that

$$P(x, t_i) - P(x, t_{i+1}) < \frac{1}{4}\eta_1, \quad x \in I_s, \quad i \leq u_1 - 1.$$

This is possible because of the continuity of P.

   We now choose a number u_2 and define

$$t_{u_1+1} := 1 + \frac{1}{u_2}, \quad t_{u_1+2} := 1 + \frac{2}{u_2}, \ldots, t_{u_1+u_2} = 2$$

so that

$$P(x, t_i) - P(x, t_{i+1}) < \frac{1}{4}\eta_2, \quad x \in I_s, \quad u_1 \leq i \leq u_2 - 1.$$

Continuing in this manner we obtain the sequence $\{t_i\}$. Finally, a sequence $\{\xi_i\}$ is defined such that

$$\xi_i > 0, \quad \xi_i \geq \xi_{i+1}, \quad \sum_{i=0}^{\infty}\xi_i < \frac{1}{8}\eta_1, \quad \sum_{i=u_1}^{\infty}\xi_i < \frac{1}{8}\eta_2, \quad \sum_{i=u_1+u_2}^{\infty}\xi_i < \frac{1}{8}\eta_3, \ldots$$

Since the function $P(x, t)$ is continuous with respect to x there exists a sequence of functions $w_i(x)$ such that

$$P(x, t_i) + \sum_{j=i}^{\infty}\xi_j - w_i(x) < \frac{1}{2}\xi_i.$$

The $w_i(x)$ are monotone decreasing, x being fixed, and may without loss of generality be assumed to be differentiable. Now let $q(t, \tau_1, \tau_2, a, b)$ be a third degree polynomial in t with real coefficients which depends on four real parameters, in fact let

$$q(t, \tau_1, \tau_2, a, b) := a + 3(b - a)\left(\frac{t - \tau_1}{\tau_2 - \tau_1}\right)^2 - 2(b - a)\left(\frac{t - \tau_1}{\tau_2 - \tau_1}\right)^3.$$

We are less interested in its explicit form than in its properties. For $\tau_1 \leq t \leq \tau_2$, q is monotone and

$$q(\tau_1, \tau_1, \tau_2, a, b) = a, \qquad q(\tau_2, \tau_1, \tau_2, a, b) = b,$$

$$\dot{q}(\tau_1, \tau_1, \tau_2, a, b) = \dot{q}(\tau_2, \tau_1, \tau_2, a, b) = 0.$$

We now define a function

$$\psi_s(x, t) := q\big(t, t_i, t_{i+1}, w_i(x), w_{i+1}(x)\big), \quad t_i \leq t \leq t_{i+1}, \quad i = 0, 1, \ldots.$$

It is defined and differentiable with respect to x at least in the domain $J_s = \{x \mid 2^{s-2} \leq |x|^2 \leq 2^{s+3}\}$. $\psi_s(x, t)$ is continuous and monotone decreasing with respect to t. The estimate

$$(48.5) \qquad 2P(x, t) \geq \psi_s(x, t) \geq \frac{1}{2}P(x, t), \quad x \in J_s, \quad t \geq 0,$$

is obtained as follows: We choose $x' \in J_s$, $t' > 0$, and then $i \geq 0$, $m \geq 0$ so that $t_i \leq t' < t_{i+1}$ and

$$\sum_{j=0}^{m} u_j \leq i < \sum_{j=0}^{m+1} u_j \quad (u_0 := 0).$$

We then have, for the difference,

$$\psi_s(x', t') - P(x', t') \leq \psi_s(x', t_i) - P(x', t_{i+1})$$

$$\leq |\psi_s(x', t_i) - P(x', t_i)| + P(x', t_i) - P(x', t_{i+1})$$

$$\leq \sum_{j=i}^{\infty} \xi_j + \frac{1}{2}\xi_i + \frac{1}{4}\eta_{m+1} < \frac{1}{8}\eta_{m+1} + \frac{1}{8}\eta_{m+1} + \frac{1}{4}\eta_{m+1} = \frac{1}{2}\eta_{m+1}$$

because of the definition of ξ_i, u_i, and m as well as η_m. We have $t_i < t_{i+1} \leq t_{u_1 + \cdots + u_{m+1}} = m + 1$. Similarly

$$P(x', t') - \psi_s(x', t') \leq P(x', t_i) - \psi_s(x', t_{i+1})$$

$$\leq P(x', t_i) - P(x', t_{i+1}) + |P(x', t_{i+1}) - \psi_s(x', t_{i+1})|$$

$$\leq \frac{1}{4}\eta_{m+1} + \sum_{j=i+1}^{\infty}\xi_j + \frac{1}{2}\xi_{i+1} \leq \frac{1}{4}\eta_{m+1} + \frac{1}{8}\eta_{m+1} + \frac{1}{8}\eta_{m+1} = \frac{1}{2}\eta_{m+1}.$$

Therefore

$$|P(x', t') - \psi_s(x', t')| \leq \frac{1}{2}\eta_{m+1} \leq \frac{1}{2}\eta_m,$$

$$P(x', t') - \frac{1}{2}\eta_m \leq \psi_s(x', t') \leq P(x', t') + \frac{1}{2}\eta_m.$$

Because of the definition of m and η_m, this implies (48.5). We now define a sequence of continuously differentiable functions $\lambda_s(v)$ of a real variable v. $\lambda_s(v)$ is to be positive in the interval $2^{s-1} < v < 2^{s+2}$ and is to vanish outside of this interval. Then if

$$\lambda(v) := \sum_{s=-\infty}^{+\infty} \lambda_s(v)$$

then

(48.6) $\qquad g(x, t) := \frac{1}{\lambda(|x|^2)} \sum_{s=-\infty}^{+\infty} \psi_s(x, t)\,\lambda_s(|x|^2), \qquad g(0, t) = 0,$

is the function whose existence is asserted in the lemma. The differentiability is obvious. Since $\psi_s(x, t)$ is monotone decreasing with respect to t, the same is true for $g(x, t)$. The property b) follows from (48.5); for in the domain J_s

$$g(x, t) = \frac{1}{\lambda(|x|^2)} \sum_{i=s-1}^{s+1} \psi_i(x, t)\,\lambda_i(|x|^2)$$

is a finite sum.

Lemma 48.4. Let $h(x, t)$ be a continuous function defined for $x \in R_n$, $t \geq 0$, with continuous first order partial derivatives for $x \neq 0$ and let $h(0, t) = 0$. Then there exists an odd, continuously differentiable function $\sigma(u)$ of the real variable u, such that 1) $\sigma'(u) > 0$, $u \neq 0$, 2) $\sigma'(u)$ is increasing for $u \geq 0$, 3) the function $\sigma(h(x, t))$ has continuous first order partial derivatives with respect to x and t for *all* $x \in R_n$, $t \geq 0$.
Proof. For $0 \leq v \leq 1$ set

$$\varrho_i(v) := \max\left(\frac{\partial h}{\partial x_j}, \left|\frac{\partial h}{\partial t}\right|\right) \quad \text{for} \quad v \leq |x|^2 \leq 1, \quad 0 \leq t \leq i,$$

and for $v \geq 0$ set

$$\chi_i(v) := \max |h(x, t)| \quad \text{for} \quad 0 \leq |x|^2 \leq v, \quad 0 \leq t \leq i.$$

Also let $\tau(v)$ be a continuously differentiable function such that $\tau(0) = 0$, $\tau'(v) > 0$ for $v > 0$, $\tau'(v)$ is increasing, and

$$\lim_{v \to 0} \tau'\big(\chi_i(v)\big)\varrho_i(v) = 0, \quad i = 1, 2, 3, \ldots.$$

The function

$$\sigma(v) := \begin{cases} \tau(v), & v \geq 0, \\ -\tau(-v), & v < 0 \end{cases}$$

has the desired properties. For if all the functions $\varrho_i(v)$ vanish for small values of v then $h(x, t)$ is constant and the theorem is trivial. Otherwise, $\lim \tau'(v) = 0$ and $\lim \sigma'(v) = 0$ as $v \to 0+$. In order to see that

$$\lim_{|x| \to 0} \frac{\partial \sigma(h(x, t))}{\partial x_j} = 0, \quad j = 1, 2, \ldots, n, \quad t \geq 0,$$

we choose an integer $i > t_0$. Then for $0 \leq t \leq i$, $0 < |\boldsymbol{x}|^2 < 1$, we have

$$\frac{\partial \sigma(h(\boldsymbol{x}, t))}{\partial x_j} = \sigma'(h(\boldsymbol{x}, t)) \cdot \frac{\partial h(\boldsymbol{x}, t)}{\partial x_j} \leq \tau'\left(\dot{\chi}_i(|\boldsymbol{x}|^2)\right) \varrho_i(|\boldsymbol{x}|^2)$$

and the right side tends to zero as $|\boldsymbol{x}| \to 0$. Thus the partial derivatives of $\sigma(h(\boldsymbol{x}, t))$ with respect to $\boldsymbol{x}$ exist everywhere and are continuous. The continuity of the partial derivative with respect to t is shown similarly.

To complete the proof of Theorem 48.2 (for differential equations in R_n) we use the auxiliary function

$$P(\boldsymbol{x}, t) = \min_{0 \leq \tau \leq t} |\boldsymbol{p}(\tau, \boldsymbol{x}, 0)|$$

which is connected with the function (48.4) by the relation

$$(48.7) \qquad\qquad P\big(\boldsymbol{p}(0, \boldsymbol{x}, t), t\big) = w(\boldsymbol{x}, t)$$

and which satisfies the hypotheses of Lemma 48.3. From P and the lemma we obtain the function $g(\boldsymbol{x}, t)$ and finally form the function $\tilde{g}(\boldsymbol{x}, t) := g(\boldsymbol{p}(0, \boldsymbol{x}, t), t)$. It is positive definite and decrescent, vanishes identically for $\boldsymbol{x} = 0$, and by construction has continuous partial derivatives with respect to $\boldsymbol{x}$ and t for $|\boldsymbol{x}| > 0$. To obtain differentiability as $\boldsymbol{x} \to 0$ we apply Lemma 48.4 which yields the function

$$v(\boldsymbol{x}, t) = \sigma\big(\tilde{g}(\boldsymbol{x}, t)\big).$$

The definiteness is not affected and since $v(\boldsymbol{x}, t)$ is non-increasing along any motion, $Dv \leq 0$. Lemma 48.4 implies the existence of continuous partial derivatives for all $\boldsymbol{x} \in R_n$ and $t \geq t_0$.

49. The Converse of Theorems on Asymptotic Stability

Theorem 49.1. If the equilibrium is equiasymptotically stable then there exists a strongly positive definite function with a non-positive derivative (converse of Theorem 42.3).

Proof. By hypothesis there exists an estimate

$$(49.1) \qquad \|\boldsymbol{p}(t, \boldsymbol{a}, t_0)\| \leq \varphi(\|\boldsymbol{a}\|; t_0)\, \sigma(t - t_0; t_0), \qquad \boldsymbol{a} \in K_h, \quad t \geq t_0.$$

The function

$$(49.2) \qquad\qquad w(\boldsymbol{x}, t) := \|\boldsymbol{p}(0, \boldsymbol{x}, t)\|\, (1 + e^{-t})$$

has a non-positive derivative because

$$w\big(\boldsymbol{p}(t, \boldsymbol{a}, t_0), t\big) = \|\boldsymbol{p}(0, \boldsymbol{p}(t, \boldsymbol{a}, t_0), t)\|(1 + e^{-t})$$
$$= \|\boldsymbol{p}(0, \boldsymbol{a}, t_0)\|(1 + e^{-t})$$

and hence

$$Dw = \|\boldsymbol{p}(0, \boldsymbol{a}, t_0)\|\, D(1 + e^{-t}) < 0.$$

(49.1) implies

$$\varphi(\|\boldsymbol{a}\|; t_0) \geq \frac{\|\boldsymbol{p}(t, \boldsymbol{a}, t_0)\|}{\sigma(t - t_0; t_0)}$$

and by taking inverses we obtain

$$\|\boldsymbol{a}\| \geq \varphi^I\left(\frac{\|\boldsymbol{p}(t, \boldsymbol{a}, t_0)\|}{\sigma(t - t_0; t_0)}; t_0\right).$$

If we set $\boldsymbol{p}(t, \boldsymbol{a}, t_0) = \boldsymbol{x}$ then we have $\boldsymbol{a} = \boldsymbol{p}(t_0, \boldsymbol{x}, t)$, and for $t_0 = 0$,

$$\|\boldsymbol{p}(0, \boldsymbol{x}, t)\| \geq \varphi^I\left(\frac{\|\boldsymbol{x}\|}{\sigma(t; 0)}; 0\right).$$

The function on the right belongs to class KK. The function $w(\boldsymbol{x}, t)$ is therefore strongly positive definite and possesses the desired properties.

Since uniform asymptotic stability is a special case of equiasymptotic stability the theorem is valid also for uniform asymptotic stability. It is clear that even for autonomous equations this construction yields a function v which explicitly depends on t.

The construction applies also to motions with a discretely varying t, so for instance to the solutions of difference equations (*cf.* sec. 50, example 8).

To construct Liapunov functions which have smoothness properties we must utilize a different procedure which was used for the first time by MASSERA [1]. A lemma, also due to Massera, is needed and we shall prove it first.

Lemma 49.2. Let the function $\beta(\tau)$ be defined for $0 \leq \tau < \infty$ and belong to class L. Let $\beta^*(\tau)$ denote a positive continuous function which satisfies $\beta^*(\tau) \leq \beta(\tau)$ for all τ. Finally let $\xi(\tau)$ be continuous, positive, and non-decreasing for $0 \leq \tau < \infty$. Then there exists a function $x(r)$ of class K, defined for $0 \leq r \leq \beta(0)$, such that the integral

$$(49.3) \qquad \int_0^\infty \alpha\left(\beta^*(\tau)\right)\xi(\tau)\, d\tau$$

converges uniformly with respect to the set of functions $\beta^*(\tau)$.

Proof. Since $\beta(\tau) \in L$ we can choose a sequence t_n such that

$$\beta(t_n) \leq \frac{1}{n+1}, \qquad n = 1, 2, \ldots.$$

We use this sequence to define a function $\eta(t)$ as follows: a) $\eta(t_n) = \frac{1}{n}$, b) between t_n and t_{n+1}, $\eta(t)$ is linear, c) in the interval $0 < t \leq t_1$, $\eta(t) = (t_1/t)^p$, where p is a positive integer chosen so large that the derivative $\eta'(t)$ has a positive jump at t_1, $\eta'(t_1 - 0) < \eta'(t_1 + 0)$. The function η belongs to class L and, for $t \geq t_1$, we have $\beta(t) < \eta(t)$. As t tends to $0+$, $\eta(t)$ grows without bound. The inverse $\eta^I(t)$ also belongs to L and also grows without bound as $t \to 0+$. Obviously

$$\eta^I\left(\beta^*\left(\tau\right)\right) \geq \eta^I\left(\beta\left(\tau\right)\right) > \eta^I\left(\eta\left(\tau\right)\right) = \tau.$$

If we define $\alpha\left(0\right) = 0$ and

$$\alpha\left(r\right) := \frac{\exp\left(-\eta^I\left(r\right)\right)}{\xi\left(\eta^I\left(r\right)\right)}, \qquad r > 0,$$

then

$$\alpha\left(\beta^*\left(\tau\right)\right) = \frac{\exp\left(-\eta^I\left(\beta^*\left(\tau\right)\right)\right)}{\xi\left(\eta^I\left(\beta^*\left(\tau\right)\right)\right)} < \frac{\exp\left(-\tau\right)}{\xi\left(\tau\right)}$$

and

$$\int\limits_{0}^{\infty} \alpha\left(\beta^*\left(\tau\right)\right)\xi\left(\tau\right)d\tau < \int\limits_{0}^{\infty} \frac{e^{-\tau}}{\xi\left(\tau\right)}\,\xi\left(\tau\right)d\tau = 1.$$

Since the bound on the right is independent of $\beta^*\left(\tau\right)$ the estimate is valid uniformly with respect to the functions $\beta^*\left(\tau\right) \leq \beta\left(\tau\right)$.

Theorem 49.3. Let the equilibrium of the family $p\left(t, a, t_0\right)$ be uniformly asymptotically stable. Then there exists a positive definite decrescent function $v\left(x, t\right)$ with a negative definite derivative.

Proof. a) We first consider the case of t varying continuously. We set

$$(49.4) \qquad z\left(x, t\right) := \int\limits_{t}^{\infty} u\left(\|p\left(\tau, x, t\right)\|\right) d\tau$$

where $u\left(r\right)$ denotes a function of class K which will be discussed later as will be the convergence of the integral (49.4). Since the integrand is independent of t along a fixed motion (*cf.* the proof of Theorem 49.1)

$$Dz = -u\left(\|p\left(t, x, t\right)\|\right) = -u\left(\|x\|\right)$$

is negative definite. By hypothesis we have for $\|a\| \leq h$ an estimate

$$(49.5) \qquad \|p\left(t, a, t_0\right)\| \leq \varphi\left(\|a\|\right)\sigma\left(t - t_0\right).$$

Therefore, the integrand of (49.4) is no larger than

$$u\left(\varphi\left(\|x\|\right)\sigma\left(\tau - t\right)\right)$$
$$\leq \left[u\left(\varphi\left(\|x\|\right)\sigma\left(0\right)\right)\right]^{1/2} \left[u\left(\varphi\left(h\right)\sigma\left(\tau - t\right)\right)\right]^{1/2}$$

provided $\|x\| \leq h$ (cf. sec. 24 A f). Accordingly

$$z\left(x, t\right) \leq \left[u\left(\varphi\left(\|x\|\right)\sigma\left(0\right)\right)\right]^{1/2} \int\limits_{t}^{\infty} \left[u\left(\varphi\left(h\right)\sigma\left(\tau - t\right)\right)\right]^{1/2} d\tau.$$

The integral can be written

$$(49.6) \qquad \int\limits_{0}^{\infty} \left[u\left(\varphi\left(h\right)\sigma\left(\tau\right)\right)\right]^{1/2} d\tau.$$

To ensure its convergence we use Lemma 49.2, setting

$$\beta(\tau) := \varphi(h)\,\sigma(\tau),\ \xi(\tau) \equiv 1$$

and

$$u(r) = \big(\alpha(r)\big)^2.$$

Because by the lemma the integral converges uniformly with respect to a set of functions dominated by $\beta(\tau)$, it follows that $z(x, t)$ is continuous with respect to $\|x\|$; for the function β depends by construction on the number h which is a bound for the norm $\|x\|$.

Since by construction of the function $\alpha(r)$, the integral (49.6) is smaller than 1 it follows that

$$z(x, t) \leq \big[u\big(\varphi(\|x\|)\,\sigma(0)\big)\big]^{1/2}$$

and since the function on the right belongs to K, $z(x, t)$ is decrescent. Now, if $z(x, t)$ happens to be positive definite we can set $v(x, t) = z(x, t)$. Otherwise, we proceed as follows.

Since the hypotheses of Theorem 48.2 are fulfilled we can construct the function $w(x, t)$ of this theorem. Then

$$v(x, t) := z(x, t) + w(x, t)$$

has the properties stated in Theorem 49.3. For we have

$$w(x, t) \leq v(x, t) \leq z(x, t) + \|x\|,$$

$$Dv = Dz + Dw \leq Dz.$$

b) If t varies discretely we define

$$(49.7) \qquad v(x, t) := \sum_{n=0}^{\infty} u\big(\|p(t + n, x, t)\|\big)$$

and choose the function $u \in K$ so that the series converges. That such a choice is possible follows from Lemma 49.2. To apply this lemma we must majorize the series by means of an integral. The definiteness of (49.7) follows from the inequality

$$v(x, t) \geq u(\|x\|).$$

The remaining properties can be discussed as above.

From the proof of Theorem 49.3 we can draw several further conclusions.

1) Frequently by means of a slight modification of (49.4) a positive definite function can be obtained, in which case the additional term $w(x, t)$ can be dispensed with. Let

$$(49.8) \qquad Z(x, t) := \int_{t}^{\infty} u\big(\|p(\tau, x, t)\|\big)\,\gamma(\tau)\,d\tau,$$

where $\gamma(\tau)$ is defined for $0 \leq \tau < \infty$, continuous, positive and non-decreasing. We shall dispose of $\gamma(\tau)$ later on. One proves as above that $Z(\boldsymbol{x}, t)$ is decrescent and has a negative definite derivative. Applying Lemma 49.2 we set

$$\beta(\tau) := \varphi(h)\,\sigma(\tau), \quad \xi(\tau) := \gamma(\tau + t).$$

We assume that in addition to the hypotheses of Theorem 49.3 the family of motions has the following property:

The function

$$(49.9) \qquad y(t) := \|\boldsymbol{p}(t, \boldsymbol{a}, t_0)\|$$

satisfies an estimate

$$(49.10) \qquad \left|\frac{dy}{dt}\right| < g(t)$$

for $\|\boldsymbol{a}\| \leq h$ where $g(t)$ is continuous, positive and independent of t_0. In order to obtain an estimate for $Z(\boldsymbol{x}, t)$ from below we first suppose that the function $y(t)$ is monotone decreasing. Introducing y in the integral (49.4) instead of τ one gets

$$Z(\boldsymbol{x}, t) = \int\limits_{y=\|\boldsymbol{x}\|}^{y=0} u(y) \left(\frac{dy}{d\tau}\right)^{-1} \gamma(\tau)\,dy \geq \int\limits_{0}^{\|\boldsymbol{x}\|} u(y) \frac{\gamma(\tau)}{g(\tau)}\,dy$$

where τ is to be considered as a function of y. Choosing $\gamma(t) \geq g(t)$ one obtains

$$Z(\boldsymbol{x}, t) \geq \int\limits_{0}^{\|\boldsymbol{x}\|} u(y)\,dy$$

which means definiteness. If y is increasing in certain intervals $t'_j < t < t''_j$, $j = 1, 2, \ldots$, we omit them and restrict the integration to the remaining t-axis. Then the substitution (49.9) and the subsequent argument can be carried through as above.

2) The condition (49.10) is satisfied, for instance, if the motion is defined by a differential equation in R_n,

$$(49.11) \qquad \dot{\boldsymbol{x}} = \boldsymbol{f}(\boldsymbol{x}, t), \quad \boldsymbol{f} \in E.$$

Then we have

$$|\boldsymbol{x}|\frac{d|\boldsymbol{x}|}{dt} = \boldsymbol{x}^T \boldsymbol{f}(\boldsymbol{x}, t)$$

and by (49.9)

$$\left|\frac{dy}{dt}\right| \leq |\boldsymbol{f}(\boldsymbol{p}(t, \boldsymbol{a}, t_0), t)|.$$

Because of the uniform stability, the right hand side can be estimated uniformly with respect to $|\boldsymbol{a}| \leq h$ and $t_0 \geq 0$, and (49.10) follows.

If $\boldsymbol{f}(\boldsymbol{x}, t)$ is bounded, the factor $\gamma(\tau)$ in (49.8) can be taken as a constant, $\gamma(\tau) = 1$. If $\boldsymbol{f}(\boldsymbol{x}, t)$ is independent of t or periodic $z(\boldsymbol{x}, t)$ is independent of t or periodic, respectively. This follows from (38.6). Since in this case asymptotic stability implies uniform stability by Theorem 38.5, we have

Theorem 49.4. If the equilibrium of a differential equation (difference equation) with constant or periodic coefficients is asymptotically stable then there exists a positive definite decrescent Liapunov function which is independent of t, respectively periodic in t, with a negative definite derivative.

MASSERA [1] was the first to prove this converse theorem in the form 49.4; however, he made further assertions about the smoothness of v (see sec. 51). The rôle played by the uniform stability was not recognized until somewhat later by MALKIN [4].

3) If the norm (49.9) satisfies a differential inequality

$$\frac{dy}{dt} \leq - m(t)\, v$$

with bounded $m(t)$, $0 < m(t) \leq c$, then, in (49.8), one can take $u(r) = r$ and $\gamma(t) = 1$. For the derivative we then have

$$Dv = Dz = - \|x\|.$$

In lieu of this equation we can write two inequalities

$$Dv \geq - \|x\| \quad \text{or} \quad Dv \leq - \|x\|,$$

and we can follow the outline of the proof of Theorem 42.4 and obtain an estimate for $\|p(t, a, t_0)\|$ from below. We obtain

Theorem 49.5. If the equilibrium is uniformly asymptotically stable and if the function $y(t) = \|p(t, a, t_0)\|$ satisfies a differential inequality

$$\frac{dy}{dt} \leq - m(t)\, y$$

where $m(t)$ is bounded, then there exists an estimate

(49.12) $$\|p(t, a, t_0)\| \geq \varphi(\|a\|)\, \sigma(t - t_0)$$

of type A' 7.

For differential equations $\dot{x} = f(x, t)$ the hypothesis on $m(t)$ in Theorem 49.5 is clearly satisfied if $f(x, t)$ is bounded with respect to t. For then (cf. above 2.)

$$\left| \frac{dy}{dt} \right| \leq |f(x, t)| .$$

4) If we apply the construction (49.4) to the case of non-uniform but equiasymptotic stability then we obtain a positive definite function $v(x, t)$ with a non-negative derivative. To prove Theorem 49.1 a little more is needed, namely the proof that the constructed function is strongly positive definite. This is difficult because the parameter t_0 appearing on the right in (49.1) must be taken into consideration in the construction

and it is not clear what rôle it plays. In many concrete cases (49.4) yields a Liapunov function which is usable also in the equiasymptotic case.

5) Let the family of motions be defined by a differential equation valid in all of R_n. Let the equilibrium be uniformly asymptotically stable in the whole and assume that the estimate (49.5) is valid for all $a \in R_n$. Then the Liapunov function defined by (49.4) is radially unbounded and a converse of Theorem 42.5 is obtained.

50. Examples for the Converse Theorems

1) The general solution of the scalar system of equations

$$\dot{x}_1 = -a^2 x_2, \quad \dot{x}_2 = b^2 x_1, \quad a \geq b,$$

is

$$x_1 = x_{10} \cos \left(a b (t - t_0) \right) - \frac{a}{b} x_{20} \sin \left(a b (t - t_0) \right),$$

$$x_2 = x_{10} \frac{b}{a} \sin \left(a b (t - t_0) \right) + x_{20} \cos \left(a b (t - t_0) \right).$$

The Liapunov function obtained from Theorem 48.1 ($t_0 = 0$) is

$$\left(x_1^2 + x_2^2 \right) \cos^2 (a b t) + \left(\frac{b^2}{a^2} x_1^2 + \frac{a^2}{b^2} x_2^2 \right) \sin^2 (a b t)$$

$$+ 2 x_1 x_2 \left(\frac{a}{b} - \frac{b}{a} \right) \sin (a b t) \cos (a b t).$$

Because of uniformity we can also apply Theorem 48.2. Here the minimum of the expression $x_1^2 + x_2^2$ is equal to

$$a^{-2} \left(b^2 x_{10}^2 + a^2 x_{20}^2 \right).$$

This is the squared length of the minor half-axis of the ellipse for which the ratio of the axes is $a : b$ and which passes through the initial point (x_{10}, x_{20}). This yields

$$v = b^2 x_1^2 + a^2 x_2^2$$

to within a constant.

2) We have frequently utilized the equation (21.14)

$$\dot{r} = r^2 \sin \left(\frac{1}{r} \right), \quad \dot{\varphi} = 1.$$

Since

$$r^{-1} = 2 \arctan \left(e^{t_0 - t} \tan \frac{1}{2 r_0} \right),$$

Theorem 48.1 yields

$$v(r, t) = \left[2 \arctan \left(e^t \tan \frac{1}{2 r} \right) \right]^{-1}.$$

The arctan is not single-valued but this is compensated for by the fact that for $t = 0$, $v = r$.

3) For the scalar equation $\dot{x} = -x$ Theorem 49.1 produces the function

$$v(x, t) = |x|(e^t + 1), \text{ or } v^2 = x^2(e^t + 1)^2,$$

so that

$$\dot{v} = -|x|, \text{ resp. } \frac{d}{dt}(v^2) = -2x^2(e^t + 1).$$

For

$$\dot{x} = -x/(1 + t)$$

we obtain

$$v^2 = (1 + t)^2(1 + e^{-t})^2 x^2, \quad \frac{d}{dt}(v^2) = -e^{-t}(1 + t)^2(1 \div e^{-t})x^2.$$

4) If

$$\dot{\boldsymbol{x}} = A\,\boldsymbol{x}$$

is a linear autonomous equation in R_n then

$$\boldsymbol{p}(t, \boldsymbol{a}, t_0) = e^{A(t-t_0)}\,\boldsymbol{a}.$$

Setting $u(r) = r^2$ in (49.4) we obtain

$$v(\boldsymbol{x}) = \boldsymbol{x}^T \left(\int_0^\infty e^{A^T \tau}\, e^{A\tau}\, d\tau \right) \boldsymbol{x}.$$

Comparison with the formulas of sec. 27 shows that this is exactly the construction described there with the special choice $C = E$. For the autonomous difference equation

$$\theta\,\boldsymbol{x} = \boldsymbol{x}(t + 1) = A\,\boldsymbol{x}$$

the corresponding construction yields

$$v(\boldsymbol{x}) = \boldsymbol{x}^T \sum_{n=0}^\infty (A^T A)^n \boldsymbol{x} = \boldsymbol{x}^T (E - A^T A)^I \boldsymbol{x}.$$

5) For the equation

$$\dot{x} = -x/(1 + t)$$

which we already considered as an example for Theorem 49.1 we can also use (49.4) with $u(r) = r^2$. Then

$$v(x, t) = (1 + t)x^2.$$

6) The example of the scalar equation

$$\dot{x} = -2tx$$

shows that in Theorem 49.5 we cannot do without the hypothesis on $m(t)$. The general solution $p(t, x_0, t_0) = x_0 \exp(t_0^2 - t^2)$ of this equation cannot be estimated by a function of the form (49.12) which is independent of t_0: From an inequality

$$|x_0| \exp(t_0^2 - t^2) \geq |x_0|\sigma(t - t_0)$$

it follows that

$$\sigma(t - t_0) \leq \exp\left((t_0 + t)(t_0 - t)\right)$$

and, for $t - t_0 = T$,

$$\sigma(T)^{-1/T} e^{-T} \geq e^{2t_0}.$$

This would have to be valid for arbitrary T and t_0, which is impossible.

7) If the equilibrium of the family $p(t, a, t_0)$ is exponentially stable (Def. 26.2) then

$$\|x\| = \|p(t, a, t_0)\| \leq \beta \|a\| \exp\left(-\alpha(t - t_0)\right)$$

for some positive α, β. The integral (49.4) converges and is positive definite if we choose $u(r) = r^2$. In this case the Liapunov function satisfies an estimate

$$v(x, t) \leq c\|x\|^2$$

(*cf.* also secs. 25 und 56).

8) The general solution of the scalar difference equation

$$\theta x = \gamma x$$

is

$$p(t, x_0, t_0) = x_0 \gamma^{t-t_0}.$$

For $0 < \gamma < 1$, Theorem 49.1 yields the function

$$w(x, t) = |x| \gamma^{-t}(1 + e^{-t})$$

for which

$$Dw = |x| \gamma^{-t}(e^{-1} - 1) e^{-t}.$$

9) The general solution of the scalar differential difference equation

$$\dot{x} = a x(t) + b x(t - \tau)$$

can be written in the form

$$p(t, w, 0) = x_0 k(t) + \int_0^t k(t - u) q(u) du \qquad (t \geq 0)$$

by sec. 12. Here

$$w(t) = x(t - \tau) \text{ for } 0 \leq t \leq \tau$$

is the initial function, $x_0 = w(\tau)$ the initial value. Furthermore $k(t)$ is the inverse Laplace transform of the transfer function $(s - a - be^{-s\tau})^{-1}$, and $q(t)$ denotes the inverse Laplace transform of

$$\int_0^\tau e^{-us} w(u) du.$$

The Liapunov function computed by (49.4) is a homogeneous quadratic function of x_0 and $w(t)$ of a rather complicated appearance.

51. Refinements of the Converse Theorems for Ordinary Differential Equations

We already pointed out that in the proofs of the converse theorems it is desirable to construct Liapunov functions which are as smooth as possible. Since the immediate construction of such functions is difficult we usually proceed in two steps. We somehow construct a continuous function with the proper type of definiteness and then subject this function to a smoothing process such as we described in sec. 48. This is illustrated in the proof of the following theorem.

Theorem 51.1.[1]) Let

$$(51.1) \qquad \dot{x} = f(x), \quad f \in E, \quad f \in C_1, \quad x \in R_n,$$

be an autonomous differential equation whose right side has continuous first order partial derivatives. Let the equilibrium be asymptotically stable and let A be the domain of attraction. Then there exist two functions $v(x)$ and $h(x)$ satisfying the hypotheses of Theorem 34.1; $h(x)$ can be so chosen that v has continuous first order partial derivatives.

The conditions of Theorem 34.1 are therefore necessary and sufficient.

Proof. We may assume that (51.1) is defined for all $x \in R_n$ and that the solutions exist for all values of the independent variable, $-\infty < t < +\infty$. This can be arranged by means of the transformation (16.9). Since the system is autonomous the solution can be written in the form $p(t, x)$ (see sec. 16). By hypothesis (*cf.* Theorem 38.5) there exists an estimate

$$(51.2) \qquad |p(t, x)| \leq \varphi(|x|) \sigma(t), \quad |x| \leq h, \quad t \geq 0.$$

We now set

$$(51.3) \qquad u(r) := r \exp\left(-\sigma^I(r)\right),$$

$$(51.4) \qquad w(x) := 1 - \exp\left(-\int_0^\infty u(|p(\tau, x)|)\, d\tau\right).$$

We first show that the integral converges. As in the proof of Massera's lemma 49.2 we choose t_n so large that

$$\varphi\left(|p(t_n, x)|\right) = \frac{1}{n}$$

and we consider the intervals of integration $(0, t_n)$ and (t_n, ∞) separately. The integral over the second interval is (*cf.* (16.8))

$$\int_{t_n}^\infty u(|p(\tau, x)|)\, d\tau = \int_0^\infty u(|p(t_n + \tau, x)|)\, d\tau = \int_0^\infty u(|p(\tau, p(t_n, x))|)\, d\tau$$

---

[1]) ZUBOV [2, 4].

The argument of the last integrand can be estimated by (51.2):

$$\left| \boldsymbol{p}\left(\tau, \boldsymbol{p}\left(t_n, \boldsymbol{x}\right)\right)\right| \leq \varphi\left(\left|\boldsymbol{p}\left(t_n, \boldsymbol{x}\right)\right|\right) \sigma(\tau) = \frac{1}{n}\, \sigma(\tau) \leq \sigma(\tau)$$

so that because of (51.3) and because

$$\sigma^I\left(\frac{1}{n}\, \sigma(\tau)\right) \geq \tau$$

we have

$$u\left(\left|\boldsymbol{p}\left(\tau, \boldsymbol{p}\left(t_n, \boldsymbol{x}\right)\right)\right|\right) \leq \frac{1}{n}\, \sigma(\tau) \exp\left(-\sigma^I\left(\frac{\sigma(\tau)}{n}\right)\right) \leq \frac{1}{n}\, \sigma(\tau)\, e^{-\tau}.$$

The integral on the second interval is therefore $O\left(\dfrac{1}{n}\right)$ and this estimate is valid uniformly in $\boldsymbol{x}$. This proves the existence of the function (51.4) as well as its continuity. Obviously $w(0) = 0$ and $0 < w(\boldsymbol{x}) < 1$ for $\boldsymbol{x} \neq 0$, $\boldsymbol{x} \in A$. Thus $w(\boldsymbol{x})$ is positive definite and since it is continuous it is also decrescent.

We also have

$$1 - w(\boldsymbol{x}) = \exp\left(-\int_0^t u\left(\left|\boldsymbol{p}\left(\tau, \boldsymbol{x}\right)\right|\right) d\tau\right) \cdot \exp\left(-\int_t^\infty u\left(\left|\boldsymbol{p}\left(\tau, \boldsymbol{x}\right)\right|\right) d\tau\right).$$

The second factor is, as we have just shown, equal to

$$\exp\left(-\int_0^\infty u\left(\left|\boldsymbol{p}\left(\tau, \boldsymbol{p}\left(t, \boldsymbol{x}\right)\right)\right|\right) d\tau\right) = 1 - w\left(\boldsymbol{p}\left(t, \boldsymbol{x}\right)\right).$$

Hence

$$(51.5) \quad \left(1 - w(\boldsymbol{x})\right) \exp\left(\int_0^t u\left(\left|\boldsymbol{p}\left(\tau, \boldsymbol{x}\right)\right|\right) d\tau\right) = 1 - w\left(\boldsymbol{p}\left(t, \boldsymbol{x}\right)\right).$$

We differentiate with respect to t, $\boldsymbol{x}$ being fixed, and obtain

$$\frac{\partial}{\partial t}\, w\left(\boldsymbol{p}\left(t, \boldsymbol{x}\right)\right) = -u\left(\left|\boldsymbol{p}\left(t, \boldsymbol{x}\right)\right|\right)\left(1 - w\left(\boldsymbol{p}\left(t, \boldsymbol{x}\right)\right)\right).$$

With an appropriate choice of the notation, this is the differential equation of Theorem 34.1.

In order to prove property d) of Theorem 34.1 we choose a point $\bar{\boldsymbol{x}}$ on the boundary of A and a sequence $\boldsymbol{x}_n \in A$ which converges to $\bar{\boldsymbol{x}}$. Let t_n be the time at which the trajectory $\boldsymbol{p}\left(t, \boldsymbol{x}_n\right)$ meets the fixed sphere $|\boldsymbol{x}| = \eta$ for the first time. By Theorem 33.2 the sequence $|t_n|$ is unbounded. (51.5) implies

$$1 - w\left(\boldsymbol{p}\left(t_n, \boldsymbol{x}_n\right)\right) = \left(1 - w(\boldsymbol{x}_n)\right) \exp\left(\int_0^{t_n} u\left(\left|\boldsymbol{p}\left(\tau, \boldsymbol{x}_n\right)\right|\right) d\tau\right).$$

On the sphere $|\boldsymbol{x}| = \eta$, $1 - w(\boldsymbol{x})$ lies between two fixed bounds a_1 and a_2. Hence

$$a_1 < 1 - w\left(\boldsymbol{p}\left(t_n, \boldsymbol{x}_n\right)\right) < a_2 \quad \text{for } \left|\boldsymbol{p}\left(t_n, \boldsymbol{x}_n\right)\right| = \eta$$

and

$$a_1 \exp\left(- \int_0^{t_n} u\big(|\boldsymbol{p}(\tau, \boldsymbol{x}_n)|\big)\, d\tau\right) < 1 - w(\boldsymbol{x}_n)$$

$$< a_2 \exp\left(- \int_0^{t_n} u\big(|\boldsymbol{p}(\tau, \boldsymbol{x}_n)|\big)\, d\tau\right).$$

Since the function $u(r)$ is positive definite the integrands are bounded away from zero,

$$u\big(|\boldsymbol{p}(\tau, \boldsymbol{x}_n)|\big) \geq a_3 > 0 ,$$

and the integrals are unbounded because $t_n \to \infty$. Therefore it follows that

$$\lim w(\boldsymbol{x}_n) = 1 \quad \text{for} \quad \boldsymbol{x}_n \to \bar{\boldsymbol{x}}.$$

It now remains to smooth out the function $w(\boldsymbol{x})$ in such a way that it retains its other properties. For this purpose we choose a positive constant γ whose exact nature will be discussed later, and form the auxiliary functions

$$u_\gamma(s) := s \exp\left(-\gamma\, \sigma^I\left(\frac{s}{\varphi(h)}\right)\right),$$

$$U_\gamma(r) := \int_0^r u_\gamma(s)\, ds ,$$

$$w_\gamma(\boldsymbol{x}) := 1 - \exp\left(- \int_0^\infty U_\gamma\big(|\boldsymbol{p}(\tau, \boldsymbol{x})|\big)\, d\tau\right).$$

The function $w_\gamma(\boldsymbol{x})$ is thus obtained from $U_\gamma(r)$ in the same way in which previously $w(\boldsymbol{x})$ was obtained from $u(r)$. It possesses the same properties as $w(\boldsymbol{x})$.

At this point we enter an auxiliary consideration. If $\boldsymbol{p}(t, \boldsymbol{a}, t_0)$ is the general solution of the equation $\dot{\boldsymbol{x}} = \boldsymbol{f}(\boldsymbol{x}, t)$ then for each i, $i = 1, 2, \ldots, n$, we have identically in t

$$\frac{d}{dt}\, p_i(t, \boldsymbol{a}, t_0) = f_i\big(\boldsymbol{p}(t, \boldsymbol{a}, t_0), t\big).$$

If $\boldsymbol{f} \in C_1$ then this identity can be differentiated and we obtain for the derivatives $q_{ij} := \partial p_i / \partial a_j$ the equation

$$(51.6) \qquad \dot{q}_{ij} = \sum_{k=1}^n \frac{\partial f_i}{\partial x_k}\, q_{kj}, \quad i, j = 1, 2, \ldots, n.$$

From this linear differential equation we obtain an estimate

$$(51.7) \qquad |q_{ij}(t)| \leq k_1 e^{k_2 t}$$

where k_1, k_2 are certain constants.

16*

Because of the hypothesis on the differential equation, the function $U_\gamma(|\boldsymbol{p}(\tau, \boldsymbol{x})|)$ has a partial derivative with respect to x_i. We have

$$\frac{\partial U_\gamma(|\boldsymbol{p}(\tau, \boldsymbol{x})|)}{\partial x_i} = \frac{dU_\gamma(r)}{dr}\bigg|_{r=|\boldsymbol{p}(\tau,\boldsymbol{x})|} \cdot \frac{1}{|\boldsymbol{p}(\tau, \boldsymbol{x})|} \sum_{k=1}^{n} \frac{\partial p_k}{\partial x_i}\, p_k$$

$$= \frac{u_\gamma(|\boldsymbol{p}(\tau, \boldsymbol{x})|)}{|\boldsymbol{p}(\tau, \boldsymbol{x})|} \sum_{k=1}^{n} \frac{\partial p_k}{\partial x_i}\, p_k$$

and from (51.7) it follows that

$$\frac{\partial U_\gamma(|\boldsymbol{p}(\tau, \boldsymbol{x})|)}{\partial x_i} \leq \frac{u_\gamma(|\boldsymbol{p}(\tau, \boldsymbol{x})|)}{|\boldsymbol{p}(\tau, \boldsymbol{x})|}\, k_1 e^{k_2 \tau} \sum_{k=1}^{n} |p_k(\tau, \boldsymbol{x})| \leq k_3 e^{k_2 \tau} u_\gamma(|\boldsymbol{p}(\tau, \boldsymbol{x})|)$$

But by (51.2)

$$u_\gamma(|\boldsymbol{p}(\tau, \boldsymbol{x})|) = |\boldsymbol{p}(\tau, \boldsymbol{x})| \exp\left(-\gamma \sigma^I\left(\frac{|\boldsymbol{p}(\tau, \boldsymbol{x})|}{\varphi(h)}\right)\right)$$

$$\leq |\boldsymbol{p}(\tau, \boldsymbol{x})| \exp\left(-\gamma \sigma^I\left(\frac{\varphi(|\boldsymbol{x}|)\, \sigma(\tau)}{\varphi(h)}\right)\right)$$

$$\leq |\boldsymbol{p}(\tau, \boldsymbol{x})|\, e^{-\gamma \tau},$$

and consequently

$$\frac{\partial U_\gamma(|\boldsymbol{p}(\tau, \boldsymbol{x})|)}{\partial x_i} \leq k_3 |\boldsymbol{p}(\tau, \boldsymbol{x})| \exp\left((k_2 - \gamma)\, \tau\right).$$

By formally differentiating the function $w_\gamma(\boldsymbol{x})$ we obtain

$$\frac{\partial w_\gamma(\boldsymbol{x})}{\partial x_i} = (1 - w_\gamma(\boldsymbol{x})) \int_0^\infty \frac{\partial U_\gamma(|\boldsymbol{p}(\tau, \boldsymbol{x})|)}{\partial x_i}\, d\tau.$$

If we choose the constant $\gamma > k_2$ then the integral on the right converges absolutely and uniformly with respect to $\boldsymbol{x}$. The formal differentiation is therefore justified: The function $w_\gamma(\boldsymbol{x})$ has first order partial derivatives. The function $v(\boldsymbol{x}) := w_\gamma(\boldsymbol{x})$ satisfies the requirements of the theorem.

If the right side of the original differential equation has second order partial derivatives then we can integrate again and construct a twice differentiable Liapunov function, etc.

The converse theorems considered so far furnish closed analytic expressions for the Liapunov function $v(\boldsymbol{x})$, respectively $v(\boldsymbol{x}, t)$, and furthermore the construction can be carried out for arbitrary motions. Under more special assumptions the smoothing process allows us to obtain Liapunov functions which have first or higher order partial derivatives. However, considerably more can be shown. The next theorem which we quote without proof is due to MASSERA [2, 3].

Theorem 51.2. Let the differential equation $\dot{\boldsymbol{x}} = \boldsymbol{f}(\boldsymbol{x}, t)$ have a uniformly asymptotically stable equilibrium and suppose that in a domain K_{h, t_0}

the right side satisfies a Lipschitz condition ($f \in C_0$). Then there exists in K_{h,t_0} a positive definite decrescent Liapunov function with a negative definite derivative, which has partial derivatives of any order desired with respect to all of its variables. If $f \in \bar{C}_0$, *i.e.* if there exists a uniform Lipschitz constant, then v can be so determined that all the partial derivatives are uniformly bounded and that, in fact, the same bound can be used everywhere.

The proof is lengthy but ingenious: We start from an expression such as (49.5) and apply a transfinite smoothing process. Clearly, this is a pure existence theorem. For further converse theorems see MASSERA [5].

There exists an entirely different type of constructive proofs for converse theorems. As an example for such a construction we discuss a theorem of KRASOVSKII [4] which gives a necessary and sufficient condition for a differential equation to have a Liapunov function with a definite derivative. It is inessential in this case whether the equilibrium is asymptotically stable or unstable. In the autonomous case the condition consists in the existence of a neighborhood of the origin which is free of closed trajectories. For a linear system this is expressed in Theorem 27.2. For a non-autonomous equation in R_n

51.8) $\dot{x} = f(x, t), \quad f \in E, \quad f \in C_0$ for $x \in G \subset R_n, \quad t \geq t_0$,

where G is a certain domain containing the origin, the corresponding property of the trajectories is somewhat more complicated.

Def. 51.1. Let $h_k \to 0$ be a null sequence and let $B_k = \{x: |x| < h_k\}$ be a sequence of balls. The trajectories of the differential equation (51.8) are said *to possess property A* in case: For each bounded sub-domain H_0 of G and for each natural number k there exists a number T_k such that no segment of the trajectory $p(t, x_0, t_0)$, with $t_0 - T_k \leq t \leq T_k + t_0$, lies entirely within H_0, provided only that $t_0 \geq T_k$ and $x_0 \in G \setminus B_k$.

If the differential equation has constant or periodic coefficients property A is equivalent with the condition that no closed sub-domain of G contains complete trajectories ($-\infty < t < +\infty$) other than the equilibrium. It is easily seen that property A precludes the existence of complete trajectories in a neighborhood of the origin. To show the converse we proceed indirectly. Assume that each closed sub-domain $H_0 \subset G$ is free of complete trajectories but that property A is not satisfied. Then there exists an $\varepsilon > 0$ and a sequence x_{0n} such that $x_{0n} \in H_0$, $|x_{0n}| > \varepsilon$. a sequence t_{0n}, and an unbounded sequence $T_n \to \infty$ such that the segment

$$p(t_{0n} + \tau, x_{0n}, t_{0n}), \quad -T_n \leq \tau \leq T_n,$$

of the trajectory belongs to H_0. Since H_0 is compact we may assume that the sequence x_{0n} converges to an accumulation point $x_0 \in H_0$. If necessary a subsequence may be chosen. Let ω be the period and assume that the numbers t_{0n} have been reduced modulo ω to yield t'_{0n}, *i.e.* $0 \leq t'_{0n} < \omega$. Let t'_0 be an accumulation point of the numbers t'_{0n}. Because of the continuity the points $p(t'_0 + \tau, x_0, t'_0)$ are arbitrarily close to the points $p(t_{0n} + \tau, x_{0n}, t_{0n})$. This implies that

$$p(t'_0 + \tau, x_0, t'_0) \in H_0, \quad -T_n \leq \tau \leq +T_n,$$

for all sufficiently large n. Since $T_n \to \infty$, the whole trajectory belongs to H_0, contradicting the assumption.

Theorem 51.3. (KRASOVSKII [4]). The differential equation

$$(51.9) \qquad \dot{x} = f(x, t), \quad f \in C_1,$$

has property A if it is possible to construct in each bounded subdomain H_0 whose closure $\overline{H_0}$ is contained in G a function $v(x, t)$ which is decrescent in H_0 and has a definite derivative.

Proof. Since $\dot{v}$ has a fixed sign in H_0 we may assume without loss of generality that $\dot{v}$ is positive. Therefore by hypothesis

$$(51.10) \qquad \dot{v} \geq \psi(|x|), \quad x \in H_0,$$

for a suitable $\psi \in K$. We next choose initial values x_0, t_0 such that $v(x_0, t_0) =: v_0 > 0$. Because of (51.10), $v(t) > v_0$ throughout. The hypothesis on $v(x, t)$ can be written in the form

$$(51.11) \qquad v(x, t) \leq \varphi(|x|), \quad x \in H_0,$$

for some function $\varphi \in K$. We obtain the differential inequality

$$\dot{v} \geq \psi\big(\varphi^I(v)\big)$$

which leads to an estimate

$$v(t) \geq \varrho(v_0)\,\lambda(t - t_0).$$

$\varrho(r)$ denotes a function of class K and $\lambda(s)$ a monotone increasing positive function. Since $v(x, t)$ is bounded for $x \in H_0$ the last inequality cannot hold for arbitrarily large t. Hence there exists a number T such that the trajectory leaves the domain H_0 in the interval $t_0 \leq t \leq t_0 + T$. T depends on the comparison functions φ and ψ as well as on v_0, that is on x_0 and t_0, and finally on the maximum assumed by $v(x, t)$ in H_0.

If $v_0 < 0$, the same argument is carried out for decreasing t. Then we have $dv/d(-t) < 0$, *i.e.* it is of the same sign as v, and as above we arrive at an estimate

$$|v(t)| \leq \varrho(|v_0|)\,\lambda(|t - t_0|).$$

The trajectory leaves the domain H_0 in the interval $t_0 - T \leq t \leq t_0$.

A simple example for Theorem 51.3 is given by the differential equation

$$\dot{x}_i = \sum_{j=1}^{n} c_{ij} x_j^{2k_j}, \quad i = 1, 2, \ldots, n,$$

where $k_j \geq 1$ is an integer. Here we choose

$$v = \sum_{i=1}^{n} b_i x_i.$$

Then

$$Dv = \sum_{i=1}^{n} b_i \sum_{j=1}^{n} c_{ij} x_j^{2k_j} = \sum_{j=1}^{n} \left(\sum_{i=1}^{n} b_i c_{ij} \right) x_j^{2k_j}$$

is a quadratic form in the variables $x_1^{k_1}, \ldots, x_n^{k_n}$, whose matrix is

$$\mathrm{diag} \left(\sum_{i=1}^{n} b_i c_{ij} \right).$$

Dv is positive definite if the b_i are so chosen that the elements of the diagonal matrix are positive. This is clearly possible if the rank of the matrix (c_{ij}) is equal to n. In that case the hypotheses of Theorem 51.3 are satisfied. (EFENDIEV [1]; the case rank $(c_{ij}) < n$ is also examined there.)

Theorem 51.3 has a converse (KRASOVSKII [4]).

Theorem 51.4. Let the right side of the differential equation (51.8) belong to C_1. In the domain G let the differential equation have property A. Then for each sub-domain H_0 whose closure lies in the interior of G there exists a Liapunov function $v(x, t)$ which is decrescent and has a definite derivative in H_0, and which has uniformly bounded first order partial derivatives $\dfrac{\partial v}{\partial x_i}, \dfrac{\partial v}{\partial t}$ in a cylindrical domain $\{(x, t) : x \in H_0,\ 0 \leq t < \infty\}$.

We first prove a lemma.

Lemma 51.5. Let H_0 be a compact sub-domain of the domain G and let x_0 be a point of H_0. Assume that in H_0 the differential equation (51.8) satisfies a Lipschitz condition with Lipschitz constant L. Let the segment of the trajectory $p(t, x_0, t_0)$ for which $0 \leq u := t - t_0 \leq \theta$ be entirely contained in H_0 for all $0 \leq T$. Then for each number $\gamma > 0$ there exists a continuous function $v(x, t)$ defined for $x \in G$, $-\infty < t < +\infty$, which has continuous first order partial derivatives and satisfies:

a) $v = 0$ for $|x - p(t, x_0, t_0)| \geq \gamma$, $\quad -\infty < t < +\infty$,

   and for $u := t - t_0 < -2\tau$, $\quad u > 2\tau + \theta$.

b) $\dot{v} > d$ for $|x - p(t, x_0, t_0)| < \alpha$, $\quad -\tau \leq u \leq \theta$.

c) $\dot{v} \geq 0$ for $x \in R_n$, $\quad -\infty < u \leq \tau + \theta$

   and for $|x - p(t_0 + \theta, x_0, t_0)| \geq \gamma$, $\quad \theta + \tau \leq u \leq \theta + 2\tau$.

d) $v > d$ for $|x - p(t, x_0, t_0)| < \alpha$, $\quad -\tau \leq u \leq \theta$.

The numbers τ, α, d appearing in the inequalities are positive numbers which are independent of the choice of the point $x_0 \in H_0$ but which depend on the domains H_0 and G as well as on T and γ. The inequalities define domains which are cylindrical neighborhoods of the motion $\boldsymbol{p}(t, \boldsymbol{x}_0, t_0)$. The second inequality in c) guarantees that $\dot{v} \geq 0$ in the entire domain H_0.

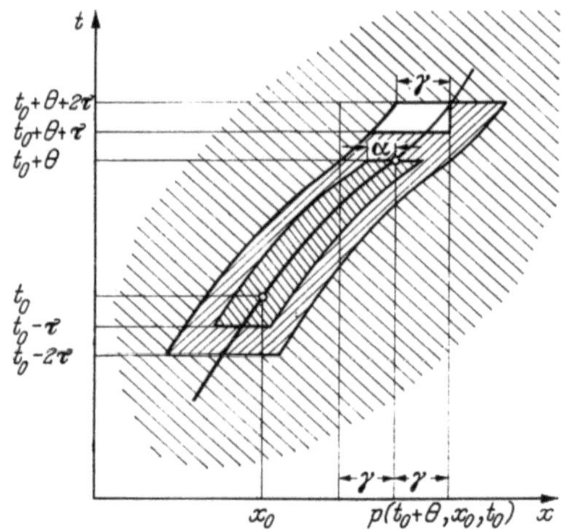

Fig. 51.1. (Lemma 51.5)

Proof. We choose numbers $\tau > 0$ and $\eta > 0$ so that the inequalities

$$|\boldsymbol{x} - \boldsymbol{p}(t, \boldsymbol{x}_0, t_0)| < \eta, \quad -2\tau \leq u \leq \theta + 2\tau,$$

define points $(\boldsymbol{x}, t) \in H_0$. The subset of these points defined by $\theta \leq u \leq \theta + 2\tau$ is to lie in the interior of the ball

$$|\boldsymbol{x} - \boldsymbol{p}(t_0 + \theta, \boldsymbol{x}_0, t_0)| < \gamma.$$

Such a choice is possible because $f \in C_1$ and therefore the velocity of the phase point along the motion is uniformly bounded. Let

$$\boldsymbol{y}(t) := \mathrm{col}\big(y_1(t), \ldots, y_n(t)\big)$$

be a vector whose components are polynomials in t so chosen that the inequalities

$$|\boldsymbol{p}(t, \boldsymbol{x}_0, t_0) - \boldsymbol{y}(t)| < \varepsilon, \quad \left|\frac{d}{dt}\boldsymbol{p}(t, \boldsymbol{x}_0, t_0) - \frac{d\boldsymbol{y}(t)}{dt}\right| < \varepsilon,$$

hold on the interval $-2\tau \leq u \leq 2\tau + \theta$; the number $\varepsilon < \eta/2$ will be discussed later. This choice is possible by the Weierstrass approximation theorem. We then define the function $v(\boldsymbol{x}, t)$ as follows:

$$\ln v(\boldsymbol{x}, t) := p \ln(u + 2\tau) + (u + 2\tau)^{-1} (u - \theta - 2\tau)^{-1}$$
$$+ \big(|\boldsymbol{x} - \boldsymbol{y}(t)|^2 e^{-2qu} - \beta^2\big)^{-1}$$

(51.12)

in the domain

(51.13) $|\boldsymbol{x} - \boldsymbol{y}(t)| < \beta\, e^{qu}, \quad -2\tau < u < \theta + 2\tau,$

and

$$v(\boldsymbol{x}, t) = 0$$

outside of the domain (51.13). In this definition

$$\beta := \frac{\eta}{2}\, \exp\left(-q(T + 2\tau)\right).$$

The constant q is larger than the Lipschitz constant L, $q > L$; p is a natural number whose magnitude will be apparent from the proof.

The function $v(\boldsymbol{x}, t)$ has the desired properties as we shall now prove in detail.

1) The continuity of $v(\boldsymbol{x}, t)$ and the continuity of the partial derivatives in the interior of the domains given in the definition follows from the construction. In order to form the derivatives at the boundary we must take the one-sided limits $u \to -2\tau + 0$, $u \to 2\tau + \theta - 0$, and $|\boldsymbol{x} - \boldsymbol{y}(t)| \to \beta e^{qu} - 0$. The values of the limits are equal to zero or are expressions of the form $\lim\limits_{r \to -\infty} r^2 e^r$, thus again zero.

2) To check a) we must show that the domain (51.13) is disjoint from the domain given in a). Consider an arbitrary $\boldsymbol{x}$ in the domain (51.13). We choose $\eta < \gamma$ and $\varepsilon \le \dfrac{\eta}{2}$ and obtain

$$|\boldsymbol{x} - \boldsymbol{p}(t, \boldsymbol{x}_0, t_0)| \le |\boldsymbol{x} - \boldsymbol{y}(t)| + |\boldsymbol{y}(t) - \boldsymbol{p}(t, \boldsymbol{x}_0, t_0)| \le \beta\, e^{qu} + \varepsilon \le \gamma$$

since

$$\beta e^{qu} = \frac{\eta}{2}\, e^{-qT - 2q\tau + qu} \le \frac{\eta}{2}\, e^{-qT - 2q\tau + q\theta + 2q\tau} = \frac{\eta}{2}\, e^{-q(T-\theta)} \le \frac{\eta}{2}.$$

3) From equation (51.12) it follows by differentiation that

(51.14) $\dfrac{\dot{v}}{v} = \dfrac{p}{u + 2\tau} + \dfrac{\theta - 2u}{(u + 2\tau)^2\,(u - \theta - 2\tau)^2} + \dfrac{2Q\, e^{-2qu}}{(|\boldsymbol{x} - \boldsymbol{y}|^2\, e^{-2qu} - \beta^2)^2}$

where

(51.15) $Q := q\,|\boldsymbol{x} - \boldsymbol{y}(t)|^2 - (\boldsymbol{x} - \boldsymbol{y}(t))^T\,(\dot{\boldsymbol{x}} - \dot{\boldsymbol{y}}(t))$

$((\boldsymbol{x} - \boldsymbol{y})^T$ is the transposed vector). (51.14) must be estimated in the domain

(51.16) $-\tau \le u \le \theta + \tau, \quad |\boldsymbol{x} - \boldsymbol{y}(t)| \le \dfrac{\beta}{2}\, e^{qu}.$

The first two summands in (51.14) are larger than

(51.17) $p(T + 3\tau)^{-1}, \quad -(T + 2\tau)\,\tau^{-4},$

respectively. Further we have

$$2(x - y)^T (\dot{x} - \dot{y}) e^{-2qu} \leq 2|x - y| |\dot{x} - \dot{y}| e^{-2qu}$$

$$\leq 2 \frac{\beta}{2} e^{qu} e^{-2qu} |\dot{x} - \dot{y}| = \beta e^{-qu} |\dot{x} - \dot{y}| \leq \beta e^{q\tau} |\dot{x} - \dot{y}|$$

and

$$|\dot{x} - \dot{y}| \leq |\dot{x} - \dot{p}(t, x_0, t_0)| + |\dot{p}(t, x_0, t_0) - \dot{y}(t)|$$

$$\leq |f(x, t) - f(p(t, x_0, t_0), t)| + \varepsilon \leq L|x - p(t, x_0, t_0)| + \varepsilon$$

$$\leq \varepsilon + L(|x - y| + |y - p(t, x_0, t_0)|) \leq \varepsilon + L\left(\varepsilon + \frac{\beta}{2} e^{qu}\right)$$

$$\leq \varepsilon(L + 1) + L \frac{\beta}{2} e^{q(T+\tau)} \leq L \frac{\beta}{2} e^{q(T+\tau)} + (L + 1) \frac{\beta}{4} e^{-q\tau}$$

if we choose $\varepsilon < \frac{\beta}{4} e^{-q\tau}$.

The denominator of the last summand of (51.14) is no smaller than $9\beta^4/16$. The summand is therefore no larger than

$$\frac{16}{9\beta^4} \beta e^{q\tau} \left(L \frac{\beta}{2} e^{q(T+\tau)} + (L + 1) \frac{\beta}{4} e^{-q\tau}\right) = \frac{8}{9} \frac{L}{\beta^2} e^{q(T+2\tau)} + \frac{4}{9} \frac{L+1}{\beta^2}.$$

In sum, it follows that

$$\frac{\dot{v}}{v} > \frac{p}{T + 3\tau} - \frac{T + 2\tau}{\tau^4} - \frac{8}{9} \frac{L}{\beta^2} e^{q(T+2\tau)} - \frac{4}{9} \frac{L+1}{\beta^2}.$$

We recognize that by an appropriate choice of p we can make the right side larger than one. Thus $\dot{v} > v$ in the domain (51.16).

Since the domain (51.16) is closed, v assumes a positive minimum there, the number d of conditions b) and d). To show that these conditions are satisfied we choose $\alpha = \frac{\beta}{4} e^{-q(T+\tau)}$ and prove that the domain (51.16) contains the domain defined in conditions b) and d). It suffices to show the inequality for an arbitrary x. We have

$$|x - y(t)| \leq |x - p(t, x_0, t_0)| + |p(t, x_0, t_0) - y(t)|$$

$$\leq \frac{\beta}{4} e^{-q(T+\tau)} + \frac{\beta}{4} e^{-q\tau} \leq \frac{\beta}{4} e^{qu} + \frac{\beta}{4} e^{qu} = \frac{\beta}{2} e^{qu}.$$

4) To check c) we consider the expression Q which was defined above. Since $q > L$, Q is bounded below in the domain

$$(51.18) \qquad -2\tau < u < \theta + \tau, \qquad \frac{\beta}{2} e^{qu} \leq |x - y(t)| \leq \beta e^{qu}$$

by

$$Q \geq (q - L)|x - y(t)|^2 - \varepsilon(L + 1)|x - y(t)| > 0$$

in case ε is sufficiently small. Since the denominator of the last summand in (51.14) is also positive this summand is positive. For $-2\tau < u \leq 0$, the first two summands of (51.14) are also positive. For $0 \leq u \leq \theta + \tau$, the bounds (51.17) can again be used. By an appropriate choice of p we can obtain that the first two terms have a positive sum for $0 \leq u \leq \theta + \tau$. Since v is non-negative in the domain (51.18) we have $\dot{v} \geq 0$ there.

This completes the proof of the lemma 51.5. The assertion that the constants are independent of the special motion is contained in the proof. It is also seen from the construction that the first order partial derivatives of v are uniformly bounded in H_0.

The lemma has been proved for a segment of the positive half-trajectory $\boldsymbol{p}(t, \boldsymbol{x}_0, t_0)$, but we can immediately formulate and prove a corresponding assertion for the segment $t_0 - \theta \leq t \leq t_0 \, (0 < \theta \leq T)$. In the proof v must be replaced by $-v$ and u by $-u$. The property b), for instance, must then be expressed by an inequality,

$$\dot{v} > d \quad \text{for} \quad |\boldsymbol{x} - \boldsymbol{p}(t, \boldsymbol{x}_0, t_0)| < \alpha, \quad -\theta + t_0 \leq t \leq t_0 + \tau.$$

Proof of Theorem 51.4. Let $\gamma < \frac{1}{3}\, \varrho\,(b\,(G), H_0)$. Set $H_0(\gamma) = \{\boldsymbol{x} : \varrho\,(\boldsymbol{x}, H_0) < \gamma\}$ and $H_0(2\gamma) = \{\boldsymbol{x} : \varrho\,(\boldsymbol{x}, H_0) < 2\gamma\}$. Let $\{h_m\}$ be the null sequence of Def. 51.1. We may assume without loss of generality that $h_1 < \varrho\,(b\,(H_0), 0)$. Also, let $H_m := H_0 \, B_m$. Because of property A there exist for each integer m a number T_m and a number $N(m)$ such that the segment of the trajectory $\boldsymbol{p}(t, \boldsymbol{x}_0, t_0)$ for which $t - t_0 \leq T_m$, lies outside of the ball $B_{N(m)}$ provided $\boldsymbol{x}_0 \in H_m$. In fact, $N(m)$ is the smallest index j for which the inequality

$$h_j \leq h_m \, e^{-LT}$$

holds. This follows from the Lipschitz condition. (Sec p. XI).

We now choose a number γ_m so that it satisfies the inequalities

$$\gamma_m \leq \frac{1}{2}\,(h_{N(m)} - h_{N(m)+1}), \quad \gamma_m < \gamma.$$

Because of property A there exists a θ, $|\theta| \leq T_m$, such that

$$\boldsymbol{p}(t_0 + \theta, \boldsymbol{x}_0, t_0) \in b\,(H_0(2\gamma)).$$

We assume that $\theta > 0$. For a negative θ the argument is analogous.

In H_0, we construct the function v of the lemma, for $\gamma = \gamma_m$. It depends on $\boldsymbol{x}_0$ and t_0 as parameters and we will therefore write it as $v(\boldsymbol{x}, t; \boldsymbol{x}_0, t_0)$. This function satisfies the conditions

$$v(\boldsymbol{x}, t; \boldsymbol{x}_0, t_0) = 0 \quad \text{for} \quad |\boldsymbol{x}| \leq h_{N(m)+1},$$

$$\dot{v} \geq 0 \quad \text{in} \quad H_0,$$

(51.19)
$$\dot{v} > d_m,$$

in the domain

$$|x - p(t, x_0, t_0)| < \alpha_m, \qquad -\tau_m < u < \tau_m.$$

The constants α_m, τ_m, and d_m depend only on m and not on the point (x_0, t_0). By making the numbers α_m and τ_m smaller we obtain that

$$(51.20) \qquad \dot{v} > d_m \text{ for } |x - x_0| < \alpha_m, \qquad -\tau_m < u < \tau_m.$$

Here we are using the fact that the derivative dx/dt is uniformly bounded on $H_0(2\gamma)$ since $f \in C_1$.

The domains in (51.19) and (51.20) have a non-void intersection if we use the same bounds. The compact set H_m can be covered by a finite number of balls

$$|x - x_{0l}| < \alpha_m, \qquad l = 1, 2, \ldots, k_m,$$

where $x_{0l} \in H_m$. Let $t_{0s} := \dfrac{s\,\tau_m}{2}$, $s = 1, 2, \ldots$. For each point (x_{0l}, t_{0s}) we construct the function $v(x, t; x_{0l}, t_{0s})$ of the lemma and set

$$v_m(x, t) := \sum_{s=1}^{\infty} \sum_{l=0}^{k_m} v(x, t; x_{0l}, t_{0s}).$$

This function is equal to zero in the domain $B_{N(m) \div 1}$. In a neighborhood of an arbitrary point of G the number of non-zero terms of the series is, for each t, smaller than a fixed number which is independent of the point. This implies that v_m is continuous and has bounded derivatives. Since each point x belongs to at least one of the neighborhoods $|x - x_{0l}| < \alpha_m$, in which (51.20) is valid, $\dot{v}_m$ is positive definite in H_m. Let P_m be a common bound for the derivatives and for v_m,

$$|v_m| < P_m, \qquad \left|\frac{\partial v_m}{\partial x_i}\right| < P_m, \qquad i = 1, \ldots, n; \qquad \left|\frac{\partial v_m}{\partial t}\right| < P_m,$$

and set

$$(51.21) \qquad v(x, t) := \sum_{m=1}^{\infty} \frac{1}{P_m 2^m} v_m(x, t).$$

This function satisfies all the conditions of the theorem; for the series and its formal derivatives converge absolutely and uniformly on G so that v has the same properties as the functions v_m.

KRASOVSKII [4] showed even more. If the right side $f(x, t)$ of the differential equation is uniformly continuous with respect to t, $0 < t < \infty$, for $x \in G$, then the function (51.21) has partial derivatives of all orders.

The mathematical idea upon which the proof is based can also be utilized for a proof of Theorem 49.3 (KRASOVSKII [4]).

52. The Converse of the Instability Theorems

As we explained in sec. 37, several instability types can be characterized by comparison functions in the same manner as stability types. Then there exists an entire neighborhood of the origin, such that all the motions beginning in this neighborhood tend away from the origin, at least to a certain minimal distance. If an inequality

$$(52.1) \qquad \| \boldsymbol{p}(t, \boldsymbol{a}, t_0) \| \geq \varphi(\|\boldsymbol{a}\|; t_0)\, \lambda(t - t_0; t_0)$$

is known, where λ is an increasing positive function, then a Liapunov function can be constructed as in sec. 49. We set

$$(52.2) \qquad v(\boldsymbol{x}, t) := \int_t^{t+T} \| \boldsymbol{p}(\tau, \boldsymbol{x}, t) \|\, d\tau.$$

The number T will be determined later. Obviously

$$v(\boldsymbol{x}, t) \geq \varphi(\|\boldsymbol{x}\|; t) \int_t^{t+T} \lambda(\tau - t; t)\, d\tau.$$

Furthermore

$$Dv = \| \boldsymbol{p}(t + T, \boldsymbol{x}, t) \| - \| \boldsymbol{p}(t, \boldsymbol{x}, t) \|,$$

$$(52.3) \qquad Dv \geq \varphi(\|\boldsymbol{x}\|; t)\, \lambda(T; t) - \|\boldsymbol{x}\|.$$

As in p. 236 we make a change of variables and obtain a majorant:

$$v(\boldsymbol{x}, t) \leq \left(\| \boldsymbol{p}(t + T, \boldsymbol{x}, t) \| - \|\boldsymbol{x}\| \right) \cdot \delta$$

where

$$(52.4) \qquad \delta := \inf_{t \leq \tau \leq t+T} \frac{d}{d\tau} \| \boldsymbol{p}(\tau, x, t) \|.$$

From these inequalities we can read off, for instance, that v is positive definite if $\varphi \in K$ and that v is decrescent if the infimum (52.4) is positive. Dv is definite in case

$$\lambda(T; t) \geq (1 + \varepsilon)\, \frac{\|\boldsymbol{x}\|}{\varphi(\|\boldsymbol{x}\|; t)}, \qquad \varepsilon > 0.$$

If the infimum is zero the construction must be modified by introducing a compensating factor in the integrand, as in (49.8). If the comparison function (52.1) is a function of class KL then the upper limit T of the integral is possibly an unbounded function. But the integrals are always finite.

The Liapunov functions constructed in this manner satisfy the hypotheses of the instability Theorems 42.6 and 42.7 but actually fulfill much stronger conditions. We shall not formulate the corresponding assertions as theorems here. We only state that Theorems 42.6 and 42.7 do have converses in the same sense that the Liapunov functions satisfy the hypothesis exactly, implying no more. We discuss a method of construction which is related to that used in sec. 51. This method has first been published by VRKOČ [1]. It gives a little more information than the mere existence of the Liapunov function, namely a statement on the domain of those initial values at which trajectories originate which tend away from the origin. We are here concerned with differential equations

$$(52.5) \qquad \dot{x} = f(x, t); \quad f \in E, \quad x \in G \subset R_n$$

with an unstable equilibrium.

Def. 52.1. Let H_0 be a bounded sub-domain of the domain of definition of (52.5), which contains the origin,

$$0 \in H_0 \subset \bar{H}_0 \subset G$$

A set $I(t_0) \subset H_0$ depending on the initial time t_0 is called a *domain of instability in H_0 for $t = t_0$* if each trajectory $p(t, x_0, t_0)$, $x_0 \in I(t_0)$, leaves the domain H_0 as $t > t_0$ increases through a finite time interval.

For an autonomous differential equation the domain of instability does not, of course, depend on t_0. The elliptic sector of the example in sec. 40 belongs to the domain of instability. If inequalities of type (52.1) exist then the domain of instability is a neighborhood of the origin.

The domain of instability is an open set in R_n. This is seen as follows: By definition there exists for each $x_0 \in I(t_0)$ a number t' such that $p(t_0 + t', x_0; t_0)$ lies on the boundary of $I(t_0)$. But then, by continuity, there exists for each x_0' in a sufficiently small neighborhood of x_0 a t'' such that $p(t_0 + t'', x_0', t_0)$ lies on the boundary of $I(t_0)$.

Theorem 52.1. Let the equilibrium of (52.5) be unstable. Then, in the notation of Def. 52.1, there exists a function $v(x, t)$ with the following properties: 1) If $x \in \bar{H}_0$ then $v(x, t) < 0$ for $t > t_0$; therefore there exists a "domain $v(x, t) < 0$" (Theorem 42.6). 2) In the domain $v < 0$, v is bounded from below. 3) In the domain $v < 0$, $\dot{v} \leq -\varphi(v)$, for some $\varphi \in K$. 4) The domain of instability $I(t_0)$ coincides with the domain $v(x, t_0) < 0$.

The last assertion goes further than the converse of the theorem of Chetaev.

Proof. (KRASOVSKII [4]) Let I be the set $\cup \{I(t_0), 0 \leq t_0 < \infty\}$. Also let $\{h_k\}$ be a monotone null sequence such that $h_1 < 1$. Let I_k denote the set of all points (x, t) with the following properties.

$$(52.6) \qquad x \in I(t), \quad h_k \leq t \leq k, \quad \varrho(x, b(I(t))) \geq h_k.$$

We may assume without loss of generality that the sets I_k are all non-void. If $(x_0, t_0) \in I_k$, then by definition of H_0, the trajectory $p(t, x_0, t_0)$ leaves the domain H_0 after a finite elapsed time. This means that there exists a number t' such that $p(t', x_0, t_0) \notin H_0$. Let

$$\gamma := \min_{0 \leq t \leq t'} \varrho(p(t, x_0, t_0), H_0 \setminus \bar{I}(t)).$$

We now use the trajectory $p(t, x_0, t_0)$ and Lemma 51.5 to construct a continuous function $v(x, t; x_0, t_0)$ in such a way that

$$(52.7) \qquad v(x, t; x_0, t_0) = 0 \text{ for } x \in \overline{H_0} \setminus I(t),$$
$$v > d \text{ and } \dot{v} > d \text{ for } |x_0 - x| < \alpha, \quad |t_0 - t| < \tau,$$

where d is a positive constant and α and τ are sufficiently small numbers; also

$$\dot{v} \geq 0 \text{ in } \overline{H_0}$$

and, finally, v has continuous first order partial derivatives in G. We can construct such a function for each point $(x_0, t_0) \in I_k$ and then choose a finite number N_k of points such that the neighborhoods in (52.7), $x - x_0| < \alpha$, $|t - t_0| < \tau$, which belong to them cover the $(n + 1)$-dimensional bounded domain I_k. We denote the corresponding functions v of the lemma by $v_{kl}(x, t)$; l runs from 1 to N_k. The function

$$(52.8) \qquad v_k(x, t) := \sum_{l=1}^{N_k} v_{kl}(x, t)$$

is defined for $x \in G$, $0 \leq t < \infty$, is continuous, and has continuous first order partial derivatives. v_k is identically zero outside of I. Furthermore, for each $k > 1$ there exists a number T_k such that $v_k(x, t) = 0$ for $t \geq T_k$; for each summand in (52.8) is different from zero only on a finite time interval. Thus the v_k are uniformly bounded and the same is true for their first order partial derivatives. The total derivative satisfies

$$(52.9) \qquad \dot{v}_k \geq \delta > 0 \text{ in } I_k, \quad \dot{v}_k \geq 0 \text{ in } \overline{H_0},$$

for a certain δ.

We define

$$w_k(x, t) := v_k(x, t) e^{t - T_k}$$

and thus provide the function (52.8) with a factor depending on time. Because of (52.9) we have

$$(52.10) \qquad \frac{dw_k}{dt} = w_k + \dot{v}_k e^{t - T_k} \geq w_k.$$

Just as v_k, w_k is bounded: $w_k \leq P_k$. The function

$$(52.11) \qquad v(x, t) := - \sum_{k=0}^{\infty} \frac{1}{2^k P_k} w_k(x, t)$$

satisfies all the hypotheses of the theorem. It is identically zero outside of $I(t)$. In $I(t)$

$$0 > v(x, t) > - \sum_{k=0}^{\infty} 2^{-k}$$

and $\dot v < v$ since because of (52.10) we have

$$(52.12) \qquad \frac{dv}{dt} = - \sum_{k=0}^{\infty} \frac{w_k(\boldsymbol{x}, t)}{2^k P_k} - \sum_{k=0}^{\infty} \frac{e^{t-T_k}}{2^k P_k} \dot v_k.$$

Hence, for a fixed t, $I(t)$ is the domain of instability.

It can be shown in addition that the function (51.11) has derivatives of all orders (VRKOČ [1]).

The relation (52.12) implies that the derivative has the form

$$\dot v = v + v_1.$$

The second summand has the same sign as the function w of Theorem 42.7. Thus the function (52.11) also satisfies the hypothesis of the second instability theorem of Liapunov (for $g = 1$) and we have

Theorem 52.2. If the equilibrium of the differential equation (52.5) is unstable then in each compact subdomain $\overline{H}_0 \subset G$ there exists the function v satisfying the requirements of the second instability theorem of Liapunov.

We conclude that Theorems 42.6 and 42.7 are equivalent for differential equations.

Chapter VII

Stability Properties of Ordinary Differential Equations

Preliminary remark. The motions to be examined in this chapter are defined by ordinary differential equations in R_n. A number of the properties discussed also apply to more general motions, but we shall not mention this explicitly in each case.

53. The Meaning of the Decrescence of Liapunov Functions

If in the hypothesis of Theorem 42.4 we omit the requirement "v is decrescent" then the stability of the equilibrium is still assured by Theorem 42.1. Since stability is already given by the condition "$\dot{v}$ is negative semi-definite" the question arises, what behavior of the motions defined by

$$(53.1) \qquad \dot{\boldsymbol{x}} = \boldsymbol{f}(\boldsymbol{x}, t), \quad (\boldsymbol{x}, t) \in K_{h, t_0},$$

is described by the conditions

$$(53.2) \quad v(\boldsymbol{x}, t) \geq \varphi(|\boldsymbol{x}|), \quad \dot{v} \leq - \psi(|\boldsymbol{x}|); \quad \varphi \in K, \quad \psi \in K.$$

This question is answered in part by the theorems of sec. 51: The trajectories must possess property A (Def. 51.1). The behavior described by (53.2) is examined somewhat more closely in

Theorem 53.1. If there exists a Liapunov function satisfying (53.2) then

$$\liminf_{t \to \infty} |\boldsymbol{p}(t, \boldsymbol{x}_0, t_0)| = 0.$$

Proof. Let us assume that the theorem is false. Then

$$\liminf_{t \to \infty} |\boldsymbol{p}(t, \boldsymbol{x}_0, t_0)| =: \eta > 0$$

and there exists for each given $\varepsilon > 0$ a T such that

$$\dot{v} \leq - \psi(\eta - \varepsilon) \quad \text{for } t \geq T > t_0.$$

This implies

$$(53.3) \quad v(t) = v_0 + \int_{t_0}^{T} \dot{v}\, dt + \int_{T}^{t} \dot{v}\, dt \leq v(T) - (t - T)\, \psi(\eta - \varepsilon)$$

and for sufficiently large t this yields a contradiction to the assumed definiteness.∎

If it is known in addition that the equilibrium of the differential equation is not asymptotically stable then we may assume that

$$\limsup_{t \to \infty} |\boldsymbol{p}(t, \boldsymbol{x}_0, t_0)| =: \gamma > 0.$$

Then there exists for each number $\gamma' < \gamma$ an unbounded sequence $\{t_n\}$ such that

$$|\boldsymbol{p}(t_n, \boldsymbol{x}_0, t_0)| \geq \gamma'; \quad n = 1, 2, \ldots .$$

In fact, there exists a sequence of intervals (t'_n, t''_n) such that

$$|\boldsymbol{p}(t, \boldsymbol{x}_0, t_0)| \geq \gamma', \quad t'_n < t < t''_n.$$

From (53.3) we conclude that

$$(53.4) \qquad v(t) \leq v_0 - \psi(\gamma') \sum_{n=1}^{\infty} (t''_n - t'_n)$$

and this is compatible with the definiteness only if the series $\sum (t''_n - t'_n)$ converges. For large arguments the function $y(t) = |\boldsymbol{p}(t, \boldsymbol{x}_0, t_0)|$ is thus almost everywhere arbitrarily small. Its graph is a sawtooth curve.

An example for this type of behavior is given by the following differential equation due to MASSERA [1]:

$$\dot{x} = \frac{\dot{g}(t)}{g(t)} x, \quad g(t) := \sum_{n=1}^{\infty} \frac{1}{1 + n^4(t - n)^2}$$

whose general solution is

$$p(t, x_0, t_0) = x_0 \frac{g(t)}{g(t_0)}.$$

Since the function

$$g(t) \leq 2 + \sum n^{-4}$$

is bounded the equilibrium is stable. For an integer argument $t = k$ we have

$$g(k) > 1$$

so that the equilibrium is not asymptotically stable. If t is not an integer and k is the largest integer less than t, $k = [t]$, and $t - k =: \eta$ then

$$g(t) = \sum_{n=1}^{k-1} \frac{1}{1 + n^4(t - n)^2} + \sum_{n=k}^{k+1} \frac{1}{1 + n^4(t - n)^2} + \sum_{n=k+2}^{\infty} \frac{1}{1 + n^4(t - n)^2}$$

$$< \sum_{n=1}^{k-1} \frac{1}{1 + n^4(t - n)^2} + \frac{1}{1 + k^4 \eta^2} + \frac{1}{1 + (k + 1)^4 (1 - \eta)^2} + \sum_{n=k}^{\infty} \frac{1}{1 + n^4}.$$

In the first sum the term for $n = 1$ is the largest, the second sum is the remainder of a convergent series, so that

$$g(t) < (k - 1) \frac{1}{1 + (k - 1)^2} + O(k^{-3}) = O(k^{-1}).$$

Hence $g(t)$ tends to zero as t grows without bound and remains bounded away from integers.

The Liapunov function constructed by (49.4)

$$v(x, t) = \int\limits_t^\infty |p(\tau, x, t)|^2 \, d\tau = \frac{x^2}{(g(t))^2} \int\limits_t^\infty (g(\tau))^2 \, d\tau$$

is positive definite but not decrescent. Its derivative $\dot{v} = -x^2$ is negative definite.

If the function $y(t) := |p(t, x_0, t_0)|$ has a bounded derivative then the series in (53.4) is divergent. For it follows from $\left|\dfrac{dy}{dt}\right| \leq M$ that $t_n'' - t_n' \geq \dfrac{\gamma'}{M}$. The derivative dy/dt is bounded if the right side of (53.1) is bounded (*cf.* sec. 49). In this case, if v is positive definite γ cannot be positive and we get

Theorem 53.2. If the right side of the differential equation (53.1) is bounded and if there exists a positive definite Liapunov function with a negative definite derivative then the equilibrium is asymptotically stable.

Differential equations which satisfy the conditions of Theorem 53.1 have been studied recently by STRAUSS [1].

54. Existence of a Liapunov Function in Case of Non-Uniform Asymptotic Stability

In the converse theorems of sec. 50 it was assumed that the asymptotic stability was uniform at least with respect to the spatial coordinates. General converse theorems for non-uniform asymptotic stability (type A 2 + A 4) are as yet unknown. The following example suggests that such general theorems exist, but that the basic assumptions on the Liapunov functions have to be weakened somewhat.

Let the differential equation

$$(54.1) \qquad \dot{r} = \frac{\dfrac{\partial}{\partial t} k(t, a)}{k(t, a)}\, r, \qquad \dot{\psi} = 0;$$

$$k(t, a) := \frac{1 + 2at^2}{1 + t + a^2 t^3}, \qquad a = \sin^2 \psi$$

be given. This is a special case of (39.3). The general solution is

$$p(t; r_0, \psi_0; t_0) = r_0 \frac{k(t, \sin^2 \psi_0)}{k(t_0, \sin^2 \psi_0)}; \qquad \psi = \psi_0.$$

The function

$$(54.2) \quad v(r, \psi, t) := \int\limits_t^\infty (p(\tau; r, \psi; t))^2 \, d\tau = \frac{r^2}{(k(t, a))^2} \int\limits_t^\infty (k(\tau, a))^2 \, d\tau$$

17*

converges for a fixed $a = \sin^2 \psi$ since for large τ the integrand behaves like τ^{-2}, but the convergence is not uniform with respect to ψ, $0 \leq \psi \leq 2\pi$. The integral

$$\int\limits_t^\infty \big(p(\tau; r, 0; t)\big)^2 \, d\tau = r^2(1 + t)^2 \int\limits_t^\infty \frac{d\tau}{(1 + \tau)^2} = r^2(1 + t)$$

exists but the limit

$$\lim_{\psi \to 0} v(r, \psi, t)$$

does not exist because we have for $a > 0$

$$\int\limits_t^\infty \left(\frac{1 + 2a\tau^2}{1 + \tau + a^2\tau^3}\right)^2 d\tau = \frac{1}{a} \int\limits_{at}^\infty \left(\frac{a + 2u^2}{a + u + u^3}\right)^2 du > \frac{1}{a} \int\limits_{at}^\infty \left(\frac{2u^2}{u + u^3}\right)^2 du$$

$$= \frac{2}{a} \left(- \frac{u}{1 + u^2} + \arctan u\right)\Big|_{at}^\infty = 2 \left(\frac{t}{1 + a^2 t^2} - \frac{\arctan (at)}{a} + \frac{\pi}{2a}\right)$$

and the right side does not have a finite limit as $a \to 0$. The function (54.2) is thus discontinuous for $\psi = 0$. The last estimate implies that $v(r, \psi, t)$ is positive definite for $\psi > 0$ since

$$v(r, \psi, t) > 2r^2 \left(\frac{1 + t + a^2 t^3}{1 + 2a t^2}\right)^2 \left(\frac{t}{1 + a^2 t^2} + \frac{1}{a}\left(\frac{\pi}{2} - \arctan (at)\right)\right).$$

The derivative Dv is, of course, negative definite.

55. Modified Stability Criteria

A. We have previously pointed out the significance of Theorem 26.2: It permits to infer asymptotic stability of the equilibrium if in addition to $v > 0$ we only know that $\dot v \leq 0$ and that the point set $\dot v = 0$ in R_n does not contain a complete positive half-trajectory. This theorem can be extended to differential equations

$$(55.1) \quad \dot x = f(x, t); \quad f(x, t + \omega) = f(x, t), \quad f \in E, \quad x \in R_n,$$

with right side periodic in t.

Theorem 55.1 (KRASOVSKII [4], LA SALLE [4]). For the differential equation (55.1) let a once continuously differentiable function $v(x, t)$ with the following properties be given:

1) $v(x, t) \geq 0$ for all (x, t), 2) $v(x, t) = v(x, t + \omega)$, 3) $\dot v \leq 0$ for all (x, t). Let M_0 denote the set of all points (y_0, t_0) for which $\dot v = 0$ and let M be the union of all trajectories $p(t, y_0, t_0)$ for which $(p(t, y_0, t_0), t) \in M_0$.

Then all bounded solutions of (55.1) tend toward M.

Proof. Let $L^+(x_0, t_0)$ be the positive limit set of $p(t, x_0, t_0)$, *i.e.* for each point $q \in L^+(x_0, t_0)$ there exists an unbounded increasing sequence $\{t_n\}$

such that $p(t_n, x_0, t_0) \to q$. If the motion $p(t, x_0, t_0)$ is bounded then it tends toward the compact set $L^+(x_0, t_0)$. Also, for integral k, let $\gamma^+(x_0, t_0)$ be the limit set of the discrete motion $p(t_0 + k\omega, x_0, t_0)$ and let $\tilde{L}^+(x_0, t_0)$ be the union of all trajectories $p(t, y_0, t_0)$ with initial points $y_0 \in \gamma^+(x_0, t_0)$. If the motion is bounded then $L^+(x_0, t_0) = \tilde{L}^+(x_0, t_0)$. For if $q \in \tilde{L}^+(x_0, t_0)$ then there exists a t^* and a y_0 such that

$$p(t^*, y_0, t_0) = q$$

and such that

$$y_0 = \lim_{n' \to \infty} p(t_0 + n'\omega, x_0, t_0),$$

where $\{n'\}$ is an increasing sequence of integers. Now

$$p\big(t^*, p(t_0 + n'\omega, x_0, t_0), t_0\big) = p(t^*, x_0, t_0 - n'\omega) = p(t^* + n'\omega, x_0, t_0)$$

and since the motion depends continuously on the initial values we have

$$q = \lim_{n' \to \infty} p(t^* + n'\omega, x_0, t_0) \in L^+(x_0, t_0)$$

so that $\tilde{L}^+(x_0, t_0) \subset L^+(x_0, t_0)$. If conversely $q \in L^+(x_0, t_0)$ and $p(t_n, x_0, t_0) \to q$ then we reduce the sequence t_n modulo ω,

$$t_n = t'_n + k(n)\,\omega, \quad 0 \le t'_n < \omega, \ k(n) \text{ an integer,}$$

and consider a point of accumulation t' of the sequence $\{t'_n\}$. Then there exists a subsequence $t_{n'}$ such that $t_{n'} - k(n')\,\omega$ tends toward t', *i.e.* we have

$$p\big(t' + k(n')\,\omega, x_0, t_0\big) \to q.$$

It follows that $L^+(x_0, t_0) \subset \tilde{L}^+(x_0, t_0)$ and hence the assertion

$$\tilde{L}^+(x_0, t_0) = L^+(x_0, t_0).$$

The function $v(t) := v(p(t, x_0, t_0), t)$ is non-increasing and bounded from below. Hence $\lim v(t) = a_\infty$ exists. For $y_0 \in \gamma^+(x_0, t_0)$ we have $p(t, y_0, t_0) \in L^+(x_0, t_0)$. Consequently, $v(p(t, y_0, t_0), t) = a_\infty$ for $y_0 \in \gamma^+(x_0, t_0)$ and for all t. Thus $\dot{v} = 0$ for $y_0 \in \gamma^+(x_0, t_0)$ and this implies that $\dot{v} = 0$ for $y_0 \in L^+(x_0, t_0)$. Hence $L^+(x_0, t_0) \subset M$ and it follows that the motion $p(t, x_0, t_0)$ tends toward M. ∎

Theorem 55.1 contains of course a statement on asymptotic stability if it is known that the set $\dot{v} = 0$ does not contain a complete half-trajectory other than the equilibrium. In that case M consists of only the origin. In addition, the hypotheses assure at least local stability of the origin and thus the existence of a neighborhood of the origin in which only bounded solutions originate.

If the set $\dot{v} = 0$ can be represented by an equation $F(x, t) = 0$ then it clearly contains no complete half-trajectory if the derivative

$$\frac{dF}{dt} = \sum_{i=1}^{n} \frac{\partial F}{\partial x_i} f_i(x, t) + \frac{\partial F}{\partial t}$$

for the equation (55.1) is different from zero; for this condition implies that the motion intersects the surface $F = 0$.

B. There also exists an instability theorem analogous to Theorem 55.1.

Theorem 55.2 (KRASOVSKII [4]). For the differential equation (55.1) let there exist a Liapunov function with the following properties: a) $v(x, t)$ is periodic in t (or independent of t in case (55.1) is autonomous), $v(x, t)$ is decrescent in the domain $x \in K_h$, $t \geq t_0$, b) there exists a domain $v > 0$ (cf. Theorem 42.6), c) $\dot{v} \geq 0$ for $x \in K_h$, $t \geq t_0$, d) the point set $\dot{v} = 0$ does not contain a complete half-trajectory. Then the equilibrium is unstable.

Proof. Suppose that the equilibrium is stable. We choose the initial point $\bar{x}_0$ in the domain $v > 0$ and consider the motion $p(t, \bar{x}_0, t_0)$, t_0 being fixed. By a) and c), the function

$$\bar{v}(t) := v\big(p(t, \bar{x}_0, t_0), t\big)$$

is non-decreasing and bounded from above. Therefore $\lim \bar{v}(t) = v_\infty$ exists. Let x_0^* be an accumulation point of the sequence $\{x_n\}$, $x_n := p(t_0 + n\omega, \bar{x}_0, t_0)$ and $\{x_{n'}\}$ a subsequence converging to x_0^*. From the continuity we get

$$v(x_0^*, t_0) = \lim_{n' \to \infty} v\big(p(t_0 + n'\omega, \bar{x}_0, t_0), t_0\big)$$

$$= \lim_{n' \to \infty} v\big(p(t_0 + n'\omega, \bar{x}_0, t_0), t_0 + n'\omega\big) = \lim_{n' \to \infty} \bar{v}(t_0 + n'\omega) = v_\infty.$$

By d), the motion $p(t, x_0^*, t_0)$ leaves the point set $\dot{v} = 0$. Therefore there exists a $t' > t_0$ such that $dv(p(t, x_0^*, t_0), t)/dt$ is positive and

$$v_1 := v\big(p(t', x_0^*, t_0), t'\big) > v(x_0^*, t_0) = v_\infty.$$

If n' is sufficiently large the values $v(p(t', x_{n'}, t_0), t')$ are arbitrarily close to v_1, and we have,

$$\lim_{n' \to \infty} v\big(p(t', x_{n'}, t_0), t'\big) > v_\infty.$$

On the other hand,

$$p(t, x_{n'}, t_0) = p\big(t, p(t_0 + n'\omega, \bar{x}_0, t_0), t_0\big) = p(t + n'\omega, \bar{x}_0, t_0)$$

and, since $v(x, t)$ is periodic,

$$v\big(p(t', x_{n'}, t_0), t'\big) = v\big(p(t' + n'\omega, \bar{x}_0, t_0), t'\big)$$

$$= v\big(p(t' + n'\omega, \bar{x}_0, t_0), t' + n'\omega\big) = \bar{v}(t' + n'\omega)$$

whence

$$\lim_{n'\to\infty} v\big(\boldsymbol{p}(t', \boldsymbol{x}_{n'}, t_0), t'\big) = v_\infty.$$

We arrive at a contradiction to the last inequality.

C. The above considerations apply to autonomous or periodic equations. In the case of the more general differential equation

$$(55.2) \qquad \dot{\boldsymbol{x}} = f(\boldsymbol{x}, t), \quad f \in E, \; f \text{ bounded},$$

we can also infer asymptotic stability if appropriate additional hypotheses are given, even though we only know that $\dot{v} \le 0$. This was shown by MATROSOV [1]. In order to formulate this theorem we need

Def. 55.1. Let $g(\boldsymbol{x})$ be continuous in K_h and let M denote the set $\{\boldsymbol{x} \in K_h: g(\boldsymbol{x}) = 0\}$. Let the scalar function $k(\boldsymbol{x}, t)$ be defined and continuous in $K_{h,0}$. $k(\boldsymbol{x}, t)$ is called *definitely non-zero on M* if for each two numbers x, γ such that $0 < \gamma < x < h$, there exist two numbers $\beta = \beta(x, \gamma)$ and $\delta = \delta(x, \gamma)$ such that $|k(\boldsymbol{x}, t)| > \beta$ for $\gamma < |\boldsymbol{x}| < x, \varrho(\boldsymbol{x}, M) < \delta$, $t \ge 0$.

Theorem 55.3. Let two functions $v(\boldsymbol{x}, t)$, $w(\boldsymbol{x}, t)$ be given which are continuous on $K_{h,0}$ and satisfy:

1) $v(\boldsymbol{x}, t)$ is positive definite and decrescent.

2) The derivative $\dot{v}$ can be estimated from above by a non-positive continuous t-independent function

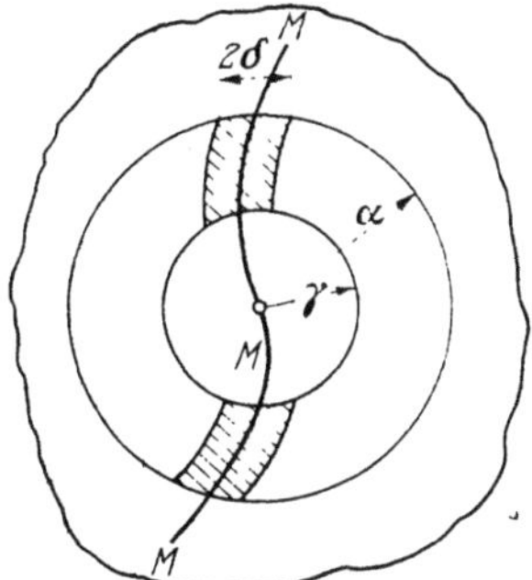

Fig. 55.1. Definition 55.1. M: the set $g(\boldsymbol{x}) = 0$. Hatched: The domain $|k(\boldsymbol{x}, t)| > \beta$

$$\dot{v} \le u(\boldsymbol{x}) \le 0.$$

3) The function $w(\boldsymbol{x}, t)$ is bounded.

4) The derivative $\dot{w}$ is definitely non-zero on the set $N := \{\boldsymbol{x}: u(\boldsymbol{x}) = 0\}$. Then the equilibrium of (55.2) is uniformly asymptotically stable.

Proof. a) The first two hypotheses imply the uniform stability, *i.e.* the existence of a function $\varphi \in K$ such that

$$|\boldsymbol{p}(t, \boldsymbol{x}_0, t_0)| < \varphi(|\boldsymbol{x}_0|), \; t \ge t_0.$$

b) Let F_i be a bound for $f_i(\boldsymbol{x}, t)$, $|f_i(\boldsymbol{x}, t)| < F_i$, for $(\boldsymbol{x}, t) \in K_{h, t_0}$. Then we have

$$|p_i(t_1, \boldsymbol{x}_0, t_0) - p_i(t_2, \boldsymbol{x}_0, t_0)| = |(t_1 - t_2)\, \dot{p}_i(t_1 + \theta(t_2 - t_1), \boldsymbol{x}_0, t_0)|$$

$$\le |t_1 - t_2|\, F_i.$$

This implies: If

$$|\boldsymbol{p}(t_1, \boldsymbol{x}_0, t_0) - \boldsymbol{p}(t_2, \boldsymbol{x}_0, t_0)| > r > 0,$$

then

$$|t_1 - t_2| > r(F_1^2 + \cdots + F_n^2)^{-1/2} =: c r.$$

c) Let $x > 0$ be given and let $|x_0| < \varphi^I(\alpha)$ so that by a)

$$|p(t, x_0, t_0)| < \alpha, \; t \geq t_0.$$

Also let $\varepsilon > 0$ be given, and set $\gamma := \varphi^I(\varepsilon)$. We must show that there exists a number T depending only on α and γ such that

(55.3)
$$|p(t_0 + T, x_0, t_0)| < \gamma.$$

Then by a) we obtain the inequality

$$|p(t, x_0, t_0)| < \varepsilon \; \text{ for } \; t \geq t_0 + T.$$

For the numbers α and γ we choose the numbers β and δ in accordance with 4) and with Def. 55.1 and we set

$$m := \sup w(x, t) \; \text{ for } \; |x| \leq \alpha, t \geq t_0,$$

$$U(r) := \{x : \gamma \leq |x| \leq \alpha, \varrho(x, N) \leq r\}.$$

Let us assume that the segment of the trajectory $p(t, x_0, t_0)$, $t_1 \leq t \leq t_2$, belongs to the domain $U(\delta)$. Then by hypotheses 3) and 4) the function

$$w(t) := w\big(p(t, x_0, t_0), t\big)$$

satisfies the inequalities

$$|w(t_2) - w(t_1)| = \left| \int_{t_1}^{t_2} \dot{w}(t)\, dt \right| \geq (t_2 - t_1)\beta,$$

$$|w(t_2) - w(t_1)| \leq 2m.$$

This implies

$$t_2 - t_1 \leq \frac{2m}{\beta}.$$

The interval on which the trajectory belongs to the domain $U(\delta)$ is therefore finite and bounded by $2m/\beta$.

There exists a time t', $t_1 < t' < t_1 + 2m/\beta$, such that $p(t', x_0, t_0)$ lies on the boundary of $U(\delta)$; but because of c) it does not lie on the part of the boundary where $|x| = \alpha$.

d) Let $\delta_1 < \delta$ and suppose $p(\tau, x_0, t_0)$ belongs to $U(\delta_1)$. There exist two possibilities. Either the trajectory leaves the domain $U(\delta)$ in the interval $\tau < t \leq \tau + 2m/\beta$ through the part of the boundary where $|x| = \gamma$, and then (55.3) is satisfied, or we have $|p(t, x_0, t_0)| > \gamma$ on this interval. By c) the trajectory must meet the boundary of $U(\delta)$ in this time interval and in particular that of $U(\delta_1)$. Since it does not pass through the part of the boundary where $|x| = \alpha$ or $|x| = \gamma$, there exists a time t'' such that

$$\varrho\big(p(t'', x_0, t_0), N\big) = \delta_1.$$

In the interval $t'' < t < t'$ we then have

$$\delta_1 \leq \varrho\big(\boldsymbol{p}(t, \boldsymbol{x}_0, t_0), N\big) \leq \delta.$$

If we denote by $-\eta$ the upper bound of $u(\boldsymbol{x})$ for $\gamma \leq |\boldsymbol{x}| \leq \alpha$, $\varrho(\boldsymbol{x}, N) \geq \delta_1$, then hypothesis 2) yields

$$\dot{v}(t) \leq u\big(\boldsymbol{p}(t, \boldsymbol{x}_0, t_0)\big) \leq -\eta, \quad t'' \leq t \leq t',$$

and further

$$v(t') - v(\tau) = \int_\tau^{t''} \dot{v}\, dt + \int_{t''}^{t'} \dot{v}\, dt \leq \int_{t''}^{t'} \dot{v}\, dt \leq -(t' - t'')\,\eta.$$

But since

$$\big|\boldsymbol{p}(t', \boldsymbol{x}_0, t_0) - \boldsymbol{p}(t'', \boldsymbol{x}_0, t_0)\big| \geq \delta - \delta_1 =: \delta_2$$

b) implies

$$t' - t'' \geq \delta_2 c,$$

and it follows that

$$v(t') \leq v(\tau) - (t' - t'')\,\eta < v(\tau) - c\,\delta_2\eta.$$

Thus the function $v(t)$ decreases in the interval (τ, t') by an amount greater than the fixed bound $a := c\,\delta_2\eta$.

e) Let

$$t_k = t_0 + \frac{2\,k\,m}{\beta}, \quad k = 1, 2, \ldots.$$

Consider the motion $\boldsymbol{p}(t, \boldsymbol{x}_0, t_0)$ in the interval $t_k \leq t \leq t_{k+1}$. If in this interval the motion remains in the region $\gamma < |\boldsymbol{x}| < \alpha$ but outside of $U(\delta_1)$ then by d),

$$v(t_{k+1}) - v(t_k) \leq -a.$$

If at some time τ, $t_k < \tau < t_{k+1}$, the motion enters the region $U(\delta_1)$ then there exists by c) a time t', $\tau < t' < t_{k+2}$, such that $\varrho\big(\boldsymbol{p}(t', \boldsymbol{x}_0, t_0), N\big) = \delta_1$ and then we have

$$v(t_{k+2}) \leq v(t') \leq v(t_k) - a.$$

But since $v(t)$ is non-increasing we have in general

$$v(t_{k+2s}) \leq v(t_k) - s\,a, \quad s = 1, 2, \ldots$$

and by hypothesis 1) we obtain a contradiction to the assumption that the motion satisfies the inequality $\gamma < |\boldsymbol{p}(t, \boldsymbol{x}_0, t_0)|$ for an arbitrarily long time. The bound for the time at which the motion leaves the annulus does not depend on the particular motion.

This completes the proof.

Mathematically, the idea in Theorem 55.2 consists in introducing the auxiliary function $u(\boldsymbol{x})$ with whose aid we estimate, so to speak, the vanishing of the derivative $\dot{v}$. This principle has many modifications and can also be applied to instability theorems.

Example. (MATROSOV [1]). Consider the equations of a symmetric gyro
in the form

$$A\dot{p} + (C - A)qr = Pz_0\gamma_2 - R_p(p, q, t),$$

$$A\dot{q} + (A - C)pr = -Pz_0\gamma_1 - R_q(p, q, t),$$

$$C\dot{r} = M(t, r),$$

$$\dot{\gamma}_1 = r\gamma_2 - q\gamma_3, \quad \dot{\gamma}_2 = p\gamma_3 - r\gamma_1, \quad \dot{\gamma}_3 = q\gamma_1 - p\gamma_2.$$

A, B, C, P are positive constants, $R = R(p, q, t)$ is homogeneous of
degree $m \geq 2$, positive definite with respect to p and q, and bounded
with respect to t. The moment M is also bounded. Finally $\gamma_1^2 - \gamma_2^2 + \gamma_3^2 = 1$.
The last equations show that r is a bounded function of t and accordingly
the system has a particular solution

$$p = q = 0, \quad r = r(t, r_0, t_0), \quad \gamma_1 = \gamma_2 = 0, \quad \gamma_3 = 1.$$

To investigate its stability we consider the functions

$$v = \frac{A}{2}(p^2 + q^2) - \frac{P}{2}z_0(\gamma_1^2 + \gamma_2^2 + (1 - \gamma_3)^2),$$

$$w = A(p\gamma_2 - q\gamma_1)$$

whose derivatives in the sense of the equation of the perturbed motion
are

$$\dot{v} = -mR,$$

$$\dot{w} = Pz_0(\gamma_1^2 + \gamma_2^2) - \gamma_1(Cpr - R_q) - \gamma_2(Cqr + R_p) + A\gamma_3(p^2 + q^2).$$

The set N of Theorem 55.3 is given by $R = 0$, *i.e.* by $p = 0, q = 0$. On
this set we have

$$\dot{w} = Pz_0(\gamma_1^2 + \gamma_2^2)$$

and for $z_0 \neq 0$ we have

$$|\dot{w}| > P|z_0|\alpha^2,$$

in case

$$p = 0, \quad q = 0, \quad 0 < \alpha^2 < \gamma_1^2 + \gamma_2^2 < A^2.$$

Thus $\dot{w}$ is definitely non-zero on N and the unperturbed motion is
asymptotically stable in case $z_0 < 0.$

D. For several problems in analytical mechanics it is convenient to
study not the vanishing of the (general) coordinates $x_1, \ldots, x_n$ but the
vanishing of the derivatives $\dot{x}_i$. If it is known that $\lim \dot{x} = 0$ we can
infer that the vector $x(t)$ approaches an equilibrium position $\bar{x}$ provided
we also know that $|\dot{x}|$ does not become too small in the vicinity of the
equilibrium, possibly that an estimate

$$(55.4) \qquad\qquad |\dot{x}| \geq \varkappa(|\bar{x} - x|), \quad \varkappa \in K,$$

holds. For then $|x - \bar{x}| \leq \varkappa^I(|\dot{x}|)$ and because $\varkappa \in K$, it follows that $\lim x = \bar{x}$. The condition $\lim \dot{x} = 0$ can sometimes be tested by means of a Liapunov function depending on x, $\dot{x}$ and t (a so-called *kinetic* Liapunov function)[1]. If there exists such a function for which

$$\varphi_1(|\dot{x}|) \leq v(x, \dot{x}, t) \leq \varphi_2(|\dot{x}|),$$

$$\frac{d}{dt} v(x, \dot{x}, t) \leq -\psi(|\dot{x}|)$$

then we see by the usual reasoning that $\lim \dot{x} = 0$.

We are here concerned with a special case of a more general concept. Let x and y be vectors of dimension n, resp. m, and let $z = \mathrm{col}(x, y)$. The vector z satisfies a differential equation

$$(55.5) \qquad \dot{z} = f(z, t), \quad f \in E,$$

with general solution

$$p(t, z_0, t_0) := \mathrm{col}\left(x(t, z_0, t_0), y(t, z_0, t_0)\right).$$

Def. 55.2. The equilibrium of (55.5) is called *stable with respect to the x-coordinates* if for each ε there exists a number $\delta = \delta(\varepsilon, t_0)$ such that

$$|x(t, z_0, t_0)| < \varepsilon \text{ for } t \geq t_0$$

provided $|z_0| < \delta$. If the number δ can be chosen independent of t_0 then the stability is *uniform*. The equilibrium of (55.5) is called *asymptotically stable with respect to the x-coordinates* if it is stable with respect to the x-coordinates and if in addition

$$\lim_{t \to \infty} x(t, z_0, t_0) = 0.$$

Again uniformity is defined in the obvious way.

The partial stability thus defined (it is occasionally also called *conditional* stability[2]) is especially important for equations (55.5) which can be decomposed into two equations of the form

$$\dot{x} = f(x, t), \quad \dot{y} = g(x, y, t).$$

It is easy to find f and g so that the equilibrium is stable only with respect to the x-coordinates but not with respect to all the coordinates.

A criterion for partial stability is given by

Theorem 55.4.[3] Let a function $v(z, t)$ be defined in $K_{h_0, t}$ such that

$$\text{a)} \qquad \varphi_1(|x|) \leq v(z, t), \qquad \varphi_1 \in K,$$

$$\text{b)} \qquad v(z, t) \leq \varphi_2(|z|), \qquad \varphi_2 \in K.$$

[1]) CHANG [1].

[2]) This term is also used with different meaning, *cf.* LEFSCHETZ [1].

[3]) CORDUNEANU [5]. Another criterion appears in ZUBOV [2].

c) The derivative of v for (55.5) satisfies a scalar differential inequality (*cf.* sec. 42)

$$\dot{v} \leq h\big(t, v(t)\big).$$

Suppose that the solution $w = 0$ of the differential equation

$$\dot{w} = h(t, w)$$

is uniformly asymptotically stable. Then the equilibrium of (55.5) is uniformly asymptotically stable with respect to the x-coordinates.

Proof. The proof proceeds as the proofs in secs. 25 and 42. From a) we obtain an inequality

$$|x| \leq \varphi_1^I(v),$$

from c) an estimate

$$v \leq q(v_0)\, \varrho(t - t_0), \qquad q \in K, \quad \varrho \in L,$$

and b) yields

$$v_0 \leq \varphi_2(|z_0|).$$

Summarizing and taking into account the formulas of sec. 24, we obtain

$$|x| \leq \varphi(|z_0|)\, \sigma(t - t_0), \qquad \varphi \in K, \quad \sigma \in L,$$

and this is the assertion.

The theorem can be modified: Condition a) and the stability of the solution $w = 0$ imply the partial stability with respect to x, etc.

E. Stability on a finite interval.

The statement "the equilibrium is stable in the sense of Liapunov" involves an assertion about the infinite interval $t_0 \leq t < \infty$ and can therefore never be tested in practice since each observation lasts only a finite time. One has attempted in many ways to overcome this difficulty. One possibility is to start with a fixed finite interval and to apply the definiton to this interval in a reasonable manner. This yields an assertion of type A 1 of the table in sec. 36. In a somewhat generalized form it says: For the motion to remain within the region $G(t_1)$ throughout the interval $t_0 \leq t \leq t_1$ it is sufficient that at time t_0 it is in a region G_0 determined by $G(t_1)$. For type A 1, sec. 36, the regions are balls. But the definition can be extended to other types of regions, for instance, to ellipsoids

$$(55.6) \qquad\qquad x^T B(t)\, x = c$$

where $B(t)$ is positive definite and symmetric for $t_0 \leq t \leq t_1$. In this case one would require that the diameter of the region $x^T B(t)\, x = c$, *i.e.* the maximal distance of two points, does not grow as t increases and that

$$p(t, x_0, t_0)^T B(t)\, p(t, x_0, t_0) \leq c, \qquad t_0 \leq t \leq t_1,$$

provided

$$x_0^T B(t_0)\, x_0 \leq c.$$

If this condition is satisfied the equilibrium is called *stable in the interval* (t_0, t_1) with respect to the region (55.6). In lieu of the quadratic form $x^T B(t) x$ we can of course use any positive definite function $u(x, t)$ provided that the diameter of the region $u(x, t) = c$ does not increase as t increases. Criteria for this stability type are obtained as follows. Let a positive definite function $v(x, t)$ be given and let

$$\alpha_1(t) \left(u(x, t)\right)^{l_1} \leq v(x, t) \leq \alpha_2(t) \left(u(x, t)\right)^{l_2},$$

$$- \beta_1(t) \left(u(x, t)\right)^{l_3} \leq \dot{v} \leq - \beta_2(t) \left(u(x, t)\right)^{l_4}$$

where $\alpha_1, \alpha_2, \beta_1, \beta_2$ are positive continuous functions and $l_1, \ldots, l_4$ are real positive numbers. As in sec. 42 we obtain differential equations from the two inequalities and thus estimates for $v(x, t)$ with whose aid we can determine whether and how long the inequality

$$u\left(\boldsymbol{p}(t, x_0, t_0), t\right) \leq c$$

holds.[1]

As an example we cite a theorem (LEBEDEV [1, 2]) which at the same time shows some of the limitations in applicability. Let

$$\dot{x} = P(t) x + f(x, t), \quad |f| = o(|x|),$$

and assume that the matrix $P(t_0)$ is stable. For a given positive definite C we can then construct (*cf.* sec. 27) a positive definite B such that $P(t_0)^T B + B P(t_0) = - C$. The derivative of the function $v(x, t) = x^T B x$ for the given equation is

$$\dot{v} = x^T \left(P(t)^T B + B P(t)\right) x + f^T B x + x^T B f.$$

If we set

$$G(t) = P(t) - P(t_0)$$

we obtain

$$\dot{v} = - x^T C x + x^T \left(G(t)^T B + B G(t)\right) x + 2 f^T B x$$

and we see immediately that this expression is negative for sufficiently small $|x|$ and sufficiently small values of $t - t_0$. Then $v(t)$ decreases and we conclude that the equilibrium is stable with respect to a region of the form $x^T B x = c$ in a certain finite interval. In a concrete case the interval and the constant can be determined. But possibly they are so small that they have no practical significance.

F. Utilization of the functional matrix.

Let

(55.7) $$\dot{x} = f(x, t), \quad f \in C_1,$$

be given and let the functional matrix

(55.8) $$J(x, t) = \frac{\partial f}{\partial x}$$

---

[1] *cf.* also KAMENKOV [1], LEBEDEV [1, 2], ZUBOV [2].

be continuous and bounded with respect to t in a certain domain K_{h,t_0}. If (55.7) is linear and autonomous then (55.8) is a constant matrix which, according to sec. 27, we can use to construct a Liapunov function for the equation. Under certain conditions we can proceed similarly in the general case.

Theorem 55.5 (KRASOVSKII [4]). Let a real symmetric positive definite constant matrix B be given such that the characteristic roots $\lambda_1, \ldots, \lambda_n$ (which depend on x and t) of the symmetric matrix

$$M = \tfrac{1}{2}\,(J^T B + B J)$$

satisfy in K_{h,t_0} an estimate

$$(55.9) \qquad\qquad \sup_i \lambda_i \leq -\delta < 0, \quad \delta \text{ a constant.}$$

Then the equilibrium of (55.8) is asymptotically stable and for fixed t_0, K_h is a subset of the domain of attraction.

Proof. We use the matrix B of the hypothesis to construct the function

$$v = x^T B x$$

and we estimate the derivative

$$\dot{v} = f^T(x, t)\,B x + x^T B f(x, t)$$

in the following way. Let y and z be n-vectors. The inner product

$$q(y) := \sum_{i,j} f_i(y, t)\,b_{ij} z_j$$

can be considered as a function of y alone. From the identity

$$q(y) = q(y) - q(0) + q(0)$$

we obtain by the mean value theorem

$$q(y) = \sum_{k=1}^{n} \frac{\partial q}{\partial y_k}\Big|_{\theta y} y_k + q(0) = \sum_{i,j,k} \frac{\partial f_i(\theta y, t)}{\partial y_k} y_k b_{ij} z_j + q(0),$$

and setting $y = x$, $z = x$,

$$f^T(x, t)\,B x = x^T J(\theta x, t)^T B x + f(0, t)^T B x,$$

and since $f(0, t) = 0$,

$$\dot{v} = 2 x^T M(\theta x, t)\,x \leq -\delta |x|^2$$

for all $(x, t) \in K_{h,t_0}$. Together with Theorem 42.4 this yields the assertion.
If $f(0, t) \not\equiv 0$ then we still have

$$\dot{v} \leq -\delta |x|^2 + \|B\|\,|x|\,|f(0, t)|,$$

and this leads to a differential inequality of the form

$$\dot{v} \leq -a v + b(t)\,v^{1/2}.$$

Under further hypotheses on $b(t)$, *i.e.* on $f(0, t)$, we can conclude that $\lim_{t\to\infty} v(t) = 0$ and hence that $\lim_{t\to\infty} x(t) = 0$. Strictly speaking, we do not have asymptotic stability in this case.[1])

If $\dot{x} = f(x)$ is autonomous then $J = J(x)$ does not depend on t. Then the hypotheses imply that the real parts of the characteristic roots of BJ which, of course, lie between the smallest and the largest characteristic root of M, are smaller than $-\delta$. This implies

$$\left| \det \big(BJ(x)\big) \right| \geq \delta^n, \quad \left| \det J(x) \right| \geq \alpha > 0 ,$$

for a suitable x. The absolute value of the Jacobian determinant is different from zero. This implies that the mapping $y = f(x)$ of the y-space into the x-space is invertible in a certain neighborhood of the origin, and further that the differential equation has an isolated singularity at $x = 0$. The quadratic form

$$(55.10) \qquad\qquad v = y^T B y = f(x)^T B f(x)$$

which is defined in the y-space can therefore also be used as a Liapunov function in the x-space; it is positive definite. Its derivative

$$\dot{v} = \dot{f}(x)^T B f(x) + f(x)^T B \dot{f}(x) = f(x)^T (J^T B + BJ) f(x)$$

is negative definite because of the hypotheses. If now the inequality

$$\left| \det J(x) \right| \geq \alpha > 0$$

holds for all $x \in R_n$ then the mapping $y = f(x)$ is everywhere invertible. The equation $f(x) = 0$ does not have a non-zero solution and the differential equation has only one finite singularity. The volume of a domain B_y in the y-space is

$$\int_{B_y} dy = \int_{B_x} \left| \det J(x) \right| dx \geq \alpha \int_{B_x} dx .$$

It grows without bound as the x-domain increases. This implies that the vector $y = f(x)$ becomes unbounded as $|x|$ increases; at least one of its components grows without bound. Thus the function (55.10) is radially unbounded and we recognize: If in the autonomous case the estimate (55.9) holds for all $x \in R_n$ then the equilibrium is asymptotically stable in the whole. KRASOVSKII [4] showed that the last assertion remains valid in the non-autonomous case also.

56. Perturbed Equations

In addition to the differential equation

$$(56.1) \qquad\qquad \dot{x} = f(x, t), \quad f \in E, \ f \in C_1,$$

---

[1]) *cf.* EZEILO [1].

let a "perturbed" differential equation

$$(56.2) \qquad \dot{x} = f(x, t) + g(x, t), \quad f + g \in E,$$

be given. Assume that the stability behavior of the equilibrium of (56.1) is known. The question arises, under which conditions on the function $g(x, t)$ the equilibrium of (56.2) shows the same stability behavior as that of (56.1), or in other words, when (56.1) and (56.2) are equivalent with respect to stability. We have treated a special case in sec. 28: If both equations are autonomous, if $f(x) = A\,x$ is linear and $g(x) = o(|x|)$, and if we are not in a critical case (all the characteristic roots of A have non-positive real parts but some of them have zero real part) then the equations are equivalent with respect to stability. This *principle of stability in the first approximation* was proved in sec. 28 with help of the direct method. An analysis of the proof shows, however, that it is valid without some of the special hypotheses.

Let a Liapunov function $v(x, t)$ with a negative definite derivative

$$(56.3) \qquad \dot{v}_{(56.1)} \leq -\,\psi(|x|)$$

for (56.1) be given. In sec. 51 we treated conditions for the existence of such a function. Because of property A (Def. 51.1) the critical case is of course excluded for linear systems also. The index of $\dot{v}$ in (56.3) emphasizes that the total derivative for (56.1) was formed. Obviously

$$(56.4) \qquad \begin{aligned} \dot{v}_{(56.2)} &= \sum_{i=1}^{n} \frac{\partial v}{\partial x_i} f_i(x, t) + \frac{\partial v}{\partial t} + \sum_{i=1}^{n} \frac{\partial v}{\partial x_i} g_i(x, t) \\ &= \dot{v}_{(56.1)} + \sum_{i=1}^{n} \frac{\partial v}{\partial x_i} g_i(x, t) \end{aligned}$$

and this expression is again definite whenever the second term can be estimated uniformly with respect to t,

$$\left| \sum_{i=1}^{n} \frac{\partial v}{\partial x_i} g_i(x, t) \right| < \chi(|x|),$$

in such a way that

$$(56.5) \qquad \chi(|x|) < \psi(|x|).$$

Then $v(x, t)$ can be used as a Liapunov function for both equations and for both equations it satisfies the conditions of the same stability theorem.

The hypotheses are easily checked if the functions $v, \dot{v}$, the partial derivatives $\partial v/\partial x_i$, and the disturbance $g(x, t)$ can be estimated by powers of $|x|$:

$$(56.6) \qquad v \leq c_1 |x|^\alpha, \quad \left| \frac{\partial v}{\partial x_i} \right| \leq c_2 |x|^{\alpha-1}, \quad |\dot{v}| \geq c_3 |x|^\beta,$$

and

$$|g_i(\boldsymbol{x})| \leq c_4 |\boldsymbol{x}|^{\gamma}$$

imply

$$\left|\frac{\partial v}{\partial x_i}\, g_i(\boldsymbol{x},\, t)\right| \leq c_5 |\boldsymbol{x}|^{\varkappa+\gamma-1} \, ;$$

and if

$$\beta < \varkappa + \gamma - 1$$

then (56.5) is satisfied for sufficiently small $|\boldsymbol{x}|$. Depending on the signs of $\dot{v}$ and v we can then infer asymptotic stability or instability.

If v is a quadratic form, as for example in the proof of the principle of stability in the first approximation in sec. 28, then $\varkappa = 2$, $\beta = 2$, and the condition is $2 < 1 + \gamma$. Thus it suffices to assume $|\boldsymbol{g}(\boldsymbol{x},t)| = o(|\boldsymbol{x}|)$. But the Liapunov function need not be a quadratic form if it can be estimated by a quadratic form and this is the case for exponential stability, as we already pointed out in sec. 50. In particular we have

Theorem 56.1. Let the right side of the differential equation $\dot{\boldsymbol{x}} = \boldsymbol{f}(\boldsymbol{x},\, t)$ have continuous first order partial derivatives bounded in K_{h,t_0}. Let the equilibrium be exponentially stable, *i.e.* let

$$(56.7) \qquad |\boldsymbol{p}(t,\, \boldsymbol{x}_0,\, t_0)| \leq a\, |\boldsymbol{x}_0|\, e^{-b(t-t_0)}, \qquad a > 0, \ b > 0.$$

Then there exists a Liapunov function $v(\boldsymbol{x},\, t)$ which satisfies estimates of the form

$$a_1|\boldsymbol{x}|^2 \leq v(\boldsymbol{x},\, t) \leq a_2|\boldsymbol{x}|^2,$$

$$\dot{v} \leq -a_3|\boldsymbol{x}|^2, \qquad \left|\frac{\partial v}{\partial x_i}\right| \leq a_4|\boldsymbol{x}|, \qquad i = 1, 2, \ldots, n,$$

for certain positive constants $a_1, \ldots, a_4$.

Proof. We use the construction (49.4) with $u(r) = r^2$, but we use for the length of the interval of integration a finite constant T to be determined later, that is we set

$$v(\boldsymbol{x},\, t) := \int_t^{t+T} |\boldsymbol{p}(\tau,\, \boldsymbol{x},\, t)|^2\, d\tau.$$

Because of the Lipschitz condition we have an estimate

$$|\boldsymbol{p}(\tau,\, \boldsymbol{x},\, t)| \geq |\boldsymbol{x}|\, e^{-b'(\tau-t)}, \qquad b' > 0,$$

and from it and (56.7) we obtain immediately the first two estimates of the assertion. Further

$$\dot{v} = -|\boldsymbol{x}|^2 + |\boldsymbol{p}(t + T,\, \boldsymbol{x},\, t)|^2$$

since the derivative of the integrand is zero. If we choose $T = \dfrac{2\ln a + \ln}{2b}$

then

$$\dot{v} \leq -\frac{1}{2}|\boldsymbol{x}|^2.$$

Finally we have

$$\frac{\partial v}{\partial x_i} = 2 \int\limits_{t}^{t+T} \sum_{j=1}^{n} p_j(\tau, \boldsymbol{x}, t) \frac{\partial p_j(\tau, \boldsymbol{x}, t)}{\partial x_i} \, d\tau.$$

The partial derivatives $\partial p_j/\partial x_i$ which are known to satisfy a certain linear differential equation (*cf.* (51.6)), increase under the given hypotheses at most exponentially so that for suitable constants we have

$$\left| \frac{\partial v}{\partial x_i} \right| \leq 2 n a \, |\boldsymbol{x}| \, k_1 \int\limits_{t}^{t+T} e^{-b(\tau-t)+k_2(\tau-t)} \, d\tau.$$

The integral is independent of t and the proof is complete.

From the discussion at the beginning of this section and using Theorem 56.1 we arrive at

Theorem 56.2. If the equilibrium of the unperturbed differential equation (56.1) is exponentially stable and if the disturbance satisfies an estimate

$$(56.8) \qquad \boldsymbol{g}(\boldsymbol{x}, t) = o(|\boldsymbol{x}|), \qquad (\boldsymbol{x}, t) \in K_{h, t_0},$$

then the equilibrium of the perturbed equation is also exponentially stable, in fact with the same number b as exponent.

Starting from an inequality

$$|\boldsymbol{p}(t, \boldsymbol{x}_0, t_0)| > a \, |\boldsymbol{x}_0| \, e^{b(t-t_0)}, \qquad a > 0, \quad b > 0,$$

we can formulate theorems parallel to Theorems 56.1 and 56.2 which make the appropriate statements about instability properties. Furthermore, if in place of (56.8) we have an inequality

$$|\boldsymbol{g}(\boldsymbol{x}, t)| \leq \alpha(t) \, |\boldsymbol{x}|$$

we can under certain circumstances infer exponential stability. For then we obtain for v a differential inequality of the form

$$\dot{v} \leq \left(-\frac{a_3}{a_1} + a_5 \alpha(t) \right) v, \qquad a_5 > 0,$$

whose solution decreases like an exponential function with $t - t_0$ in the exponent, under suitable hypotheses on $\alpha(t)$. The equilibrium of the perturbed differential equation is also exponentially stable in this case but we might have to use a smaller constant in the exponent.[1]) A sufficient condition, for example, is an inequality

$$\alpha(t) \leq c < \frac{a_3}{a_1 a_5}.$$

The above assumptions on the differential equation (56.2) and on the perturbation function are reasonable from a mathematical point of view but they are not quite realistic if the differential equation describes a

---

[1]) CORDUNEANU [1].

concrete physical system. In general the term $g(x, t)$ describes the effect of small disturbances caused by friction, current heat, etc.; but often it also takes care of inexact measurements and similar factors.

Frequently, such small disturbances do not allow exact numerical measurement and only their order of magnitude is known, so that it is actually not permissible to characterize these variables by giving a "function" $g(x, t)$. In particular the assumption $g(0, t) = 0$ has no justification whatsoever. But its omission denies the hypothesis that (56.2) even admits the solution $x = 0$. In order to express none the less the fact that the unperturbed system which is described by (56.1) is not essentially affected by the additional forces, provided these remain sufficiently small, we introduce the concept of *total stability*.

Def. 56.1. In the domain K_{h, t_0} let

$$(56.9) \qquad \dot{x} = f(x, t), \quad f \in E,$$

and

$$(56.10) \qquad \dot{x} = f(x, t) + g(x, t)$$

be given. Let the general solution of (56.10) exist and be denoted by $p(t, x_0, t_0)$. The equilibrium of the differential equation (56.9) is called *totally stable* if for each $\varepsilon > 0$ there exist two positive numbers $\delta_1(\varepsilon)$ and $\delta_2(\varepsilon)$ such that

$$(56.11) \qquad |p(t, x_0, t_0)| < \varepsilon \quad \text{for} \quad t \geq t_0$$

provided only that

$$|x_0| < \delta_1$$

and that in the domain K_{ε, t_0} the inequality

$$(56.12) \qquad |g(x, t)| < \delta_2$$

is satisfied.

In place of the term *totally stable* one also uses *stable under persistent disturbances*.

The property given in Def. 56.1 can be characterized by means of the general stability concept (sec. 36) if the solution is considered not in dependence on the initital values but on the right side and if accordingly the solutions of (56.9) and (56.10) are denoted by $p(t; f_1)$ resp. $p(t; f_2)$, for identical initial values. Total stability then amounts to saying that an appropriately defined norm

$$\|p(t; f_1) - p(t; f_2)\|$$

can be estimated by an expression $\varphi(\|f_1 - f_2\|)$.

Theorem 56.3. If a positive definite function $v(x, t)$ exists whose partial derivatives $\partial v / \partial x_i$ are bounded in a region G and whose derivative formed

18*

for (56.9) is negative definite then the equilibrium of (56.9) is totally stable.

Proof. By Theorem 41.1, v is decrescent. Therefore there exist inequalities

$$\varphi_1(|\boldsymbol{x}|) \leq v(\boldsymbol{x}, t) \leq \varphi_2(|\boldsymbol{x}|),$$

$$\dot{v} \leq -\psi(|\boldsymbol{x}|)$$

and

$$\varphi_2^I(v) \leq |\boldsymbol{x}| \leq \varphi_1^I(v).$$

$v = \dot{v}(\boldsymbol{x}, t)$ denotes the derivative of v for (56.9). Let $\varepsilon > 0$ be given and choose a number β such that $0 < \beta < \varphi_1(\varepsilon)$. Then for a point $\boldsymbol{x}'$ which satisfies the equation $v(\boldsymbol{x}', t) = \beta$, we have

$$(56.13) \qquad \varphi_2^I(\beta) \leq |\boldsymbol{x}'| < \varepsilon, \quad \dot{v}(\boldsymbol{x}', t) \leq -\psi\left(\varphi_2^I(\beta)\right).$$

The derivatives of v for (56.9) and for (56.10) differ by the summand $\sum\limits_{i=1}^{n} \dfrac{\partial v}{\partial x_i} g_i$, cf. (56.4). Since the partial derivatives are bounded by hypothesis and since (56.12) is valid this summand can be made so small that the derivative of v for (56.10) is negative. We now choose $\delta_1(\varepsilon)$ so that $\delta_1(\varepsilon) < \varepsilon$ and that $v_0 := v(\boldsymbol{x}_0, t_0) < \beta$ for $|\boldsymbol{x}_0| < \delta_1$. Then (56.11) holds throughout; for v decreases along a trajectory of (56.10). Since $v_0 < \beta$, $v < \beta$ throughout, and this implies by (56.13) the inequality

$$|\boldsymbol{x}| = |\boldsymbol{p}(t, \boldsymbol{x}_0, t_0)| < \varphi_1^I(\beta) < \varepsilon.$$

From the converse theorems of secs. 49 and 51 we see that if the equilibrium of (56.9) is uniformly asymptotically stable then there exists a Liapunov function which satisfies the hypotheses of Theorem 56.3. Therefore we have

Theorem 56.4. If the equilibrium is uniformly asymptotically stable then it is also totally stable.

The concept of total stability admits many modifications. We mention without going into the details the concept of *integral stability* introduced by VRKOČ [2]: The equilibrium of (56.9) is called integrally stable if there exists a function $\varphi(r) \in K$ such that

$$|\boldsymbol{p}(t, \boldsymbol{x}_0, t_0)| < \varphi(\delta)$$

provided

$$|\boldsymbol{x}_0| < \delta \quad \text{and} \quad \int\limits_{t_0}^{\infty} \sup_{|\boldsymbol{x}| < \varphi(\delta)} |\boldsymbol{g}(\boldsymbol{x}, t)| \, dt < \delta.$$

Here we estimate a certain mean value of $\boldsymbol{g}(\boldsymbol{x}, t)$ and it is permissible that $\boldsymbol{g}(\boldsymbol{x}, t)$ assumes rather large values in certain small intervals, whereas in Def. 56.1 formula (56.12) requires that $\boldsymbol{g}$ remains uniformly small.

If, as above, we consider total stability as stability in a function space then integral stability differs from the concept defined in Def. 56.1 only in the choice of the norm.

Vrkoč [1] has shown that necessary and sufficient conditions for integral stability can be given by means of Liapunov functions: $v(x, t)$ must be positive definite and radially unbounded; it must have a non-positive derivative, and satisfy a Lipschitz condition

$$|v(x, t) - v(y, t)| < L |x - y|.$$

He also defined asymptotic integral stability and gave conditions for it.

If the differential equation (56.1) is replaced by a difference equation

$$(56.14) \qquad x(t + \tau) - x(t) = \tau f(x(t), t)$$

which can be used to gain approximate solutions, then the question arises whether the equilibrium of the difference equation has the same stability behavior as that of the differential equation. This question can be attacked as follows.

Let $v(x, t)$ be a Liapunov function for (56.1). Its total difference formed for (56.14) is

$$v(x(t + \tau), t + \tau) - v(x(t), t) = v(x + \tau f(x, t), t + \tau) - v(x, t)$$

$$(56.15) \qquad = \sum_{i=1}^{n} \tau \left(\frac{\partial v}{\partial x_i}\right)_0 f_i(x, t) + \tau \left(\frac{\partial v}{\partial t}\right)_0.$$

The index 0 signifies that in the partial derivatives the arguments $x + \theta \tau f(x, t)$ and $t + \theta \tau, 0 < \theta < 1$ have to be substituted. The question is whether this form is negative definite when the derivative of v for (56.1) is negative definite. If $f(x, t)$ and the partial derivatives of v are bounded with respect to t we can make (56.15) negative by an appropriate choice of the step length τ. The choice of τ, of course, depends on the bounds for f and the derivatives. As Theorem 51.2 indicates, the following conditions are sufficient for the existence of these bounds: 1) $f(x, t)$ satisfies in a domain K_{h, t_0} a Lipschitz condition with uniform Lipschitz constant. 2) The equilibrium of (56.1) is uniformly asymptotically stable.

A simple example shows that the boundedness of $f(x, t)$ is essential. The equilibrium of the scalar equation $\dot{x} = -tx$ is exponentially stable; $v = x^2$ can be used as the Liapunov function. The associated difference equation $x(t + \tau) = x(t)(1 - \tau t)$ however, has for its solution

$$x(t) = (-\tau^2)^{\frac{1}{\tau}(t - t_0)} \frac{\Gamma\left(\dfrac{t}{\tau} - \dfrac{1}{\tau^2}\right)}{\Gamma\left(\dfrac{t_0}{\tau} - \dfrac{1}{\tau^2}\right)}$$

which does not tend to zero as t increases. The total difference of the Liapunov function x^2 is $x^2(\tau^2 t^2 - 2\tau t)$. It is positive for large values of t.

By modifying Def. 56.1, a reasonable definition of the concept of *practical stability*, which we already mentioned in secs. 2 and 26, can be obtained. From a practical point of view a concrete system will be considered as stable if the deviations of the motions from the equilibrium remain within certain bounds determined by the physical situation, in case the initial values and the disturbance are bounded by suitable constants. One does not, however, insist on the narrower stability property; that is one will not require that the deviation from zero can be made arbitrarily small by a suitable choice of constants. Analytically this has the following form suggested by LaSalle and Lefschetz [1]:

Def. 56.2. The equilibrium of equation (56.9) is called *practically stable* if there exist two constants k_1, k_2 with the following property: If $|x_0| < k_1$ and if the function $g(x, t)$ in (56.10) satisfies an inequality

$$|g(x, t)| < k_2, \quad x \in R_n, \quad t \geq t_0,$$

then the general solution of (56.10) admits an estimate

$$|p(t, x_0, t_0)| < k_3$$

where the constant k_3 depends on the particular physical situation.

A systematical study of this concept, the development of criteria, etc., have not yet been undertaken.

57. Equations with Homogeneous Right Side [1]

In this section we consider equations of the form

$$(57.1) \qquad \dot{x} = f(x, t),$$

whose right sides are homogeneous in x, *i.e.* satisfy a relation

$$(57.2) \qquad f(cx, t) = c^k f(x, t), \quad k \geq 1.$$

In the case of uniform asymptotic stability we can exactly describe for equations of this type the rate with which the motions tend to zero. The main goal of this investigation is to prove that an estimate of the form

$$(57.3) \qquad |p(t, x_0, t_0)| \leq a\,|x_0|^{1/2}(t - t_0)^{-1/2(k-1)}, \quad (k > 1)$$

respectively an exponential estimate

$$(57.4) \qquad |p(t, x_0, t_0)| \leq a\,|x_0|\,e^{-b(t-t_0)}, \quad (k = 1)$$

is possible. In the case (57.4) we speak of *exponential stability* (*cf.* Def. 26.2).

---

[1] Hahn [5].

For the right side of (57.1) we assume that it is continuous in x, $x \in R_n$, and in t, $0 < t < \infty$. The equation $f(0, t) = 0$, $0 < t < \infty$, is a consequence of (57.2). But we do require that $x = 0$ is the only singular point, *i.e.* that $f(x_0, t)$ is not identically zero for any $x_0 \neq 0$.

We assume that the number k, the so called degree of homogeneity, is either an integer or a rational number with odd denominator, *i.e.* that $k = \frac{p}{2q+1}$ ($p > 0$ an integer, $q \geq 0$ an integer, p and $2q+1$ relatively prime). For if k is not of this form then $f(x, t)$ cannot be real valued in a neighborhood of $x = 0$, as is seen from (57.2) if we set $c = -1$. If $k < 1$ then the solution of (57.1) is not uniquely determined by the initial value $x_0 = 0$. For example, the solution of the equation

$$\dot{x} = x^{2/3}$$

is

$$\left(x_0^{1/3} + \frac{1}{3}(t - t_0)\right)^3$$

which passes through the origin if x_0 is negative.

As in sec. 17 we can obtain auxiliary equations for r and y (*cf.* (17.4)) by the transformation $x = ry$, $r = |x|$. If we set

$$y(t)^T f\left(y(t), t\right) := g(t) \, ,$$

we obtain

$$(57.5) \qquad\qquad \dot{r} = g(t)\, r^k$$

$$(57.6) \qquad \dot{y} = r^{k-1}\left(f(y, t) - g(t)\, y\right), \quad |y(t)| = 1.$$

The equation for r can be integrated in closed form:

$$(57.7) \qquad r(t, r_0, t_0) = [r_0^{1-k} + (1 - k)\, s(t, t_0)]^{1/1-k}, \quad k > 1,$$

resp.

$$(57.8) \qquad\qquad r(t, r_0, t_0) = r_0\, e^{s(t,t_0)}, \quad k = 1,$$

where

$$(57.9) \qquad\qquad s(t, t_0) := \int_{t_0}^{t} g(\tau)\, d\tau.$$

Obviously the equilibrium of (57.1) is stable, respectively attractive, if and only if the same is true for the variable $r(t, r_0, t_0)$, as long as the side condition $|y(t)| = 1$ is satisfied. We recognize immediately:

1) The equilibrium is stable if and only if $s(t, t_0)$ is bounded above for a fixed t_0; the equilibrium is uniformly stable if and only if $s(t, t_0)$ is smaller than a fixed number which is independent of t and t_0. 2) The equilibrium is uniformly attractive with respect to r_0 if and only if

$s(t, t_0)$ tends to $-\infty$ as t increases. 3) For the equilibrium to be uniformly attractive with respect to t_0 there must be an estimate

$$r(t, r_0, t_0) < \sigma(t - t_0), \qquad \sigma \in L.$$

If we take (57.7) and (57.8) into consideration then this implies an estimate

$$|s(t_0 + T, t_0)| \geq \varkappa(T)$$

which holds for $T \geq T_1$ and is uniform with respect to t_0. $\varkappa(T)$ is an unbounded monotone increasing function, in fact, $\varkappa(T) \geq cT$. If this were not true we had a relation

$$|s(t_0 + T, t_0)| \leq \varkappa_1(T)$$

where $\varkappa_1(T) = o(T)$ monotone increasing, and for suitable T_n, τ_n,

$$\varkappa(\tau_n) \leq |s(t_0 + T_n + \tau_n, t_0 + T_n)| = |s(t_0 + T_n + \tau_n, t_0) - |s(t_0 + T_n, t_0)| \leq \varkappa_1(T_n + \tau_n) - \varkappa_1(T_n).$$

Since $\varkappa_1(T) = o(T)$ the right side is bounded and we arrive at a contradiction. From

$$|s(t_0 + T, t_0)| \geq cT, \qquad T \geq T_1,$$

we get with fitting constants

$$r(t, r_0, t_0) \leq a r_0 e^{c(t_0 - t)}, \qquad k = 1,$$
$$r(t, r_0, t_0) \leq (a r_0^{1-k} - (1 - k) c(t - t_0))^{1/1-k}, \qquad k > 1.$$

Returning to the initial equation, we have:

Theorem 57.1. Let the above assumptions on $f(x, t)$ and k be satisfied for the equation (57.1). Let the equilibrium be uniformly asymptotically stable. Then we have for suitable $a > 0$, $c_1 > 0$, $b_1 > 0$,

$$(57.10) \quad |p(t, x_0, t_0)| \leq (a|x_0|^{1-k} + c_1(t - t_0))^{1/1-k}, \qquad k > 1.$$
$$(57.11) \quad |p(t, x_0, t_0)| \leq a|x_0| e^{-b_1(t-t_0)}, \qquad k = 1.$$

As we shall see below, the hypothesis on the stability contains a further assumption on k. If the right side of (57.10) is estimated by means of a product (*cf.* sec. 24 A) then (57.3) results.

Theorem 57.2. In addition to the hypotheses of Theorem 57.1 let $f(x, t)$ be bounded for $|x| = 1$. Then the following inequalities hold:

$$(57.12) \quad |p(t, x_0, t_0)| \geq (|x_0|^{1-k} + c_2(t - t_0))^{1/1-k}, \qquad k > 1,$$
$$(57.13) \quad |p(t, x_0, t_0)| \geq |x_0| e^{-b_2(t-t_0)}, \qquad k = 1.$$

Proof. The hypothesis says that

$$g(t) \leq |y(t)| |f(y, t)| \leq K.$$

This implies

$$|s(t, t_0)| \leq K(t - t_0)$$

and the assertion follows$\bullet$

Under certain circumstances we can do without the requirement that the solution exists in a full neighborhood of the origin. This is illustrated in the example of the system of scalar equations

$$\dot x = -x, \quad \dot y = -y + \sqrt{xy}, \quad \sqrt{xy} \geq 0.$$

This system is real valued only in the region $xy \geq 0$. The solution (for $t_0 = 0$)

$$x = x_0 e^{-t}, \quad y = e^{-t}\left(y_0 + t\sqrt{x_0 y_0} + \tfrac{x_0}{4}t^2\right)$$

is real for $x_0 y_0 \geq 0$, remains entirely in the region $xy \geq 0$, and satisfies an exponential estimate there.

In general this can be stated as follows: 1) If $f(x, t)$ is real valued in certain sectors, 2) if each solution which starts in such a sector remains in it, and 3) if these solutions tend to zero uniformly with respect to t_0, then there exist estimates (57.10) resp. (57.11).

The estimates (57.11) and (57.13) are especially applicable to linear equations (*cf.* chapter VIII). An example of a non-linear equation with degree of homogeneity 1 is given by the scalar system

$$\dot x = -y - x\,h(x, y), \quad \dot y = x - y\,h(x, y); \quad h(x, y) := \frac{a^2 x^2 + b^2 y^2}{x^2 + y^2}.$$

whose solution is (for $t_0 = 0$)

$$x = r\cos\varphi, \quad y = r\sin\varphi,$$

where

$$r = r_0 \exp\left(-\frac{a^2 + b^2}{2}t - \frac{a^2 - b^2}{4}\left(\sin 2(t + \varphi_0) - \sin 2\varphi_0\right)\right),$$

$$\varphi = t + \varphi_0.$$

Starting with the estimates which we proved we can immediately find a Liapunov function for (57.1) by the method of sec. 49. If $|f(x, t)|$ is bounded for $|x| = 1$ then

$$(57.14) \quad v(x, t) = \int_t^\infty |p(\tau, x, t)|^{m(k-1)}\, d\tau, \quad k > 1, \quad m > 2,$$

resp.

$$(57.15) \quad v(x, t) = \int_t^\infty |p(\tau, x, t)|^2\, d\tau, \quad k = 1,$$

satisfies all the requirements, and in fact

$$\frac{1}{(m-1)c_2}|x|^{(k-1)(m-1)} < v(x, t) < \frac{1}{(m-1)c_1}|x|^{(k-1)(m-1)}, \quad k > 1,$$

resp.

$$\frac{1}{2b_2}|x|^2 < v(x, t) < \frac{1}{2b_1}|x|^2, \quad k = 1.$$

For the derivative we conclude

$$\dot{v} = - \, |x|^{m(k-1)} \text{ resp. } \dot{v} = - \, |x|^2.$$

If $f(x, t)$ has partial derivatives with respect to x which are uniformly bounded with respect to t then we obtain an estimate for the partial derivatives $\partial v/\partial x_i$ in the following manner. We first consider $k > 1$ and have

(57.16)

$$\frac{\partial v}{\partial x_i} = m(k-1) \int_t^\infty |p(\tau, x, t)|^{m(k-1)-2} \sum_{j=1}^n p_j(\tau, x, t) \frac{\partial p_j(\tau, x, t)}{\partial x_i} \, d\tau.$$

The partial derivatives of p_j satisfy the linear system of equations (51.6). In the present case

$$\left| \frac{\partial f_r}{\partial x_i} \right| < K \, |x|^{k-1} \; (i, r = 1, 2, \ldots, n).$$

If we replace x by $p(t, x, t_0)$, as is required in forming (51.6), and if we observe (57.10) we see that the elements of the matrix of (51.6) are no larger than a fixed constant times t^{-1} in absolute value. In this case the solutions of the linear equation do not grow faster than a fixed power t^s (cf. Theorem 58.2). If the number m in (57.14) is chosen larger than $s + 2$ then the integral (57.16) converges uniformly with respect to x. Hence the partial derivative exists and (57.10) implies an estimate

$$\left| \frac{\partial v}{\partial x_i} \right| < c_3 \, |x|^{(m-1)(k-1)-1}.$$

The argument is valid for $k = 1$ as well. But then the convergence of the integral which is analogous to (57.16) needs no special justification.

We can now utilize sec. 56 and obtain

Theorem 57.3. Let the hypothesis of Theorems 57.1 and 57.2 be given. Let the function $f(x, t)$ have continuous first order partial derivatives with respect to x which are uniformly bounded with respect to t. In addition to (57.1) let the perturbed equation

(57.17)
$$\dot{x} = f(x, t) + g(x, t),$$

where

$$g(x, t) = o(|x|^k),$$

be given. Then the equilibrium of (57.17) is uniformly asymptotically stable and the solution satisfies estimates of the types (57.10) etc., respectively.

If equation (57.1) is autonomous or has coefficients which are periodic in t then the function defined by (57.14), respectively (57.15), is independent of t, respectively periodic in t. This follows directly from the relation

$$p(t + \omega, x_0, t_0 + \omega) = p(t, x_0, t_0),$$

which is satisfied in the autonomous case for arbitrary ω and in the periodic case for the period, and which implies $v(x, t + \omega) = v(x, t)$. Furthermore we have

Theorem 57.4. If the differential equation (57.1) is autonomous then the function $v(x)$ defined by (57.14) is a homogeneous function of degree $(m - 1)(k - 1)$.

Proof. We apply (17.2) and observe that the function constructed in (57.14) does not depend on t. We then have the following computation:

$$v(c\,x) = \int_0^\infty |p(\tau, c\,x, 0)|^{m(k-1)}\, d\tau$$

$$= c^{m(k-1)} \int_0^\infty |p(c^{k-1}\,\tau, x, 0)|^{m(k-1)}\, d\tau$$

$$= c^{(m-1)(k-1)} \int_0^\infty |p(\tau, x, 0)|^{m(k-1)}\, d\tau = c^{(m-1)(k-1)}\, v(x)\,\bullet$$

If $k = \dfrac{2p}{2q+1}$ then by Theorem 57.4 the function $v(x)$, for $m = 2q + 2$, is homogeneous of the odd order $2(p - q) - 1$ and hence is clearly not definite. But its derivative is negative definite. Thus we obtain a contradiction: The assumption that the equilibrium is asymptotically stable is false. In this case we cannot form the function (57.14) at all since the integral does not converge.

For the function $v(x)$ of Theorem 57.4 we have

$$\dot{v} = \sum_{i=1}^n \frac{\partial v}{\partial x_i}\, f_i(x) = - |x|^{m(k-1)}$$

and

$$\sum_{i=1}^n \frac{\partial v}{\partial x_i}\, x_i = (m - 1)(k - 1)\, v(x).$$

In simple cases the function $v(x)$ can be computed from these two equations.[1]

These ideas can be extended somewhat. Let

$$(57.18) \qquad \dot{x}_i = f_i(x_1, \ldots, x_n; t), \quad i = 1, 2, \ldots, n,$$

and assume that the right sides are such that

$$f_i(c^{m_1} x_1, \ldots, c^{m_n} x_n; t) = c^{k+m_i-1}\, f_i(x_1, \ldots, x_n; t).$$

If we introduce the variable

$$\varrho := \sum_{i=1}^n |x_i|^{1/m_i} = \sum_{i=1}^n (x_i\, \operatorname{sgn}\, x_i)^{1/m_i}$$

---

[1] *cf.* ZUBOV [3].

and set

$$x_i = \varrho^{m_i} y_i, \quad i = 1, 2, \ldots, n,$$

we obtain

$$\sum_{i=1}^{n} |y_i|^{1/m_i} = 1, \quad \sum_{i=1}^{n} \frac{1}{m_i} |y_i|^{1/m_i} y_i^{-1} \dot{y}_i = 0$$

and

$$\dot{x}_i = \varrho^{m_i} \dot{y}_i + m_i \dot{\varrho} \, \varrho^{m_i-1} y_i = \varrho^{k+m_i-1} f_i(\mathbf{y}, t),$$

$$\dot{y}_i + m_i \frac{\dot{\varrho}}{\varrho} y_i = \varrho^{k-1} f_i(\mathbf{y}, t), \quad i = 1, \ldots, n,$$

$$\frac{\dot{\varrho}}{\varrho} \sum_{i=1}^{n} |y_i|^{1/m_i} = \varrho^{k-1} \sum_{i=1}^{n} \frac{1}{m_i} f_i(\mathbf{y}, t) |y_i|^{1/m_i} y_i^{-1},$$

$$\dot{\varrho} = \varrho^k \sum_{i=1}^{n} \frac{1}{m_i} |y_i|^{1/m_i} y_i^{-1} f_i(\mathbf{y}, t).$$

We thus obtain for ϱ a scalar differential equation of type (57.5) and a representation (57.7), respectively (57.8). If ϱ tends to zero as t increases then the same is true for $\mathbf{p}(t, \mathbf{x}_0, t_0)$ and as in the proof of Theorem 57.1 the order of magnitude of $|\mathbf{p}(t, \mathbf{x}_0, t_0)|$ can be estimated with the help of ϱ. In case of uniform asymptotic stability the solution of (57.18) again decreases like a power of $(t - t_0)^{-1}$. But now the exponent depends on k and on the m_i. It also makes sense to extend the remaining results.

It follows from Theorems 57.1 and 57.2 that there exist differential equations whose solutions decrease like a given power. We finally mention that there exist equations whose solutions decrease more slowly than any power. A simple example is

$$\dot{x} = - x^3 \exp(- x^{-2})$$

whose solution is

$$p(t, x_0, t_0) = [\ln\left(\exp(x_0^{-2}) + 2(t - t_0)\right)]^{-1/2}.$$

Chapter VIII

Linear Differential Equations

58. The General Solution of a Linear Homogeneous Differential Equation

In this section we consider the vector equation

$$(58.1) \qquad \dot{x} = A(t)\, x.$$

We assume that the elements $a_{ik}(t)$ of the matrix A are continuous functions of t defined for $t \geq t_0$. Equations with a constant A, treated in sec. 4, are special cases of (58.1). On the other hand, (58.1) is a special case of the equations with homogeneous right side which were examined in sec. 57.

The theory of linear equations is well-known and can be reviewed in many sources. We only cite here those results and notations which we shall need and we refer the reader to the literature for proofs, etc.[1]

An important property of the solutions of (58.1) is that they obey the principle of superposition (sec. 3); if $x^{(1)}$ and $x^{(2)}$ are solutions then $c_1 x^{(1)} + c_2 x^{(2)}$ is also a solution. The number of linearly independent solutions of (58.1) is equal to n, the order of the matrix $A(t)$, *i.e.* the solutions of (58.1) determine a vector space of dimension n.

The n solution vectors $x^{(1)}, \ldots, x^{(n)}$ are linearly independent if and only if their determinant

$$D(t) := \det(x^{(1)}, \ldots, x^{(n)})$$

is different from zero. Now it is well-known that

$$(58.2) \qquad \begin{aligned} \dot{D}(t) &= \operatorname{Tr}\big(A(t)\big)\, D(t) \\ D(t) &= D(t_0) \exp\left(\int_{t_0}^{t} \operatorname{Tr}\big(A(\tau)\big)\, d\tau \right). \end{aligned}$$

$D(t)$ is therefore different from zero if $D(t_0) \neq 0$, that is, if the initial vectors $x^{(k)}(t_0)$, $k = 1, \ldots, n$, are linearly independent. A *fundamental system* of solutions of equation (58.1) is a system of n linearly independent

---

[1] *cf.* for instance CESARI [1], CODDINGTON and LEVINSON [1], KAMKE [1].

solutions $x^{(1)}, \ldots, x^{(n)}$. These also form a basis in the space of solutions. It is convenient to use the columns $x^{(1)}, \ldots, x^{(n)}$ to form a matrix X which then satisfies the matrix equation

$$\dot{X} = A(t)\, X.$$

Between two different bases X and Y we have the relation

$$(58.3) \qquad Y = XC, \quad \det C \neq 0,$$

which expresses the fact that each solution belonging to one basis can be expressed as a linear combination of elements of the other basis,

$$y^{(k)} = \sum_{j=1}^{n} c_{jk}\, x^{(j)}, \qquad k = 1, \ldots, n.$$

The particular basis whose matrix for $t = t_0$ is the unit matrix is called the *standard basis*. It is denoted by

$$K(t, t_0) \quad \text{resp.} \quad K\big(t, t_0; A(t)\big),$$

the second in case the dependence on the matrix $A(t)$ is to be emphasized. Thus we have

$$\frac{\partial K(t, t_0; A(t))}{\partial t} = A(t)\, K\big(t, t_0; A(t)\big),$$

$$K\big(t_0, t_0; A(t)\big) = E.$$

From (58.3) we see that

$$(58.4) \qquad X(t) = K\big(t, t_0; A(t)\big)\, C, \ \det C \neq 0,$$

is a representation of the general fundamental system of (58.1). The general solution of the equation therefore has the form

$$(58.5) \qquad p(t, x_0, t_0) = K(t, t_0)\, x_0,$$

which again immediately shows the validity of the superposition principle. From the general fact

$$p(t, x_0, t_0) = p\big(t, p(t_1, x_0, t_0), t_1\big)$$

we immediately obtain the important relation

$$(58.6) \qquad K(t, t_0) = K(t, t_1)\, K(t_1, t_0).$$

It is valid for any three values t, t_0, t_1 taken from the interval of continuity of (58.1). In particular we have

$$K(t, t_0) = K(t_0, t)^I,$$

$$K(t, t_0) = K(t, 0)\, K(0, t_0) = K(t, 0)\, K(t_0, 0)^I.$$

The last formula shows that the two parameter family of matrices $K(t, t_0)$ can be expressed in terms of the elements of the one parameter family $K(t, 0)$ and their inverses.

To derive a closed expression for the standard solution we solve (58.1), respectively the equivalent integral equation

$$(58.7) \qquad x(t) = x(t_0) + \int_{t_0}^{t} A(\tau)\, x(\tau)\, d\tau$$

by successive approximations. Define

$$\Re_A z := \int_{t_0}^{t} A(\tau)\, z(\tau)\, d\tau$$

and let the sequence x_i be given by

$$x_{i+1} = \Re_A x_i, \qquad i = 0, 1, 2, \ldots.$$

The series

$$x_0 + x_1 + x_2 + \cdots$$

converges and represents a solution of (58.1), namely $p(t, x_0, t_0)$.

We do not give the proof here and only mention that it leads to the representation

$$K(t, t_0; A(t)) = E + \int_{t_0}^{t} A(t_1)\, dt_1 + \int_{t_0}^{t} A(t_1) \int_{t_0}^{t_1} A(t_2)\, dt_2\, dt_1 + \cdots.$$

In case the matrices $A(t)$ and $\int_{t_0}^{t} A(\tau)\, d\tau$ commute,

$$K(t, t_0; A(t)) = \exp\left(\int_{t_0}^{t} A(\tau)\, d\tau\right);$$

and, in particular, we have for a constant A

$$K(t, t_0; A) = e^{A(t-t_0)}.$$

The proof also yields the estimate

$$(58.8) \qquad |K(t, t_0; A(t))\, x_0| \leq e^{mn(t-t_0)}\, |x_0|$$

where m is an upper bound for $|a_{ik}(\tau)|$, $t_0 \leq \tau \leq t$. We obtain

Theorem 58.1. If the elements $a_{ik}(t)$ are bounded then the norm of $K(t, t_0; A(t))$ increases at most as fast as an exponential function with $t - t_0$ in the exponent.

We also have

Theorem 58.2. If the elements of $A(t)$ satisfy an estimate

$$|a_{ik}(t)| < k t^{-1}, \qquad k \text{ constant,}$$

then the norm of $K(t, t_0; A(t))$ increases at most as fast as a power of t.

Proof. We introduce the new variable $s = \ln t$. Then (58.1) becomes

$$\frac{d\boldsymbol{x}}{ds} = e^s A\left(e^s\right)\boldsymbol{x}.$$

By hypothesis, the elements of $e^s A\left(e^s\right)$ are bounded. The solution therefore grows at most as fast as an exponential function of s or as a power of t.

From the representation (58.4) and from our discussion on $K\left(t, t_0\right)$ we obtain

Theorem 58.3. The equilibrium of (58.1) is stable if and only if $K\left(t, t_0; A\left(t\right)\right)$ is bounded with respect to t. It is uniformly stable if $K\left(t, t_0; A\left(t\right)\right)$ is bounded with respect to t and t_0.

The equilibrium is attractive if and only if $K\left(t, t_0; A\left(t\right)\right)$ tends to zero as t increases. If the equilibrium is asymptotically stable then it is also equiasymptotically stable in the whole.

A different characterization of equations with stable, respectively attractive, equilibrium is possible with the aid of the function

$$(58.9) \qquad \mu\left(A\left(t\right)\right) : = \lim_{h \to 0} \frac{\|E + h A(t)\| - 1}{h}$$

If we observe that $\|E + hA\|$ is equal to the square root of the largest root of the equation

$$\det\left(\lambda E - E - h(A + A^T) - h^2 A^T A\right) = 0$$

and that this square root can be written in the form $1 + \frac{h}{2}\gamma + O\left(h^2\right)$, where γ denotes the largest characteristic root of $(A + A^T)$, then we see that $\mu(A)$ is equal to the largest characteristic root of $(A + A^T)/2$.

Theorem 58.4.[1]) The solution of (58.1) admits an estimate

$$|\boldsymbol{x}_0| \exp\left(-\int_{t_0}^{t} \mu\left(-A\left(u\right)\right) du\right) \leq \left|\boldsymbol{p}\left(t, \boldsymbol{x}_0, t_0\right)\right|$$

$$\leq |\boldsymbol{x}_0| \exp\left(\int_{t_0}^{t} \mu\left(A\left(u\right)\right) du\right).$$

The function

$$(58.10) \qquad \left|\boldsymbol{p}\left(t, \boldsymbol{x}_0, t_0\right)\right| \exp\left(-\int_{t_0}^{t} \mu\left(A\left(u\right)\right) du\right)$$

is non-increasing with respect to t; the function

$$(58.11) \qquad \left|\boldsymbol{p}\left(t, x_0, t_0\right)\right| \exp\left(\int_{t_0}^{t} \mu\left(-A\left(u\right)\right) du\right)$$

is non-decreasing.

---

[1]) *cf.* COPPEL [2] and the literature cited there.

Proof. Let $y(t) := |\boldsymbol{x}(t)|$. By (41.4),

$$\dot{y}(t) = \lim_{h \to 0} \frac{1}{h}\left(|\boldsymbol{x}(t) + h\dot{\boldsymbol{x}}(t)| - |\boldsymbol{x}(t)|\right).$$

Now

$$|\boldsymbol{x}(t) + h\dot{\boldsymbol{x}}(t)| = |(E + hA(t))\,\boldsymbol{x}(t)| \leq \|E + hA(t)\|\,|\boldsymbol{x}(t)|$$

and thus

$$\dot{y}(t) \leq \lim_{h \to 0} \frac{1}{h}\left(\|E + hA(t)\| - 1\right) y(t) = \mu(A(t))\,y(t).$$

From this differential inequality we conclude that (58.10) has a non-positive derivative and hence does not increase. Changing over to $-t$ we obtain the result for (58.11) and the proof is complete.

Theorem 58.4 implies: The equilibrium of (58.1) is unstable or stable according as

$$\liminf_{t \to \infty} \int_{t_0}^{t} \mu(-A(u))\,du = -\infty \quad \text{or} \quad \limsup_{t \to \infty} \int_{t_0}^{t} \mu(A(u))\,du < +\infty.$$

The stability is uniform if $\mu(A(t)) \leq 0$ for $t \geq t_0$. If

$$\lim \int_{t_0}^{t} \mu(A(u))\,du = -\infty$$

then the equilibrium is asymptotically stable and uniformly so in case $\mu(A(t)) \leq -\alpha < 0$ for $t \geq t_0$.

A further criterion due to COPPEL [1, 2] depends on

Theorem 58.5. Let $Q(t)$ be a nonsingular matrix which is continuous for $t \geq t_0$ and let P_0, P_1 be *supplementary projections*, that is constant matrices satisfying the equations

$$P_0^2 = P_0, \quad P_1^2 = P_1, \quad P_0 + P_1 = E.$$

Suppose that there exists a positive constant $\varkappa$ such that for $t \geq t_0$ we have

$$(58.12) \qquad \int_{t_0}^{t} \|Q(t)\,P_0\,Q^I(s)\|\,ds + \int_{t}^{\infty} \|Q(t)\,P_1\,Q^I(s)\|\,ds \leq \varkappa.$$

Then as $t \to \infty$,

$$\lim \|Q(t)\,P_0\| = 0,$$

$$\limsup \|Q(t)\,P_1\| = \infty, \quad \text{if} \quad P_1 \neq 0.$$

Proof. We have identically in t,

$$(t - t_0)\,Q(t)\,P_0 = \int_{t_0}^{t} Q(t)\,P_0\,Q^I(s)\,Q(s)\,P_0\,ds$$

and hence

$$(t - t_0) \, \|Q(t) \, P_0\| \leq \varkappa \sup_{t_0 \leq s \leq t} \|Q(s) \, P_0\|.$$

Furthermore

$$\|Q(t) \, P_0\| = O(t^{-1}).$$

Similarly we obtain from the identity

$$\frac{1}{2}(t - t_0)^2 \, Q(t) \, P_0 = \int_{t_0}^{t} Q(t) \, P_0 \, Q^I(s) \int_{t_0}^{s} Q(s) \, P_0 Q^I(u) \, Q(u) \, P_0 \, du \, ds$$

the inequality

$$\frac{1}{2!}(t - t_0)^2 \, \|Q(t) \, P_0\| \leq \varkappa^2 \sup_{t_0 \leq s \leq t} \|Q(s) \, P_0\|,$$

and continuing in this manner we arrive at

$$\exp\left(\frac{t - t_0}{\varkappa}\right) \|Q(t) \, P_0\| \leq \sup_{t_0 \leq s \leq t} \|Q(s) \, P_0\|,$$

which implies the first statement.

To prove the second statement we set $g(t) := |Q(t) \, P_1 x|^{-1}$, and from the identity

$$\int_{t}^{T} g(s) \, ds \, Q(t) \, P_1 x = \int_{t}^{T} Q(t) \, P_1 Q^I(s) \, Q(s) \, P_1 x g(s) \, ds,$$

taking absolute values, we obtain the estimate

$$(g(t))^{-1} \int_{t}^{T} g(s) \, ds \leq \int_{t}^{T} \|Q(t) \, P_1 Q^I(s)\| \, (g(s))^{-1} g(s) \, ds \leq \varkappa \quad \text{for} \quad T > t.$$

Thus $g(s)$ is integrable on (t, ∞) whence $\liminf g(t) = 0$.

If we apply (58.12) with $Q(t) = K(t, t_0)$, we obtain

Theorem 58.6. Let two supplementary projections P_0, P_1 and a positive constant $\varkappa$ be given such that

$$\int_{t_0}^{t} \|K(t, t_0) \, P_0 K(t_0, s)\| \, ds + \int_{t}^{\infty} \|K(t, t_0) \, P_1 K(t_0, s)\| \, ds \leq \varkappa.$$

If $P_1 \neq 0$ then the equilibrium is unstable. If $P_1 = 0$ then the equilibrium is asymptotically stable. If the constant $\varkappa$ can be chosen independent of t_0 then the equilibrium is exponentially stable in case $P_1 = 0$.

Theorem 58.6 implies

Theorem 58.7. If the equilibrium of a linear equation is uniformly asymptotically stable then it is also exponentially stable.

This theorem is also a special case of Theorem 57.1. We give a further proof here which is based on the linearity of the operator $K(t, t_0)$. Let $t_k = t_0 + kT$, $T > 0$ fixed, $k = 0, 1, 2, \dots$. It follows from (58.6) that

$$K(t_k, t_0) = K(t_k, t_{k-1}) \, K(t_{k-1}, t_{k-2}) \cdots K(t_1, t_0).$$

By hypothesis we have an estimate

$$\|K(t, t_0)\| \leq \sigma(t - t_0), \qquad \sigma \in L.$$

Hence

$$\|K(t_i, t_{i-1})\| \leq \sigma(T)$$

and

$$\|K(t_k, t_0)\| \leq (\sigma(T))^k, \quad k = 0, 1, 2, \ldots.$$

As $\sigma(t) \in L$, T can be chosen so large that $\ln \sigma(T)$ is equal to a pre-assigned negative number $-\alpha$, and since $k = (t_k - t_0)/T$, we obtain, for sufficiently large t_k,

$$\|K(t_k, t_0)\| \leq \exp\left(-\alpha \frac{t_k - t_0}{T}\right)$$

and this is our assertion.

The proof is based only on the properties of the operator $K(t, t_0)$ and not on the fact that this operator defines the solution of a differential equation. Theorem 58.7 is therefore valid also for linear functional equations whose solutions can be written in the form (58.5), provided an inequality

$$\|K(t_2, t_0)\| \leq \|K(t_2, t_1)\|\,\|K(t_1, t_0)\| \quad \text{for } t_0 < t_1 < t_2$$

exists.

In addition to equation (58.1) we often consider the *adjoint equation*

(58.13) $$\dot{y} = -A^T y.$$

If $Y(t)$ is a fundamental system for (58.13) and $X(t)$ is one for (58.1) then

$$\frac{d}{dt}(X^T Y) = \dot{X}^T Y + X^T \dot{Y} = X^T A^T Y - X^T A^T Y = 0$$

and hence $X^T Y$ is constant.

Let $K^*(t, t_0; A(t)) := K(t, t_0; -A^T(t))$ be the matrix of the standard basis of (58.13). Then the equation

$$K^T(t, t_0)\, K^*(t, t_0) = C = \text{const.}$$

implies that $C = E$ for $t = t_0$ and that

(58.14) $$K(t, t_0; -A^T(t)) = K^*(t, t_0; A(t)) = K^{TI}(t, t_0; A(t)).$$

59. The Nonhomogeneous Linear Equation

The general nonhomogeneous equation has the form

(59.1) $$\dot{x} = A(t)\, x + z(t)$$

where x and z are vectors. The nth order scalar equation

(59.2) $$a_0(t)\, y^{(n)} + a_1(t)\, y^{(n-1)} + \cdots + a_n(t)\, y = w(t)$$

19*

is equivalent to (59.1). $y(t)$ and $w(t)$, and the coefficients $a_i(t)$ are scalar functions. In general one assumes that the elements of the matrix A as well as the coefficients $a_i(t)$ in (59.2) are continuous for $t > t_0$. As in the autonomous case (sec. 4) it can be shown that every n^{th} order scalar equation can be written as a vector equation of the form (59.1) and that, quite formally, n^{th} order scalar equations can be derived from equation (59.1). But we already pointed out in sec. 4 that this formal procedure is not always justified and that additional assumptions must be made. This is of course especially true in the more general non-autonomous case. In addition to the conditions mentioned in sec. 4, further conditions involving the differentiability of the $a_{ik}(t)$ must be satisfied. For most applications it suffices to know that (59.2) can be considered as a special case of (59.1) and henceforth we will almost exclusively treat systems of equations.

The general solution of (59.1) can be represented in the form

$$(59.3) \qquad x(t) = K(t, t_0; A)\, x_0 + \int_{t_0}^{t} K(t, u; A)\, z(u)\, du.$$

This is verified immediately; by differentiation we obtain

$$\dot{x} = A\, K(t, t_0)\, x_0 + A \int_{t_0}^{t} K(t, u)\, z(u)\, du + K(t, t)\, z(t) = A\, x + z.$$

(For the standard basis, the argument A is omitted). In (59.3) the superposition principle again appears. The first component depends only on the initial value x_0 and the second only on the external force $z(u)$. If the equilibrium of (58.1) is asymptotically stable then the first component becomes arbitrarily small as t increases. The behavior of the solution (59.3) for large values of t is then determined by $z(t)$. We are interested here in the question whether the expression

$$(59.4) \qquad \Re\, z := \int_{t_0}^{t} K(t, u)\, z(u)\, du$$

remains bounded for bounded $z(t)$ or not. It is easy to see that (59.4) is not necessarily bounded for an asymptotically stable equilibrium of (58.1). As an example consider the scalar differential equation

$$\dot{x} = \frac{-1}{1+t}\, x.$$

Its standard solution

$$K(t, t_0) = \frac{1+t_0}{1+t}$$

tends to zero as t increases. None the less $\Re z$ is unbounded for $z(u) = 1$, because the integral

$$\int_{t_0}^{t} \frac{1+u}{1+t}\, du = \frac{1}{2}\, \frac{(1+t)^2 - (1+t_0)^2}{1+t}$$

is unbounded.

On the other hand $\Re z$ is clearly bounded for a bounded function $z(u)$, if the equilibrium of (58.1) is uniformly asymptotically stable. In this case

$$\| K(t,\, t_0) \| \le a\, e^{-b(t-t_0)}, \qquad a > 0,\ b > 0,$$

because of Theorem 58.7, and for a bounded $z(u)$, $|z(u)| \le m$, we have

$$|\Re z| \le a \int_{t_0}^{t} e^{-b(t-u)}\, |z(u)|\, du \le a\, m\, e^{-bt} \int_{t_0}^{t} e^{bu}\, du = \frac{a\,m}{b}\, (1 - e^{b(t_0-t)}).$$

Theorem 59.1. If the equilibrium of the homogeneous equation (58.1) is uniformly asymptotically stable, if $z(t)$ is bounded and $\lim\limits_{t\to\infty} z(t) = 0$ then all the solutions of the nonhomogeneous equation (59.1) tend to zero.

Proof. For each given $\varepsilon > 0$ we can find a t_1, such that $|z(t)| < \varepsilon$ for $t > t_1$. For $t_0 < t \le t_1$, $|z(t)| \le m$. Thus for $t > t_1$,

$$|\Re z| \le \int_{t_0}^{t_1} \| K(t,\, u) \|\, |z(u)|\, du + \int_{t_1}^{t} \| K(t,\, u) \|\, |z(u)|\, du$$

$$\le a\, m \int_{t_0}^{t_1} e^{-b(t-u)}\, du + \varepsilon\, a \int_{t_1}^{t} e^{-b(t-u)}\, du$$

$$= \frac{a\,m}{b}\, e^{-bt} (e^{bt_1} - e^{bt_0}) + \varepsilon\, \frac{a}{b}\, (1 - e^{b(t_1-t)}).$$

Hence (59.4) becomes arbitrarily small as t increases and ε decreases, and the assertion follows from (59.3).

If equation (59.1) is of such a nature that for each bounded input $z(t)$ we obtain a bounded output then the expression

$$\int_{t_0}^{t} \| K(t,\, s) \|\, ds$$

is bounded for a fixed t_0, and as in the proof of Theorem 58.7 we can conclude that the solutions fade exponentially. We can thus obtain a certain converse of the boundedness statement. The relationship between the stability of the homogeneous equation and the boundedness of the solutions of the nonhomogeneous equation has been investigated very carefully by MASSERA and SCHÄFFER [1].

Theorem 59.1 can be generalized:

Theorem 59.2. Let the equilibrium of

$$(59.5) \qquad \dot{x} = f(x,\, t), \quad f \in E,\ f \in C_1,\ (x,\, t) \in K_{h,t'},\ t' > 0,$$

be exponentially stable and let G be the domain of attraction (*cf.* sec. 36). Let the vector $g(t)$ be continuous and bounded and of such a nature that the equation

$$(59.6) \qquad\qquad \dot{x} = f(x,\, t) + g(t)$$

also belongs to class E. Further assume that $\lim \boldsymbol{g}(t) = 0$ $(t \to \infty)$. Then there exist two numbers c and τ such that each solution of (59.6) tends to zero provided that its initial values satisfy the inequalities $|\boldsymbol{x}_0| < c$, $t_0 > \tau$. These numbers depend on G and on $\boldsymbol{g}(t)$.

Proof. By Theorem 56.1 we can construct a function $v(\boldsymbol{x}, t)$ which satisfies the estimates

$$(59.7) \qquad a_1 |\boldsymbol{x}|^2 \le v(\boldsymbol{x}, t) \le a_2 |\boldsymbol{x}|^2,$$

$$\sum_{i=1}^{n} \frac{\partial v}{\partial x_i} f_i(\boldsymbol{x}, t) + \frac{\partial v}{\partial t} \le - a_3 |\boldsymbol{x}|^2, \quad \left| \frac{\partial v}{\partial x_i} \right| \le \frac{1}{n} a_4 |\boldsymbol{x}| \quad (i = 1, 2, \ldots, n)$$

for $|\boldsymbol{x}| \le h_1 < h$, $\boldsymbol{x} \in G$. If we form the derivative of v for (59.6) we see that

$$\dot{v}_{(59.6)} = \sum_{i=1}^{n} \frac{\partial v}{\partial x_i} f_i(\boldsymbol{x}, t) + \frac{\partial v}{\partial t} + \sum_{i=1}^{n} \frac{\partial v}{\partial x_i} g_i(t) \le - a_3 |\boldsymbol{x}|^2 + a_4 |\boldsymbol{x}| |\boldsymbol{g}(t)|.$$

Since $\boldsymbol{g}(t)$ tends to zero there exists a function $\sigma(t)$ of class L such that $|\boldsymbol{g}(t)| \le \sigma(t)$. Thus

$$\dot{v}_{(59.6)} \le - a_3 |\boldsymbol{x}|^2 + a_4 |\boldsymbol{x}| \sigma(t) \le - \frac{1}{2} a_3 |\boldsymbol{x}|^2 + \frac{1}{2} a_3^{-1} a_4^2 \sigma(t)^2.$$

A direct computation shows the validity of the last inequality. Using inequalities (59.7), we finally obtain

$$\dot{v}_{(59.6)} \le - \frac{a_3}{2 a_2} v + \frac{1}{2} \frac{a_4^2}{a_3} \sigma(t)^2.$$

If we choose the initial values so that $|\boldsymbol{x}_0| \le h_1$ and $a_3^2 v_0 > a_2 a_4^2 \sigma(t_0)^2$, i.e. $a_3 \sqrt{a_1} |\boldsymbol{x}_0| > a_4 \sqrt{a_2} \sigma(t_0)$ then at the beginning of the motion $\boldsymbol{p}(t, \boldsymbol{x}_0, t_0)$, $\dot{v}_{(59.6)}$ is negative and later on it can be positive only if $|\boldsymbol{x}|$ is smaller that $a_4 a_3^{-1} |\boldsymbol{g}(t)|$, i.e. decreases.

The inequalities (59.7) and the differential inequality for v remain valid throughout. From Theorem 59.1 it follows that $v(t)$ tends to zero as t increases, completing the proof.

Besides (59.5), we consider

$$(59.8) \qquad \dot{\boldsymbol{y}} = \boldsymbol{h}(\boldsymbol{y}, t), \quad \boldsymbol{h} \in E, \quad \boldsymbol{h} \in C_1 \text{ for } (\boldsymbol{y}, t) \in K_{h, t'}.$$

Let $C(t)$ be a bounded matrix, $\|C(t)\| \le \gamma$. If we interpret (59.5) and (59.8) as the equations of the motions of two nonlinear transfer units (sec. 3) then (59.5) together with

$$(59.9) \qquad \dot{\boldsymbol{y}} = \boldsymbol{h}(\boldsymbol{y}, t) + C(t) \boldsymbol{x}$$

describes a generalized arrangement in series. Let $w(\boldsymbol{y}, t)$ be a Liapunov function and let a_i', $i = 1, 2, 3, 4$, be the constants from Theorem 56.1 for (59.8). Then it follows from the proof of Theorem 59.2 that the

y-motions tend to zero in case the hypersurface $w(y, t) = w(y_0, t_0)$ lies in the domain of attraction of (59.8) and if also $\gamma \, |x| < \sqrt{\dfrac{a_1'}{a_2'} \dfrac{a_3'}{a_4'}} \, |y_0|$. On the other hand, the general solution of (59.5) can be estimated. We have

$$|x(t, x_0, t_0)|^2 \leq a_1^{-1} a_2 \, |x_0|^2,$$

provided x_0 lies in the domain of attraction. Thus if the initial values x_0, y_0 lie in the pertinent domains of attraction and if

$$\gamma \sqrt{\frac{a_2}{a_1}} \, |x_0| < \sqrt{\frac{a_1'}{a_2'}} \cdot \frac{a_3'}{a_4'} \, |y_0|$$

then the motion of the total system determined by x_0, y_0 tends to zero. If both systems individually are exponentially stable for arbitrary initial values then the same is true for the combined system.

A system of equations

$$\text{(59.10)} \qquad \begin{aligned} \dot{x} &= f(x, t) + B(t) \, y, \\ \dot{y} &= h(y, t) + C(t) \, x, \end{aligned}$$

where $\|B(t)\| \leq \beta$, can be interpreted as a generalized feedback arrangement. Similar to the above, we find (for sufficiently small initial values) the differential inequalities

$$\dot{v} \leq -a_3 \, |x|^2 + a_4 \, |x| \, \beta \, |y| \leq -\tfrac{1}{2} a_3 \, |x|^2 + \tfrac{1}{2} a_3^{-1} (a_4 \beta)^2 \, |y|^2,$$

$$\dot{v} \leq -\frac{1}{2} a_2^{-1} a_3 v + \frac{1}{2} a_3^{-1} (a_1')^{-1} (a_4 \beta)^2 \, w$$

and analogously

$$\dot{w} \leq -\frac{1}{2} (a_2')^{-1} a_3' w + \frac{1}{2} a_1^{-1} (a_3')^{-1} (a_4' \gamma)^2 \, v.$$

If v and w are to tend to zero as t increases then the corresponding linear system of equations must have an asymptotically stable equilibrium. This leads to the inequality

$$a_1 a_1' (a_3 a_3')^2 > a_2 a_2' (a_4 a_4')^2 \, \beta^2 \gamma^2.$$

If the individual units are each exponentially stable for arbitrary initial values then the last inequality is sufficient for the system (59.10) to be asymptotically stable in the whole. Otherwise we must use Liapunov functions to obtain bounds for the initial values, similarly as above.

The preceding considerations permit us in certain cases to test the stability behavior of a composite nonlinear system by investigating the stability of the individual members. It is assumed however that the units are composed linearly.[1]

---

[1] cf. BAILEY [1].

60. Linear Equations with Periodic Coefficients

We have an important special case of (58.1) if the matrix $A(t)$ has periodic elements. Many mechanical and electrical systems with transfer units which change periodically (e.g. rotate) are described by such differential equations.[1] Frequently the equations are nonlinear; then the equation of the first approximation belongs to the type under consideration:

$$(60.1) \qquad \dot{x} = P(t)\, x, \quad P(t + \omega) = P(t).$$

It follows from the periodicity of P that by translating the variable t in the argument of the solution by one period ω, another solution is obtained:

$$p(t + \omega, x_0, t_0) = K\big(t + \omega, t_0; P(t)\big)\, x_0$$

is again a solution. Hence $K(t + \omega, t_0)$ is matrix of a basis and by (58.4) there exists a constant matrix C such that

$$K(t + \omega, t_0) = K(t, t_0)\, C.$$

Now

$$K(t + \omega, t_0 + \omega) = K(t, t_0)$$

and on the other hand by (58.6),

$$K(t + \omega, t_0) = K(t + \omega, t_0 + \omega)\, K(t_0 + \omega, t_0)$$
$$= K(t, t_0)\, K(t_0 + \omega, t_0),$$

which implies

$$C = K(t_0 + \omega, t_0).$$

From an arbitrary basis

$$X(t) = K(t, t_0)\, Q_X,$$

we again obtain a basis if we replace t by $t + \omega$,

$$X(t + \omega) = K(t, t_0)\, C\, Q_X = X(t)\, Q_X^I\, C\, Q_X.$$

Therefore if we substitute $t + \omega$ for t then this has the effect of multiplying the basis $X(t)$ by a matrix $Q_X^I C Q_X$ which is similar to C. Since

$$C = K(t_0 + \omega, t_0) = K(t_0, 0)\, K(\omega, 0)\, K(t_0, 0)^I,$$

we have: The transformation matrices Q_X are all similar to the matrix $K(\omega, 0)$.

Making the substitution $t \to t + \omega$ repeatedly we obtain

$$(60.2) \qquad K(t + n\omega, t_0) = K(t, t_0)\, C^n, \quad n = 1, 2, \ldots,$$

where

$$C^n = K(t_0, 0)\, K(\omega, 0)^n\, K(t_0, 0)^I.$$

---

[1] ARSCOTT [1], MALKIN [3, 5], MINORSKY [1], STOKER [1].

We can assume without loss of generality that the variable t in (60.2) ranges over the interval $t_0 \leq t \leq t_0 + \omega$ and thus we see that the behavior of $K(t, t_0)$ for large values of t is determined by the behavior of C^n, respectively of $K(\omega, 0)^n$. In particular, the equilibrium of (60.1) is stable if C^n is bounded for all n and it is asymptotically stable if C^n tends to zero. By Theorem 38.5 the stability is uniform each time.

In order to study the behavior of C^n we look at the characteristic roots of the matrix C, respectively of the similar matrix $K(\omega, 0)$: These roots as well as the equation

$$(60.3) \qquad \det(E\mu - C) = 0$$

depend not on the particular basis, but only on the matrix $P(t)$. The roots $\mu_1, \mu_2, \ldots, \mu_n$ of (60.3) are called the *characteristic multipliers* of the differential equation (60.1). From the relation (58.2) we obtain

$$\det K(\omega, 0) = \exp\left(\int_0^\omega \operatorname{Tr}\big(P(t)\big)\, dt\right),$$

and by (60.3) we have

$$(60.4) \qquad \mu_1 \mu_2 \cdots \mu_n = \exp\left(\int_0^\omega \operatorname{Tr}\big(P(t)\big)\, dt\right).$$

The numbers $\varrho_k := \dfrac{1}{\omega} \ln \mu_k$ are called the *characteristic exponents* of the differential equation. They are only determined up to integer multiples of $2\pi i/\omega$. Relation (60.4) applied to the scalar differential equation

$$(60.5) \qquad \ddot{x} + q_1(t)\, \dot{x} + q_2(t)\, x = 0$$

has the form

$$\mu_1 \mu_2 = \exp\left(- \int_0^\omega q_1(t)\, dt\right).$$

This implies (for the principal values of the characteristic exponents)

$$(60.6) \qquad \varrho_1 + \varrho_2 = - \frac{1}{\omega} \int_0^\omega q_1(t)\, dt.$$

If it were possible to find a second rational expression relating ϱ_1 and ϱ_2, which is independent of (60.6), these values could be calculated. So far such a second relation is not known and because of the following situation, discovered by MARKHASHOV [1], it is doubtful that such a relation exists at all. Let $f(t, y_1, \ldots, y_{2k})$ be a function which is continuous and once differentiable with respect to all variables and which is periodic in t with ω but otherwise arbitrary. Then the mean value formed analogous to (60.6),

$$(60.7) \qquad \frac{1}{\omega} \int_0^\omega f\big(t, q_1(t), q_2(t), \dot{q}_1(t), \dot{q}_2(t), \ldots, q_2^{(k)}(t)\big)\, dt,$$

is of the form $c_1(\varrho_1 + \varrho_2) + c_2$, that is, is linearly dependent on (60.6). Since this is true for arbitrary functions f and arbitrary k a calculation of ϱ_1, ϱ_2 from mean values of the type (60.7) is not possible.

Let the constant matrix R be so chosen that $R^I C R =: J$ is a Jordan normal form for C. The power

$$C^n = R J^n R^I$$

is bounded for increasing n if and only if J^n is bounded and this is the case if and only if the inequality $|\mu_i| \leq 1$ holds for all the characteristic multipliers. If furthermore $|\mu_i| < 1$, $i = 1, \ldots, n$, then $\lim J^n = 0$ and $\lim C^n = 0$. Therefore the stability behavior of the equilibrium of (60.1) can be described by means of the characteristic multipliers similarly as was done with the characteristic roots of A in the autonomous case.

Theorem 60.1. Let $\dot{x} = P(t)\, x$ be a differential equation with periodic coefficients and let $\mu_1, \ldots, \mu_n$ be the characteristic multipliers, *i.e.* the characteristic roots of the matrix $K(\omega, 0)$. If for all i, $i = 1, 2, \ldots, n$, $|\mu_i| < 1$ then the equilibrium is asymptotically (in fact exponentially) stable. If there exists a j such that $|\mu_j| > 1$ then the equilibrium is unstable. If $|\mu_i| \leq 1$, $i = 1, \ldots, n$, and if for some j, $|\mu_j| = 1$ then the equilibrium is stable but not asymptotically stable in case only simple elementary divisors belong to the μ_j for which $|\mu_j| = 1$; it is unstable if there exists a multiplier of absolute value 1 with a multiple elementary divisor.

The last statement on instability follows from the fact that the nth power of the Jordan normal form has a sub-matrix with at least two rows containing the element $(n - 1)\, \mu^{n-1}$, which is unbounded in case $|\mu| = 1$.

We point out that we can also look at this theorem from the point of view of sec. 15. The substitution $t \rightarrow t + \omega$ maps the totality of the solutions into itself and thus defines a linear operator L in the space of solutions which is represented analytically as multiplication by $K(t + \omega, t)$. A characteristic vector for this operator satisfies the equation

$$L x = \lambda x,$$

and for x, being a solution and of the form (58.5), we have

$$L x = L K(t, t_0)\, x_0 = K(t + \omega, t)\, K(t, t_0)\, x_0 = \mu K(t, t_0)\, x_0 = \mu x.$$

The number μ is a characteristic value of the operator L and the stability behavior can thus be characterized by means of the characteristic values of a linear operator. If we use the constant matrix R defined above to make the substitution

$$x = R z,$$

we obtain the equation

(60.8) $$\dot{z} = R^I P R z.$$

The matrix

$$Q(t) = R^I P(t) R$$

is also periodic and we have

$$K(t, t_0; Q) = R^I K(t, t_0; P) R,$$

$$K(\omega, 0; Q) = J.$$

Equation (60.8) is therefore of such a nature that the matrix $K(\omega, 0; Q)$ which is used to define the characteristic multipliers appears in Jordan normal form.

Let $Z(t) := K(t, 0; Q)$. Then by the above,

(60.9) $$Z(t + \omega) = Z(t) J.$$

The matrix J is composed of submatrices J_r of the form

$$\begin{pmatrix} \mu & 1 & 0 & \cdots & 0 & 0 \\ 0 & \mu & 1 & \cdots & 0 & 0 \\ 0 & 0 & \mu & \cdots & 0 & 0 \\ & & \cdots & \cdots & \cdots & \\ 0 & 0 & 0 & \cdots & \mu & 1 \\ 0 & 0 & 0 & \cdots & 0 & \mu \end{pmatrix}.$$

The submatrix of the matrix Z which corresponds to J_r is denoted by Z_r,

$$Z_r = \left(z_i^{(k)}(t)\right), \quad i = 1, \ldots, r \text{ rows}, \quad k = 1, \ldots, r \text{ columns}.$$

Equation (60.9) gives rise to the system

(60.10) $$z^{(1)}(t + \omega) = \mu z^{(1)}(t),$$

$$z^{(2)}(t + \omega) = \mu z^{(2)}(t) + z^{(1)}(t),$$

$$\cdots \cdots \cdots \cdots \cdots \cdots \cdots$$

$$z^{(r)}(t + \omega) = \mu z^{(r)}(t) + z^{(r-1)}(t).$$

It differs from the one considered in sec. 14 only by the fact that t varies continuously. The general solution is therefore a linear combination of terms μ^k, the coefficients being periodic functions of t, however. It is easily seen by direct verification that the following functions are the solutions of (60.10) $\left(\varrho := \dfrac{\ln \mu}{\omega}\right)$:

$$(60.11) \quad z^{(1)}(t) = e^{\varrho t} p^{(1)}(t),$$

$$z^{(2)}(t) = e^{\varrho t}\left(p^{(2)}(t) + \frac{1}{\mu}\binom{\frac{t}{\omega}}{1} p^{(1)}(t)\right),$$

$$\cdots\cdots\cdots\cdots\cdots\cdots\cdots\cdots\cdots\cdots\cdots$$

$$z^{(r)}(t) = e^{\varrho t}\left(p^{(r)}(t) + \frac{1}{\mu}\binom{\frac{t}{\omega}}{1} p^{(r-1)}(t)\right.$$

$$\left. + \frac{1}{\mu^2}\binom{\frac{t}{\omega}}{2} p^{(r-2)}(t) + \cdots + \frac{1}{\mu^{r-1}}\binom{\frac{t}{\omega}}{r-1} p^{(1)}(t)\right),$$

where the vectors $p^{(i)}(t)$, $i = 1, \ldots, n$ have periodic components. For we have

$$z^{(k)}(t + \omega) - \mu z^{(k)}(t)$$

$$= \mu e^{\varrho t}\left(p^{(k)}(t) + \frac{1}{\mu}\binom{\frac{t}{\omega}+1}{1} p^{(k-1)}(t) + \cdots + \frac{1}{\mu^{k-1}}\binom{\frac{t}{\omega}+1}{k-1} p^{(1)}(t)\right.$$

$$\left. - p^{(k)}(t) - \frac{1}{\mu}\binom{\frac{t}{\omega}}{1} p^{(k-1)}(t) - \cdots - \frac{1}{\mu^{k-1}}\binom{\frac{t}{\omega}}{k-1} p^{(1)}(t)\right)$$

$$= e^{\varrho t}\left(p^{(k-1)}(t) + \frac{1}{\mu}\binom{\frac{t}{\omega}}{1} p^{(k-2)}(t) + \cdots + \frac{1}{\mu^{k-2}}\binom{\frac{t}{\omega}}{k-1} p^{(1)}(t)\right)$$

$$= z^{(k-1)}(t).$$

The explicit representation (60.11) shows that the matrix Z_r, whose columns are the vectors (60.11), has the form

$$(60.12) \qquad Z_r = e^{\varrho t} S_r(t), \quad S_r(t + \omega) = S_r(t).$$

S_r is constructed from the vectors $p^{(k)}$. The fundamental system Z is composed of the submatrices Z_r. If we reverse the transformation $x = R z$ we obtain

Theorem 60.2. The general solution of the differential equation (60.1) is a linear combination of terms of the form

$$e^{\varrho_k t}\, q^{(k,j)}(t), \qquad \begin{array}{l} k = 1, \ldots, r;\ j = 1, \ldots, n_k,\\[4pt] \varrho_k = \dfrac{\ln \mu_k}{\omega}. \end{array}$$

The numbers $\mu_1, \ldots, \mu_r$ are the distinct characteristic roots of the matrix $K(\omega, 0)$; n_j denotes the multiplicity of the elementary divisor belonging to μ_j. The components of the vector $\boldsymbol{q}^{(k,j)}(t)$ are polynomials in t with periodic coefficients. The degree of each polynomial is at most $n_j - 1$.

Theorem 60.2 is known as *Floquet's theorem*. From the Floquet representation we can, of course, derive the stability properties expressed in Theorem 60.1 directly.

If the number μ in (60.10) is a root of unity, $\mu^s = 1$, then

$$\boldsymbol{z}^{(1)}(t + s\omega) = \boldsymbol{z}^{(1)}(t).$$

The differential equation has a periodic solution of period $s\omega$ in this case. If $s = 1$ then $\boldsymbol{z}^{(1)}(t)$ has the same period as the differential equation itself and this condition is also necessary. If the equation is to have a solution with period ω then the linear operator L introduced above must have the number 1 as a characteristic value, *i.e.* 1 must be a characteristic root of $K(\omega, 0)$.

Theorem 60.3. For (60.1) to have a periodic solution of period ω it is necessary and sufficient that there exists a characteristic multiplier 1, (resp. that there exists a characteristic exponent $2k\pi i$, k an integer).

The adjoint equation for (60.1),

(60.13) $$\dot{\boldsymbol{y}} = - P^T(t)\,\boldsymbol{y},$$

also has periodic coefficients of period ω. By (58.14) we have

$$K(\omega, 0; - P^T) = K^{TI}(\omega, 0; P).$$

This implies that the characteristic multipliers $\mu_1^*, \ldots, \mu_n^*$ of (60.13) which satisfy the equation

$$\det\left(E\,\mu^* - K^{IT}(\omega, 0; P)\right) = 0$$

are exactly the reciprocals $\mu_1^{-1}, \ldots, \mu_n^{-1}$. Corresponding elementary divisors have the same multiplicity. If there exists a multiplier μ such that $\mu^s = 1$ then there also exists a multiplier μ^* such that $(\mu^*)^s = 1$. Hence (60.1) and the adjoint equation (60.3) have the same number of (linearly independent) periodic solutions.

Under further assumptions on the matrices of the equation we can of course assert more. This is true especially for the so-called *canonical equations* which play a rôle in mechanics. They are of the form (*cf.* sec. 25)

(60.14) $$\dot{x}_i = \frac{\partial H}{\partial y_i}, \quad \dot{y}_i = - \frac{\partial H}{\partial x_i}, \quad i = 1, 2, \ldots, n,$$

where H is a quadratic form in the $2n$ variables $x_1, \ldots, x_n, y_1, \ldots, y_n$, with continuous coefficients. In practical problems H is usually the kinetic energy. The x_i are position coordinates, the y_i are impulses. If we write $\boldsymbol{x}$ and $\boldsymbol{y}$ together as a $2n$-vector, $\boldsymbol{z} = \mathrm{col}(\boldsymbol{x}, \boldsymbol{y})$, and if we set

$$H(\boldsymbol{x}, \boldsymbol{y}, t) = \boldsymbol{z}^T A\,\boldsymbol{z},$$

where

$$A = \begin{pmatrix} A_{11} & A_{12} \\ A_{12}^T & A_{22} \end{pmatrix}, \quad A_{11} \text{ and } A_{22} \text{ symmetric,}$$

then the system (60.14) can be written as a single equation for z:

$$\dot{z} = Pz, \quad P := \begin{pmatrix} A_{12}^T & A_{22} \\ -A_{11} & -A_{12} \end{pmatrix}.$$

The matrix of the adjoint equation is

$$-P^T = \begin{pmatrix} -A_{12} & A_{11} \\ -A_{22} & A_{12}^T \end{pmatrix}.$$

It is similar to P,

$$-P^T = QPQ^I, \quad Q := \begin{pmatrix} 0 & -E \\ E & 0 \end{pmatrix}.$$

If H is periodic in t then the equation for z and its adjoint equation have the same characteristic multipliers. On the other hand, these multipliers are reciprocals of each other (*cf.* above). Hence, for each multiplier μ, the reciprocal multiplier μ^{-1} also appears. Multipliers $\mu = \pm 1$ always have even multiplicity.

The equilibrium of (60.14) can thus never be asymptotically stable. For its simple stability it is necessary that all the multipliers have absolute value 1. If (60.14) is the equation of the first approximation for a nonlinear equation then the equilibrium of that equation is either unstable or we have a critical case.

61. The Liapunov Reducibility Theorem

Let the matrix $K(\omega, 0)$ be written in the form

$$K(\omega, 0) = e^{\omega D}$$

and let

$$S(t) := K(t, 0) e^{-tD}.$$

S is nonsingular and periodic, for

$$S(t + \omega) = K(t + \omega, 0) e^{-(t+\omega)D} = K(t, 0) K(\omega, 0) e^{-\omega D} e^{-tD} = S(t),$$

and $K(t, 0)$ admits a representation

$$K(t, 0) = S(t) e^{tD}.$$

Let the vector $y(t)$ be defined by

$$(61.1) \qquad\qquad y(t) := S(t)^I x(t)$$

and put $y_0 := y(t_0)$. From

$$S(t)\, y(t) = K(t, t_0)\, x_0 = K(t, 0)\, K(0, t_0)\, x_0 = S(t)\, e^{tD}\left(S(t_0)\, e^{t_0 D}\right)^I x_0$$

it follows that

$$y(t) = e^{tD}\, e^{-t_0 D}\, S(t_0)^I\, x_0 = e^{tD}\, e^{-t_0 D}\, y_0$$

and

$$\dot{y}(t) = D\, y(t).$$

The vector $y(t)$ defined in (61.1) satisfies an autonomous differential equation.

Theorem 61.1. Each linear differential equation with periodic coefficients can be transformed into an autonomous differential equation by means of a nonsingular substitution with periodic coefficients.

Def. 61.1 A differential equation $\dot{x} = A(t)\, x$ is called *reducible* if it can be transformed into an autonomous differential equation $\dot{y} = Dy$ by means of a bounded substitution $x = S(t)\, y$ with a bounded inverse $S^I(t)$.

Theorem 61.1 can now be stated as follows: Every equation with periodic coefficients is reducible.

Differentiating the equation of the substitution we obtain

$$\dot{x} = \dot{S}(t)\, y + S(t)\, \dot{y} = A(t)\, S(t)\, y,$$

that is

(61.2) $$\dot{y} = \left(S^I A S - S^I \dot{S}\right) y, \quad D = S^I A S - S^I \dot{S}.$$

The last equation defines explicitly the connection between the matrices D, S, and A, resp. P.

In the periodic case the characteristic roots of the matrix D are, of course, the numbers $\varrho_1, \varrho_2, \ldots, \varrho_n$, i.e. they are the characteristic exponents of the original equation (60.1), respectively the principal values of the logarithms.

Def. 61.2. Two matrices $A(t)$ and $D(t)$ are *kinematically similar* if there exists a bounded matrix $S(t)$ with a bounded inverse such that

$$D(t) = S^I A S - S^I \dot{S}.$$

Thus an equation $\dot{x} = A(t)\, x$ is reducible if its matrix is kinematically similar to a constant matrix. Obviously we have

Theorem 61.2. Equations with kinematically similar matrices have the same stability behavior. In particular, the equilibrium of a reducible equation is exponentially stable if the kinematically similar autonomous equation has an asymptotically stable equilibrium.

Def. 61.1 can be extended immediately to difference equations: The equation

$$x(t + 1) = A(t)\,x(t) \qquad (t = t_0, t_0 + 1, \ldots) \tag{61.3}$$

is called *reducible* if it can be transformed into an equation $y(t + 1) = Dy$, with constant D, by means of a substitution $x = S(t)\,y$. The matrix $S(t)$ is defined for discrete arguments $t = t_0 + k$, $k = 0, 1, 2, \ldots$, is bounded and has a bounded inverse. Obviously

$$D = S(t + 1)^I\,A(t)\,S(t).$$

It is easy to show that every difference equation (61.3) with periodic coefficients, $A(t + p) = A(t)$, is reducible. For we see as in sec. 60 that a basis $X(t)$ is subjected to a constant substitution, $X(t + p) = X(t)\,Q_X$, when t is replaced by $t + p$. The matrices Q_X depend on the basis but they are all similar to each other. The matrix

$$S(k) := X(t_0 + k)\,Q_X^{-k/p}$$

is periodic since $S(k + p) = X(t_0 + k)\,Q_X Q_X^{(-k+p/p)} = S(k)$. The substitution $x(k) = S(k)\,y(k)$ transforms (61.3) into an autonomous equation, for

$$S(k + 1)^I\,A(k)\,S(k) = Q^{(k+1)/p}\,X(t_0 + k + 1)^I\,A(k)\,X(t_0 + k)\,Q^{-k/p}$$
$$= Q^{(k+1-k)/p} = Q^{1/p}$$

is constant.[1])

62. Stability Criteria for Special Linear Differential Equations

In a number of the preceding sections (5, 6, 27, 57, 60) we proved stability theorems which were valid specifically for linear equations. But among these only the theorems of sec. 5 and the equivalent theorems of sec. 27 were stability criteria in the narrow sense of the word. Only for autonomous linear differential equations a knowledge of the coefficients of the equation is sufficient for the solution of the stability problem; it can be solved, in that case, through finite algebraic processes. Already for the slightly more complicated periodic case the stability theorems which we gave (sec. 60) are not exactly criteria, nor is the reducibility theorem 61.1. For in order to apply them several things must already be known about the solutions, or the characteristic exponents must be known.

Many papers are dedicated to the problem of determining the characteristic exponents, and many individual results are known. We point to the literature listed in CESARI [1] and STARZHINSKII [1] and consider here only the special equation (60.5) and sketch one possible way for computing the characteristic exponents.

---

[1]) KOVAL' [1].

First of all we can make a substitution

$$x = y \exp\left(-\frac{1}{2} \int^t q_1(u)\, du\right)$$

and thereby change the equation to

$$(62.1) \qquad \ddot{y} + p(t)\, y = 0, \quad p(t) := q_2(t) - \frac{1}{4}\left(q_1(t)\right)^2 - \frac{1}{2}\dot{q}_1(t).$$

This is known as *Hill's differential equation*. The quantity (60.6) is equal to zero here. Hence the constant term of the characteristic equation is equal to 1 and the equation can be written in the form

$$(62.2) \qquad \lambda^2 - 2A\lambda - 1 = 0.$$

The multipliers satisfy the relations

$$\mu_1 \mu_2 = 1, \quad \mu_1 + \mu_2 = 2A.$$

If one of the multipliers has an absolute value smaller than 1 then the absolute value of the other one is larger than 1. Then there exists an unbounded solution and the equilibrium is unstable. This happens if $A^2 > 1$. If $A^2 = 1$ then, by sec. 60, there exists a periodic solution. In case $A^2 < 1$, the μ_i have absolute value 1 and we have stability. The problem has thus been reduced to investigating A.

Let $g(t)$ and $h(t)$ be the solutions determined by

$$g(0) = 1, \quad \dot{g}(0) = 0,$$

$$h(0) = 0, \quad \dot{h}(0) = 1.$$

Together with their derivatives they form the standard matrix $K(t, 0)$ belonging to (62.1). Therefore

$$K(\omega, 0) = \begin{pmatrix} g(\omega) & h(\omega) \\ \dot{g}(\omega) & \dot{h}(\omega) \end{pmatrix},$$

$$A = \frac{1}{2}\left(g(\omega) + \dot{h}(\omega)\right).$$

In addition to (62.1) we consider the equation

$$(62.3) \qquad \ddot{z} - \gamma p(t)\, z = 0,$$

which depends on a parameter, and its solution $z = z(t; \gamma)$. Then $z(t; -1) = y(t)$ is the solution of (62.1). We represent the solution z as a power series in γ

$$(62.4) \qquad z(t; \gamma) = \sum_{i=0}^{\infty} \gamma^i z_i(t)$$

and compare coefficients to obtain

$$\ddot{z}_0 = 0, \quad \ddot{z}_i = p(t)\, z_{i-1}, \quad i = 1, 2, \ldots.$$

We solve these equations for the initial conditions

$$z_0(0) = 1, \quad \dot{z}_0(0) = 0, \quad z_i(0) = \dot{z}_i(0) = 0, \atop \tilde{z}_0(0) = 0, \quad \dot{\tilde{z}}_0(0) = 1, \quad \tilde{z}_i(0) = \dot{\tilde{z}}_i(0) = 0,} \quad i = 1, 2, \ldots,$$

whence

(62.5)

$$z_0(t) = 1, \; \tilde{z}_0(t) = t, \; z_1(t) = \int_0^t \int_0^\tau p(u) \, du \, d\tau, \; \tilde{z}_1(t) = \int_0^t \int_0^\tau u \, p(u) \, du \, d\tau,$$

$$z_i(t) = \int_0^t \int_0^\tau p(u) \, z_{i-1}(u) \, du \, d\tau, \quad \tilde{z}_i(t) = \int_0^t \int_0^\tau p(u) \, \tilde{z}_{i-1}(u) \, du \, d\tau,$$

and obtain two solutions $z(t; \gamma)$, $\tilde{z}(t; \gamma)$ for (62.3) which for $\gamma = -1$ are just the functions $g(t)$ and $h(t)$ defined above.

We have

(62.6)
$$A = \frac{1}{2} \left(z(\omega; -1) + \dot{\tilde{z}}(\omega; -1) \right)$$

$$= 1 + \frac{1}{2} \sum_{n=1}^\infty (-1)^n \left(z_n(\omega) + \dot{\tilde{z}}_n(\omega) \right),$$

and if we have further information on $p(t)$ we can at times determine whether A is greater or smaller than 1. For instance, the terms of the series are all positive if $p(t) \leq 0$, $p(t) \not\equiv 0$, $0 \leq t \leq \omega$. This follows from (62.5). In this case the equilibrium of (62.1) is thus unstable.

If $p(t) \geq 0$ ($p(t) \not\equiv 0$) and if in addition

$$\omega \int_0^\omega p(t) \, dt \leq 4,$$

then we see again by discussing the series in (62.6) that $A^2 < 1$ so that we have stability.[1]) This criterion is due to Liapunov. Further criteria are listed in CESARI [1] and STARZHINSKII [1].

The special equation

(62.7)
$$\ddot{x} + (\alpha + \beta \cos 2t) \, x = 0$$

is called the *Mathieu equation*. Its stability depends on the two parameters α and β, the second of which indicates how strongly the equation deviates from an autonomous equation. For small values of β we start with an expression similar to (62.4) with β as the parameter and try to calculate the number (62.6).[2])

Stability criteria for general linear equations are unknown. The criteria listed below concern certain special cases of the equation $\dot{x} = A(t) \, x$.

---

[1]) MALKIN [3].　　[2]) ARSCOTT [1], MALKIN [3].

a) Consider a matrix $A(t)$ of the form

$$A(t) = A + Q(t),$$

let the matrix A be constant and stable and let $Q(t)$ tend to zero as t increases. We can then construct a matrix B by the rule (27.3). The derivative of the Liapunov function $v(x) = x^T B x$ is

$$- x^T C x + x^T (Q^T B + B Q)\, x.$$

It is negative definite for $t > t_0$ if t_0 is chosen sufficiently large. Hence the equilibrium is asymptotically stable for large t_0.

b) Let $\alpha_1, \ldots, \alpha_n$ be the characteristic roots of the matrix

$$M(t) = \frac{1}{2}\left(A^T(t) + A(t)\right) \text{ and let } \operatorname{Re}\alpha_i \leq \delta,\ i = 1, \ldots, n;\ t \geq t_0.$$

Then we have: If the number δ is equal to zero then the equilibrium is uniformly stable; if $\delta < 0$ the equilibrium is exponentially stable.

The proof uses the direct method. The derivative of the Liapunov function $v(x) = |x|^2 = x^T x$ is $\dot{v} = 2x^T M(t)\, x$, and thus $\dot{v} \leq 2\delta\, |x|^2$. Therefore the equilibrium is uniformly stable, respectively uniformly asymptotically stable, by Theorem 42.2, resp. 42.4, and because of Theorem 58.7 it is exponentially stable in the latter case.

c) It is not sufficient for asymptotic stability that the matrix $A(t)$ have characteristic roots $\alpha_i(t)$, $i = 1, \ldots, n$, with negative real parts for all $t \geq t_0$, even if the real parts are all smaller than a fixed negative number. A simple counterexample is due to VINOGRAD [1]. The matrix of the scalar system

$$\dot{x} = (-1 - 9 \cos^2 6t + 6 \sin 12t)\, x + \left(12 \cos^2 6t + \frac{9}{2} \sin 12t\right) y,$$

$$\dot{y} = \left(-12 \sin^2 6t + \frac{9}{2} \sin 12t\right) x - (1 + 9 \sin^2 6t + 6 \sin 12t)\, y$$

has constant negative characteristic roots -1 and -10. The characteristic equation is $\lambda^2 + 11\lambda + 10 = 0$. But two linear independent solutions of the system are

$$x_1(t) = e^{2t}(\cos 6t + 2 \sin 6t), \quad y_1(t) = e^{2t}(2 \cos 6t - \sin 6t),$$

$$x_2(t) = e^{-13t}(\sin 6t - 2 \cos 6t), \quad y_2(t) = e^{-13t}(2 \sin 6t + \cos 6t),$$

one of which is not bounded. Thus the equilibrium is unstable. Of course, the criterion in b) is not satisfied here. The determinant of the matrix $M(t)$ is constant and equal to the negative number -26, *i.e.* one α_i is positive.

However, the information that $\operatorname{Re}\alpha_i(t) < 0$ can be utilized if additional conditions are satisfied. Then we can at least solve the matrix equation

$$A^T(t) B(t) + B(t) A(t) = -C, \quad C \text{ constant and positive},$$

20*

in such a way that $B(t)$ is positive definite for a fixed t. Let $\beta_1(t), \ldots, \beta_n(t)$ denote the characteristic roots of B and suppose that $\beta_1(t)$ is always the smallest characteristic root. Assume that we can choose a continuous, bounded, and continuously differentiable function $\varphi(t)$ such that $\varphi(t)^2 \beta_1(t)$ is always larger than a fixed positive number γ. The function

$$v(\boldsymbol{x}, t) = \varphi(t)^2 \, \boldsymbol{x}^T B(t) \, \boldsymbol{x}$$

is then positive definite. Its derivative is

$$\dot{v} = \boldsymbol{x}^T \left(- \varphi(t)^2 C + 2\varphi(t)\dot{\varphi}(t) B(t) + \varphi(t)^2 \dot{B}(t) \right) \boldsymbol{x}$$

$$= - \varphi(t)^2 \, \boldsymbol{x}^T \left(C - \dot{B}(t) - 2\frac{\dot{\varphi}(t)}{\varphi(t)} B(t) \right) \boldsymbol{x}.$$

The matrix $C - \dot{B}(t) - 2\dfrac{\dot{\varphi}(t)}{\varphi(t)} B(t)$ is non-negative if $2\dfrac{\dot{\varphi}(t)}{\varphi(t)}$ is smaller than the smallest characteristic root of $(C - \dot{B}) B^I$, respectively smaller than the smallest root $\mu(t)$ of the equation

$$\det \left(C - \dot{B}(t) - \mu B(t) \right) = 0,$$

and the inequality

$$\varphi(t) < \varphi(t_0) \exp \left(\int_{t_0}^{t} \mu(t)\, dt \right)$$

implies that $\dot{v} \leq 0$. We can therefore infer stability of the equilibrium at least if a function $\varphi(t)$ can be found which satisfies the given inequalities. If we assume more we can assert more. If, for example, $\operatorname{Re} x_i(t) \leq \alpha < 0$ then $B(t)$ is positive definite for all t and the first condition for $\varphi(t)$ becomes unnecessary. If in addition $\int^{\infty} \mu(\tau)\, d\tau$ diverges then we can choose $\varphi \equiv 1$ and actually deduce asymptotic stability.

63. The Order Numbers of a Differential Equation

In sec. 35 we described the different types of asymptotic stability by means of comparison functions. The corresponding inequalities (A 4 through A 7) say that the variable $\|\boldsymbol{p}(t, \boldsymbol{x}_0, t_0)\|$ decreases at least as fast as the comparison function. In the theory of *order numbers* we try to make the statement "at least as fast" more precise. Let $f(t)$ be a real or complex-valued function of the real variable t, defined for $t > 0$ and bounded for finite t.

Def. 63.1. The order number of $f(t)$ is the expression

$$(63.1) \qquad\qquad \pi(f) := \limsup_{t \to \infty} \frac{\ln |f(t)|}{t}.$$

We are comparing the function with an exponential function. For each number $\alpha < \pi(f)$ there exists a sequence $t_n \to \infty$ such that $f(t_n)\, e^{-\alpha t_n} > 1$. If $\beta > \pi(f)$ then for sufficiently large values of t, $f(t)\, e^{-\beta t} < 1$.

Examples:

$$\pi(t^x) = 0, \quad \pi(e^{\alpha t}) = \alpha, \quad \pi(e^{\sqrt{t}}) = 0,$$

$$\pi(e^{t^2}) = \infty, \quad \pi(e^{t \sin t}) = 1,$$

$$\pi(c) = 0.$$

LIAPUNOV [1], to whom this concept is due, had a somewhat different definition: He calls a quantity $-\pi(f)$ the *characteristic number* of the function. Def. 63.1 is due to PERRON [2], who developed the theory of order numbers independent of Liapunov.

A few simple computational rules and inequalities follow directly from the definition.

$$(63.2) \qquad\qquad \pi(f g) \leq \pi(f) + \pi(g),$$

since $\ln(f g) = \ln f + \ln g$ and $\limsup (a + b) \leq \limsup a + \limsup b$. In particular we have for constant c,

$$\pi(c f) = \pi(f).$$

$$(63.3) \qquad \pi(f + g) = \max\big(\pi(f), \pi(g)\big), \ \text{if} \ \pi(f) \neq \pi(g),$$

$$(63.4) \qquad \pi(f + g) \leq \pi(f), \ \text{if} \ \pi(f) = \pi(g).$$

For if $\pi(f) \geq \pi(g)$ then (63.3), resp. (63.4), follows from (63.2) applied to the product $f \cdot \left(1 + \frac{g}{f}\right)$.

We also have

$$(63.5) \qquad\qquad \pi\left(\int\limits_{t_0}^{t} f(u)\, du \right) \leq \pi(f),$$

$$\pi\left(\int\limits_{t}^{\infty} f(u)\, du \right) \leq 0, \ \text{in case the integral exists,}$$

$$(63.6)$$

$$\pi\left(\int\limits_{t}^{\infty} f(u)\, du \right) \leq \pi(f), \ \text{if} \ \pi(f) < 0.$$

The first inequality follows from

$$\int\limits_{t_0}^{t} f(u)\, du \leq (t - t_0) \max_{t_0 \leq u \leq t} |f(u)|.$$

In order to prove (63.6) we apply (63.3) to the sum

$$\int\limits_{t}^{t_1} f(u)\, du + \int\limits_{t_1}^{\infty} f(u)\, du$$

and choose t_1 so large that the second term becomes smaller than some fixed number. This is possible because the integral converges.

The order number of a vector $\boldsymbol{x} = \mathrm{col}(x_1, \ldots, x_n)$ is equal to the order number of its norm:

$$(63.7) \qquad\qquad \pi(\boldsymbol{x}) := \pi(|\boldsymbol{x}|).$$

The definition

$$(63.8) \qquad\qquad \pi(\boldsymbol{x}) := \max_i \pi(x_i)$$

is equivalent.

Theorem 63.1. Every solution of a linear differential equation

$$\dot{\boldsymbol{x}} = A(t)\, \boldsymbol{x}$$

with bounded coefficients has a finite order number.

Proof. The statement is a consequence of Theorem 58.1, which asserts that each solution grows at most like an exponential function. We give a second proof here which furnishes an estimate for the order numbers. Let $\boldsymbol{z} = e^{\lambda t}\boldsymbol{x}$, λ a parameter, and let

$$B(t, \lambda) := A(t) + A(t)^T + 2\lambda E.$$

Let the numbers β_1, β_2, and $\alpha > 0$ be so chosen that the matrices $B(t, \beta_1)$ and $B(t, \beta_2)$ have only negative, respectively only positive, characteristic roots for $t \geq 0$ and that

$$2\alpha\,|\boldsymbol{z}|^2 < \boldsymbol{z}^T B(t, \beta_2)\,\boldsymbol{z},$$

$$-\,2\alpha\,|\boldsymbol{z}|^2 > \boldsymbol{z}^T B(t, \beta_1)\,\boldsymbol{z}.$$

Since

$$\frac{d}{dt}\,|\boldsymbol{z}|^2 = \boldsymbol{z}^T B(t, \lambda)\,\boldsymbol{z},$$

we have for $\lambda = \beta_1$

$$\frac{d}{dt}\,|\boldsymbol{z}|^2 < -\,2\alpha\,|\boldsymbol{z}|^2$$

and

$$|\boldsymbol{z}| < |\boldsymbol{z}_0|\,e^{-\alpha t}, \quad |\boldsymbol{x}| < |\boldsymbol{x}_0|\,e^{-\alpha t - \beta_1 t}.$$

Analogously we obtain

$$|\boldsymbol{x}| > |\boldsymbol{x}_0|\,e^{\alpha t - \beta_2 t}$$

and we recognize that the order numbers lie in the interval $(\alpha - \beta_2,\ -\alpha - \beta_1)$.

If A is not bounded the theorem is false. A solution of $\dot{x} = 2\,tx$ is $x = e^{t^2}$ whose order number is ∞.

Let $x^{(1)}, \ldots, x^{(n)}$ be a basis, *i.e.* a fundamental system of solutions of the differential equation and let

$$(63.9) \qquad\qquad y = \sum_{j=1}^{n} c_j x^{(j)}$$

be an arbitrary solution.

Def. 63.2. A basis is called *normal* if the order number of each solution y is one of the $\pi(x^{(j)})$.

Theorem 63.2. Each equation, with bounded or with unbounded coefficients, has a normal basis.

Proof. Let X be the matrix of the basis and let $\sigma(X)$ denote the number of the distinct order numbers $\pi(x^{(j)})$, $j = 1, \ldots, n$. Since $\sigma(X) \leq n$ there exists a maximum m of $\sigma(X)$ for all possible bases X. If $m = n$ and if y is an arbitrary solution then by (63.8) and (63.3) the order number $\pi(y)$ is equal to the maximum of the order numbers $\pi(x^{(j)})$, the maximum being taken among those $x^{(j)}$ which actually appear in the expression (63.9). Hence X is normal. If $\pi(y)$ is not one of the $\pi(x^{(j)})$ then at least two solutions $x^{(i)}$, $x^{(i')}$ must have equal order numbers, this number being larger than $\pi(y)$. In this case we replace one of the solutions $x^{(i)}$, $x^{(i')}$ by y, which increases the number σ by 1. If the new basis is still not normal we continue the process. It breaks off when $\sigma = n$, at the latest.

From this proof and from Theorem 63.1 we obtain

Theorem 63.3. The basis of an equation with bounded coefficients is normal if and only if the sum

$$(63.10) \qquad\qquad \sum_{j} \pi(x^{(j)})$$

is as small as possible.

Def. 63.3. The order numbers of a normal basis arranged in order of magnitude are called the *order numbers of the differential equation*.

Let $S(t)$ be a bounded matrix with differentiable elements. Assume that the matrices $S(t)^I$ and $\dot{S}(t)$ are also bounded. Then the transformation $y = S(t) x$ is called a *Liapunov transformation*. It transforms an equation with bounded coefficients into an equation with bounded coefficients (*cf.* (61.2)). Since S and S^I are bounded it follows from $y = S x$ and $x = S^I y$ that $\pi(y) \leq \pi(x)$ and $\pi(x) \leq \pi(y)$. Hence $\pi(y) = \pi(x)$: A Liapunov transformation does not change the order numbers of the equation. Special cases of Liapunov transformations are the transformations (4.11), which transform an autonomous equation into its Jordan form, and (61.1) which transforms a periodic equation into an autonomous equation. As an application of the last remark we obtain

Theorem 63.4. The order numbers of an autonomous equation are the real parts of the characteristic roots of its matrix. The order numbers of a periodic equation are the real parts of the characteristic exponents.

If all the characteristic numbers are negative then the solutions decrease like exponential functions and the equilibrium is, of course, asymptotically stable. But it need not be uniformly stable and certainly not exponentially stable. For example, the solution

$$x = x_0 \exp\left(t \cos t - t_0 \cos t_0 + 2(t_0 - t)\right)$$

of the equation

(63.11) $$\dot{x} = (\cos t - t \sin t - 2)\, x$$

which we considered in sec. 39, has order number -1. But its equilibrium is not uniformly stable. For if we set $t = t_0 + T$, then

$$x = x_0 \exp\left[t_0\left(\cos(t_0 + T) - \cos t_0\right) + T\left(\cos(t_0 + T) - 2\right)\right]$$

and the exponent can become arbitrarily large as t_0 increases.

Let $A(t)$ be bounded. Equation (58.2) implies

(63.12) $$\exp\left(\int_0^t \operatorname{Tr} A(\tau)\, d\tau\right) = c \ \det\ X(t),$$

and since by (63.2) through (63.4), the order number of the determinant of the fundamental system is certainly no larger than the sum (63.10), it follows that

(63.13) $$\sum_{j=1}^{n} \pi(x^{(j)}) \geq \pi\left(\exp\left(\int_0^t \operatorname{Tr} A(\tau)\, d\tau\right)\right).$$

If we have equality in (63.13) then the corresponding fundamental system is normal; this follows from the definition. The converse, however, is not correct: There exist normal bases for which the sum of the order numbers does not attain the bounds in (63.13). An example for this was given by LIAPUNOV [1]. It has the form

$$\begin{aligned}\dot{x} &= x \cos \ln t + y \sin \ln t, \\ \dot{y} &= x \sin \ln t + y \cos \ln t.\end{aligned} \qquad t \geq 1,$$

In this case

$$\int_1^t \operatorname{Tr} A(\tau)\, d\tau = 2 \int_1^t \cos \ln \tau\, d\tau = t(\sin \ln t + \cos \ln t) - 1$$

and the order number of $\exp\left(\int_1^t \operatorname{Tr} A(\tau)\, d\tau\right)$ is $\sqrt{2}$. A fundamental system is given by

$$\begin{aligned}x^{(1)} &= \exp(t \sin \ln t), & y^{(1)} &= \exp(t \sin \ln t), \\ x^{(2)} &= \exp(t \cos \ln t), & y^{(2)} &= -\exp(t \cos \ln t).\end{aligned}$$

It has order numbers 1 and 1. The system is normal but in (63.13) we have a strict inequality.

At times differential equations with discontinuous right sides have to be considered. In that case a finite or infinite sequence $0 < t_1 < t_2 < \dots$ is given. The elements $a_{ij}(t)$ of the matrix $A(t)$ are bounded for $t \geq 0$. $|a_{ij}(t)| < L$, and they are continuous on the intervals $t_r < t < t_{r+1}$. At $t = t_r$ finite jumps may occur. The differential equation must therefore be accompanied by requirements on the behaviour of the solutions at the discontinuities; for example, by giving a sequence of bounded nonsingular matrices T_r, $r = 1, 2, \dots$, and requiring that the solutions satisfy the condition

$$(63.14) \qquad \mathbf{x}(t_r + 0) = T_r \mathbf{x}(t_r - 0).$$

Within the intervals of continuity we have

$$(63.15) \qquad |\mathbf{x}(t_r + 0)| \, e^{-nL(t-t_r)} \leq |\mathbf{x}(t)| \leq |\mathbf{x}(t_r - 0)| \, e^{nL(t-t_r)}.$$

From (63.14) follows

$$(63.16) \qquad \mu_r |\mathbf{x}(t_r - 0)| \leq |\mathbf{x}(t_r + 0)| \leq m_r |\mathbf{x}(t_r - 0)|.$$

The constants μ_r^2 and m_r^2 are the smallest and the largest characteristic root, respectively, of the matrix $T_r^T T_r$. From (63.15) and (63.16) we obtain an estimate

$$(63.17) \qquad |\mathbf{x}(0)| \, \mu(t) \, e^{-nLt} \leq |\mathbf{x}(t)| \leq |\mathbf{x}(0)| \, m(t) \, e^{nLt},$$

which is valid for all $t > 0$. The jump functions $\mu(t)$ and $m(t)$ are defined by

$$(63.18) \qquad
\begin{aligned}
&\mu(t) := 1, \; 0 \leq t < t_1; \; \mu(t) := \mu_1 \mu_2 \cdots \mu_r, \; t_r \leq t < t_{r+1}, \\
&m(t) := 1, \; 0 \leq t < t_1; \; m(t) := m_1 m_2 \cdots m_r, \; t_r \leq t < t_{r+1}, \\
&\qquad\qquad r = 1, 2, \dots.
\end{aligned}$$

As a supplement for Theorem 63.1 we obtain:

Theorem 63.5.[1] Each solution of a piecewise continuous differential equation $\dot{\mathbf{x}} = A(t)\,\mathbf{x}$ satisfying the condition (63.14) has a finite order number if the coefficients of the differential equation are bounded and the functions (63.18) have a finite order number.

The place of equation (63.12) is taken in the discontinuous case by

$$\det X(t) = \det X(0) \, \Pi(t) \exp\left(\int_0^t \operatorname{Tr} A(\tau) \, d\tau \right).$$

On the right we again have a jump function

$$\Pi(t) := 1, \; 0 \leq t < t_1; \; \Pi(t) := \det(T_1 T_2 \cdots T_r), \; t_r \leq t < t_{r+1}.$$

---

[1] LIVARTOVSKII [1].

The discontinuous case can be reduced to the continuous case, for we can construct a linear transformation

$$(63.19) \qquad\qquad x = L(t)\, y,$$

so that the equation is transformed into a continuous equation

$$\dot{y} = P(t)\, y, \quad P(t) := L(t)^I \left(A(t)\, L(t) - \dot{L}(t) \right),$$

with bounded coefficients, and (63.14) is observed. L, L^I, and $\dot{L}$ are continuous except at $t = t_r$, $t = 1, 2, \ldots$. We first define

$$(63.20) \qquad\qquad L(t_r - 0) = E, \quad \dot{L}(t_r - 0) = 0,$$

$$L(t_r + 0) = T_r, \quad \dot{L}(t_r + 0) = A(t_r + 0)\, T_r - T_r A(t_r - 0),$$

and thereby make $P(t)$ continuous at $t = t_r$. If we require in addition to the hypothesis, that $|\det T_r| \geq \gamma > 0$, $r = 1, 2, \ldots$, then $L^I(t)$ is certainly bounded at the discontinuities. In order to arrange for L to map the interval $t_r \leq t \leq t_{r+1}$ continuously and that L, L^I and $\dot{L}$ remain bounded we proceed as follows. We consider the columns of T_r as vectors in R_n and change these vectors, keeping their length constant, so that the parallel-epiped spanned by them becomes rectangular. Then the edge length of the rectangular parallelepiped is changed uniformly so that it becomes a cube. The cube is rotated so that its edges point in the direction of the axes and finally the edges are all normalized to have length 1. These operations are executed so that each step takes up $1/4$ of the time $t_{r+1} - t_r$ and $|l_{ik}(t)|$ remains under a fixed bound while always $|\det L(t)| \geq \gamma$. Finally the variables $l_{ik}(t)$ are altered so that $\dot{L}$ satisfies the conditions in (63.20). This can certainly be done in such a way that the new values also remain below a fixed bound and that $|\det L(t)| \geq \gamma' > 0$, for some $\gamma' \leq \gamma$.

64. Regular Differential Equations

We again consider equation (58.1) and its adjoint equation

$$(64.1) \qquad\qquad \dot{y} = - A(t)^T\, y, \quad t \geq 0,$$

and we denote the order numbers of these two equations by

$$\lambda_1 \leq \lambda_2 \leq \cdots \leq \lambda_n, \quad \text{resp.} \quad \mu_1 \geq \mu_2 \geq \cdots \geq \mu_n.$$

If A is constant then $\lambda_i = -\mu_i$, $i = 1, 2, \ldots, n$. In general this is not true. For example, $+3$ is the order number of the solution

$$y_0 \exp\left(t_0 \cos t_0 - t \cos t + 2(t - t_0) \right)$$

of the adjoint of equation (63.11).

Def. 64.1. Equation (58.1) is called *regular* if its order numbers λ_i and the order numbers μ_i of its adjoint equation are related by

$$(64.2) \qquad \lambda_i + \mu_i = 0, \quad i = 1, 2, \ldots, n.$$

Theorem 64.1. Equation (58.1) is regular if and only if

$$(64.3) \qquad \pi\left(\exp\left(\int_0^t \operatorname{Tr} A(\tau)\, d\tau\right)\right) + \pi\left(\exp\left(-\int_0^t \operatorname{Tr} A(\tau)\, d\tau\right)\right) = 0$$

and simultaneously

$$(64.4) \qquad \lambda_1 + \lambda_2 + \cdots + \lambda_n = \pi\left(\exp\left(\int_0^t \operatorname{Tr} A(\tau)\, d\tau\right)\right).$$

Proof. Since $\operatorname{Tr} A = \operatorname{Tr} A^T$, we have by (63.13),

$$\lambda_1 + \cdots + \lambda_n \geq \pi\left(\exp\left(\int_0^t \operatorname{Tr} A(\tau)\, d\tau\right)\right),$$

$$\mu_1 + \cdots + \mu_n \geq \pi\left(\exp\left(-\int_0^t \operatorname{Tr} A(\tau)\, d\tau\right)\right).$$

If the equation is regular, *i.e.* if (64.2) holds then

$$(64.5) \qquad 0 \geq \pi\left(\exp\left(\int_0^t \operatorname{Tr} A(\tau)\, d\tau\right)\right) + \pi\left(\exp\left(-\int_0^t \operatorname{Tr} A(\tau)\, d\tau\right)\right).$$

On the other hand, we have by (63.2) for any arbitrary function f,

$$\pi(f) + \pi\left(\frac{1}{f}\right) \geq 0.$$

The left side of (64.3) is thus clearly nonnegative and by (64.5) equality must obtain.

Now let (64.3) and (64.4) be satisfied. Let X and Y be the matrices of normal fundamental systems for (58.1), resp. (64.1) Then we have

$$X^T Y = C = \text{constant}.$$

If we apply (63.2) to the order numbers of the individual inner products $x^{(i)T} y^{(i)}$ then

$$\lambda_i + \mu_i \geq 0$$

follows. Furthermore we consider the elements of $Y = X^{T I} C$ as linear combinations of the elements of $X^{T I}$. These are the $(n-1)^{\text{st}}$ order minors of X divided by $\det X$. The solution $x^{(j)}$ never appears in the jth column of $X^{T I}$. The order number of the jth column is therefore, again by (63.2), at most as large as

$$\sum_{i=1}^{n} \lambda_i - \lambda_j + \pi\left(\frac{1}{\det X}\right).$$

Observing (63.12) and (64.4) we obtain

$$\lambda_j + \mu_j \leq 0$$

and hence (64.2).

Since Theorem 64.1 contains a necessary and sufficient condition it can also be used to define the concept regular: in fact, this was the original definition of Liapunov. The definition given above is a characterization due to PERRON [2].

Autonomous and periodic equations are obvious examples of regular equations, and so are all reducible equations (Def. 61.1). Furthermore we have

Theorem 64.2. Let (58.1) be an equation whose matrix is triangular i.e. $a_{ij} = 0$ for $i < j$, and has bounded elements. Then it is regular if and only if the limits

$$(64.6) \qquad \lim_{t \to \infty} \frac{1}{t} \int_0^t a_{ii}(\tau)\, d\tau = : c_i, \quad i = 1, 2, \ldots, n,$$

exist.

Proof. 1) If the limits (64.6) exist then

$$\pi \left(\exp \left(\int_0^t a_{ii}(\tau)\, d\tau \right) \right) = c_i, \quad \pi \left(\exp \left(- \int_0^t a_{ii}(\tau)\, d\tau \right) \right) = - c_i$$

and (64.3) is satisfied. Then (64.4) is valid because

$$\pi \left(\exp \left(\int_0^t \operatorname{Tr} A(\tau)\, d\tau \right) \right) = \limsup \frac{1}{t} \int_0^t \operatorname{Tr} A(\tau)\, d\tau = c_1 + \cdots + c_n.$$

2) Since the system $\dot{x}_i = \sum_{k=1}^{n} a_{ik} x_k$, $i = 1, 2, \ldots, n$, is triangular it can be solved step for step. We have

$$x_1(t) = x_1(0) \exp \left(\int_0^t a_{11}(\tau)\, d\tau \right).$$

By (64.3), the limit c_1 exists whence

$$\pi(x_1) = c_1, \quad \text{if } x_1(0) \neq 0.$$

$$x_2(t) = x_2(0) \exp \left(\int_0^t a_{22}(\tau)\, d\tau \right) + x_1(0) \exp \left(\int_0^t a_{22}(\tau)\, d\tau \right)$$

$$\times \int_0^t a_{21}(\tau) \exp \left(\int_0^\tau \big(a_{11}(u) - a_{22}(u) \big)\, du \right) d\tau$$

$$= \exp \left(\int_0^t a_{22}(\tau)\, d\tau \right) \left(x_2(0) + x_1(0) \right.$$

$$\times \left. \int_0^t a_{21}(\tau) \exp \left(\int_0^\tau \big(a_{11}(u) - a_{22}(u) \big)\, du \right) d\tau \right).$$

From (64.3), the existence of the limit c_2 follows.

The order number of the first summand is c_2 in case $x_2(0) \neq 0$. That of the second summand is, because of (64.4), at most equal to c_1 in case $x_1(0) \neq 0$. If $n = 2$ then the solutions

$$\mathrm{col}\left(0,\ \exp\left(\int_0^t a_{22}(\tau)\,d\tau\right)\right), \quad \mathrm{col}\left(\exp\left(\int_0^t a_{11}(\tau)\,d\tau\right),\ \psi(t)\right)$$

form a basis with order numbers c_2 and c_1, $\psi(t)$ denoting the coefficient of $x_1(0)$ in $x_2(t)$. If $n > 2$ then the process can be continued.

For further development of the theory the following theorem is of importance.

Theorem 64.3. Every equation $\dot{x} = A(t)\,x$ with bounded coefficients can be transformed into an equation with triangular matrix by means of a Liapunov transformation (sec. 63).

The theorem is due to PERRON [1]. The proof given here, which is simpler than Perron's, is due to DILIBERTO [1], respectively VINOGRAD [2].

1) Let X be the matrix of a real, respectively complex basis of the original equation. Starting from the solution $x^{(1)}$ and proceeding step for step, we procure n linearly independent mutually orthogonal unit vectors. For this purpose we set

$$u^{(1)} = c_{11}\,x^{(1)}, \quad c_{11} = |x^{(1)}|^{-1},$$

$$u^{(2)} = c_{21}\,u^{(1)} + c_{22}\,x^{(2)}, \quad \text{so that} \quad \bar{u}^{(1)T}\,u^{(2)} = 0, \quad |u^{(2)}| = 1;$$

i.e.

$$c_{21} + c_{22}\,\bar{u}^{(1)T}\,x^{(2)} = 0, \quad 1 = c_{21}^2 + 2\,c_{21}c_{22}\,\bar{u}^{(1)T}\,x^{(2)} + c_{22}^2\,|x^{(2)}|^2,$$

etc. The vectors $u^{(1)}$, $u^{(2)}$, ... form the columns of an orthogonal, respectively unitary, matrix U which is related to X by the equation

$$(64.7) \qquad\qquad U = XR.$$

R is nonsingular and triangular. The elements below the main diagonal are zero. By differentiating the identity $\bar{U}^T U = E$ we obtain

$$\dot{\bar{U}}^T U + \bar{U}^T \dot{U} = 0$$

and hence

$$(64.8) \qquad\qquad U^I \dot{U} = -\left(\overline{U^I \dot{U}}\right)^T.$$

The matrix $U^I \dot{U}$ is scew-symmetric, respectively scew-hermitian. If we set

$$x = U\,y, \quad X = U\,Y,$$

then we obtain (*cf.* (61.2))

$$\dot{Y} = \left(U^I A U - U^I \dot{U}\right) Y.$$

On the other hand,

$$Y = U^I X = (X R)^I X = R^I,$$
$$\dot{Y} = - R^I \dot{R} R^I = - \left(R^I \dot{R}\right) Y,$$

and, comparing with the above, we obtain

$$(64.9) \qquad\qquad U^I A U - U^I \dot{U} = - R^I \dot{R}.$$

The right side is a triangular matrix since both R^I and $\dot{R}$ are triangular. Hence the elements below the main diagonal on the left side are zero and this implies that the elements of $U^I \dot{U}$ below the main diagonal are bounded. From (64.8) we conclude that $U^I \dot{U}$ has only zeros on the main diagonal, all of which implies that $\dot{U}$ is bounded. U is bounded since it is a unitary matrix and thus Y satisfies an equation whose matrix is triangular and U is a Liapunov transformation.

Theorem 64.3 is of theoretical interest only since in order to find the transformation U one must know a fundamental system of the equation under consideration. The last two theorems together imply

Theorem 64.4. If the equation (58.1) is regular then for each solution x the limit

$$(64.10) \qquad\qquad \lim_{t \to \infty} \frac{1}{t} \ln |x(t)|$$

exists and clearly is the order number.

Proof. Theorem 64.3 is used to put the equation into triangular form. It remains regular. The existence of the limit follows from the step for step solution described in the proof of Theorem 64.2 and for the diagonal elements the limit (64.6) exists. We further see that the limit (64.10) exists for the functions $x_i(t)$ constructed in the proof of Theorem 64.2. For x_1 this is immediately clear. Next,

$$\ln x_2 = \int_0^t a_{22}(\tau)\, d\tau$$
$$+ \ln \left(x_2(0) + x_1(0) \int_0^t a_{21}(\tau) \exp\left(\int_0^\tau a_{11}(u) - a_{22}(u)\, du\right) d\tau\right).$$

By (64.6) the inner integrand of the second summand behaves like $\tau(c_1 - c_2)$, and since $a_{21}(t)$ is bounded the outer integral of the second summand lies between two bounds of the form

$$k_i e^{t(c_1 - c_2)}, \quad i = 1, 2, \ldots.$$

This implies the existence of the limit (64.10) for $x_2(t)$ and in the same way we show its existence for the other components of $x(t)$.

65. Stability in the First Approximation

If the right side of the differential equation

$$(65.1) \qquad \dot{x} = f(x, t), \ f \in E,$$

admits a power series expansion in a neighborhood of the origin the equation can be written in the form

$$(65.2) \qquad \dot{x} = A(t)\, x + g(x, t), \ |x| \le h, \ t \ge t_0.$$

If $f(x, t)$ is bounded with respect to t then

$$(65.3) \qquad g(x, t) = O(|x|^2).$$

As in the case of autonomous equations (*cf.* sec. 28) we seek conditions under which the stability behavior of (65.1) is equivalent with that of the *equation in the first approximation*

$$(65.4) \qquad \dot{x} = A(t)\, x.$$

If it is known that for (65.4) there exists a Liapunov function which permits a quadratic estimate

$$a_1\, |x|^2 \le v(x, t) \le a_2\, |x|^2,$$

then the proof of Theorem 28.1 applies directly. Such a Liapunov function can be constructed as in sec. 56 if the equilibrium of (65.4) is exponentially stable. If in addition we observe Theorem 58.7 we have

Theorem 65.1. If the equilibrium of the equation in the first approximation is uniformly asymptotically stable then so is the equilibrium of the complete equation.

In place of (65.3) it suffices to assume that

$$(65.5) \qquad g(x, t) = o(|x|).$$

Theorem 65.1 can also be considered as a special case of Theorem 56.2.

Another theorem is based on the concepts introduced in Theorem 58.6.

Theorem 65.2. (COPPEL [1, 2]) Let two supplementary projections P_0, P_1 and two constants $\varkappa$ and γ with $\varkappa\gamma < 1$ be given such that

$$\int_0^t \| K(t, t_0)\, P_0 K(t_0, s) \|\, ds + \int_t^\infty \| K(t, t_0)\, P_1 K(t_0, s) \|\, ds \le \varkappa, \ t \ge t_0,$$

and such that

$$|g(x, t)| \le \gamma\, |x|.$$

The equilibrium of (65.2) is asymptotically stable if $P_1 = 0$ and unstable if $P_1 \ne 0$.

Proof. First assume $P_1 = 0$ and $P_0 = E$. A solution $x(t)$ of (65.2) satisfies the integral equation

$$x(t) = K(t, t_0)\, x(t_0) + \int_{t_0}^t K(t, s)\, g(x(s), s)\, ds.$$

Since by Theorem 58.6 the equilibrium of the linear equation (65.4) is asymptotically stable there exists a bound μ for $\|K(t, t_0)\|$. From the hypothesis we obtain

$$|\boldsymbol{x}(t)| \leq \mu\,|\boldsymbol{x}(t_0)| + \gamma\varkappa \sup_{s \geq t_0} |\boldsymbol{x}(s)|,$$

$$|\boldsymbol{x}(t)| \leq (1 - \gamma\varkappa)^{-1}\mu\,|\boldsymbol{x}(t_0)|.$$

Thus the equilibrium of (65.2) is stable. Next, assume that

$$\alpha := \limsup_{t \to \infty} |\boldsymbol{x}(t)| > 0.$$

A number β, $\varkappa\gamma < \beta < 1$, and a number t_1, are so chosen that $|\boldsymbol{x}(t)| < \beta^{-1}\alpha$ for $t \geq t_1$. Then for $t \geq t_1$ the integral equation and the assumptions imply

$$|\boldsymbol{x}(t)| \leq \|K(t, t_0)\|\,|\boldsymbol{x}(t_0)| + \gamma\varkappa\beta^{-1}\alpha$$

$$+ \|K(t, t_0)\| \int_{t_0}^{t_1} K(t, s)\,\boldsymbol{g}\big(\boldsymbol{x}(s)\,s,\big)\,ds .$$

As t increases the right side tends to $\gamma\varkappa\beta^{-1}\alpha < \alpha$ and a contradiction is obtained. Therefore $\alpha = 0$ and the equilibrium is asymptotically stable.

Next suppose $P_1 \neq 0$. Now the integral equation has the form

$$\boldsymbol{x}(t) = K(t, t_0)\,P_0\boldsymbol{x}(t_0) + \int_{t_0}^{t} K(t, t_0)\,P_0 K(t_0, s)\,\boldsymbol{g}\big(\boldsymbol{x}(s), s\big)\,ds$$

$$- \int_{t}^{\infty} K(t, t_0)\,P_1 K(t_0, s)\,\boldsymbol{g}\big(\boldsymbol{x}(s), s\big)\,ds.$$

If the equilibrium is assumed to be stable then $|\boldsymbol{x}(t)|$ is arbitrarily small for sufficiently small values of $|\boldsymbol{x}(t_0)|$, and as above we would obtain an estimate of the form

$$|\boldsymbol{x}(t)| \leq (1 - \gamma\varkappa)^{-1}\mu\,|P_0\boldsymbol{x}(t_0)|.$$

If we now choose $\boldsymbol{x}(t)$ so that $P_0\boldsymbol{x}(t_0) = 0$ (this is possible because $P_0 \neq E$), we arrive at a contradiction.

A theorem based on yet a different foundation is

Theorem 65.3. In (65.2) let

$$|\boldsymbol{g}(\boldsymbol{x}, t)| \leq c\,|\boldsymbol{x}|^m, \quad m > 1,\ c > 0,$$

and assume that the equation (65.4) of the first approximation is regular and has only negative order numbers. Then the equilibrium of (65.2) is asymptotically stable.

Before proving this theorem which is due to Liapunov, we note that Theorems 56.2 and 65.3 are not equivalent. The collection of regular equations with negative order numbers and the collection of equations with exponentially stable equilibrium are not the same. This is shown in the following examples.[1])

1) The system

$$\dot{x}_1 = - \left(2 - \sin \ln (t + 1)\right) x_1 ,$$

$$\dot{x}_2 = x_1 - x_2 , \quad t \geq 0 ,$$

has negative order numbers and its equilibrium is exponentially stable. But the equation is not regular because

$$\frac{1}{t} \int_0^t \left(\sin \ln (\tau + 1) - 2\right) d\tau = - 2 + \frac{1}{2} \left(\sin \ln (t + 1) - \cos \ln (t + 1)\right)$$

$$+ \frac{1}{2t} \left(1 + \sin \ln (t + 1) - \cos \ln (t + 1)\right)$$

does not approach a limit as $t \to \infty$, so that the criterion of Theorem 64.2 is not satisfied.

2) The equation

$$\dot{x} = - \left(1 + 2\pi \cos \sqrt{t}\,\right) x$$

satisfies the hypotheses of Theorem 64.2 since

$$- \frac{1}{t} \int_0^t (1 + 2\pi \cos \sqrt{\tau}\,) d\tau = - 1 + \frac{1}{t} \left(4\pi - 4\pi \sqrt{t} \sin \sqrt{t} - 4\pi \cos \sqrt{t}\,\right)$$

tends to $- 1$. It has a negative order number. But the equilibrium is not uniformly stable; for we have

$$p(t, x_0, t_0) = x_0 \exp \left[- (t - t_0) - 4\pi (\cos \sqrt{t} - \cos \sqrt{t_0})\right.$$

$$\left. - 4\pi (\sqrt{t} \sin \sqrt{t} - \sqrt{t_0} \sin \sqrt{t_0})\right].$$

For $t_0 = \left(2 k\pi + \dfrac{\pi}{2}\right)^2$, $t = (2 k\pi + \pi)^2$, we have

$$p(t, x_0, t_0) = x_0 \exp \left[- \left(4 k\pi + \frac{3}{2} \pi\right) \frac{\pi}{2} + 4\pi + 4\pi \left(2 k\pi + \frac{\pi}{2}\right)\right]$$

$$> x_0 \exp(6 k\pi^2) .$$

3) The solution of the equations

$$\dot{x}_1 = - a x_1 ,$$

$$\dot{x}_2 = (\sin \ln t + \cos \ln t - 2 a) x_2 , \quad t \geq 1$$

is

$$x_1 = x_1 (1) e^{-a(t-1)} , \quad x_2 = x_2 (1) e^{t \sin \ln t - 2a(t-1)} .$$

---

[1]) MALKIN [3].

In case $a > 1/2$, both order numbers are negative and the equilibrium is exponentially stable. But the system of equations is not regular because the expression

$$\frac{1}{t} \int_1^t (\sin \ln \tau + \cos \ln \tau - 2a) \, d\tau = \sin \ln t - 2a + \frac{2a}{t}$$

has no limit as $t \to \infty$.

This system of equations furnishes an example to show that Theorem 65.3 is false without the hypothesis "(65.4) is regular". If we replace the equation for $\dot{x}_2$ by the nonlinear equation

$$\dot{x}_2 = (\sin \ln t + \cos \ln t - 2a) \, x_2 - x_1^2,$$

we obtain

$$x_2 = e^{t \sin \ln t - 2a(t-1)} \left(x_2(1) + x_1(1)^2 \int_1^t e^{-\tau \sin \ln \tau} \, d\tau \right).$$

But for $t = t_n := \exp((2n + 1/2)\,\pi)$,

$$\int_1^t e^{-\tau \sin \ln \tau} \, d\tau > \int_{t_n e^{-\pi}}^{t_n e^{-2\pi/3}} e^{-\tau \sin \ln \tau} \, d\tau > \exp\left(\frac{1}{2} t_n e^{-\pi} \int_{t_n e^{-\pi}}^{t_n e^{-2\pi/3}} d\tau \right)$$

$$= t_n (e^{-2\pi/3} - e^{-\pi}) \exp\left(\frac{1}{2} t_n e^{-\pi} \right)$$

and

$$e^{t \sin \ln t - 2at} \int_1^t e^{-\tau \sin \ln \tau} \, d\tau > ct \exp\left(\left(1 + \frac{1}{2} e^{-\pi} - 2a \right) t \right).$$

Hence $x_2(t)$ is unbounded in case $x_1(1) \neq 0$ and $1 < 2a < 1 + \frac{1}{2} e^{-\pi}$.

Proof of Theorem 65.3 (from ČETAEV [2]). Let $x^{(j)}, y^{(j)}$ be two normal bases of (65.4) and (64.1) which are normalized so that the matrices X and Y^T are inverses of each other,

$$X Y^T = E.$$

By hypothesis the order numbers of (65.4) satisfy

$$\lambda_1 \leq \lambda_2 \leq \cdots \leq \lambda_n < 0.$$

If we set

$$D := \operatorname{diag}(e^{\lambda_1 t}, \ldots, e^{\lambda_n t})$$

and

$$(65.6) \qquad z := D Y^T e^{\beta t} x, \quad 0 < \beta < |\lambda_n|,$$

then because of the regularity and by (63.2) we have

$$\pi(z) \leq \pi(x) + \beta.$$

But also

$$x = e^{-\beta t} X D^I z,$$

and this implies

$$\pi(x) \leq \pi(z) - \beta,$$

hence

$$\pi(z) = \pi(x) + \beta.$$

The transformation (65.6) puts (65.2) into the form

$$(65.7) \qquad \dot{z} = \left(\dot{D} D^I + \beta E\right) z + D Y^T e^{\beta t} g(x, t).$$

The derivative of the Liapunov function

$$v = \frac{1}{2} |z|^2$$

for this differential equation is

$$(65.8) \quad \dot{v} = z^T \left(\dot{D} D^I + \beta E\right) z + z^T \left(e^{\beta t} D Y^T g(e^{-\beta t} X D^I z, t)\right).$$

The second term is of order $1 + m$ with respect to z. If z is fixed the term tends to zero if t increases. This can be seen as follows. The order numbers of the elements of $D Y^T$ and $X D^I$ are zero (we already used this fact). Therefore

$$\pi\left(z^T e^{\beta t} D Y^T g(e^{-\beta t} X D^I z, t)\right) \leq \pi(z) + \beta + 0 + m\,\pi(e^{-\beta t} X D^I z)$$

$$\leq \pi(z) + \beta + m\left(-\beta + \pi(z)\right)$$

$$= (m + 1)\,\pi(z) + (1 - m)\,\beta.$$

For fixed z, the order number of the term in question is at most $(1 - m)\,\beta < 0$.

By hypothesis, $\dot{v}$ is negative definite and for a given $\alpha > 0$ two numbers δ and t_1 can be so chosen that

$$\left|z^T \left(e^{\beta t} D Y^T g(x, t)\right)\right| < \alpha\,|z|^2$$

provided only that $|z| < \delta$ and $t > t_1$. Then we have

$$\dot{v} \leq \sum_{i=1}^{n} (\lambda_i + \beta)\,|z_i|^2 + \alpha\,|z|^2 < (\lambda_n + \beta + \alpha)\,|z|^2$$

and because of (65.8),

$$|z| \leq |z(t_1)|\,\exp\left((\lambda_n + \beta + \alpha)\,(t - t_1)\right).$$

If the initial value z_0 is chosen so that the solution $z(t, z_0, t)$ of (65.7, remains in the domain $|z| < \delta$ on the interval $t_0 \leq t \leq t_1$, then for $t > t_1$)

$$|z(t, z_0, t_0)| \leq |z(t_1, z_0, t_0)|\,\exp\left((\lambda_n + \beta + \alpha)\,(t - t_1)\right).$$

It follows that $\pi(z) \leq \lambda_n + \beta + \alpha$, $\pi(x) \leq \lambda_n + \alpha$. Therefore, if we choose $\alpha < -\lambda_n$ then the asymptotic stability of the equilibrium of (65.2) is assured.

21*

Theorem 65.4. Let the equations of Theorem 65.3 be given. Let the equation (65.4) be regular and have at least one positive order number. Then the equilibrium of (65.2) is unstable.

Proof. As above we form the vector z, by (65.6) but this time we set $\beta = 0$. By (65.7) the component $\dot{z}_n$ satisfies the equation

$$\dot{z}_n = \lambda_n z_n + \sum_{i=1}^{n} e^{\lambda_n t} y_i^{(n)}(t)\, g_i\big(x(t), t\big),$$

which implies

$$z_n = z_n(0)\, e^{\lambda_n t} + e^{\lambda_n t} \int_0^t \sum_{i=1}^{n} y_i^{(n)}(\tau)\, g_i\big(x(\tau), \tau\big)\, d\tau.$$

If now the equilibrium is assumed to be stable then the functions $x_i(t)$, consequently also the functions $g_i(x(t), t)$, are bounded for sufficiently small initial values. Since the order number of $y^{(n)}$ is equal to $-\lambda_n < 0$ by hypothesis, the order number $-\lambda_n$ of the integrand and the order number of the second summand are non-positive. Hence

$$\pi(z_n) = \lambda_n > 0,$$

and since $\pi(x) \geq \pi(z)$, a contradiction is obtained to the assumed stability.

Both theorems can be extended to the nonregular case under additional assumptions. For each case

$$\lambda_1 + \lambda_2 + \cdots + \lambda_n + \pi\left(\exp\left(-\int_0^t \operatorname{Tr} A(\tau)\, d\tau\right)\right) =: \gamma$$

is a non-negative number and in the notation of the proof of Theorem 65.3 we have throughout

$$\pi(x) + \beta \leq \pi(z) \leq \pi(x) + \beta + \gamma.$$

In calculating the second term of $\dot{v}$ in (65.8) we must however take into consideration that the order number of the elements of DY^T may assume the value γ. Therefore β must be chosen so large that $(1-m)\beta + \gamma$ is negative. But since $\lambda_n + \beta < 0$ is necessary, as otherwise $\dot{v}$ is not negative definite, we must require $\lambda_n < -\dfrac{\gamma}{m-1}$.

The theorems which we listed furnish only sufficient criteria for stability in the first approximation. Other sufficient criteria are available in which the equation in the first approximation is assumed to satisfy an estimate

$$\|K(t, t_0)\| \leq h\, e^{a t_0}\, e^{-bt}, \quad a > 0,\ b > 0,$$

and the nonlinear additional term an estimate

$$|g(x, t)| \leq f(t)\, |x|^m, \quad m \geq 1.$$

For m, a, b we have inequalities $m \geq \frac{a}{b}$, $a \geq b$; $f(t)$ satisfies certain integrability conditions. For example, MALKIN[2] has proved that the condition "$f(t)$ bounded, $m > 1$, $a < 2mb$" is sufficient. The most comprehensive criterion to date of this type is found in REGHIŞ [1].

Closely connected with the theory of stability in the first approximation is the so-called *stability of the order numbers* of a linear equation. By this we mean the following: Let $\lambda_1, \ldots, \lambda_n$ and $\lambda'_1, \ldots, \lambda'_n$, respectively, be the order numbers of the equations $\dot{x} = A(t) x$ and $\dot{x} = B(t) x$. The order numbers are called *stable* if for each ε there exists a δ such that $|\lambda'_i - \lambda_i| < \varepsilon$, $i = 1, 2, \ldots, n$, provided only that $|a_{ij}(t) - b_{ij}(t)| < \delta$, $i, j = 1, 2, \ldots, n$, $t \geq 0$.

There is a variety of criteria for stability of order numbers. We refer the reader to the book by MALKIN [3]. We note in closing that not even for regular systems the stability is assured without additional assumptions. VINOGRAD [3] has shown that the order numbers of the regular system

$$\dot{x} = 0, \quad \dot{y} = f(t) y,$$

where

$$f(t) = \begin{cases} -1 & \text{for } (2n)^2 \leq t < (2n+1)^2, \\ +1 & \text{for } (2n+1)^2 \leq t < (2n+2)^2, \end{cases}$$

are both zero whereas the perturbed system

$$\dot{x} = \varepsilon y, \quad \dot{y} = \varepsilon x + f(t) y$$

possesses, for arbitrarily small ε, one order number $\geq \frac{1}{2}$ and one order number $\leq -\frac{1}{2} + \varepsilon$. The largest and the smallest order number of the example are thus unstable.

The theory of the first approximation can also be extended to differential equations

$$(65.9) \qquad\qquad \dot{x} = f(x, t)$$

with discontinuous right side, provided the solutions are continuous. The term "discontinuous" is made precise as follows. Let $\tilde{x}(t)$ be a fixed continuous solution of (65.9). Its integral curve is considered as the "axis" of a cylinder C in the $(n + 1)$-dimensional (x, t)-space. In addition to the equation a sequence of continuous functions $F_r(x, t)$, $r = 1, 2, \ldots$ is given such that for each r, the surface

$$(65.10) \qquad\qquad F_r(x, t) = 0, \quad r = 1, 2, \ldots$$

intersects the curve $\tilde{x}(t)$ at the point $P_r := (x_r, t_r)$. In a neighborhood of P_r, (65.10) is smooth. No two of the surfaces (65.10) intersect within the cylinder C. C is decomposed into domains H_r by the surfaces. The planes $t = t_r$ decompose the domains H_r into *central domains* which each contain a segment of the curve $\tilde{x}(t)$, and *corner domains* which do not. In addition, the functions f and F_r are assumed to satisfy the following conditions:

1) The length of the intervals (t_r, t_{r+1}) is bounded below:

$$(65.11) \qquad t_{r+1} - t_r \geq T > 0, \quad r = 1, 2, \ldots.$$

2) The function $f(x, t)$ is continuous in the interior of H_r and of such a nature that uniquely determined solutions of (65.9) exist there.

3) In passing through the surface (65.10) the components of $f(x, t)$ may have jumps of height ξ_{ir}, these jumps being uniformly bounded with respect to r: There exists a number ξ such that

$$(65.12) \qquad \begin{aligned} |\xi_r| &< \xi, \quad r = 1, 2, \ldots, \\ \xi_r &:= \operatorname{col}(\xi_{1r}, \xi_{2r}, \ldots, \xi_{nr}). \end{aligned}$$

4) Within the central domains we have

$$f(x, t) - f(\tilde{x}(t), t) = A(t)\,(x - \tilde{x}(t)) + g(x, t),$$

in fact, $A(t)$ is continuous in each interval $t_r \leq t \leq t_{r+1}$ and uniformly bounded for $t \geq 0$. The remainder $g(x, t)$ satisfies an estimate

$$(65.13) \qquad |g(x, t)| \leq a\,|y|, \quad y := x - \tilde{x}(t).$$

5) If the point (x, t) approaches P_r we have in each corner domain

$$f(x, t) - f(\tilde{x}(t), t) \to \xi_r,$$

resp.

$$f(x, t) - f(\tilde{x}(t), t) \to -\xi_r,$$

depending on whether (x, t) lies "below" or "above" the plane $t = t_r$. The approach is to be uniform with respect to r.

6) The total derivative of the function $F_r(x, t)$ along the curve $\tilde{x}(t)$ is different from zero for $t = t_r - 0$. In passing through the point P_r, i.e. in passing from $P_r^+ := (x_r, t_r + 0)$ to $P_r^- := (x_r, t_r - 0)$,

$$\frac{dF_r}{dt}\bigg|_{P_r^+} : \frac{dF_r}{dt}\bigg|_{P_r^-} \geq \gamma > 0$$

is valid. For the surface defined by

$$G_r(y, t) := F_r(y + \tilde{x}(t), t) = 0$$

we obtain from (65.13) an approximative representation: We introduce the vectors

$$h_r^+ := \frac{\operatorname{grad} F_r(x, t)|_{P_r}}{\dfrac{dF_r}{dt}\bigg|_{P_r^+}} \;;\quad h_r^- := \frac{\operatorname{grad} F_r(x, t)|_{P_r}}{\dfrac{dF_r}{dt}\bigg|_{P_r^-}}.$$

Then the approximate equations are

$$(65.14) \qquad t_r - t = (\boldsymbol{h}_r^-)^T \boldsymbol{y} + o(|\boldsymbol{y}|)\,,$$

$$t - r_r = (\boldsymbol{h}_r^+)^T \boldsymbol{y} + o(|\boldsymbol{y}|)\,,$$

depending on whether the surface lies below the plane $t = t_r$ or above it.

7) The vectors $\boldsymbol{h}_r^+$ and $\boldsymbol{h}_r^-$ are uniformly bounded. The quantities denoted by o in (65.14) tend to zero uniformly as $\boldsymbol{y}$ tends to zero.

Under these hypotheses the equation

$$(65.15) \qquad \dot{\boldsymbol{y}} = A(t)\,\boldsymbol{y}$$

together with the jump condition

$$(65.16) \qquad \boldsymbol{y}(t_r + 0) = T_r\,\boldsymbol{y}(t_r - 0)\,, \quad T_r := E + \boldsymbol{\xi}_r\,(\boldsymbol{h}_r^-)^T\,,$$

which follows from (65.14), is to be called the equation of the first approximation (for the equation of the perturbed motion). It is of the type discussed at the end of sec. 63.

Theorem 65.5 (AIZERMAN and GANTMACHER [3]). If the equilibrium of equation (65.15) and (65.16) is exponentially stable then the solution $\boldsymbol{x}(t)$ of (65.9) is asymptotically stable provided the constant a in (65.13) is sufficiently small.

Proof. The discontinuous linear equation is transformed by a transformation $\boldsymbol{y} = L(t)\,\boldsymbol{z}$ of type (63.19) into a continuous equation $\dot{\boldsymbol{z}} = P(t)\,\boldsymbol{z}$. This maintains the exponential stability and by Theorem 56.1 we can construct a Liapunov function $v(\boldsymbol{z}, t)$ such that

$$(65.17) \qquad b_1\,|\boldsymbol{z}|^2 \le v(\boldsymbol{z}, t) \le b_2\,|\boldsymbol{z}|^2\,, \quad \left|\frac{\partial v}{\partial z_i}\right| \le b_3\,|\boldsymbol{z}|\,,$$

$$(\operatorname{grad} v)^T\,P(t)\,\boldsymbol{z} + \frac{\partial v}{\partial t} = -\,|\boldsymbol{z}|^2.$$

We have to examine the derivative of this function for the nonlinear equation (65.9), which has been mapped by the transformation into

$$(65.18) \qquad \dot{\boldsymbol{z}} = \boldsymbol{q}(\boldsymbol{z}, t) := L(t)^I \left[\boldsymbol{g}(\boldsymbol{x}(t) + L(t)\,\boldsymbol{z}, t) - \boldsymbol{f}(\boldsymbol{x}(t), t) - L(t)\,\boldsymbol{z}\right].$$

In the central domains the argument differs in no way from the argument in the continuous case: Since the derivative of v for (65.18) and for $\dot{\boldsymbol{z}} = P(t)\,\boldsymbol{z}$ differ only by a term $O(|\boldsymbol{z}|^2)$ we obtain a differential inequality $\dot{v} \le -cv$ as in the proof of Theorem 28.1. If the motion remains in a given central domain between the times $t = t'$ and $t = t''$ then

$$(65.19) \qquad v(t'') \le v(t')\,\exp\left(-c(t'' - t')\right),$$

where c does not depend on r.

In the corner domains we must argue somewhat differently. If $\boldsymbol{z}$ tends to zero within a corner domain then t tends to t_r and by hypothesis 5) and because of the discontinuity of L, the vector $\boldsymbol{q}(\boldsymbol{z}, t)$ tends toward a finite limit $\boldsymbol{\eta}_r$ which is equal to $L(t_r - 0)^I \boldsymbol{\xi}_r$ in the lower corner domain and is equal to $-L(t_r + 0)^I \boldsymbol{\xi}_r$ in the upper corner domain. We therefore have in the domain, for sufficiently small $|\boldsymbol{z}|$,

$$\frac{dv}{dt} = (\mathrm{grad}\ v)^T \boldsymbol{\eta}_r + O(|\boldsymbol{z}|).$$

The first term is also $O(|\boldsymbol{z}|)$ and since also $\sqrt{v} = O(|\boldsymbol{z}|)$, $\dot{v}/\sqrt{v}$ is smaller than a constant k so that

$$\sqrt{v(t'')} - \sqrt{v(t')} < k\,|t'' - t'|$$

provided the motion remains in the same corner domain between $t = t''$ and $t = t'$. Now

$$(65.20) \qquad \frac{\sqrt{v(t')}}{\sqrt{v(t'')}} = 1 + \frac{\sqrt{v(t')} - \sqrt{v(t'')}}{\sqrt{v(t'')}} \le 1 + k\,\frac{t' - t''}{\sqrt{v(t'')}},$$

and since by (65.14)

$$|t' - t''| = O(|\boldsymbol{z}|) = O(\sqrt{v}),$$

it follows that

$$\frac{\sqrt{v(t')}}{\sqrt{v(t'')}} \le k_1, \quad v(t') \le k_2\,v(t''),$$

where k_2 does not depend on r (hypotheses 5 and 7). (65.20) is valid for any two times between which the motion remains in the same corner domain.

If $|\boldsymbol{z}|$ is sufficiently small the values of $v(\boldsymbol{z}, t)$ at the points at which the integral curve cuts the plane $t = t_r$, respectively the surface $G_r(L(t)\,\boldsymbol{z}, t)) = 0$, differ by an arbitrarily small amount. This can be expressed by an inequality

$$(65.21) \qquad e^{-\delta} < \frac{v(\boldsymbol{z}_r, t_r)}{v(\boldsymbol{z}', t')} < e^{\delta}.$$

$(\boldsymbol{z}', t')$ is the point of intersection with the surface, δ an arbitrarily small quantity.

(65.20) indicates that the function $v(t)$ can increase along the motion within a corner domain. It is our problem to show that its decrease, assured in a central domain by (65.19), definitely exceeds the increase. We have to take into consideration here that as $|\boldsymbol{z}|$ decreases, $i.e.$ as the radius of the "cylinder" C decreases, the integral curve remains in the central domain for a relatively longer and in the corner domain for a shorter time.

We first choose $\varepsilon > 0$ so small that the planes $t = t'_r$ through the points $t'_r = \frac{1}{2}(t_r + t_{r+1})$ do not meet any corner domain in the interior of the cylinder $|\boldsymbol{z}|^2 = \varepsilon$. Then we choose $(\boldsymbol{z}_0, t'_1)$ as the initial point of the integral curve, where

$$|\boldsymbol{z}_0|^2 < \frac{\varepsilon\,b_1}{k_2\,b_2}.$$

The b_i are the numbers from (65.17). Then we have $v(z_0, t_1') < \dfrac{\varepsilon\, b_1}{k_2}$ and since the growth coefficient of v in the interval (t_1', t_2') is smaller than k_2, $v(t)$ is smaller than εb_1 in this interval and because of (65.17), $|z(t)|^2 < \varepsilon$. Let $\varDelta t$ be the length of the time interval during which the motion remains in the corner domain. Applying (65.11), (65.19), and (65.21) we obtain

$$v(t_2') < v(t_1') \exp\left(-c(T - \varDelta t) + \delta\right).$$

We now choose ε, and thus $\varDelta t$ and δ, so small that the exponent in the last inequality becomes negative, say $\leq -\varkappa$. Then $v(t_2') < v(t_1')\, e^{-\varkappa}$. Repeating the argument we find that

$$v(t_r') \leq v(t_1')\, e^{-(r-1)\varkappa}$$

and this implies the desired assertion: $v(t) \to 0$ as $t \to \infty$. For in the interval $t_r' < t < t_{r+1}'$ we certainly have $v(t) < v(t_r')\, k_2$. From the estimates (65.17) we see then that $|z| \to 0$. This completes the proof.

Chapter IX

The Liapunov Expansion Theorem

66. Families of Solutions Depending on a Parameter

Let a linear autonomous equation

$$(66.1) \qquad \dot{x} = A\,x$$

be given. Let the first k of the characteristic roots $\alpha_1, \ldots, \alpha_n$ of the matrix A have negative real parts, $\operatorname{Re}\alpha_j \leq -\eta < 0$, $j = 1, \ldots, k$, and the remaining characteristic roots have non-negative real parts. By a suitable linear transformation we can put A into the form $\operatorname{diag}(A_1, A_2)$ in such a way that $\alpha_1, \ldots, \alpha_k$ are the characteristic roots of A_1. We shall thus assume that A has this form. Let a k-vector c and an n-vector x_0 be given, where

$$(66.2) \qquad c = \operatorname{col}(c_1, \ldots, c_k), \quad x_0 = \operatorname{col}(c, 0).$$

The solution

$$(66.3) \qquad x(t) = p(t, x_0, 0) = e^{At}\,x_0 = \operatorname{col}(e^{A_1 t}\,c, 0),$$

depends on the k components of the vector c and therefore constitutes a k-parameter family of solutions of the differential equation (66.1). Its components are linear combinations of terms of the form

$$(66.4) \qquad t^{r_{ij}}\, e^{\alpha_j t}\, c_k, \quad 0 \leq r_{ij} \leq n - 1.$$

If the characteristic roots α_j are all distinct then the integers r_{ij} are all equal to zero. An expression of the form

$$t^s\, e^{-\xi t}, \quad t > 0, \quad \xi > 0, \quad s > 0,$$

assumes its maximum when $t = \dfrac{s}{\xi}$, so that

$$(66.5) \qquad t^s\, e^{-\xi t} \leq \left(\frac{s}{\xi}\right)^s e^{-s}.$$

For $s = 0$ the maximum is equal to 1. Consequently, if the characteristic roots α_i are all distinct then (66.3) satisfies an estimate

$$|x(t)| \leq \gamma\, e^{-\eta t}\,|c|, \quad t \geq 0, \quad \gamma > 0.$$

If not all of the α_j are distinct we must formulate the assertion somewhat more carefully because of (66.5): For each $\varepsilon > 0$ there exists a constant $\gamma = \gamma(\varepsilon)$ such that

$$(66.6) \qquad |\boldsymbol{x}(t)| \leq \gamma(\varepsilon)\, e^{-(\eta-\varepsilon)t}\, |\boldsymbol{c}|, \qquad t \geq 0.$$

The constant $\gamma(\varepsilon)$ increases without bound as ε decreases. This follows from (66.5).

A similar estimate for a family of solutions depending on a parameter can also be given for the perturbed system

$$(66.7) \qquad \dot{\boldsymbol{x}} = A\boldsymbol{x} + \boldsymbol{f}(\boldsymbol{x}, t), \qquad \boldsymbol{x} \in K_h, \qquad t \geq t_0 \geq 0,$$

provided the function $\boldsymbol{f}(\boldsymbol{x}, t)$ remains small in a way to be explained later and provided the length of the parameter vector $\boldsymbol{c}$ is no larger than a certain constant depending on equation (66.7). The nonlinear function in (66.7) is assumed to satisfy: a) for fixed $\boldsymbol{x} \in K_h$, it is continuous and bounded for $t \geq t_0$, and b) with respect to $\boldsymbol{x}$ it satisfies a strong Lipschitz condition

$$(66.8) \qquad |\boldsymbol{f}(\boldsymbol{x}, t) - \boldsymbol{f}(\boldsymbol{y}, t)| \leq L\, |\boldsymbol{x} - \boldsymbol{y}|\, \max(|\boldsymbol{x}|^{\beta}, |\boldsymbol{y}|^{\beta}),$$

where L and β are fixed positive numbers. No generality is lost if the matrix A is assumed to have the form $\mathrm{diag}\,(A_1, A_2)$. Then there exists a family of solutions $\boldsymbol{x}(t, \boldsymbol{c})$ which depends on the real k-vector $\boldsymbol{c}$ (cf. (66.2)), such that for sufficiently small $\boldsymbol{c}$, $|\boldsymbol{c}| < \xi$, an estimate of type (66.6) obtains (cf. below, Theorem 66.1).

The proof is essentially based on the construction of the desired solution by means of successive approximations.[1] For this purpose it is desirable to rewrite the differential equation (66.7) as an integral equation. We introduce a Green's matrix

$$(66.9) \qquad G(s) := \begin{cases} \mathrm{diag}\,(e^{A_1 s}, 0), & s > 0, \\ \mathrm{diag}\,(0, -e^{-A_2 s}), & s < 0. \end{cases}$$

It is continuous for $s \neq 0$, has a jump at $s = 0$,

$$(66.10) \qquad G(s + 0) - G(s - 0) = E,$$

and satisfies the differential equations

$$(66.11) \qquad \frac{\partial G(t - u)}{\partial t} = A\, G(t - u), \qquad \frac{\partial G(t - u)}{\partial u} = -\, G(t - u)\, A,$$

for $t \neq u$. Because of the form of A as a partitioned matrix, $G(s)$ can be interchanged with A. By (66.4) the elements of $G(s)$ are composed of

---

[1] cf. ZUBOV [4].

terms $s'^{ij} e^{\alpha_i s}$ to within a constant and they can be estimated. We find:
For each $\delta > 0$ there exists a $\xi = \xi(\delta)$ such that

(66.12)
$$\| G(s) \| \leq \xi e^{-\eta s} e^{\delta s}, \quad s > 0$$
$$\| G(s) \| \leq \xi e^{\delta s}, \quad s < 0.$$

The first estimate corresponds precisely to the estimate (66.6). As in
that case, $\xi(\delta)$ can grow without bounds as δ decreases if there exist
multiple characteristic roots. The second estimate follows from the fact
that the characteristic roots $\alpha_{k+1}, \ldots, \alpha_n$ have nonnegative real parts by
hypothesis.

The integral equation

$$(66.13) \quad x(t) = G(t - t_0) c + \int_{t_0}^{\infty} G(t - u) f(x(u), u) \, du, \quad t \geq t_0,$$

is equivalent to equation (66.7) in the sense that each solution of the
integral equation also satisfies the differential equation and that each
solution of the differential equation which satisfies (66.6) also satisfies
the integral equation. This is seen as follows: We multiply the differential
equation by $G(t - u)$ and obtain

$$G(t - u) \dot{x}(u) - A G(t - u) x(u) = G(t - u) f(x(u), u).$$

Because of (66.11) the left side can be written as a derivative:

$$\frac{d}{du} G(t - u) x(u) = G(t - u) f(x(u), u).$$

We integrate from $u = t_0$ to $u = \infty$, but on the left we first integrate
from t_0 to t then from t to ∞, and we obtain

$$G(t - u) x(u) \Big|_{t_0}^{t} + G(t - u) x(u) \Big|_{t}^{\infty} = \int_{t_0}^{\infty} G(t - u) f(x(u), u) \, du.$$

The integral on the right converges because of (66.12) and (66.6). In
evaluating the left side we must observe (66.10); its value is

$$- G(t - t_0) x(t_0) + x(t) \big(G(0+) - G(0-) \big) = x(t) - G(t - t_0) c.$$

The solution $x(t)$ thus satisfies the integral equation. If we start, on the
other hand, from (66.13) and write the integral on the right as a sum of
the integrals from t_0 to t and from t to ∞, differentiate, and observe
(66.10) we obtain the differential equation.

The remainder of the argument is based on the integral equation. We
shall consider its right side as a nonlinear operator

$$(66.14) \quad H(x(t); c) := G(t - t_0) c + \int_{t_0}^{\infty} G(t - u) f(x(u), u) \, du.$$

It is defined on vectors $\boldsymbol{x} = \boldsymbol{x}(t)$ with continuous components which satisfy an estimate

$$(66.15) \qquad\qquad |\boldsymbol{x}(t)| \leq \gamma_0\, e^{-(\eta-\varepsilon)t}.$$

γ_0 and $\varepsilon < \eta$ are fixed positive numbers; η is the number in (66.6). In (66.14) t_0 is considered as a fixed number and $\boldsymbol{c}$ as a parameter vector whose length will be discussed later.

We first prove a lemma

Lemma 66.1. For each sufficiently small $\varepsilon > 0$ there exist two positive numbers γ_0 and $\varkappa$ such that the mapping $\tilde{\boldsymbol{x}} = H(\boldsymbol{x}; \boldsymbol{c})$ takes the domain characterized by (66.15) into itself provided only that

$$(66.16) \qquad\qquad |\boldsymbol{c}| < \varkappa;$$

i.e. the mapping is contracting.

Proof. Let ε be given, $0 < \varepsilon < \eta$. (66.14) yields

$$|H(\boldsymbol{x}; \boldsymbol{c})| \leq \|G(t - t_0)\|\, |\boldsymbol{c}| + \int_{t_0}^{\infty} \|G(t - u)\|\, |\boldsymbol{f}(\boldsymbol{x}(u), u)|\, du.$$

From (66.8) we obtain for $\boldsymbol{x} = \boldsymbol{x}(u)$, $\boldsymbol{y} = 0$,

$$|\boldsymbol{f}(\boldsymbol{x}(u), u)| \leq L\, |\boldsymbol{x}(u)|\, |\boldsymbol{x}(u)|^{\beta} \leq L\, \gamma_0^{1+\beta}\, e^{(1+\beta)(\varepsilon-\eta)t}.$$

To further estimate the integral we again decompose the interval into the subintervals (t_0, t) and (t, ∞). In the first interval we use the first inequality in (66.12) and in the second the second inequality; $\delta > 0$ is assumed given. The absolute value of the integral is smaller than

$$\xi(\delta)\, L\, \gamma_0^{1+\beta}\left[\int_{t_0}^{t} e^{\delta(t-u)-\eta(t-u)}\, e^{(1+\beta)(\varepsilon-\eta)u}\, du + \int_{t}^{\infty} e^{\delta(t-u)}\, e^{(1+\beta)(\varepsilon-\eta)u}\, du\right].$$

The first integral in the brackets has the value

$$I_1 = \frac{1}{p}\, e^{(\delta-\eta)t}\, (e^{pt} - e^{pt_0}),$$

where $p := \eta - \delta + (1 + \beta)(\varepsilon - \eta)$.

The value of the second integral is

$$I_2 = \frac{1}{\eta - p}\, e^{(\delta-\eta)t + pt}.$$

δ is chosen so that the two inequalities

$$0 < \delta < \varepsilon, \quad \delta > (1 + \beta)(\eta - \varepsilon) + \eta,$$

are satisfied. Then $p < 0$. Hence

$$I_1 + I_2 = e^{(\delta-\eta)t} \left[e^{pt} \left(\frac{1}{p} + \frac{1}{\eta - p} \right) - \frac{1}{p} e^{pt_0} \right]$$

$$= e^{(\delta-\eta)t} \left[- e^{pt} \frac{\eta}{|p|(\eta - p)} + \frac{1}{|p|} e^{pt_0} \right]$$

$$< e^{(\delta-\eta)t} \frac{e^{pt_0}}{|p|} < C e^{(\varepsilon-\eta)t}, \quad C := \frac{e^{pt_0}}{|p|}.$$

We use this inequality to obtain

$$\left| \int_{t_0}^{\infty} G(t - u) f(x(u), u) \, du \right| < L \gamma_0^{1+\beta} \xi(\delta) C e^{(\varepsilon-\eta)t}$$

and further

$$|H(x; c)| < (L C \gamma_0^{1+\beta} \xi(\delta) + \xi(\delta) |c|) e^{(\varepsilon-\eta)t}.$$

If we choose

$$(66.17) \qquad |c| < \varkappa := \gamma_0^{1+\beta},$$

we obtain

$$|H(x; c)| < \varkappa \, \xi(\delta) (L C + 1) e^{(\varepsilon-\eta)t}.$$

We now only have to choose γ_0 so small that, for example,

$$(66.18) \qquad \varkappa \, \xi(\delta) (L C + 1) < \frac{1}{2} \gamma_0,$$

whence

$$\gamma_0^{\beta} \, \xi(\delta) (L C + 1) < \frac{1}{2}.$$

Then, analogous to (66.15), we have

$$|H(x; c)| < \frac{1}{2} \gamma_0 \, e^{(\varepsilon-\eta)t},$$

and the lemma is established.

We now solve the integral equation (66.13) which can be written in the form

$$(66.19) \qquad H(x; c) = x$$

by successive approximation. For this purpose we form the sequence of functions

$$x^{(1)} = G(t - t_0) \, c,$$

$$x^{(i)} = H(x^{(i-1)}; c), \quad i = 2, 3, \ldots,$$

and the series

$$(66.20) \qquad x(t; c) := \sum_{i=1}^{\infty} (x^{(i+1)} - x^{(i)}).$$

To check the convergence of this series we introduce the abbreviations

$$y_i(u) := \max\left(|x^{(i)}(u)|^\beta, \ |x^{(i-1)}(u)|^\beta\right), \quad i = 2, \ldots,$$

$$z_i(t) := \max_{v \geq t}\left(|x^{(i)}(v) - x^{(i-1)}(v)|\right), \quad i = 2, \ldots,$$

$$z_1(t) := \max_{v \geq t}|x^{(1)}(v)|.$$

From (66.12) and (66.8) it appears that

$$z_1(t) \leq |c|\,\xi(\delta)\,e^{(\varepsilon-\eta)t}$$

and

$$(66.21) \qquad z_{i+1}(t) \leq \max_{v \geq t}\int_{t_0}^{\infty} \|G(v-u)\|\,L\,z_i(u)\,y_i(u)\,du, \quad i = 1, 2, \ldots.$$

We also have for all i,

$$y_i(u) < (\gamma_0\,e^{(\varepsilon-\eta)u})^\beta.$$

Proceeding by induction, we assume that there exists a q such that

$$(66.22) \qquad z_i(u) < |c|\,\xi(\delta)\,q^{i-1}\,e^{(\varepsilon-\eta)u}.$$

Then (66.21) implies

$$z_{i+1}(t) \leq \max_{v \geq t}\int_{t_0}^{\infty} \|G(v-u)\|\,L\,\xi(\delta)\,|c|\,q^{i-1}\,\gamma_0^\beta\,e^{(\varepsilon-\eta)(1+\beta)u}\,du$$

$$\leq L\,|c|\,\xi(\delta)\,q^{i-1}\,\gamma_0^\beta\,\max_{v \geq t}\int_{t_0}^{\infty} \|G(v-u)\|\,e^{(\varepsilon-\eta)(1+\beta)u}\,du$$

$$\leq L\,|c|\,\xi(\delta)\,q^{i-1}\,\gamma_0^\beta\,\xi(\delta)\,C\,e^{(\varepsilon-\eta)t}$$

and this in turn implies, analogous to (66.22),

$$z_{i+1}(t) < |c|\,\xi(\delta)\,q^i\,e^{(\varepsilon-\eta)t},$$

if we set

$$q = L\,\gamma_0^\beta\,\xi(\delta)\,C.$$

By (66.18) this number is smaller than 1 and the uniform convergence of the series (66.20) follows.

In order to show that (66.20) is a solution of the integral equation we consider the difference $x - H(x; c)$. Then

$$x - H(x; c) = x - x^{(i+1)} + H(x^{(i)}; c) - H(x, c).$$

The norm of the right side is no larger than

$$|x - x^{(i+1)}| + |H(x^{(i)}; c) - H(x; c)|$$

$$= |x - x^{(i+1)}| + \left|\int_{t_0}^{\infty} G(t-u)\,(f(x^{(i)}(u), u) - f(x(u), u))\,du\right|$$

$$\leq |x - x^{(i+1)}| + \left|\int_{t_0}^{T} G(t-u)\,(f(x^{(i)}(u), u) - f(x(u), u))\,du\right|$$

$$+ \left|\int_{T}^{\infty} G(t-u)\,f(x^{(i)}(u), u)\,du \ + \int_{T}^{\infty} G(t-u)\,f(x(u), u)\,du\right|.$$

The number T is to be chosen so large that the last two integrals become smaller than a preassigned arbitrarily small number $\frac{\alpha}{4}$ in absolute value. This is possible because of the uuiform convergence in t and i of the integral. Next, i is chosen so large that the first two summands each become smaller than $\frac{\alpha}{4}$. It follows that

$$|x - H(x; c)| < \alpha.$$

Since the difference is independent of i and α is arbitrary we see that $x = H(x; c)$. The series (66.20) is therefore a solution of the integral equation. We show that this solution is unique. If there were two different solutions x and $\bar{x}$ then we would have

$$|x(t) - \bar{x}(t)| \leq \left| \int_{t_0}^{\infty} G(t - u) \left(f(x(u), u) - f(\bar{x}(u), u)\right) du \right.$$

$$\leq \gamma_0^\beta L \int_{t_0}^{\infty} \|G(t - u)\| \, e^{\beta(\varepsilon - \eta)u} \, |x(u) - \bar{x}(u)| \, du.$$

We set

$$m := \sup_{t > t_0} |x(t) - \bar{x}(t)|$$

and it follows that

$$m \leq \gamma_0^\beta L m \sup_{t \geq t_0} \int_{t_0}^{\infty} \|G(t - u)\| \, e^{\beta(\varepsilon - \eta)u} \, du < m C L \, \gamma_0^\beta \, \xi(\delta).$$

This is a contradiction for $m \neq 0$ because

$$L \, \xi(\delta) \, \gamma_0^\beta \, C < 1.$$

We have thus completely solved the integral equation. If we now observe the connection between (66.13) and (66.7) we obtain
Theorem 66.2. Let

$$\dot{x} = A x + f(x, t)$$

be given. Suppose that the matrix A has exactly k (not necessarily distinct) characteristic roots $\alpha_1, \ldots, \alpha_k$, such that $\operatorname{Re} \alpha_i \leq -\eta < 0$. Let the function $f(x, t)$ be continuous and bounded with respect to t and satisfy a Lipschitz condition

$$|f(x, t) - f(y, t)| \leq L \, |x - y| \, \max \left(|x|^\beta, |y|^\beta\right), \quad \beta > 0,$$

for $t \geq t_0$, $x \in K_h$, $y \in K_h$. Then there exists a k-parameter family of solutions $x(t, c)$, $c = \operatorname{col}(c_1, \ldots, c_k)$, with the following property: For each ε there exists a $\gamma(\varepsilon)$ such that for sufficiently small $|c|$

$$|x(t, c)| \leq \gamma(\varepsilon) \, |c| \, e^{-(\eta - \varepsilon)t}, \quad t > t_0.$$

This theorem can immediately be extended to nonautonomous systems

$$\dot{x} = A(t)\,x + f(x, t)$$

provided the corresponding homogeneous equation is reducible (*cf.* sec. 61). For, a linear transformation with bounded coefficients transforms the function $f(x, t)$ into a bounded function which again satisfies the strong Lipschitz condition. However, the reducibility is not an essential assumption here. Essential is the estimate (66.12) of the Green's matrix, and this estimate can be made analogously in the nonautonomous case, provided the equation which corresponds to (66.1) is regular (*cf.* sec. 64). The number $-\eta$ must here be replaced by the largest negative order number. Lemma 66.1, which makes the successive approximations possible, can be proved even in the nonregular case. Therefore Theorem 66.2 can be stated quite generally but with an estimate for the k-parameter family which is not as good as (66.6) (*cf.* also ZUBOV [4]). Theorem 66.2 implies a theorem on stability in the first approximation. If all the characteristic roots of A have negative real parts then there exists an n-parameter family of solutions satisfying (66.6). In fact, in a sufficiently small neighborhood of the origin this is the general solution since no more than n parameters may occur. The matrix A_2 in (66.9), in the definition of G, does not appear in this case. The limits of integration in (66.13) are t_0 and t only, and c is the initial value for $t = t_0$. This stability theorem is not quite as "good" as Theorem 56.2, since there we only require that $|f(x,t)| = o(|x|)$ whereas here we need the strong Lipschitz condition. On the other hand, for the present theorem the generalization to nonautonomous systems involves weaker assumptions on the linear part of the equation. (*cf.* also sec. 65.)

67. The Liapunov Expansion Theorem

The proof of Theorem 66.2 depends on successive approximations and is accordingly a constructive proof which permits us to compute the solutions as closely as desired. If the assumptions on $f(x, t)$ are strengthened it is possible to give the family of solutions $x(t, c)$ of Theorem 66.2 in closed form so that its dependence on the parameters becomes obvious. In order to state both the hypotheses and the result effectively we introduce the symbol

$$(m) = (m_1, \ldots, m_k), \qquad m = m_1 + m_2 + \ldots + m_k; \qquad m_i \geq 0,$$

$$1 \leq k \leq m,$$

for a decomposition of the integer m into k nonnegative summands, as well as the symbol

(67.1) $$x_{(m)} := x_1^{m_1} x_2^{m_2} \cdots x_k^{m_k}$$

Furthermore the symbol $p^{(m)}(t)$ denotes a vector which depends on the decomposition (m) and has n continuous and bounded components. The function $f(x, t)$ is assumed to be analytic in a certain neighborhood K_h of the origin and representable as a power series

$$(67.2) \qquad f(x,t) = \sum_m \sum_{(m)} p^{(m)}(t) \, x_{(m)}, \qquad m = 2, 3, \ldots,$$

without linear or constant terms. The inner sum runs over all possible decompositions (m); the $p^{(m)}(t)$ are uniformly bounded. Then we have

Theorem 67.1. Let the differential equation (66.7) be given. Let the function $f(x, t)$ be of the form (67.2). Then the family of solutions in Theorem 66.2 can be represented in the form

$$(67.3) \qquad x(t, c) = \sum_{r=1}^{\infty} \sum_{(r)} q^{(r)}(t) \, c_{(r)} \, e^{(\varkappa_1 r_1 - \cdots - \varkappa_k r_k)t}.$$

The symbol $c_{(r)}$ has a definition analogous to (67.1), and (r) runs over all decompositions of r into nonnegative summands. The coefficients $q^{(r)}(t)$ are either bounded or increase like powers of t; in either case, they have nonpositive order numbers.

The validity of the theorem is seen a follows. For $|c| \le c_0$ the series obtained by successive approximations and representing $x(t, c)$, can be majorized by a convergent series. This, after all, is the basis for the entire construction. Therefore the function $x(t, c)$ is analytic in c and a representation by a series of the form

$$x(t, c) = \sum_{r=1}^{\infty} \sum_{(r)} k^{(r)}(t) \, c_{(r)}, \qquad |c| \le c_0,$$

clearly exists. The inner summands

$$g_r(t, c) := \sum_{(r)} k^{(r)}(t) \, c_{(r)}$$

are homogeneous in c of degree r. The successive approximations permit us to derive defining equations for these expressions. For we have

$$x(t, c) = \sum_r g_r(t, c),$$

and from (66.13) it follows that

$$\sum_{r=1}^{\infty} g_r(t, c) = G(t - t_0) \, c + \int_{t_0}^{\infty} G(t - u) f\left(\sum_{s=1}^{\infty} g_s(u, c), u \right) du.$$

We apply (67.2) and expand the integrand in a series whose terms are homogeneous in c of degrees $r = 1, 2, \ldots$, and we compare coefficients. First we obtain

$$g_1(t, c) = \mathrm{col} \, (e^{A_1(t - t_0)} \, c, 0).$$

The components of this vector are of the form $p_i(t)\, e^{\alpha_i(t-t_0)}$. The coefficients $p_i(t)$ are either bounded or powers of t. Then we obtain a representation

$$g_2(t,\, c) = \int\limits_{t_0}^{\infty} G(t - u)\, h_2(t,\, c)\, du,$$

where in the computation of the function $h_2(t,\, c)$ we must combine the second degree terms of the series. These arise by multiplying two components each of $g_1(t,\, c)$. When we integrate, only integrals of the type

$$\int\limits_{t_0}^{t} e^{\alpha\mu(t-u)}\, p(u)\, e^{(\lambda_i + \lambda_j)(u-t_0)}\, du, \qquad \text{resp.}$$

(67.4)

$$\int\limits_{t}^{\infty} e^{\beta\nu(t-u)}\, q(u)\, e^{(\lambda_i + \lambda_j)(u-t_0)}\, du,$$

appear because of the special form of the Green's matrix (66.9). $p(t)$ and $q(t)$ are functions which grow at most as fast as powers of t, and $\beta_1, \ldots, \beta_{n-k}$ are the characteristic roots of A_2; these have nonnegative real parts. In both cases the integration yields an expression of the form

$$r(t)\, e^{(\alpha_i + \alpha_j)t}, \qquad r(t) \ \text{bounded},$$

(We need not emphasize the dependence on the number t_0 which was fixed in the very beginning). The second degree terms which make up $g_2(t,\, c)$ are thus of the form

$$c_{(2)}\, e^{(\alpha_i + \alpha_j)t}\, P(t),$$

where $P(t)$ denotes a polynomial in t. The next equation is

$$g_3(t,\, c) = \int\limits_{t_0}^{\infty} G(t - u)\, h_3(u,\, c)\, du,$$

in which h_3 is formed using $g_1(t,\, c)$ and $g_2(t,\, c)$, etc. The sth integration leads to expressions of the form:

"$c_{(s)}$ times a polynomial in t times an exponential function with s summands $\alpha_i(t)$ in the exponent".

If equation (66.7) is autonomous and if the numbers $\alpha_1, \ldots, \alpha_k$ are all distinct then the coefficients $p^{(m)}$ in (67.2) are constant and $\exp(A_1 t)$ is a diagonal matrix whose elements are pure exponential functions. The integrations in (67.4) then again furnish pure exponential functions if we exclude the possibility that

$$\alpha_\mu = \alpha_i + \alpha_j.$$

A corresponding statement is true for the calculation of the further terms. We obtain

Theorem 67.2. Let the differential equation (66.7) be autonomous and let the function $f(x)$ be of the form (67.2) with constant $p^{(m)}$. Let the

22*

numbers $\alpha_1, \ldots, \alpha_k$ be all distinct and suppose that there exist no linear relations

$$\alpha_i = \sum_{j \neq i} m_j \alpha_j$$

for nonnegative integers m_j, $\sum m_j > 1$. Then the coefficients $q^{(r)}$ in (67.3) are constant.

If equation (66.7) is periodic then the coefficients in (67.3) are also periodic if 1) all the characteristic exponents $\varrho_1, \ldots, \varrho_k$ of the linear part which have absolute value less than 1 are distinct, and 2) these numbers do not satisfy a relation of the form

$$\sum_{j \neq i} m_j \varrho_j - \frac{2 \pi i}{\omega} N = \varrho_\mu, \qquad N \text{ an integer.}$$

Theorem 67.1 is called the *Expansion Theorem of Liapunov*. It is the core of the *first method of Liapunov* whose purpose is to make stability assertions on the basis of the explicit form of the solutions. The original proof by Liapunov differs somewhat (*cf.* for instance LEFSCHETZ [1]) from the proof given here which is due to ZUBOV [4].

Because of (66.9), the vector relation

$$\boldsymbol{x}(t_0, \, \boldsymbol{c}) = \mathrm{col}(\boldsymbol{c}, \, 0) - \int_{t_\bullet}^{\infty} \mathrm{diag}\,(0, \, e^{A_2(t_0 - u)})\,\mathrm{col}(0, \, \ldots, \, 0, f_{k+1}, \, \ldots, \, f_n)\,du$$

(67.5)

follows from the integral equation (66.13),

$$\boldsymbol{x}(t, \, \boldsymbol{c}) = G(t, \, t_0)\,\boldsymbol{c} + \int_{t_0}^{t} G(t, \, u)\,\boldsymbol{f}\big(\boldsymbol{x}(u), \, u\big)\,du$$

$$+ \int_{t}^{\infty} G(t, \, u)\,\boldsymbol{f}\big(\boldsymbol{x}(u), \, u\big)\,du,$$

for $t \to t_0 + 0$. If we introduce

$$\boldsymbol{x}_0 = \mathrm{col}(x_{01}, x_{02}, \, \ldots, \, x_{0n}) : = \boldsymbol{x}(t_0, \, \boldsymbol{c})$$

and denote the components of the integral term in (67.5) by $g_{k+1}, g_{k+2}, \ldots$ $\ldots, g_n$, relation (67.5) can be written as

$$(67.6) \qquad\qquad x_{0i} = c_i, \qquad i = 1, 2, \ldots, k,$$

$$(67.7) \qquad\qquad x_{0r} = g_r(\boldsymbol{x}_0), \qquad r = k + 1, \ldots, n,$$

since the quantities g_r depend on the components of $\boldsymbol{x}_0$, *i.e.* of the parameters $c_1, \ldots, c_k$. The functions g_r are at least of degree 2 with respect to the c_i. Equations (67.6), (67.7) say that the parameters $c_1, \ldots, c_k$ can be expressed in terms of the initial values $x_{01}, \ldots, x_{0n}$ provided that the relations (67.7) exist between these initial values. The expansion theorem yields: The general solution

$$\boldsymbol{p}\,(t,\ \boldsymbol{x}_0,\ t_0)$$

of the differential equation (66.7) tends to zero in case $|\boldsymbol{x}_0|$ is sufficiently small and the components of $\boldsymbol{x}_0$ satisfy the relations (67.7). In such a case we speak of *conditional* (asymptotic) stability (*cf.* p. 267). It is conditional because of the relation (67.7). If $k = n$ then we have unconditional asymptotic stability (*cf.* also sec. 55).

Chapter X

The Critical Cases for Differential Equations

68. General Remarks Concerning Critical Cases; Subsidiary Results

In secs. 13 and 43 we defined the term *critical behavior*. For a differential equation

$$(68.1) \qquad \dot{x} = A(t)\,x + g(x, t), \quad g(x, t) = O(|x|^2),$$

we have a *critical case* if the stability of the equilibrium is significantly influenced by the terms of higher order and cannot be discussed by means of the reduced equation

$$(68.2) \qquad \dot{x} = A(t)\,x.$$

For autonomous equations the idea can be positively stated: The critical case is given if the matrix A has characteristic roots with negative as well as zero real parts but none with positive real parts. For periodic equations a similar characterization can be made with the aid of the characteristic exponents.

Both times the characterization of the noncritical case amounts to: Either all solutions decrease like exponential functions $e^{-\alpha(t-t_0)}$, $\alpha > 0$, or there exists at least one solution which increases like $e^{\beta(t-t_0)}$, $\beta > 0$. This statement appears somewhat unsymmetric: The "either" part of the theorem appears to be a statement about all solutions but the "or" part only about one. For the purpose of investigating the stability of the equilibrium the distinction is adequate. If one wishes to study the growth behavior then the concept of *dichotomy* of the phase space introduced by MASSERA and SCHÄFFER [1] is useful. A linear subspace U of the phase space, of R_n in the case of the equation (68.2), generates an *exponential dichotomy* (of R_n) if there exist positive constants a, α, b, β with the following properties:

$$|p(t, x_0, t_0)| \leq a e^{-\alpha(t-t_0)} \text{ for } x_0 \in U,$$

$$|p(t, x_0, t_0)| \geq b e^{\beta(t-t_0)} \quad \text{for } x_0 \notin U.$$

For an autonomous equation with critical characteristic roots no such subspace U exists. Massera and Schäffer extended the definition of an exponential dichotomy to differential equations in a general normed linear space[1]) and have studied the effect of the existence of a dichotomy-generating subspace on the solutions of the perturbed equation (68.1). For example, the following is true: If there exists a subspace U which generates an exponential dichotomy then there exists a subspace U_1 such that each solution of (68.1), for $x_0 \in U_1$, exists for $t \geq t_0$ and is bounded. Theorems of this nature, however, lead us beyond the topic of critical cases in the narrow sense.

The general theory of the critical cases is quite involved and at this time the investigations have by no means been completed. The most comprehensive result is due to MALKIN [3]. The two principal theorems proved by him permit the treatment of a great number of special cases; many of them had their origin in concrete problems. We shall not give complete proofs of the theorems of Malkin but will only work out the leading ideas which essentially go back to Liapunov.

The first step consists in extending the concept of stability in the first approximation. The requirement that the stability behavior depends exclusively on the first degree terms and is entirely independent of the terms of second and higher degree is replaced in a natural manner by the requirement that the stability depends only on terms whose degree is less than or equal to N. This presupposes, of course, that the right side of the differential equation admits a Taylor expansion up to and including the Nth degree. Usually it is even assumed that the right side is analytic.

Accordingly, let y be a p-vector which satisfies an equation

$$(68.3) \qquad \dot{y} = g_m(y, t) + g_{m+1}(y, t) + \cdots + g_N(y, t) + \gamma(y, t)$$

for $|y| \leq h, t \geq t_0 \geq 0$. The components of $g_\mu(y, t), \mu = m, m+1, \ldots, N$, are homogeneous polynomials of degree μ with continuous and bounded coefficients. The degree of the term $\gamma(y, t)$ which is to be considered as variable, is larger than N.

Def. 68.1. The equilibrium $y = 0$ is called *stable in the Nth approximation* if for each $\varepsilon > 0$ there exists a $\delta > 0$ such that the general solution satisfies the inequality

$$|p(t, y_0, t_0)| < \varepsilon \ \text{ for } \ t \geq t_0,$$

whenever $|y_0| < \delta$. The term $\gamma(y, t)$ is arbitrary as long as it satisfies an estimate

$$(68.4) \qquad |\gamma(y, t)| < a|y|^{N+1}, \ a > 0.$$

The number δ is allowed to depend on ε and on the bound a; the definition implies uniformity with respect to t_0. If the equilibrium is stable in

---

[1]) An extension to difference equations is found in PANOV [1].

the Nth approximation and if $p(t, y_0, t_0)$ tends to zero in case y_0 belongs to a certain ball $|y_0| < \eta$, then the equilibrium is called *asymptotically stable in the Nth approximation*.

Def. 68.2. The equilibrium is called *unstable in the Nth approximation* if under the hypothesis of Def. 68.1 there exists an $\varepsilon_0 > 0$ and in each neighborhood of the origin there exist initial values y_0, such that $|p(t, y_0, t_0)|$ attains the value ε_0 on a finite time interval for each choice of $\gamma(y, t)$ for which (68.4) is valid. The number ε_0 is allowed to depend on a.

Theorem 68.1. The equilibrium of a scalar differential equation

$$(68.5) \qquad \dot{y} = g\,y^m + \gamma(y, t)\,, \quad g \text{ constant, } m \geq 1 \text{ an integer,}$$

$$|\gamma(y, t)| = o(|y|^m)$$

is asymptotically stable in the mth approximation if m is odd and $g < 0$. It is unstable in the mth approximation if m is odd and $g > 0$ or if m is even.

Proof. For an odd m we choose the Liapunov function $v = y^2$ whose derivative

$$\dot{v} = 2g\,y^{m+1} + 2y\,\gamma(y, t)$$

is positive or negative definite, depending on the sign of g. If m is even we use the indefinite function $v = -gy$ and its negative definite derivative $\dot{v} = -g^2 y^m - g\gamma(y, t)$ and apply Theorem 25.4.∎

For systems of two or more scalar equations no simple criteria for stability in the Nth approximation are known.

We now prove a generalization of Theorem 59.2 which is important for the study of critical cases.

Theorem 68.2. Let the equilibrium of the differential equation

$$(68.6) \qquad \dot{z} = f(z, t), \quad z \in R_n, \quad f \in C_0, \quad |z| \leq h, \quad t \geq t_0 \geq 0\,,$$

be uniformly asymptotically stable and let $g(t)$ be a continuous bounded n-vector, $|g(t)| < \varkappa$ for $t \geq t_0$, $\lim\limits_{t \to \infty} g(t) = 0$. Then the general solution of the differential equation

$$(68.7) \qquad\qquad\qquad \dot{z} = f(z, t) + g(t)$$

tends to zero with increasing t for sufficiently small initial values z_0 and sufficiently large values t_0.

Proof. We first consider the scalar differential equation

$$(68.8) \qquad \dot{y} = -\psi(y) + h(t), \quad \psi \in K, \quad h(t) \text{ bounded, } h(t) \underset{t \to \infty}{\to} 0,$$

for nonnegative values of y. For suitable initial values, $\dot{y}(t_0) < 0$ and y decreases. If for $t \geq t_0$, $\dot{y}$ is permanently negative then $\lim y = 0$ since the equation can not have an equilibrium different from zero. The other possibility is $\limsup \dot{y} \geq 0$ and accordingly $\limsup y =: b > 0$. Since $h(t)$ is bounded, b is finite. There exists a sequence $t_n \to \infty$, such that $y(t_n)$ differs from b by an arbitrarily small amount and such that $\dot{y}(t_n) \geq 0$. On the other hand $\dot{y}(t_n)$ tends to $-\psi(b)$ because $h(t_n) \to 0$, and under the assumption that $b > 0$ we reach a contradiction. Hence $y(t)$ tends to zero.

In the general case we apply Theorem 51.2 or 51.4 and choose for (68.6) a Liapunov function $v(z, t)$ which satisfies the following estimates:

$$\varphi_1(|z|) \leq v(z, t) \leq \varphi_2(|z|), \quad \varphi_1 \in K, \quad \varphi_2 \in K,$$

$$\sum_{i=1}^{n} \frac{\partial v}{\partial z_i} f_i(z, t) + \frac{\partial v}{\partial t} \leq -\varphi_3(|z|), \quad \varphi_3 \in K, \quad \sum_{i=1}^{n} \left| \frac{\partial v}{\partial z_i} \right| \leq c, \quad i = 1, \ldots, n.$$

Let $w(t)$ denote the function $v(z(t), t)$ along a solution of (68.7) so that

$$\dot{w} = \sum_{i=1}^{n} \frac{\partial v}{\partial z_i} f_i(z, t) + \frac{\partial v}{\partial t} + \sum_{i=1}^{n} \frac{\partial v}{\partial z_i} g_i(t)$$

and

$$\dot{w} \leq -\varphi_3(|z|) + \sqrt{n}\, c\, |g(t)| \leq -\varphi_3(\varphi_2^I(w)) + \sqrt{n}\, c\, |g(t)|.$$

With this differential inequality we can associate a differential equation of type (68.8) and conclude that $w(t)$ tends to zero. This implies $\lim z(t) = 0$.

If we omit the hypothesis $\lim g(t) = 0$ and only maintain that $|g(t)| \leq \varkappa$ then the inequality

$$\dot{w} \leq -\varphi_3(|z|) + \sqrt{n}\, c\, \varkappa,$$

implies that $w(t)$ decreases at least as long as $\varphi_3(|z|) > \sqrt{n}\, c\, \varkappa$. This furnishes the bound

$$|z(t)| \leq \varphi_1^I(v(z, t)) \leq \varphi_1^I(v(z(t_0), t_0)).$$

We cannot do without the uniformity of the asymptotic stability. This is shown by the equation

$$\dot{x} = -\frac{x}{1+t} + \frac{1}{1+t},$$

whose solution

$$x = 1 - \frac{1+t_0}{1+t}(1 - x_0)$$

does not tend to zero.

69. The Principal Theorems of Malkin

We first consider an equation of a special form. Let

$$y = \mathrm{col}(y_1, \ldots, y_k), \quad z = \mathrm{col}(z_1, \ldots, z_m), \quad x = \mathrm{col}(y, z).$$

The vector x satisfies a differential equation

$$(69.1) \qquad \dot{x} = f(x, t) = \mathrm{col}\left(p(y, z, t),\ q(y, z, t)\right),$$

$$|y| \leq h,\ |z| \leq h,\ t \geq t_0.$$

The k-vector $p(y, z, t)$ has the form

$$_2p(y, t) + {}_3p(y, t) + \cdots + {}_Np(y, t) + r(y, z, t).$$

The components of $_sp(y, t)$ are homogeneous polynomials in y of degree s with bounded coefficients. The components of $r(y, z, t)$, which may also depend on z, are at least of degree $N + 1$ with respect to y. The m-vector q has the form

$$q(y, z, t) = Q(t)\, z + {}_1q(y, t) + {}_2q(y, z, t);$$

$_1q$ and $_2q$ are at least of degree 2 in y, resp. in y and z. We further assume that the equilibrium of the differential equation

$$(69.2) \qquad \dot{y} = {}_2p(y, t) + \cdots + {}_Np(y, t),\ |y| \leq h,\ t \geq t_0,$$

is stable, asymptotically stable, or unstable, in the Nth approximation and that

$$(69.3) \qquad \dot{z} = Q(t)\, z$$

has a uniformly asymptotically stable, *i.e.* an exponentially stable equilibrium. Under these assumptions the equilibrium of (69.1) has the same stability behavior in the usual sense as the equilibrium of (69.2) has in the Nth approximation.

Proof. We replace the vector z in $r(y, z, t)$ by an arbitrary vector $\varphi(t)$ such that $|\varphi(t)| \leq h$. Then

$$(69.4) \qquad |r(y, \varphi, t)| \leq c\, |y|^{N+1}$$

and the constant c does not depend on φ. In turn we replace the vector y in $q(y, z, t)$ by an arbitrary vector $\psi(t)$ such that $|\psi(t)| < \eta$. The vector $_2q(\psi, z, t)$ satisfies an estimate

$$|{}_2q(\psi, z, t)| < a\, |z|^2,$$

where $a = a(\eta)$ depends on η and becomes arbitrarily small as $\eta \to 0$. By Theorem 65.1 the equilibrium of the equation

$$(69.5) \qquad \dot{z} = Q(t)\, z + {}_2q(\psi(t), z, t)$$

remains exponentially stable for sufficiently small values of a, *i.e.* of η. The vector $_1q(y, t)$ becomes smaller than an arbitrary given number $\varkappa$ if η is chosen sufficiently small:

$$|{}_1q(y, t)| < \varkappa\ \text{ for }\ |y| < \eta.$$

We first assume that the equilibrium of (69.2) is stable in the Nth approximation and we choose for a given ε the number $\delta = \delta(\varepsilon, c)$ so that

$$|y(t)| < \varepsilon \quad \text{for} \quad t \geq t_0$$

in case $|y_0| < \delta$ and (69.4) holds. For the equation (69.5) we construct the Liapunov function $v(z, t)$ of the proof of Theorem 68.2. For this purpose the number ε has to be so small that the equilibrium of (69.5) is exponentially stable for $|\psi(t)| < \varepsilon$. The initial value z_0 of z is chosen so small that

$$q_1^I\big(v(z_0), t_0)\big) < \varepsilon$$

and simultaneously

$$q_3\big(|z_0|\big) < c_4 \varkappa.$$

φ_1, φ_3 are defined in Theorem 68.2 and $\varkappa$ is defined above. The remark following the proof of Theorem 68.2 shows that always $|z(t)| < \varepsilon$ and hence also $|y(t)| < \varepsilon$. The equilibrium of (69.1) is thus stable.

If the equilibrium of (69.2) is asymptotically stable in the Nth approximation then $y(t)$ and $_1q(y, t)$ tend to zero. Then Theorem 68.2 applies directly. $x(t) \to 0$ follows and hence the asymptotic stability for (69.1).

Finally let the equilibrium of (69.2) be unstable in the Nth approximation. If the equilibrium of (69.1) is stable with respect to the variable z (secs. 55, 67) we can arrange that $|z(t)|$ remains arbitrarily small by properly choosing the initial values z_0 and y_0. We consider the solutions $z_1(t), \ldots, z_m(t)$ substituted as arbitrary functions of time in $r(y, z, t)$ and we utilize the assumed instability in the Nth approximation: There exist arbitrarily small values y_0 such that the solution of

$$\dot{y}(t) = p\big(y(t), z(t), t\big)$$

becomes equal to ε in norm at some time. But then this is also valid for $x = \mathrm{col}\,(y(t), z(t))$; i.e. the equilibrium of (69.1) is unstable. The proof is now complete•

The result may be formulated as follows. The behavior of the equilibrium of (69.1) is completely determined by the behavior of the equilibrium of (69.2) in the Nth approximation, and equation (69.2) is equivalent, with respect to stability, to the reduced equation

$$\dot{y} = p(y, 0, t),$$

which differs from (69.2) only by the terms of degree $N + 1$.

Equation (69.1) has a rather special form and the question arises how in the critical case the general equation can be transformed into the special form by a suitable transformation of variables. We illustrate this procedure with the equations

$$(69.6) \qquad \dot{y} = R(t)\,z + p(y, z, t),$$

$$(69.7) \qquad \dot{z} = Q(t)\,z + q(y, z, t),$$

which satisfy the following hypotheses. 1) Equation (69.3) has a uniformly asymptotically stable equilibrium. 2) The nonlinear terms $\boldsymbol{p}(\boldsymbol{y}, \boldsymbol{z}, t)$ and $\boldsymbol{q}(\boldsymbol{y}, \boldsymbol{z}, t)$ are analytic functions of $\boldsymbol{y}$ and $\boldsymbol{z}$ defined for $|\boldsymbol{y}| \leq h$, $|\boldsymbol{z}| \leq h$, and bounded for $t \geq t_0$, whose expansion as a power series in the respective variables begins with terms of at least second degree. 3) The expansion of $\boldsymbol{q}(\boldsymbol{y}, 0, t)$ begins with terms of at least $(N^* + 1)$ st degree; the number $N^* \geq 2$ will be defined later (because of the weakening of 3), see below). We introduce a new variable $\boldsymbol{\eta}$ by setting

$$(69.8) \qquad \boldsymbol{y} = \boldsymbol{\eta} + \boldsymbol{u}(\boldsymbol{z}, t).$$

$\boldsymbol{u} = \operatorname{col}(u_1, \ldots, u_k)$ is an unknown vector such that $\boldsymbol{u}(0, t) = 0$. $\boldsymbol{u}$ is to be determined in such a way that the equation in the new variable $\boldsymbol{\eta}$ obtained from (69.6) satisfies the hypothesis for (69.2). N^* is set equal to the number N defined by the properties of that equation. Then we have

$$\dot{\boldsymbol{y}} = \dot{\boldsymbol{\eta}} + \frac{\partial \boldsymbol{u}}{\partial \boldsymbol{z}}\, \dot{\boldsymbol{z}} + \frac{\partial \boldsymbol{u}}{\partial t}$$

and it follows from (69.6) and (69.7) that

(69.9)

$$\dot{\boldsymbol{\eta}} + \frac{\partial \boldsymbol{u}}{\partial \boldsymbol{z}}\, Q(t)\, \boldsymbol{z} + \frac{\partial \boldsymbol{u}}{\partial \boldsymbol{z}}\, \boldsymbol{q}(\boldsymbol{\eta} + \boldsymbol{u}, \boldsymbol{z}, t) + \frac{\partial \boldsymbol{u}}{\partial t} = R(t)\, \boldsymbol{z} + \boldsymbol{p}(\boldsymbol{\eta} + \boldsymbol{u}, \boldsymbol{z}, t).$$

If $\boldsymbol{u}$ is a solution of the partial differential equation

$$(69.10) \qquad \frac{\partial \boldsymbol{u}}{\partial t} + \frac{\partial \boldsymbol{u}}{\partial \boldsymbol{z}}\left(Q(t)\, \boldsymbol{z} + \boldsymbol{q}(\boldsymbol{u}, \boldsymbol{z}, t)\right) = R(t)\, \boldsymbol{z} + \boldsymbol{p}(\boldsymbol{u}, \boldsymbol{z}, t),$$

then all the terms of (69.9) which are independent of $\boldsymbol{\eta}$ vanish. We must solve equation (69.10). For this purpose we first set quite formally

$$(69.11) \qquad \boldsymbol{u}(\boldsymbol{z}, t) = {}_1\boldsymbol{u}(\boldsymbol{z}, t) + {}_2\boldsymbol{u}(\boldsymbol{z}, t) + \cdots.$$

${}_p\boldsymbol{u}(\boldsymbol{z}, t)$ is a homogeneous polynomial in $\boldsymbol{z}$ of degree p with bounded coefficients. These polynomials satisfy the equations

$$\frac{\partial}{\partial t}\, {}_1\boldsymbol{u}(\boldsymbol{z}, t) + \frac{\partial}{\partial \boldsymbol{z}}\, {}_1\boldsymbol{u}(\boldsymbol{z}, t)\, Q(t)\, \boldsymbol{z} = R(t)\, \boldsymbol{z},$$

$$\frac{\partial}{\partial t}\, {}_2\boldsymbol{u}(\boldsymbol{z}, t) + \frac{\partial}{\partial \boldsymbol{z}}\, {}_2\boldsymbol{u}(\boldsymbol{z}, t)\, Q(t)\, \boldsymbol{z} = -\, {}_2U(\boldsymbol{z}, t),$$

$$\cdots \cdots \cdots \cdots \cdots \cdots \cdots \cdots \cdots \cdots \cdots \cdots$$

$$\frac{\partial}{\partial t}\, {}_p\boldsymbol{u}(\boldsymbol{z}, t) + \frac{\partial}{\partial \boldsymbol{z}}\, {}_p\boldsymbol{u}(\boldsymbol{z}, t)\, Q(t)\, \boldsymbol{z} = -\, {}_pU(\boldsymbol{z}, t).$$

The right side of the pth equation depends on ${}_1\boldsymbol{u}, \ldots, {}_{p-1}\boldsymbol{u}$ and is known if the first $p - 1$ equations have been solved. The solution of the pth equation can be written in closed form: If $K(t, t_0)\, \boldsymbol{z}_0$ is the general solution of

$$\dot{\boldsymbol{z}} = Q(t)\, \boldsymbol{z},$$

then

$$_p\boldsymbol{u}(\boldsymbol{z}, t) = \int\limits_t^\infty {}_p\boldsymbol{U}\left(K(\tau, t)\,\boldsymbol{z}, \tau\right) d\tau .$$

Because of the assumed exponential stability for (69.3) the integral converges. We must still show that the series (69.11) converges and actually is a solution of the partial differential equation. MALKIN [1] showed this; another proof is given in LEFSCHETZ [1].

In the new equation for $\boldsymbol{\eta}$, therefore, all the terms on the right depend on $\boldsymbol{\eta}$. By a further transformation of variables

$$(69.12) \qquad \boldsymbol{\eta} = \boldsymbol{\theta} + \sum \eta_1^{m_1} \cdots \eta_k^{m_k}\, v^{(m_1,\ldots,m_k)}(\boldsymbol{z}, t),$$

$$m_1 + m_2 + \cdots m_k = 1,$$

we produce an equation in the variable $\boldsymbol{\theta}$, whose right side is free of all terms which depend on $\boldsymbol{z}$ and which involve $\boldsymbol{\theta}$ to the first power. Repetition of the transformation with $m_1 + \cdots + m_k = 2, 3, \ldots$ eventually leads to the desired special form. At each step we must apply the existence theorem for partial differential equations which we mentioned.

Since the expressions $\boldsymbol{q}(\boldsymbol{y}, 0, t)$ begin with terms of at least $(N + 1)$st degree the transformations (69.8), resp. (69.12), leave the terms which are independent of $\boldsymbol{z}$ and of degree less than or equal to N unchanged. This is seen as follows. After the first transformation, (69.6) has become

$$\dot{\boldsymbol{\eta}} = \boldsymbol{g}(\boldsymbol{\eta}, \boldsymbol{z}, t)$$

and by (69.9) and (69.10) we have

$$\boldsymbol{g}(\boldsymbol{\eta}, \boldsymbol{z}, t) = -\frac{\partial \boldsymbol{u}}{\partial t} - \frac{\partial \boldsymbol{u}}{\partial \boldsymbol{z}}\left(Q(t)\,\boldsymbol{z} + \boldsymbol{q}(\boldsymbol{\eta} + \boldsymbol{u}, \boldsymbol{z}, t)\right)$$

$$+ \boldsymbol{p}(\boldsymbol{\eta} + \boldsymbol{u}, \boldsymbol{z}, t) + R(t)\,\boldsymbol{z}.$$

In this equation we must express $\boldsymbol{u}$ in terms of $\boldsymbol{z}$. But $\boldsymbol{u}(0, t) = 0$ and therefore

$$\boldsymbol{g}(\boldsymbol{\eta}, 0, t) = \boldsymbol{p}(\boldsymbol{\eta}, 0, t) - \frac{\partial \boldsymbol{u}}{\partial \boldsymbol{z}}\bigg|_{\boldsymbol{z}=0} \boldsymbol{q}(\boldsymbol{u}, 0, t).$$

By the hypotheses on $\boldsymbol{q}(\boldsymbol{y}, 0, t)$, $\boldsymbol{g}(\boldsymbol{y}, 0, t)$ and $\boldsymbol{p}(\boldsymbol{y}, 0, t)$ agree except for terms of degree at least $N + 1$. The same is true for the succeeding transformations. In summary we obtain the first theorem of MALKIN [3].

Theorem 69.1. Let the equation

$$\dot{\boldsymbol{x}} = \boldsymbol{f}(\boldsymbol{x}, t)$$

be given, where $\boldsymbol{x} = \operatorname{col}(\boldsymbol{y}, \boldsymbol{z})$ and suppose: $\boldsymbol{y}$ and $\boldsymbol{z}$ satisfy the equations (69.6) and (69.7); the equilibrium of $\dot{\boldsymbol{z}} = Q(t)\,\boldsymbol{z}$ is exponentially stable;

the expansion of $q(y, 0, t)$ begins with terms of degree at least $N + 1$; the equilibrium $y = 0$ of the reduced equation

$$(69.13) \qquad \dot{y} = p(y, 0, t)$$

is stable, asymptotically stable, or unstable, in N^{th} approximation. Then the equilibrium $x = 0$ is, respectively, stable, asymptotically stable, or unstable.

If equation (69.6) also contains linear terms in y on the right then the existence theorem may no longer hold. Theorem 69.1 is then valid only under the additional assumption that the partial differential equations appearing in the reduction process have bounded analytic solutions.

If the function $q(y, 0, t)$ involves terms of degree $1, 2, \ldots, N$ then Theorem 69.1 can only be applied after the terms of lower degree in y have been removed by a transformation of variables. The transformation replaces z and has the form

$$(69.14) \qquad z = \zeta + u(y, t).$$

The components of u are polynomials in y with bounded coefficients. We shall not carry out the detailed computations here nor formulate Malkin's second theorem but direct the reader to the book by MALKIN [3]. We also mention the work of PLISS [3, 4] and the book by ZUBOV [4], sec. 13.

70. Simple Critical Cases for Autonomous Equations

If equation (68.1) is autonomous then the observations of the preceding section become simpler in many ways. We can, for instance, apply a linear transformation and put the matrix A into the form $\mathrm{diag}(A_1, A_2)$, where the critical characteristic roots are the characteristic roots of A_1 and where A_2 is stable. If there is one critical characteristic root we can without loss of generality write the equation in the form

$$(70.1) \qquad \dot{y} = g y^m + \gamma(y, z),$$
$$(70.2) \qquad \dot{z} = Q z + q(y, z).$$

The expansions of the nonlinear parts begin with terms of at least second degree; Q is stable. Furthermore $\gamma(y, 0) = 0$. The stability of the reduced equation $\dot{y} = g y^m$ can be determined from Theorem 68.1 in case $g \neq 0$. The transformation (69.14) which raises the degree of $q(y, z)$ with respect to y to $m + 1$ is easily found. We solve the equation

$$Q z + q(y, z) = 0$$

in terms of z. This we are able to do because the determinant of the Jacobian matrix $\partial(Q z + q)/\partial z$ is equal to $\det Q \neq 0$ when $z = 0$. z is obtained in the form

$$(70.3) \qquad z = u(y)$$

as an analytic function of y; $u(0) = 0$ and $\dfrac{du}{dy}\Big|_{y=0} = 0$. The first relation follows from the fact that $q(0, z) = 0$. If we differentiate this equation we obtain

$$Q\frac{dz}{dy} + \frac{\partial q}{\partial y} + \frac{\partial q}{\partial z}\frac{dz}{dy} = 0.$$

If we set $y = 0$ then all the elements of the vector $\dfrac{\partial q}{\partial y}$ vanish since q is at least of second degree in y and z and since z vanishes for $y = 0$. But since $Q + \dfrac{\partial q}{\partial z}$ is not equal to zero for $y = 0$, the derivative dz/dy must vanish for $y = 0$. If in (70.3) we make the substitution

$$z = \zeta + u(y),$$

then because of the behavior of $u(y)$ at $y = 0$, we obtain the desired increase in the degree provided $g \neq 0$. In this case we can apply Theorem 69.1.

If no number $g \neq 0$ can be found then the right side of (70.1) vanishes identically in y for $z = 0$. We then have a *singular* case. The reduced equation $y = 0$ has a stable but not an asymptotically stable equilibrium. And the same is true for the complete equation. It can be shown that every motion with sufficiently small initial values tends to a constant value as time increases (MALKIN [3]).

If there are two critical characteristic roots then the matrix which was denoted above by A_1 can be put into one of the forms

$$\begin{pmatrix} 0 & 0 \\ 0 & 0 \end{pmatrix}, \quad \begin{pmatrix} 0 & 0 \\ 1 & 0 \end{pmatrix}, \quad \begin{pmatrix} 0 & -1 \\ 1 & 0 \end{pmatrix}$$

according as we have a double characteristic root equal to zero with two simple, respectively one double, elementary divisors, or a pair of imaginary characteristic roots. Usually the transformation (69.14) already requires considerable computation. For the last case of a pair of purely imaginary characteristic roots MALKIN [3] showed a relatively easy way.

A critical case is given, in principle, whenever a concrete system is located on the stability boundary (sec. 11). We assume that the autonomous equation (68.1) depends on a parameter α which varies in a parameter space, and we write

$$(70.4) \qquad \dot{x} = A(\alpha)\, x + g(x; \alpha).$$

A particular value α_0 is to be chosen so that $A(\alpha_0)$ is critical. Then, in the notation of sec. 11, one of the equations (11.2) must be satisfied for $\alpha = \alpha_0$. Let the value α_1 lie within the stability domain, the value α_2 without. Accordingly, the equation

$$\dot{x} = A(\alpha_2)\, x + g(x; \alpha_2)$$

has an unstable equilibrium. We write the last equation in the form

$$(70.5) \qquad \dot{x} = A(\alpha_0)\, x + g(x; \alpha_0) + h(x; \alpha_0, \alpha_2).$$

For fixed x, the last term is of the order of magnitude $\|\alpha_0 - \alpha_2\|$, provided this norm is sufficiently small. If the equilibrium of (70.4) is asymptotically stable at the point $\alpha = \alpha_0$ of the stability boundary then Theorem 56.5 can be applied to (70.5). The equilibrium is then totally stable and the deviation $|x(t)|$ from the equilibrium can be held arbitrarily small by an appropriate choice of $\|\alpha_0 - \alpha_2\|$. This has the following interpretation. If the parameter α of a concrete system described by (70.4) is varied continuously so that the system moves from an asymptotically stable into an unstable state and if it is asymptotically stable on the boundary of stability then the system remains "practically" stable even outside of the stability boundary for small deviations in the parameter: It executes undamped oscillations of arbitrarily small amplitude or it approaches an equilibrium which deviates arbitrarily little from zero. If, on the other hand, the equilibrium is unstable at the point $\alpha = \alpha_0$ then the situation is quite different. Then if we move even an arbitrarily small distance beyond the stability boundary, oscillations of arbitrarily large amplitude may occur.

Accordingly, a point α_0 of the stability boundary, at which (70.4) has an asymptotically stable equilibrium, respectively a domain consisting of such points, is termed *safe* since there is no danger in passing through such a segment of the boundary (the terminology was developed for airplane control systems). In the opposite case we speak of *dangerous* segments. In principle, the exact determination of the safe and the dangerous segments is possible by the methods sketched in sec. 69. In practice, we usually use approximation procedures which frequently also give information about the "admissible" deviations of α.[1]

---

[1]) Aizerman [2], Magnus [1].

Chapter XI

Periodic and Almost Periodic Motions

71. General Remarks on Periodic Motions

By Def. 38.1 a motion described by $p(t, a, t_0)$ is called *periodic with period ω* if for all $t \geq t_0$ the relation

$$(71.1) \qquad p(t + \omega, a, t_0) = p(t, a, t_0)$$

is satisfied. Periodic motions in R_n which are described by differential equations or difference equations are exceedingly important in practice: The motion of planets can be described by differential equations and so can the operating behavior of an electric motor or steam engine. This explains the great importance of the theory of periodic motions and the numerous publications in this area. Strictly speaking, however, most of the "periodic" motions are actually not periodic but almost periodic (*cf.* also sec. 73) and the purely periodic motion is only approached as a limit. Still the study of this type of motions is indispensible for understanding many phenomena.

In this chapter we consider periodic motions which are defined by differential equations. The character of such a motion depends significantly on whether the differential equation is autonomous,

$$(71.2) \qquad \dot{x} = f(x)$$

or whether it involves t explicitly and is periodic in t,

$$(71.3) \qquad \dot{x} = f(x, t), \quad f(x, t + \omega) = f(x, t).$$

In the first case we speak of *self-excited oscillations*. It is clear that nothing can be said about their period without closer examination. In the second case we are interested in solutions whose period is equal to the external period or at least stands in a rational ratio to it. Depending on the physical situation we speak of *forced* or *parameter-excited* oscillations. The equation

$$\ddot{x} + a\dot{x} + bx = A \sin \frac{2\lambda}{\omega} t$$

describes a system which is excited by an external periodic force, whereas

$$\ddot{x} + \left(1 + a \cos \frac{2\lambda}{\omega} t\right) x = 0$$

suggests a system within which periodic changes occur, for instance a pendulum whose point of suspension vibrates. From a purely mathematical point of view this distinction is not essential. The methods for studying the existence and stability of the periodic motions are the same in both cases.

Determining a periodic motion is simplest if the general law of the motion, *i.e.* the expression $p(t, x_0, t_0)$ is known. For from the relation (71.1), the equation

$$(71.4) \qquad p(t_0 + \omega, x_0, t_0) = x_0$$

follows for $t = t_0$. If the period ω is known, as for example in the case of a non-autonomous equation, we can regard (71.4) as a system of n defining equations for the n unknown components of the initial vector $x_0 = \hat{x}_0$ of the periodic solution. If the differential equation is autonomous then the period ω is not known. The system of equations contains one unknown too many and the period as well as the variable $\hat{x}_0$ is a function of a parameter. If we think of the parameter as fixed, for example in such a way that the nth component x_{0n} of the initial vector assumes a fixed value γ, then the period and the components $\hat{x}_{01}, \ldots, \hat{x}_{0,n-1}$ are determined. This is so because in the autonomous case the parameter t_0 in (71.1) is superfluous since actually the general solution has the form $p(t - t_0, x_0)$. If $p(t, \hat{x}_0)$ is periodic of period ω then its nth component $p_n(t, \hat{x}_0)$ ranges over a certain interval. We choose for γ a number from this interval and as our initial instant t_0 a value t' for which

$$p_n(t', \hat{x}_0) = \gamma.$$

The initial vector for $t = t'$ is $p(t', \hat{x}_0)$ and the last component has the required value γ. In case $n = 2$ the phase trajectory of the periodic motion is a closed curve in the phase plane and because of what has just been said we can always choose the initial instant so that the curve happens to pass through the x-axis at time $t = 0$.

Equation (71.4) has also another interpretation. We consider the general solution $p(t, x_0, t_0)$ and regard the point

$$x^{(1)} = p(t_0 + \omega, x_0, t_0)$$

as the image of x_0 under a mapping of R_n to itself which depends on t_0 and which is defined by the motion. Then we can write, starting with $x^{(0)} := x_0$,

$$(71.5) \qquad x^{(k)} = \mathfrak{P}\, x^{(k-1)}, \quad k = 1, 2, 3, \ldots$$

A solution of equation (71.4) corresponds to a fixed point of the mapping (71.5) and the problem of determining periodic solutions has been reduced to studying the mapping (71.5) which is often more accessible than the expression $p(t, x_0, t_0)$ of the general solution.

Equation (71.5) can be interpreted as a difference equation. Each of its solutions $x^{(k)}$, $k = 0, 1, 2, \ldots$, agrees with a solution of (71.3) or (71.2) at the points where $t = k\omega$, $k = 0, 1, \ldots$.

Frequently it is not possible to solve the equations (71.4) exactly and we must be satisfied with an approximation procedure. Often perturbation methods are used which are based on the following idea. In addition to the differential equation (71.3) a perturbed equation

$$(71.6) \qquad \dot{x} = f(x, t) + g(x, t)$$

is considered and the general solutions are denoted by

$$p(t, x_0, t_0; f) \quad \text{and} \quad p(t, x_0, t_0; f + g).$$

Let a periodic solution of (71.3) be given, $i.e.$ let $\bar{x}_0$ be an initial vector such that $p(t, \bar{x}_0, t_0; f)$ is periodic. This vector satisfies the equation $p(t_0 + \omega, \bar{x}_0, t_0; f) = \bar{x}_0$; the notation is to express the fact that there exists a functional connection between f and the periodic solution. We could also write

$$(71.7) \qquad p(t, \bar{x}_0, t_0; f) = \Re f$$

in order to indicate that the space of functions f is mapped to the space of periodic functions. If the function f satisfies certain hypotheses, for example if it is continuous and satisfies a Lipschitz condition, then the solution depends continuously on the right side of the differential equation. The mapping (71.7) is continuous with respect to a suitably chosen norm.

We seek a periodic solution $p(t, \tilde{x}_0, t_0; f + g)$ of (71.6) which approaches $p(t, \bar{x}_0, t_0; f)$ as $g \to 0$. The defining equation for the initial vector $\tilde{x}_0$ is

$$(71.8) \qquad p(t_0 + \omega, \tilde{x}_0, t_0; f + g) = \tilde{x}_0$$

and it tends to (71.4) as $g \to 0$. We now concentrate on the dependence on g, $i.e.$ we consider the solution $\tilde{x}_0 = \varphi(g)$ as a function defined in the g-space. Since the value $\varphi(0) = \bar{x}_0$ is known it seems appropriate to write

$$\tilde{x}_0 = \varphi(g) = \varphi(0) + h(g) = \bar{x}_0 + h(g).$$

The term h is a function defined in the g-space which vanishes for $g = 0$ and it is our problem to determine this function for "small" arguments g, at least approximatively. The discussion is usually limited to special perturbances

$$(71.9) \qquad g(x, t) = \mu k(x, t; \mu).$$

$k(x, t)$ is a fixed given function and μ a small parameter. Under suitable assumptions on $f(x, t)$ and $k(x, t; \mu)$ we can work with power series expansions, etc., $cf.$ sec. 76ff.

23*

If we start with a differential equation with real coefficients we may assume that (71.4) is a real equation and that only its real solutions are of interest. As is well known, there are several possibilities for the solution of such an equation. They can be characterized by the behavior of the Jacobian determinant of the function $\boldsymbol{p}(t_0 + \omega, \boldsymbol{x}_0, t_0) - \boldsymbol{x}_0$. Let ω be fixed and let

$$J(\boldsymbol{x}) := \frac{\partial \boldsymbol{p}(t_0 + \omega, \boldsymbol{x}, t_0)}{\partial \boldsymbol{x}}.$$

If the determinant

$$(71.10) \qquad \det\left(J(\boldsymbol{x}_0) - E\right)$$

is different from zero for the solution $\boldsymbol{x}_0 = \bar{\boldsymbol{x}}_0$ of equation (71.4) then this solution and hence the periodic solution $\boldsymbol{p}(t, \bar{\boldsymbol{x}}_0, t_0)$ are uniquely determined. Since there exists a neighborhood of $\bar{\boldsymbol{x}}_0$ which contains no further solutions of (71.4) we also speak of an *isolated periodic solution* of the differential equation. A different situation is encountered if (71.10) vanishes identically for a certain $\bar{\boldsymbol{x}}_0$-set with an accumulation point. In that case (71.4) has a one or multi-parameter family of solutions and the same is true for the differential equation. If $\bar{\boldsymbol{x}}_0$ is an isolated zero of the Jacobian determinant then we are dealing with a singular point requiring special attention. For autonomous differential equations the period must also be considered as a variable. In that case we form the partial derivatives with respect to ω and the $n - 1$ variable components of $\bar{\boldsymbol{x}}_0$; *cf.* above.

The Jacobian determinant $\det\left(J(\boldsymbol{x}) - E\right)$ for the perturbed equation (71.6) depends, of course, also on $\boldsymbol{g}(\boldsymbol{x}, t)$. So it can happen that the determinant (71.10) formed with the solution $\tilde{\boldsymbol{x}}_0$ of (71.8), vanishes for certain perturbations. By what was said above, this means that an isolated periodic solution goes over into a family of solutions. The discussion of this case is quite difficult and it has been treated satisfactorily only for the special case (71.9) with a small parameter.

A periodic solution of a differential equation is physically realizable only if it is stable in the sense of Liapunov or orbitally stable.

Even more important is asymptotic, respectively orbitally asymptotic, stability. It means that the concrete system in time tends to a steady state, the periodic solution. In any case, no matter how the periodic solution was obtained, we must examine its stability behavior. This can be done, for example, with the help of the differential equation of the perturbed motion introduced in sec. 35. It has periodic coefficients and one would try to apply the principle of stability in the first approximation. Going over to the differential equation of the perturbed motion corresponds to applying a transformation to the motion space which takes the curve of the motion into the time axis. In addition, there is a transformation of coordinates which is specially suited for periodic motions: We construct an n-dimensional system of axes whose initial point moves along the (necessarily closed) trajectory of the motion with constant speed and use it as the new coordinate system, $cf.$ also sec. 80. In testing for stability we can also start from the fact that the periodic solution corresponds to an equilibrium of the difference equation (71.5) and study its stability behavior. We have

Theorem 71.1. The periodic solution of (71.3), $f \in C_0$, with initial value defined by (71.4) is asymptotically stable in the sense of Liapunov if and only if the equilibrium $\bar{x}_0$ of the difference equation

$$x^{(k+1)} = p(t_0 + \omega, x^{(k)}, t_0), \quad k = 0, 1, 2, \ldots,$$

is asymptotically stable in the sense of Liapunov.

Proof. Since the differential equation of the perturbed motion is periodic the assumed stability is uniform ($cf.$ sec. 38). We have

$$|p(t, \bar{x}_0, t_0) - p(t, x_0, t_0)| \leq \varphi(|\bar{x} - x_0|)\, \sigma(t - t_0),$$

where $\bar{x}_0$ is the initial value of the periodic solution, x_0 that of a neighboring solution. The inequality is valid for all $t \geq t_0$ and in particular for the sequence $t_k = t_0 + k\omega$, $k = 0, 1, 2, \ldots$. Therefore

$$(71.11) \qquad |p(t_k, x_0, t_0) - \bar{x}_0| \leq \varphi(|\bar{x}_0 - x_0|)\, \sigma(k\omega).$$

Obviously the equation

$$(71.12) \qquad p(t_k, x_0, t_0) = x^{(k)}$$

holds for $k = 1$. Its validity in general, for $k = 2, 3, \ldots$, is shown by induction: By definitiou,

$$x^{(k+1)} = p(t_0 + \omega, x^{(k)}, t_0) = p(t_{k+1}, x^{(k)}, t_k)$$

and if (71.12) is valid for k then it follows that

$$x^{(k+1)} = p(t_{k+1}, p(t_k, x_0, t_0), t_k) = p(t_{k+1}, x_0, t_0).$$

Hence we obtain from (71.11),

$$(71.13) \qquad |x^{(k)} - \bar{x}_0| \leq \varphi(|\bar{x}_0 - x_0|)\, \sigma(k\omega)$$

$i.e.$ the asymptotic stability of the equilibrium of the difference equation.

On the other hand we have for $k = 1, 2, \ldots$ and for arbitrary τ,

$$p(t_k + \tau, x_0, t_0) = p(t_k + \tau, x^{(k)}, t_k) = p(t_0 + \tau, x^{(k)}, t_0),$$

$$p(t_k + \tau, \bar{x}_0, t_0) = p(t_k + \tau, \bar{x}_0, t_k) = p(t_0 + \tau, \bar{x}_0, t_0)$$

and hence

$$|p(t_k + \tau, x_0, t_0) - p(t_k + \tau, \bar{x}_0, t_0)|$$

$$= |p(t_0 + \tau, x^{(k)}, t_0) - p(t_0 + \tau, \bar{x}_0, t_0)| \leq |x^{(k)} - \bar{x}_0|\, K.$$

The constant K is essentially equal to $e^{L\tau} \leq e^{L\omega}$ (L is a Lipschitz constant). Combining with (71.13) we obtain the asymptotic stability of the periodic solution.

If the initial equation is autonomous then the periodic solution is never asymptotically stable in the sense of Liapunov (Theorem 81.1) and it is useless to try to apply Theorem 71.1. But another difference equation can be introduced in this case. For this purpose we consider a hypersurface, for example a hyperplane in R_n, positioned in such a way that it is intersected by the closed phase trajectory of the periodic solution. We are interested in a neighborhood of the point of intersection on the surface. The trajectories define a mapping of this neighborhood into another one and it can again be written in the form (71.5); the points $x^{(k)}$ are on the hypersurface. A fixed point of this mapping corresponds to a periodic solution. But the times t_k at which the trajectory passes through the points $x^{(k)}$ are no longer equal to $t_0 + k\omega$. If we compare the periodic solution $p(t, \bar{x}_0, t_0)$ with the solution which originates at $x^{(k)}$ at the same instant then we obtain as above

$$|p(t, x^{(k)}, t_0) - p(t, \bar{x}_0, t_0)| < |x^{(k)} - \bar{x}_0| K_1.$$

K_1 depends on the duration of the comparison. If the comparison is limited to the interval between two consecutive passes through the hypersurface then the asymptotic stability of the equilibrium of the difference equation implies that the perturbed trajectory comes arbitrarily close to the trajectory of the periodic solution. But because the two motions have different phases this means merely orbital asymptotic stability of the periodic solution; cf. also sec. 81.

Similar to the asymptotically stable equilibrium, an orbitally asymptotically stable periodic solution possesses a *domain of attraction*. It consists of the set of all those points from which motions originate which with increasing time come arbitrarily close to the periodic solution: If M denotes the set of points on the periodic solution then x_0 belongs to the domain of attraction just in case

$$\lim_{t \to \infty} \varrho\big(p(t, x_0, t_0), M\big) = 0.$$

Since the periodic solution is an invariant set the definition of its domain of attraction is in agreement with the definitions of sec. 33 and possesses the general properties shown there. It is, for instance, an open set.

We further mention that the characterization of the domain of attraction of the equilibrium which was given in sec. 34 applies to the domain of attraction of a periodic solution. The hypotheses must be brought into the proper context: In place of the hypothesis "$v(x)$ positive definite" we now have $v(x) > \varphi(\varrho(x, M))$ etc. Observing this we can repeat the proofs of sec. 34 almost verbatim. See also ZUBOV [2], sec. 5.

In the next sections we treat a few especially simple cases in which the periodic solutions can be given explicitly and their stability studied. We are dealing with exceptions here: Generally it is very difficult to obtain the periodic solutions and to discuss their stability in practical problems, and such discussion can be carried out approximatively at best.

72. Nonhomogeneous Linear Equations with Periodic External Force

In the general discussion of sec. 59 the external force $z(t)$ was restricted by no assumption other than that it be bounded. Henceforth we consider an equation

$$(72.1) \qquad \dot{x} = P(t)\,x + z(t)$$

under the following more special hypotheses: a) $P(t)$ is periodic with period ω or constant. b) $z(t)$ is periodic with period ω. Such periodic external forces appear in many problems in practice. We are then usually interested in periodic solutions and particularly in those which are stable. They describe well-determined physical states of the system and are called *forced oscillations* (*cf.* sec. 71).

To determine these solutions we start from the general representation (59.3) and observe that the periodic solution must satisfy condition (71.1). We can choose $t_0 = 0$ without loss of generality since the property (71.1) is independent of the initial time. From (59.3) we obtain

$$(72.2) \qquad x(\omega) = K(\omega, 0)\,x_0 + \int_0^\omega K(\omega, u)\,z(u)\,du = x_0$$

and because $K(\omega, u) = K(\omega, 0)\,K(0, u)$ and $K(\omega, 0) = K(0, \omega)^I$,

$$(72.3) \qquad \big(K(0, \omega) - E\big)\,x_0 = \int_0^\omega K(0, u)\,z(u)\,du.$$

This equation can be considered as a defining equation for the vector x_0, that is, for the initial vector which determines the solution which begins at $t = 0$.

In solving the equation we must distinguish several cases. 1) The homogeneous equation

$$(72.4) \qquad \dot{x} = P(t)\,x$$

belonging to (72.1) has no periodic solutions. By Theorem 60.3 this means that the matrix $K(\omega, 0)$ has no characteristic root equal to 1. The same is then true for the matrix $K(0, \omega)$. Therefore

$$\det\big(K(0, \omega) - E\big) \neq 0$$

and (72.3) can be solved immediately. The solution vector

$$x_0 = \left(K(0, \omega) - E\right)^{\mathrm{I}} \int_0^{\omega} K(0, u)\, z(u)\, du$$

is the initial vector of the periodic solution and the solution itself has the form

$$(72.5) \qquad \varphi(t) = K(t, 0)\left(K(0, \omega) - E\right)^{\mathrm{I}} \int_0^{\omega} K(0, u)\, z(u)\, du$$

$$+ \int_0^{t} K(t, u)\, z(u)\, du.$$

Since $\varphi(t)$ is periodic t can be restricted to the interval $0 \leq t \leq \omega$. With the aid of the Green's matrix

$$(72.6) \quad G(t, u) := \begin{cases} K(t, u) + K(t, 0)\left(K(0, \omega) - E\right)^{\mathrm{I}} K(0, u), & u \leq t, \\ K(t, 0)\left(K(0, \omega) - E\right)^{\mathrm{I}} K(0, u), & t < u, \end{cases}$$

the solution $\varphi(t)$ can be written as a definite integral, as is easily verified by differentiation:

$$\varphi(t) = \int_0^{\omega} G(t, u)\, z(u)\, du = \int_t^{t+\omega} \left(K(u, 0)(K(\omega, 0)^{\mathrm{I}} - E)\, K(t, 0)^{\mathrm{I}}\right)^{\mathrm{I}} z(u)\, du$$

$$(72.7)$$

$$= K(t, 0)\left(K(0, \omega) - E\right)^{\mathrm{I}} \int_t^{t+\omega} K(0, u)\, z(u)\, du.$$

2) If the homogeneous equation has periodic solutions then

$$\det\left(K(0, \omega) - E\right) = 0$$

and equation (72.3) can be solved only if the right side satisfies certain conditions. An equation

$$B p = q, \quad \det B = 0,$$

can be solved for p if and only if each vector v which satisfies the equation $v^T B = 0$, also satisfies $v^T q = 0$.[1] Applied to (72.3) this result says: Each vector v for which

$$(72.8) \qquad\qquad v^T\left(K(0, \omega) - E\right) = 0$$

must also satisfy

$$(72.9) \qquad\qquad v^T \int_0^{\omega} K(0, u)\, z(u)\, du = 0.$$

---

[1] *cf.* for instance SCHMEIDLER [1].

(72.8) implies that $K(0, \omega)^T \boldsymbol{v} = \boldsymbol{v}$ and this says that the vector

$$\boldsymbol{y}(t) := K(0, t)^T \boldsymbol{v}$$

is a periodic solution for the adjoint of equation (72.4) $(cf.$ (58.21)),

$$(72.10) \qquad \qquad \dot{\boldsymbol{y}} = -P(t)^T \boldsymbol{y}.$$

Equation (72.9) therefore assumes the form

$$(72.11) \qquad \qquad \int\limits_0^\omega \boldsymbol{y}^T(u)\,\boldsymbol{z}(u)\,du = 0.$$

If (72.1) has periodic solutions, that is if (72.3) can be solved, then (72.11) must be satisfied for all the periodic solutions of (72.10). These solutions form a linear space and it suffices to require that (72.11) is satisfied for a basis $\boldsymbol{\psi}_1, \ldots, \boldsymbol{\psi}_r$ of this space:

$$(72.12) \qquad \int\limits_0^\omega \boldsymbol{\psi}_j(u)^T\,\boldsymbol{z}(u)\,du = 0, \quad j = 1, 2, \ldots, r.$$

These conditions are the so-called *solvability conditions*. If the conditions (72.12) are not satisfied then there exists at least one periodic solution $\boldsymbol{y}(t)$ of (72.10) such that

$$\int\limits_0^\omega \boldsymbol{y}(u)^T\,\boldsymbol{z}(u)\,du \neq 0\,.$$

Then (72.1) cannot have a periodic solution since each solution is unbounded. For the vector

$$\boldsymbol{h} := \int\limits_0^\omega K(\omega, u)\,\boldsymbol{z}(u)\,du$$

has at least one component different from zero since the rows of the matrix $K(\omega, 0)$ consist of the initial vectors of periodic solutions which (72.10) might have, and since by hypothesis at least one of the corresponding integrals does not vanish. For a solution $\boldsymbol{x}(t)$ of (72.1) we have

$$\boldsymbol{x}(\omega) = K(\omega, 0)\,\boldsymbol{x}_0 + \boldsymbol{h},$$

$$\boldsymbol{x}(k\omega) = K(\omega, 0)^k\,\boldsymbol{x}_0 + \sum_{\varrho=0}^{k-1} K(\omega, 0)^\varrho\,\boldsymbol{h}, \quad k = 2, 3, \ldots.$$

Since $K(\omega, 0)$ has by hypothesis a characteristic root equal to 1, the sum is unbounded as k iucreases. $\boldsymbol{x}(k\omega)$ becomes arbitrarily large. In this case we speak of *resonance*.

If the adjoint equation (72.10) has exactly s linearly independent solutions all of which satisfy the solvability conditions (72.12) then the defining equation (72.3) has s linearly independent solutions $\boldsymbol{x}_0^{(\sigma)}$,

$\sigma = 1, \ldots, s$, and (72.1) has an s-parameter family of periodic solutions. The general member of this family has the form

$$(72.13) \qquad x(t) = \varphi(t) + \sum_{\sigma=1}^{s} c_\sigma x^{(\sigma)}(t)$$

where $\varphi(t)$ is a particular solution and the $x^{(\sigma)}$ are the linearly independent periodic solutions of (72.4).

All these results agree with the general theory discussed in sec. 71. The Jacobian determinaut (71.10) belonging to equation (72.3) is simply

$$\det\left(K(0, \omega) - E\right).$$

If it is different from zero we have an isolated solution. If it is equal to zero and there exist solutions at all then there are families of solutions.

Formula (72.2) is, in fact, the difference equation (71.5). If we set

$$x^{(m)} := x(m\omega), \quad Q := K(\omega, 0), \quad h := \int_0^\omega K(\omega, u)\, z(u)\, du,$$

then the equation has the form

$$x^{(m+1)} = Q\, x^{(m)} + h.$$

By sec. 14, the stability depends on the characteristic roots of the matrix Q: The equilibrium, and hence the periodic solution, is asymptotically stable if Q has only characteristic roots with absolute value less than one. The adjoint differential equation (72.10) has no periodic solutions in this case. If the absolute values of the characteristic roots are no larger than one then the equilibrium of the difference equation is at least stable. The same is true for the families of periodic solutions.

If Q has characteristic roots of absolute value larger than one then the equilibrium is unstable and the periodic solutions are also unstable.

The same result is obtained from the arguments at the beginning of sec. 44 as a statement on system stability of the differential equation (72.1).

The Green's matrix (72.6) is familiar from boundary value problems. Since the computation of a periodic solution can be considered as a boundary value problem (the boundary condition is $x(0) = x(\omega)$) the appearance of the Green's matrix in the formula is only natural.

We formulate these results for the special equation

$$(72.14) \qquad \dot{x} = A\, x + z(t),$$

where A is constant and $z(t)$ is periodic. The adjoint equation

$$\dot{y} = - A^T y$$

has periodic solutions if and only if the matrix A has characteristic values of the form $\lambda = 2\pi i/\omega$. If this is not the case then (72.14) has exactly one periodic solution, namely

$$\varphi(t) = e^{At}(E - e^{A\omega})^I \int_0^\omega e^{-A\omega}\, z(u)\, du + \int_0^t e^{A(t-u)} z(u)\, du.$$

This time the Green's matrix is

$$G(t, u) := \begin{cases} e^{A(t-u)}\left(E + (e^{-A\omega} - E)^I\right), & u \le t, \\ e^{A(t-u)}(e^{-A\omega} - E)^I, & t < u, \end{cases}$$

so that we can write

$$\varphi(t) = \int_0^\omega G(t, u) z(u)\, du = e^{At}(e^{-A\omega} - E)^I \int_t^{t+\omega} e^{-Au}\, z(u)\, du.$$

If A has characteristic roots of the form $2\pi i/\omega$ then the solvability conditions

$$y_0^T \int_0^\omega e^{-Au}\, z(u)\, du = 0$$

must be satisfied; y_0 is a solution of

$$(72.15) \qquad\qquad e^{-A^T\omega} y_0 = y_0.$$

If not all the solvability conditions are satisfied then (72.14) has no periodic solution; we have resonance in this case. Otherwise there exists a family of periodic solutions which depends on as many parameters as there are linear independent solutions for (72.15).

For the scalar equation

$$(72.16) \qquad\qquad \ddot{x} + a^2 x = z(t), \quad z(t + \omega) = z(t),$$

which has no damping term and is therefore in principle capable of periodic solutions, we can state the result even more simply. If $a^2 \ne (2\pi/\omega)^2$ then the equation has exactly one periodic solution. In particular, for $z = b e^{i\alpha t}$ we have

$$(72.17) \qquad\qquad x(t) = x_0 \frac{b}{a^2 - \alpha^2} e^{i\alpha t}, \quad a^2 \ne \alpha^2.$$

If $a^2 = (2\pi/\omega)^2$ then there exists a periodic solution if and only if the Fourier expansion of the function $z(t)$ in terms of $\sin(2k\pi/\omega)$ and $\cos(2k\pi/\omega)$ contains no terms with $\sin(2\pi/\omega)$ and $\cos(2\pi/\omega)$. The equations

$$\int_0^\omega z(u) \sin\frac{2\pi u}{\omega}\, du = 0, \quad \int_0^\omega z(u) \cos\frac{2\pi u}{\omega}\, du = 0$$

which express the fact that the first two Fourier coefficients vanish, are precisely the solvability conditions if the scalar equation (72.16) is written as an equation for a 2-vector.

The stability of a forced oscillation corresponds to the stability of the equilibrium of the homogeneous equation; for (72.4) is the equation of the perturbed motion for a periodic solution of (72.1). If it has an asymptotically stable equilibrium then the general solution of (72.1) tends toward a periodic steady state which is characterized by (72.5), resp. (72.7). If the equilibrium is unstable then the periodic solution cannot be realized physically.

73. Forced Almost Periodic Oscillations

From a practical point of view, a periodic motion in a physical system of higher order is a rare limiting case. For instance, a conservative system with two degrees of freedom (it is described by a fourth order differential equation) has a general solution of the form

$$a_1 \cos \alpha_1 t + b_1 \sin \alpha_1 t + a_2 \cos \alpha_2 t + b_2 \sin \alpha_2 t$$

and this expression is periodic if and only if the numbers α_1 and α_2 are commensurable, *i.e.* if the fraction α_1/α_2 is a rational number. Otherwise the expression is an *almost periodic function* in t.

Def. 73.1. A function $f(t)$ is called *almost periodic* if it satisfies the following conditions:[1]

　　a) $f(t)$ is continuous and defined on $-\infty < t < +\infty$.

　　b) For each $\varepsilon > 0$ there exists a number $L = L(\varepsilon)$ so large that in each interval $t_0 \leq t \leq t_0 + L$ there exists a number $\tau(\varepsilon)$ for which

$$|f(t + \tau) - f(t)| < \varepsilon.$$

Since for finitely many almost periodic functions $z_1, \ldots, z_n$ we can find common numbers $L(\varepsilon)$ and $\tau(\varepsilon)$ (for instance by using the theorem that the sum of finitely many almost periodic functions is again almost periodic) it makes sense to talk of an almost periodic vector $z(t) = \mathrm{col}(z_1(t), \ldots, z_n(t))$.

Finite trigonometric sums are important special cases of almost periodic functions,

$$(73.1) \qquad f(t) = a_0 + \sum_{i=1}^{N} (a_i \cos \alpha_i t + b_i \sin \alpha_i t).$$

We shall try to extend the results of sec. 72 to almost periodic functions. That is, we shall try to answer the question, when a differential equation (72.1) with an almost periodic external force $z(t)$ has almost periodic solutions. For the sake of simplicity we assume that the matrix A is constant. This does not limit the generality of the theory since by sec. 61 any equation with periodic coefficients can be transformed into an equa-

---

[1] *cf.* for instance BESICOVITCH [1].

tion with constant coefficients by a linear transformation which maps $z(t)$ again to an almost periodic vector.

We first prove

Theorem 73.1. Let

$$(73.2) \qquad \dot{x} = A\,x + z(t)$$

be given. Let all the characteristic roots of A have a nonzero real part. Let $z(t)$ be a vector with almost periodic components, and let sup $|z(t)| = M$. Then the equation has exactly one almost periodic solution $x(t)$ such that

$$(73.3) \qquad |x(t)| < c\,M.$$

The constant c depends on A only.

Proof. 1) The equation is scalar,

$$\dot{x} = \alpha\,x + g(t).$$

If $\operatorname{Re}\alpha < 0$ then the solution

$$x(t) = e^{\lambda t}\int\limits_{-\infty}^{t} e^{-\lambda u} g(u)\,du$$

is almost periodic. For we have

$$x(t+\tau) - x(t) = \int\limits_{-\infty}^{t+\tau} e^{\lambda(t+\tau-u)} g(u)\,du - \int\limits_{-\infty}^{t} e^{\lambda(t-u)} g(u)\,du$$

$$= \int\limits_{-\infty}^{t} e^{\lambda(t-u)}\big(g(u+\tau) - g(u)\big)\,du.$$

If we now choose the number τ so that

$$|g(u+\tau) - g(u)| < |\operatorname{Re}\alpha|\,\varepsilon \quad \text{for} \quad -\infty < u < +\infty$$

then

$$|x(t+\tau) - x(t)| \le |\operatorname{Re}\alpha|\,\varepsilon \int\limits_{-\infty}^{t} e^{\operatorname{Re}x(t-u)}\,du = \varepsilon.$$

The estimate (73.3) follows immediately from the definition of $x(t)$. If $\operatorname{Re}\alpha > 0$ we take the solution

$$x(t) = -\,e^{\alpha t}\int\limits_{t}^{\infty} e^{-\alpha u} g(u)\,du.$$

If we introduce the symbol

$$\varkappa := \operatorname{sgn}\operatorname{Re}\alpha$$

then the two cases can be stated together: If $\operatorname{Re}\alpha \neq 0$ then the almost periodic solution has the form

$$x(t) = -\,e^{\lambda t}\int\limits_{t}^{\varkappa\cdot\infty} e^{-\lambda u} g(u)\,du.$$

2) Assume that the equation is not scalar and the real parts of the characteristic roots of A all have the same sign. Then

$$x(t) = - e^{At} \int_t^{\infty \cdot \infty} e^{-Au} z(u)\, du$$

is the desired solution. This is shown in the same way as above. 3) If A has some characteristic roots with positive and some with negative real parts we apply a linear transformation which takes A into the matrix $\mathrm{diag}\,(A_1, A_2)$, where the characteristic roots of A_1 have negative real parts and those of A_2 have positive real parts. And then we apply the argument of 2) to the two matrices separately. The solution can be written in closed form if we introduce the Green's matrix (66.9) defined above. The almost periodic solution is

$$x(t) = \int_{-\infty}^{+\infty} G(t - u)\, z(u)\, du\,.$$

If the matrix A has characteristic roots with vanishing real parts then the situation is similar to the resonance case for a periodic external force. Almost periodic solutions exist only if the external force satisfies certain conditions. We also speak of a critical case here.

The general solution of the scalar equation is

$$x(t) = e^{\alpha t} x_0 + e^{\alpha t} \int_0^t e^{-\alpha u} g(u)\, du\,.$$

If $\operatorname{Re} \alpha = 0$, $e^{\alpha t}$ is periodic. The condition necessary and sufficient for the second term to be almost periodic is

$$(73.4) \qquad \int_0^t e^{-\alpha u} g(u)\, du \ \text{bounded,}$$

the relation

$$(73.5) \qquad \lim_{T \to \infty} \frac{1}{2T} \int_{-T}^{+T} e^{-\alpha u} g(u)\, du = 0,$$

being necessary. (The function $a(t)$ in (38.8) shows that the two conditions are not equivalent.) If (73.4) is satisfied then all solutions are almost periodic. If the equation is not scalar we apply a transformation which takes the matrix A into a Jordan normal form. A sub-system corresponding to the submatrix

$$J_r := \begin{pmatrix} \alpha & 0 & 0 & \cdots & 0 & 0 \\ 1 & \alpha & 0 & \cdots & 0 & 0 \\ 0 & 1 & \alpha & \cdots & 0 & 0 \\ & & \cdots & & & \\ 0 & 0 & 0 & \cdots & 1 & \alpha \end{pmatrix},$$

consists of scalar equations

$$\dot{y}_1 = \alpha\, y_1 + w_1(t),$$

$$\dot{y}_2 = \alpha\, y_2 + y_1 + w_2(t),$$

$$\cdot\ \cdot\ \cdot\ \cdot\ \cdot\ \cdot\ \cdot\ \cdot\ \cdot\ \cdot\ \cdot$$

(73.6)

$$\dot{y}_r = \alpha\, y_r + y_{r-1} + w_r(t).$$

The first equations has almost periodic solutions if the condition (73.4) is satisfied for $w_1 = g$. Substituting such a solution in the second equation, we obtain a scalar equation with external force

$$y_1(t) + w_2(t) = y_1(0)\, e^{\alpha t} + \int_0^t e^{\alpha(t-u)}\, w_1(u)\, du + w_2(t),$$

which must satisfy the condition (73.5). This equation defines $y_1(0)$. We can proceed in this manner and obtain a sequence of conditions which the $w_i(t)$ must satisfy in order that the system (73.6) have a 1-parameter family of almost periodic solutions. The parameter is the initial value $y_r(0)$ for the last equation. In this manner we see that the complete equation has a family of almost periodic solutions depending on as many parameters as the Jordan normal form has submatrices belonging to characteristic roots with zero real part.

Let the external force be of the special form

(73.7)
$$z(t) = a_0 + \sum_{\mu=1}^{N} (a_\mu \cos \alpha_\mu t + b_\mu \sin \alpha_\mu t),$$

in which case (73.4) and (73.5) are equivalent, and assume that the matrix A has $2m + s$ characteristic roots with vanishing real parts and is such that the equation

(73.8)
$$\dot{y} = - A^T y$$

possesses the $2m$ linearly independent periodic solutions

(73.9)
$$\psi_\nu = p_\nu \cos \omega_\nu t - q_\nu \sin \omega_\nu t,$$
$$\psi_{\nu+m} = p_\nu \sin \omega_\nu t + q_\nu \cos \omega_\nu t,$$
$$\nu = 1, 2, \ldots, m,$$

and the s linearly independent constant solutions

(73.10)
$$\psi_\nu = c_\nu, \quad \nu = 2m + 1, \ldots, 2m + s.$$

The functions (73.9) and (73.10) form a basis for the almost periodic solutions of (73.8). We have

Theorem 73.2. If $z(t)$ is of the form (73.7) then the equation (73.2) has almost periodic solutions if and only if the functions (73.9) and (73.10) satisfy the conditions

$$(73.11) \qquad \lim_{t \to \infty} \frac{1}{t} \int_0^t \boldsymbol{\psi}_\nu^T(u)\, z(u)\, du = 0, \qquad \nu = 1, 2, \ldots, 2m + s.$$

In place of the functions (73.9) and (73.10) any arbitrary basis for the almost periodic solutions of (73.8) can be used.

Proof.[1]) a) Let $\boldsymbol{x}(t)$ be an almost periodic solution of (73.2) and let $\boldsymbol{\psi}(t)$ be a member of the basis. Then

$$\frac{d}{dt}\left(\boldsymbol{\psi}(t)^T \boldsymbol{x}(t)\right) = \dot{\boldsymbol{\psi}}^T \boldsymbol{x} + \boldsymbol{\psi}^T \dot{\boldsymbol{x}} = - \boldsymbol{\psi}^T A \boldsymbol{x} + \boldsymbol{\psi}^T A \boldsymbol{x} + \boldsymbol{\psi}^T z$$

and by integration from 0 to t we obtain

$$\lim_{t \to \infty} \frac{1}{t} \int_0^t \boldsymbol{\psi}(u)^T z(u)\, du = \lim \frac{1}{t}\left(\boldsymbol{\psi}(t)^T \boldsymbol{x}(t) - \boldsymbol{\psi}^T(0)\, \boldsymbol{x}(0)\right).$$

The expression in parantheses on the right side is bounded since $\boldsymbol{\psi}(t)$ is periodic and $\boldsymbol{x}(t)$ is almost periodic. The value of the limit is therefore zero. Condition (73.11) is necessary. b) The equation

$$(73.12) \qquad \dot{\boldsymbol{x}} = A\boldsymbol{x} + \boldsymbol{a}_\mu \cos \alpha_\mu t + \boldsymbol{b}_\mu \sin \alpha_\mu t$$

which involves one of the terms of (73.7) clearly has a periodic solution of period $2\pi/\alpha_\mu$ if the number α_μ is different from the characteristic roots ω_ν. This follows from sec. 72. If α_μ is equal to one of the characteristic roots ω_ν, say ω_1, then (73.12) has a periodic solution only if the solvability conditions (72.11) are satisfied, *i.e.* if

$$(73.13) \qquad \int_0^{2\pi/\alpha_\mu} \boldsymbol{\psi}_\nu(u)^T \left(\boldsymbol{a}_\mu \cos \alpha_\mu u + \boldsymbol{b}_\mu \sin \alpha_\mu u\right) du = 0,$$

$$\nu = 1, m + 1.$$

We now have the relation

$$\lim_{t \to \infty} \frac{1}{t} \int_0^t \boldsymbol{\psi}_\nu(u)^T \left(z(u) - \boldsymbol{a}_\mu \cos \alpha_\mu u - \boldsymbol{b}_\mu \sin \alpha_\mu u\right) du = 0$$

for $\nu = 1, m + 1$, because the integrand contains only finitely many trigonometric terms. If we assume that (73.11) is satisfied then it follows that

$$\lim_{t \to \infty} \frac{1}{t} \int_0^t \boldsymbol{\psi}_\nu(u)^T \left(\boldsymbol{a}_\mu \cos \alpha_\mu u + \boldsymbol{b}_\mu \sin \alpha_\mu u\right) du = 0$$

$$\nu = 1, m + 1,$$

and hence (73.13).

---

[1]) *cf.* MALKIN [5].

Analogously we conclude that the last s conditions in (73.11) guarantee the existence of a constant solution of the equation

$$(73.14) \qquad \dot{y} = - A^T y + a_0.$$

We think of the solutions of (73.2) as formed by superposition of equations (73.12), $\mu = 1, 2, \ldots, N$, and equation (73.14) and we see that (73.11) is sufficient for the existence of an almost periodic solution.

The stability behavior of the almost periodic solution of (73.2) is determined by the behavior of the equilibrium of the homogeneous equation $\dot{x} = A x$, for this homogeneous equation is the "equation of the perturbed motion". We can, of course, again interpret the behavior of the almost periodic solution as system stability (*cf.* also Defs. 35.7 and 35.8).

The question whether there exist almost periodic solutions for non-homogeneous linear equations with almost periodic coefficients is considerably more difficult to answer.[1]

74. Piecewise Linear Equations

In this section we discuss a class of differential equations which are important for many applications and which, in addition, are distinguished by general solutions which can be written explicitly. It is thus possible to study the behavior of periodic solutions by discussing equation (71.1). We first give a few simple examples for equations in this class (*cf.* also sec. 22).

1) The motion of a particle of mass m held be a spring and sliding on a rough surface is essentially damped only by the so-called dry friction, which depends on the direction of the motion and not on the velocity. Damping proportional to the velocity is in general so small compared to frictional damping that it can be neglected. The equation of the motion is therefore

$$(74.1) \qquad \begin{aligned} m\ddot{x} + k^2 x &= - \varkappa \quad (\dot{x} > 0), \\ m\ddot{x} + k^2 x &= + \varkappa \quad (\dot{x} < 0); \quad \varkappa > 0. \end{aligned}$$

The positive number $\varkappa$ is a measure of the friction; $k^2 x$ is the linear restoring force, for example the spring tension computed by Hooke's law. Equation (74.1) is linear for positive as well as for negative velocities but otherwise it does not satisfy all the hypotheses which are usually required for differential equations; for the right side is discontinuous at $\dot{x} = 0$ and the second derivative is not even defined at this point. The concept of a solution must therefore be made somewhat more general.

---

[1] *cf.* MASSERA and SCHÄFFER [1], MASSERA [4].

The extension of this concept is most easily explained if we interpret the solution of (74.1) in a phase plane (sec. 18). Incidentally, (74.1) can also be written in the form of a single equation

$$(74.2) \qquad m\ddot{x} + k^2 x + \alpha \operatorname{sgn} \dot{x} = 0 \quad (\dot{x} \neq 0)$$

or as a system

$$\dot{x} = y, \quad \dot{y} = -\frac{k^2}{m} x - \frac{\alpha}{m} \operatorname{sgn} y \quad (y \neq 0).$$

In the upper half-plane the trajectories of (74.2) are ellipses with equations

$$m\dot{x}^2 + k^2(x + \alpha k^{-2})^2 = \text{const}$$

whose center is the point $(-\alpha k^{-2}, 0)$. Since $\dot{x} > 0$, they are traversed in the negative sense. In the lower half-plane we have ellipses with center $(\alpha k^{-2}, 0)$ which are also traversed in the negative sense.

If we require that the transition from one family of ellipses to the other across the x-axis is to be continuous, which is a reasonable assumption in view of the practical problem, then each motion has a continuous phase trajectory with a continuous tangent. At the points of intersection with the x-axis these curves have a discontinuous second derivative.

In the interval $I := (-\alpha k^{-2}, \alpha k^{-2})$ it is impossible to cross the x-axis because the motions of the upper family meet this segment in a direction opposite to that of the lower family. Physically this corresponds to the fact that the particle of mass comes to rest in the segment I after moving back and forth a finite number of times. Accordingly, equation (74.2) must be supplemented by an equation which determines the motions taking their origin in this interval. The supplementary equation

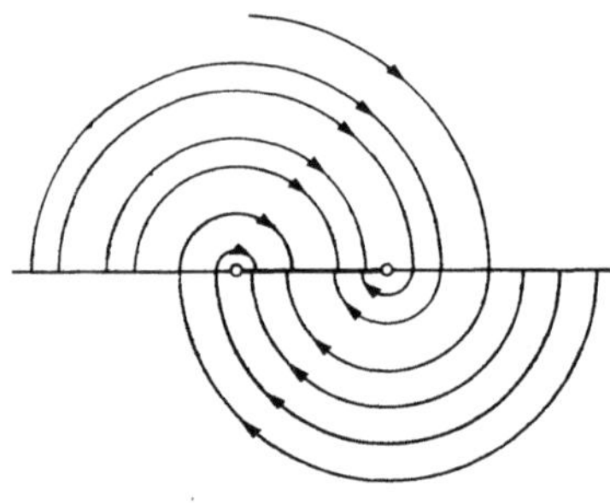

Fig. 74.1. To (74.1)

$$(74.3) \qquad\qquad \dot{x} = 0, \quad x \in I,$$

is appropriate for the given practical problem. It expresses the fact that the points of the interval I are singular points of the equation of the motion. We recognize that the set of these singular points is stable and attractive: The distance of the phase point from this set is decreasing. It follows immediately that periodic motions cannot occur since each motion ends in I.

The choice of the supplementary equation (74.3) depended exclusively on physical considerations. From a mathematical point of view, a variety of different supplements are admissible, for example

$$(74.4) \qquad \dot{x} = - x, \quad x \in I.$$

This motion approaches the origin along the x-axis. The origin is the only equilibrium point and is asymptotically stable. As the phase curve enters I it has a discontinuous tangent. Another supplement is

$$(74.5) \qquad \dot{x} = - x + \alpha k^{-2}, \quad x \in I.$$

Here the motions approach the right end point of the interval I asymptotically.

2) The scalar equation

$$(74.6) \qquad m\ddot{x} + k^2 x = - \alpha \operatorname{sgn} x \quad (x \neq 0)$$

describes the motion of an undamped oscillator which is subjected to a restoring force proportional to the distance and, in addition, to a further restoring force which is constant but whose sign depends on x (*cf.* figure 22.5). As in the preceding example, the individual trajectories in the phase plane are composed of elliptic arcs. The centers are the same as before, that is $(-\alpha k^{-2}, 0)$ and $(\alpha k^{-2}, 0)$ but this time the two arcs meet on the x-axis. We require that the transition is continuous but we cannot insist on the continuity of the tangent. The symmetry of the equation of the trajectories shows immediately that all the motions are periodic. The origin is singular. Since (74.6) is not defined for $x = 0$ we must add the definition

$$x \equiv 0 \quad \text{when} \quad x_0 = \dot{x}_0 = 0.$$

A supplementary equation of type (74.4) or (74.5) is not indicated here.

The two examples (74.1) and (74.6) belong among the differential equations

$$(74.7) \qquad \dot{\boldsymbol{x}} = \boldsymbol{f}(\boldsymbol{x})$$

which we already treated in secs. 16, 63, and 65, and whose right side has the following property. The domain of definition $B \subset R_n$ of $\boldsymbol{f}$ can be decomposed into a number of subdomains $B_i \subset B$, such that in each subdomain $\boldsymbol{f}(\boldsymbol{x})$ is linear,

$$\boldsymbol{f}(\boldsymbol{x}) = a_{i0} + a_{i1} x_1 + \cdots + a_{in} x_n, \quad \boldsymbol{x} \in B_i.$$

The subdomains are separated from each other by hypersurfaces which are usually called *switching spaces*. Usually $\boldsymbol{f}(\boldsymbol{x})$ has a jump at a point $\boldsymbol{x}$ in a switching space. Within each subdomain B_i the solution of (74.7) is uniquely determined by the initial point and can be written in closed form as a solution of a linear equation. The solution is to be continuous at points of transition through the switching spaces. If a transition is not possible (*cf.* example (74.1)) then we speak of an *endpoint* of the motion.

24*

If $f(x)$ is explicitly defined in the switching spaces, an endpoint can become an initial point of a motion which lies entirely within a switching space. We do not intend to investigate the differential equations (74.7) in complete generality.[1] We shall only concern ourselves with a few further examples which stem from practical problems and which have been chosen to illustrate the various methods used to determine the equilibria and periodic solutions.

3) *Oscillations in a vacuum tube generator with Z-shaped character-istic*.[2] The process is approximated by the scalar equation

$$(74.8) \qquad \ddot{x} + a\dot{x} + bx = \frac{1}{2} b(1 + \operatorname{sgn} \dot{x}) = \begin{cases} 0 \ (\dot{x} < 0) \\ b \ (\dot{x} > 0) \end{cases}$$

in which a and b are positive numbers.

Let α_1 and α_2 denote the roots of the characteristic equation

$$\lambda^2 + \alpha\lambda + b = 0$$

and assume that these roots are complex and have negative real parts:

$$\alpha_j = \varrho \pm i\omega, \quad j = 1, 2; \quad \varrho < 0.$$

In the lower half-plane ($\dot{x} < 0$), the solution has the form

$$x(t) = \frac{1}{\alpha_2 - \alpha_1} \left(x_0(\alpha_2 e^{\alpha_1(t-t_0)} - \alpha_1 e^{\alpha_2(t-t_0)}) - \dot{x}_0(e^{\alpha_1(t-t_0)} - e^{\alpha_2(t-t_0)}) \right).$$

The phase trajectories are spirals since the origin is a focus. If we choose $t_0 = 0$ so that $\dot{x}_0 = 0$ then the next intersection of the trajectory with the x-axis occurs at a time t_1 defined by $\dot{x}(t_1) = 0$. We have

$$e^{\alpha_1 t_1} = e^{\alpha_2 t_1}, \qquad \text{hence} \qquad t_1 = \frac{\pi}{\omega}$$

and

$$x_1 := x(t_1) = x_0 e^{\alpha_1 t_1} = -x_0 e^{\frac{\pi}{\omega}\varrho}.$$

In the upper half-plane the point $(1, 0)$ is singular. It is also a focus. The solution has the form

$$x(t) = 1 + \frac{1}{\alpha_2 - \alpha_1} \left((x_0 - 1)(\alpha_2 e^{\alpha_1(t-t_0)} - \alpha_1 e^{\alpha_2(t-t_0)}) - \dot{x}_0(e^{\alpha_1(t-t_0)} - e^{\alpha_2(t-t_0)}) \right).$$

The trajectory which begins at $x = x_1$, $\dot{x} = 0$, has its next intersection with the x-axis at

$$(74.9) \qquad x_2 := 1 + (x_1 - 1)e^{\alpha_1 t_1} = 1 + e^{\frac{\pi}{\omega}\varrho} + x_0 e^{\frac{2\pi}{\omega}\varrho}.$$

---

[1] *cf.* for instance ANDRÉ and SEIBERT [1], FILIPOV [1], FLÜGGE-LOTZ [1, 2], REISSIG [1].

[2] ANDRONOV, WITT and KHAIKIN [1].

A periodic solution is determined by $x_2 = x_0$. For its initial value x_0 we find

$$(74.10) \qquad x_0 = \left(1 - e^{\frac{\pi}{\omega}\varrho}\right)^{-1}.$$

(74.9) corresponds to the difference equation (71.5). Its equilibrium is stable since the coefficient $e^{2\pi\varrho/\omega}$ of x_0 is smaller than one, because $\varrho < 0$. Accordingly, the periodic solution defined by (74.10) is orbitally asymptotically stable.

Equation (74.8) does not describe the behavior of the motion at the origin. If we introduce the additional equation

$$x \equiv 0 \quad \text{when} \quad x_0 = \dot{x}_0 = 0$$

then the system receives an unstable solution $x = 0$ (we could just as well define $x = \beta$ for some β, $0 \le \beta \le 1$, for $\dot{x} = 0$). If the tube has an initial grid voltage, we must replace (74.8) by

$$\ddot{x} + a\dot{x} + bx = \begin{cases} 0 \ (\dot{x} < h) \\ b \ (\dot{x} > h) \end{cases}.$$

This time the *switching line* is the straight line $\dot{x} = h$. Since for sufficiently small initial values this line is never reached the corresponding trajectories satisfy the first equation throughout and the origin is a stable focus. The equation for t_1 can be put in the form

$$(x_0 b - h\varrho) \sin \omega t_1 - h\omega \cos \omega t_1 = h\omega\, e^{-\varrho t_1}.$$

In general, it has two positive roots; these lead to two distinct periodic solutions, one of which must be unstable. As h tends to zero the unstable periodic solution degenerates to a point: The unstable periodic solution and the stable singular point become the unstable singular point of the previous case. We have a typical branch phenomenon here (secs. 22, 82).

As already indicated, (74.8) is an approximative equation: a function $f(\dot{x})$ whose graph has, say, the form of the curve in figure 74.2 is approximated by a step function. Characteristics of the kind illustrated by the curve in figure 74.2 occur quite frequently and one usually attempts to apply to them the method of the example.

4) *A control system with a single non-linearity.* We first discuss the occurrence of periodic motions in a control system of the form (31.1). We assume that the characteristic roots of the matrix A are all simple and have

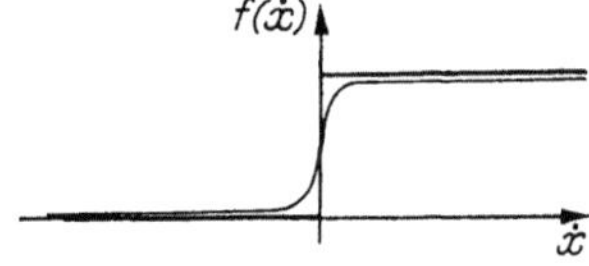

Fig. 74.2. Z-shaped characteristic

negative real parts. This assumption is not absolutely essential but it simplifies the subsequent computations. We also assume that the equation has already been put into the form (31.10) by the canonical transformation described in sec. 31. We again write this equation with a slightly changed notation,

$$(74.11) \qquad \dot{z}_i = \alpha_i z_i + f(s), \quad i = 1, 2, \ldots, n,$$

$$(74.12) \qquad \dot{s} = c_1 z_1 + \cdots + c_n z_n - h f(s), \quad h > 0.$$

The numbers α_i and the parameters $c_1, \ldots, c_n, h$, are known, fixed quantities. The function $f(s)$ is defined as follows:

$$
\begin{aligned}
f(s) &= -1 \quad &&\text{for } s < -\gamma \\
& &&\text{and for } -\gamma < s < -\beta,\ \dot{s} > 0, \\
f(s) &= 0 \quad &&\text{for } -\beta < s < \beta \\
(74.13)\qquad& &&\text{and for } -\gamma < s < -\beta,\ \dot{s} < 0 \\
& &&\text{and for } \beta < s < \gamma,\ \dot{s} > 0 \\
f(s) &= +1 \quad &&\text{for } s > \gamma \\
& &&\text{and for } \beta < s < \gamma,\ \dot{s} < 0.
\end{aligned}
$$

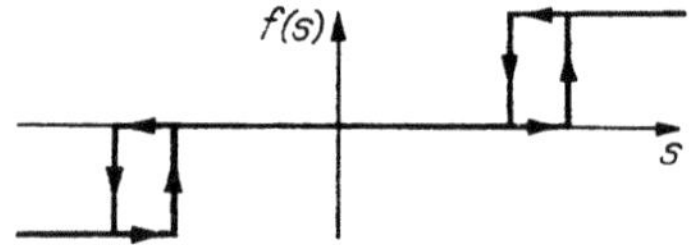

Fig. 74.3. Characteristic with hysteresis and clearance (74.13)

In these equations $0 \leq \beta \leq \gamma$.

The function (74.13) is the characteristic of a relay with a dead zone, $-\beta < s < +\beta$, and a hysteresis effect. By specializing we can simplify this characteristic. If $\beta = 0$, the dead zone does not occur. If $\beta = \gamma = 0$ we have the characteristic of (74.1). Our task is to determine periodic states of the control system, i.e. to construct periodic solutions of (71.11), (71.12) and to discuss their stability.

We begin by observing the system at time $t_0 = 0$, when

$$s(t_0) := s_0 = \gamma; \quad \dot{s}(t_0) < 0.$$

The function $f(s)$ takes a value in the upper right corner of the loop at this time (figure 74.3). The equations of the motion are

$$\dot{z}_i = \alpha_i z_i + 1, \ i = 1, 2, \ldots, n; \quad \dot{s} = c_1 z_1 + \cdots + c_n z_n - h.$$

This implies

$$z_i = (z_{i0} + \beta_i) e^{\alpha_i t} - \beta_i, \quad i = 1, 2, \ldots, n; \quad \beta_i := \alpha_i^{-1},$$

$$(74.14) \quad s = \sum_{k=1}^{n} c_k \beta_k (z_{k0} + \beta_k)(e^{\alpha_k t} - 1) - \delta t + \gamma$$

$$\delta := c_1 \beta_1 + \cdots + c_n \beta_n + h.$$

These equations are valid as long as $s > \beta$. If we define the time t_1 by

$$s(t_1) = \beta$$

then for $t > t_1$ equations (74.11), (74.12) are valid with $f(s) = 0$. In this case

(74.15)
$$z_i = z_{i1} e^{\alpha_i(t-t_1)}, \quad i = 1, 2, \ldots, n,$$
$$s = \sum_{k=1}^{n} c_k \beta_k z_{k1} (e^{\alpha_k(t-t_1)} - 1) + \beta.$$

A further switching occurs when $s(t)$ assumes the value $-\gamma$ or the value $+\gamma$.

We now define the time t_2 by

(74.16)
$$s(t_2) := -\gamma.$$

A periodic motion is obtained, in fact one for which the origin is a point of symmetry, if the state at time t_2 is symmetric to the state at time t_0, *i.e.* if

(74.17)
$$z_i(t_2) = : z_{i2} = - z_{i0}.$$

This equation corresponds to the defining equation (71.4) for the half-period. If we observe (74.14) and (74.15) then (74.17) together with a short calculation yields

(74.18)
$$\bar{z}_{i0} = - \beta_i \frac{1 - e^{-\alpha_i t_1}}{1 + e^{-\alpha_i t_2}} \; ; \quad \bar{z}_{i1} = - \beta_i \frac{1 - e^{\alpha_i t_1}}{1 + e^{\alpha_i t_2}} = - e^{\alpha_i(t_1-t_2)} \, \bar{z}_{i0}.$$

The bar indicates that the values belong to the periodic solution. Utilizing the expressions for s we obtain

(74.19)
$$\sum_{k=1}^{n} c_k \beta_k (\bar{z}_{k0} + \beta_k)(e^{\alpha_k t_1} - 1) - \delta t_1 + \gamma - \beta = 0,$$
$$\sum_{k=1}^{n} c_k \beta_k \bar{z}_{k1} (e^{\alpha_k(t_2-t_1)} - 1) + \beta + \gamma = 0.$$

Substituting the expressions (74.18) in (74.19) we obtain two defining equations for t_1 and t_2. The quantities (74.18) are the initial values of the periodic solution; the period is $2t_2$. If the Jacobian determinant of the left side of (74.19), considered as a function of t_1, t_2, is different from zero then we have an isolated periodic solution. In order that a real periodic motion correspond to this solution two further conditions must be satisfied. First we must have

$$0 < t_1 < t_2,$$

and secondly

(74.20)
$$\dot{s}(t_1 - 0) < 0, \quad \dot{s}(t_2 + 0) > 0.$$

These equations say that the relay actually switches. If they are not satisfied then s cannot leave the dead zone and remains constant (*cf.* example (74.1)) and we have $\dot{s} \equiv 0$. The equations of the motion are

$$\dot{z}_i = \alpha_i z_i + f(s), \quad i = 1, \ldots, n; \quad c_1 z_1 + \cdots + c_n z_n - h f(s) = 0,$$

and by eliminating $f(s)$ we obtain a linear system of equations which we may have to discuss further. Physically this state of the motion corresponds to a control through the feedback while the controller is not operating (fig. 74.4).

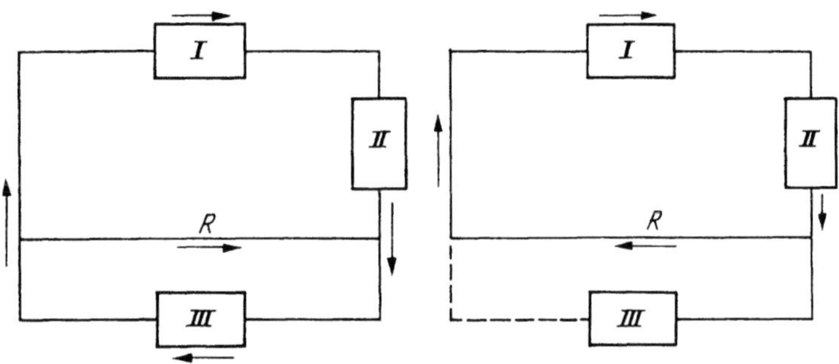

Fig. 74.4a. Condition (74.20): $\dot{s} \neq 0$
Fig. 74.4b. Condition (74.20): $\dot{s} \equiv 0$

There is no loss of generality in our choice of the initial point. For, as we have already pointed out in sec. 71, we may assign a fixed value to one of the variables in an autonomous equation. In the present case we have chosen $s(0) = \gamma$.

(74.16) is only one of two possibilities for defining the time t_2 at which the second switch occurs. Formally, there exists another possibility,

$$s(t_2) = +\gamma.$$

To it corresponds a periodic solution for which s varies periodically between β and γ, and which therefore remains entirely on the right loop of the characteristic. It is distinguished by

$$\tilde{z}_{i0} = \tilde{z}_{i2}.$$

As above, we find

$$\tilde{z}_{i0} = -\beta_i \frac{1 - e^{-\alpha_i t_1}}{1 - e^{-\alpha_i t_2}}; \quad \tilde{z}_{i1} = e^{\alpha_i(t_1 - t_2)} \tilde{z}_{i0}.$$

But there are also other possibilities. It can be that $z_{i0} \neq z_{i2}$. The variable s traverses the right loop of the characteristic on the time interval $t_2 < t < t_3$ and attains the switching value $-\gamma$ at a time $t_4 > t_3$. By evaluating the condition $z_{i4} = -z_{i0}$ we are again led to a symmetric solution which, however, is more complicated than the one previously considered; it remains to investigate whether this solution can be realized. In principle, even more complicated periodic solutions are possible. On the basis of our previous discussion the conditions can be stated for any numerically given equation. But there is little value in attempting this for the general case.[1])

Equations (74.18) and (74.19) can be interpreted as mappings in the sense of sec. 71. We consider the trajectories in an $(n + 1)$-dimensional space of the variables $z_1, \ldots, z_n, s$ and examine the mapping of the plane $s = \gamma$ onto the plane $s = -\gamma$ which is effected by the trajectories. Actually the plane $s = \gamma$ is first mapped to the plane $s = \beta$ which is then mapped further. A point which is symmetric to its image is invariant if the mapping is applied twice. It is thus a fixed point of the mapping of $s = \gamma$ into itself and hence is an initial point of a periodic solution.

The stability of the periodic solution defined by (74.18) and (74.19) can be tested by means of the variational equation. This is, however, inconvenient and it is simpler to introduce the variations into the defining equations for the initial values and the period. This amounts to testing the stability of the fixed point of the mapping just mentioned, or the stability of the equilibrium of the difference equation.

To do so we replace the exact defining values $\bar{z}_0, \bar{t}_1, \bar{t}_2$ of the periodic solution by the incremented values $\bar{z}_0 + \delta z_0, \bar{t}_1 + \delta t_1, t_2 + \delta t_2$ and derive a system of equations for the variations $\delta z_0, \delta t_i$. We use equations (74.11) and (74.12) for this system but remove all terms in which the exponent on the variations is greater than one. Computationally this looks as follows: By means of the equations for z_{i1} derived from (74.14) (these have been used before) we form the variations

$$\delta z_{i1} = \delta z_{i0} e^{\alpha_i \bar{t}_1} + \alpha_i e^{\alpha_i \bar{t}_1} \delta t_1 (\bar{z}_{i0} + \beta_i), \quad i = 1, 2, \ldots, n,$$

and similarly (74.15) is used to form

$$\delta z_{i2} = e^{\alpha_i (\bar{t}_2 - \bar{t}_1)} \delta z_{i1} + \bar{z}_{i1} \alpha_i e^{\alpha_i (\bar{t}_2 - \bar{t}_1)} (\delta t_2 - \delta t_1)$$
$$= e^{\alpha_i (\bar{t}_2 - \bar{t}_1)} \delta z_{i1} - \alpha_i \bar{z}_{i0} (\delta t_2 - \delta t_1).$$

Then we form the total variations of the equations for $s(t_1)$ and $s(t_2)$ i.e. (74.19). By the definition of $\bar{t}_1$ and $\bar{t}_2$ these variations are zero.

From these $2n + 2$ equations we eliminate the variations

$$\delta t_1, \quad \delta t_2, \quad \delta z_{11}, \ldots, \delta z_{n1}$$

---

[1]) A discussion of "higher" periodic motions is found in ANDRONOV, WITT and KHAIKIN [1].

and obtain a system of equations of the form

$$\delta z_{i2} = e^{\alpha_i \bar{t}_2}\, \delta z_{i0} + p_i\, U + q_i\, V, \quad i = 1, 2, \ldots, n.$$

$p_1, \ldots, p_n$ and $q_1, \ldots, q_n$ denote certain constants which depend on $\bar{z}^0$ and on the parameters of the system, whereas U and V are linear forms in $\delta z_{10}, \ldots, \delta z_{n0}$ which do not depend on i and therefore can be written in vector notation as

$$U =: \boldsymbol{u}^T \delta \boldsymbol{z}_0, \quad V =: \boldsymbol{v}^T \delta \boldsymbol{z}_0.$$

If we now introduce the notations

$$\boldsymbol{p} := \mathrm{col}\,(p_1, \ldots, p_n), \quad \boldsymbol{q} := \mathrm{col}\,(q_1, \ldots, q_n),$$

and

$$D(\boldsymbol{\alpha}) := \mathrm{diag}\,(e^{\alpha_1 \bar{t}_2}, \ldots, e^{\alpha_n \bar{t}_2}),$$

then the system of equations assumes the form

$$(74.21) \quad \delta \boldsymbol{z}_{m+2} = \left(D(\boldsymbol{\alpha}) + \boldsymbol{p}\,\boldsymbol{u}^T + \boldsymbol{q}\,\boldsymbol{v}^T\right) \delta \boldsymbol{z}_m, \quad m = 0, 2, 4, \ldots,$$

and the stability can be discussed by the rules of sec. 14. Equation (74.21) is the equation of the first approximation for the difference equation (71.5) which, however, has been stated in terms of the half period here.

If the characteristic roots of the matrix of (74.21) are all smaller than one in absolute value then the variations $\delta \boldsymbol{z}_m$ tend to zero, *i.e.* the perturbed motion tends to the periodic motion: The periodic motion is orbitally asymptotically stable. It is not asymptotically stable in the sense of Liapunov because of the effect of the time variations δt_1 and δt_2 which depend linearly on $\delta \boldsymbol{z}$. During the first half cycle the perturbed motion lags behind the periodic motion by δt_2: The second switch occurs at time $\bar{t}_2$ for the periodic motion and at time $\bar{t}_2 + \delta t_2$ for the perturbed motion (the variation can of course be negative). This lag is repeated during each half cycle. The lags of all the cycles add and after r half cycles the total lag is

$$(74.22) \qquad \delta t_2 + \delta t_{2,2} + \cdots + \delta t_{2,r-2}.$$

However, this sum cannot become arbitrarily large: The variation $\delta t_{2,m}$ depends linearly on the components of $\delta \boldsymbol{z}_m$ and these tend exponentially to zero because of (74.21). The terms of (74.22) can thus be estimated by the terms of a geometric progression and the total lag tends toward a constant limit. In this case we speak of *asymptotic orbital stability with an asymptotically constant phase difference* (*cf.* sec. 81).

The solution procedure which we described is theoretically interesting and informative but it leads to complicated computations and is unsuitable for practical problems. Therefore other methods were developed for studying concrete systems of relays. They depend essentially on the following ideas.

Let $F(s)$ be the transfer function of the linear part and $u(t)$ its transient response (sec. 8). Let the equation of the relay be $x_O = c \operatorname{sgn} x_I$. As the system executes a symmetric periodic oscillation of the type described by (74.18) and (74.19) the input of the linear part consists of jumps of absolute value $2c$ with alternating sign. Accordingly, the steady state output is given by

$$(74.23) \quad v_O(\tau) = 2c\left(u(\tau) - u(\tau + T) + u(\tau + 2T) - \cdots\right), \; 0 \leq \tau < T.$$

$2T$ denotes the length of the period. We now open the circuit and use as input for the open circuit the variable

$$x_I = c \sin \frac{2\pi\tau}{2T}$$

which has period $2T$ and amplitude c, and determine the corresponding output $\tilde{v}_O(\tau)$. If (74.23) actually describes the periodic steady state then the action of the open circuit must correspond to that of the closed circuit, *i.e.* the output $\tilde{v}_O$ must correspond to the output v_O, and we must have $\tilde{v}_O = -\operatorname{sgn} x_I$. The practical procedure amounts to deriving closed expressions for (74.23) which can be used to test the "closed circuit condition" which we just discussed. *cf.* also LUR'E [1], ZYPKIN [1].

75. A System with Several Discontinuity Types

We consider a further example in which several typically "nonlinear" phenomena can be observed. We are dealing with a control circuit with a relay amplifier. The measuring unit has a dead zone caused by dry friction. The equations of the motion, the parameters being simplified as much as possible[1]), are (*cf.* figure (74.4)):

$$\dot{x} = -y \quad \text{(controlled system)},$$

$$\left.\begin{array}{l} v = x - a \operatorname{sgn} \dot{v} \text{ for } \dot{v} \neq 0 \\[2mm] \dot{v} = 0 \;\text{ for } |x - v| < a \end{array}\right\} \text{(measuring unit)},$$

$$(75.1)$$

$$s = v - y \quad \text{(feedback)},$$

$$\dot{y} = \operatorname{sgn} s \text{ for } s \neq 0 \quad \text{(motor)}.$$

If $s = 0$, the control is through the feedback and the motor is not operating (*cf.* condition (74.20)). In this case the last two equations do not apply and the equation for the controlled system has the form

$$\dot{x} = -v.$$

The motion for this system has four different states which depend on the size of the initial values. In discussing the equations we again use the

---

[1]) PETROV and ULANOV [1].

fact that we can fix one of the initial values of an autonomous system arbitrarily. Here we set $s(t_0) = s_0 = 0$.

I. Let $s_0 = 0$, $y_0 = v_0 = 0$. Then $\dot{x}_0 = 0$. Also let $\dot{v}_0 = 0$. The system is then completely at rest and, in fact, we must have $|v_0 - x_0| = |x_0| < a$.

II. $s_0 = 0$, $y_0 = v_0 \neq 0$, $|v_0 - x_0| < a$. As long as $|v - x| < a$, $\dot{v} = 0$ and the measuring unit is at rest. It only begins to function when $|v - x| = a$, and this does happen since

$$\dot{x} = - v_0, \quad x = - v_0 t + x_0.$$

When the measuring unit is working we have

$$\dot{v} = \dot{x} = - v, \quad v = v_0 e^{-t}, \quad x = - v_0 e^{-t} \pm a, \quad y = v_0 e^{-t}.$$

So far the motor has been at rest. As t increases, y and v tend to zero, x to $\pm a$.

III. In studying the remaining states of the motioo we may neglect the variable x, since $|v - x| = a$ throughout. It suffices to consider the projection of the trajectory into the (y, v)-plane and, in particular, its intersection with the *switching line* L whose equation is

$$s = v - y = 0.$$

Again we take $s_0 = 0$ but assume that $|x_0 - v_0| = a$, i.e. $\dot{v}_0 \neq 0$. Furthermore $y_0 = v_0 \neq 0$ and without loss of generality we can place the initial point (y_0, v_c) in the upper half plane. On the right of L we have $s > 0$. Since $\dot{v} = \dot{x} = -y$, v is decreasing in the upper half plane: In the first quadrant the projection crosses the line L from above. The equations of the motion are

$$(75.2) \qquad \dot{y} = - 1, \quad \dot{v} = - y, \quad x = v - a, \quad \dot{x} = \dot{v}.$$

For $t > 0$ we have

$$y = y_0 - t, \quad v = y_0(1 - t) + \frac{t^2}{2}$$

and this holds until either $\dot{v} = 0$ or $s = 0$. In the first case the measuring unit enters the *insensitive zone* and in the second case the relay turns off. Let the corresponding times be t_1, resp. t_2; the same indices are used to denote the values of the variables.

If $\dot{v}_1 := \dot{v}(t_1) = 0$ then $y_1 = 0$, $t_1 = y_0$, $v_1 = \frac{y_0}{2}(2 - y_0)$, $x_1 = v_1 - a$. For $s(t_2) =: s_2 = 0$, (75.2) implies

$$(75.3) \qquad y_2 = v_2, \quad t_2 = 2(y_0 - 1).$$

If $t_1 < t_2$ then $y_0 > 2$. Thus if $y_0 \leq 2$ then $t_2 < t_1$, i.e. the relay switches first.

If $0 < y_0 \leq 1$ then (75.3) implies the inequality $t_2 < 0$, which is impossible. In this case, therefore, we must have $s = 0$. We are in state II.

If $y_0 > 2$ then $t_1 < t_2$. For $t > t_1$ the motion is described by the equations

$$\dot{x} = -y, \quad \dot{v} = 0, \quad \dot{y} = -1,$$

(75.4)

$$x = \frac{1}{2}(t - t_1)^2 + x_1, \quad y = t_1 - t, \quad v = v_1.$$

The time t_3 at which the measuring unit leaves the insensitive zone and again begins to function is given by the equation

$$|x_3 - v_1| = a.$$

Again two possibilities must be distinguished.

a) If $t_3 < t_2$ then

$$x_3 = \frac{1}{2}(t_3 - t_1)^2 + x_1 = v_1 + a,$$

$$(t_3 - t_1)^2 = 4a; \quad t_3 = t_1 + 2\sqrt{a} = y_0 + 2\sqrt{a},$$

$$y_3 = -2\sqrt{a}, \quad v_3 = v_1 = y_0\left(1 - \frac{1}{2}y_0\right).$$

For $t > t_3$ the equations (75.2) are again valid until time t_2, that is until again $s = 0$ and $y = v$. We find that

$$y_2 = y_3 - (t_2 - t_3), \quad v_2 = \frac{1}{2}(t_2 - t_3)^2 - y_3(t_2 - t_3) + v_3,$$

which implies

$$t_2 = t_3 + y_3 - 1 + \sqrt{y_3^2 + 1 - 2v_3} = y_0 - 1 + \sqrt{(y_0 - 1)^2 + 4a},$$

(75.5)

$$y_2 = 1 - \sqrt{(y_0 - 1)^2 + 4a}.$$

b) If $t_2 < t_3$ then the relay switches and the state described by (75.4) is entered again. We obtain consecutively

$$y_2 = t_1 - t_2 = v_2,$$

$$t_2 = t_1 + y_0\left(\frac{1}{2}y_0 - 1\right) = \frac{1}{2}y_0^2,$$

$$x_2 = \frac{1}{2}(t_2 - t_1)^2 + x_1 = \frac{1}{2}v_1^2 + x_1 = \frac{1}{2}v_1^2 + v_1 - a.$$

For $t > t_2$, y and v are constant. The motion is represented by a point of the (y, v)-plane. During this time the variable x satisfies

$$x = -y_2(t - t_2) + x_2$$

and this holds until $x = v_1 + a$, *i.e.* until the measuring unit begins working again. This happens at the time t_3 at which

$$x_3 = - y_2(t_3 - t_2) + x_2 = v_1 + a,$$

respectively

$$- v_1(t_3 - t_2) + \frac{1}{2} v_1^2 + v_1 - a = v_1 + a.$$

This implies

$$t_3 = - \frac{2a}{v_1} + \frac{1}{2} v_1 + t_2 = \frac{4a}{y_0(y_0 - 2)} + \frac{2y_0 + y_0^2}{4}.$$

This state can only occur when

$$|x_2 - v_1| < a.$$

Observing the signs, we are led to the inequalities

$$x_2 - v_1 = \frac{1}{2} v_1^2 - a < a, \quad i.e. \quad v_1^2 < 4a,$$

$$y_0^2(2 - y_0^2) < 16a,$$

$$y_0 < 1 + \sqrt{1 + 4\sqrt{a}}.$$

If $y_0 > 1 + \sqrt{1 + 4\sqrt{a}}$ then $t_3 < t_2$.

Thus we have four distinct states for the motion, according as

$$0 \leq y_0 < 1, \quad 1 < y_0 < 2,$$

$$2 < y_0 < 1 + \sqrt{1 + 4\sqrt{a}}, \quad \text{or} \quad 1 + \sqrt{1 + 4\sqrt{a}} < y_0.$$

In addition, there are the border line cases in which equality holds. We shall return to them presently. The projections of the trajectories map the ray $y = v > 0$ to the ray with the opposite direction, $y = v < 0$. The condition for a periodic motion is obviously $|y_2| = y_0$. If $0 < y_0 < 2$ then this cannot occur. These motions (state I and II) fade. If $2 < 1 + \sqrt{1 + 4\sqrt{a}}$ then

$$(75.6) \qquad\qquad y_2 = v_1 = \frac{1}{2} y_0(2 - y_0),$$

that is

$$\left|\frac{y_2}{y_0}\right| = \frac{1}{2}|2 - y_0| = \left|1 - \frac{1}{2} y_0\right|.$$

For $y_0 < 4$ this expression is smaller than one. If $y_0 > 4$ then $|y_2| > y_0$. The sequence $y_0, |y_2|, y_4, \ldots$ is increasing; y_{2n} is the abscissa of the nth point of intersection of the trajectory with the switching line. After finitely many steps

$$|y_{2n}| > 1 + \sqrt{1 + 4\sqrt{a}}$$

and a motion of the fourth state begins. The inequalities

$$y_0 > 4, \quad y_0 < 1 + \sqrt{1 + 4\sqrt{a}}$$

are compatible only if $a > 4$. If $a < 4$ then the motion will always fade since then $|y_2| < y_0$. In the fourth state we obtain from the relation (75.5) that

$$(y_2 - y_0)(y_2 + y_0 - 2) = 4a$$

and we see immediately that $y_0 = a$, $y_2 = -a$ furnishes a periodic solution. If $y_0 > a$ then $|y_2| < y_0$. The motion tends toward the periodic motion from the outside. If, on the other hand, $1 + \sqrt{1 + 4\sqrt{a}} < y_0 < a$ then $|y_2| > y_0$ and the motion tends toward the periodic motion from within. This implies immediately that the periodic solution is orbitally asymptotically stable with respect to y and v.

If $a > 4$ then there exists a second periodic motion for $y_0 = 4$. It satisfies the equation

$$y_2 = y_0 \left(1 - \frac{1}{2} y_0\right) = - y_0$$

(cf. (75.6)) and is obviously unstable in the (y, v)-plane. If $a = 4$ then the two periodic motions coincide. In summary we have the following result. If $a < 4$ then the motion fades for arbitrary initial values: The set of equilibrium points is attractive. In general the fading occurs in several stages. If $a > 4$ then there exist two periodic motions which correspond to the initial values $y_0 = v_0 = 4$, $x_0 = 4 - a$, resp. $y_0 = v_0 = a$, $x = 0$. As initial instant we use here the instant when the relay switches for the first time.

A motion with initial values $v_0 = y_0 < 4$, $|v_0 - x_0| = a$, tends toward one of the equilibrium positions. If $v_0 = y_0 > 4$, $|x_0 - v_0| = a$, then the motion tends toward the periodic solution. If $a = 4$ then there exists only one periodic solution; motions with $y_0 > 4$ and suitable values v_0, x_0 tend toward the periodic motion. If $y_0 < 4$ then the motion fades.

In the (y, v, x)-space the two periodic motions are unstable: The motions for $y_0 = v_0 = a$, $x_0 = \varepsilon > 0$ tend toward the equilibrium of the state II although their initial point may lie arbitrarily close to the trajectory of the periodic motion. This is possible because the curves in the (y, v)-plane are not phase trajectories but projections and the theory of sec. 18 does not apply directly. But on the switching line itself periodic solutions, respectively their projections, behave like separatrices and separate the domains of different stability behavior.

76. Perturbed Linear Equations

If we wish to carry out the program outlined immediately after (71.6) we must start with an unperturbed system (71.3) whose periodic solution is known, for example with the nonhomogeneous linear equation

$$(76.1) \qquad \dot{x} = A(t)\,x + f(t).$$

The matrix $A(t)$ and the external force $f(t)$ are periodic with period ω.

We first assume that we do not have resonance, *i.e.* that the homogeneous equation $\dot{x} = A(t)\,x$ does not possess a periodic solution with period ω. By (72.7) the uniquely determined periodic solution of (76.1) can be written in the form

$$\varphi(t) = \Re f := \int_0^\omega G(t,u)\,f(u)\,du.$$

This implies an estimate for the norm

$$\|\varphi(t)\| := \sup_{0 \leq t \leq \omega} |\varphi(t)|,$$

given by

$$\|\varphi(t)\| \leq \|f(t)\| \sup_{0 \leq t \leq \omega} \int_0^\omega \|G(t,u)\|\,du,$$

respectively

$$(76.2) \qquad \|\Re f\| \leq K\|f(t)\|.$$

The constant

$$K := \sup_{0 \leq t \leq \omega} \int_0^\omega \|G(t,u)\|\,du$$

depends only on the Green's matrix $G(t,u)$, that is only on the matrix $A(t)$.

In addition to (76.1) we consider the nonlinear equation

$$(76.3) \qquad \dot{x} = A(t)\,x + f(t) + g(x,t)$$

for sufficiently small x. For a fixed x the nonlinear function $g(x,t)$ is periodic in t. We assume that the usual hypotheses on the existence and uniqueness of the solution are satisfied. The problem of determining a periodic solution $\psi(t)$ is equivalent to a boundary value problem and thus to solving an integral equation

$$(76.4) \qquad x(t) = \Re f(t) + \Re g\big(x(t),\,t\big).$$

As in the case of (66.13), the equivalence can again be proved directly. We shall attempt to solve the nonlinear integral equation by successive approximations, starting with the periodic solution $\varphi(t)$ of equation (76.1) as the zeroth iterate. Accordingly we define the functions

$$(76.5) \qquad \begin{aligned} x^{(0)} &= \Re f, \\ x^{(i)} &= \Re f + \Re g(x^{(i-1)},\,t), \quad i = 1, 2, \ldots, \end{aligned}$$

and then construct the desired solution by means of the series $x^{(0)} + (x^{(1)} - x^{(0)}) + \cdots$ (*cf.* sec. 66). In order to guarantee the convergence of the series the differential equation must be restricted by certain assumptions. We shall assume that for all x and y in a certain domain $B \subset R_n$ a Lipschitz condition

$$(76.6) \qquad |g(x, t) - g(y, t)| \leq L\,|x - y|, \quad x \in B, \; y \in B,$$

is satisfied. We also assume that the function $g(x, t)$ remains below a fixed bound,

$$(76.7) \qquad |g(x, t)| < M(\delta) \; \text{ for } \; |x - x^{(0)}(t)| < \delta, \; 0 \leq t \leq \omega,$$

for all x in a δ-neighborhood about the function $x^{(0)}(t)$ defined above, and for all t. This second condition assures that together with $x^{(i)}$, $x^{(i+1)}$ also belongs to B, so that B is mapped into itself. For we have

$$\|x^{(i)} - x^{(0)}\| = \|\Re\,g(x^{(i-1)}, t)\| \leq K\,\|g(x^{(i-1)}, t)\|$$

and because of (76.7),

$$\|x^{(i)} - x^{(0)}\| \leq K M(\delta).$$

If the inequality

$$(76.8) \qquad\qquad K M(\delta) < \delta$$

is satisfied then

$$\|x^{(i)} - x^{(0)}\| < \delta.$$

If, for a fixed δ, we define the domain B by the inequality

$$\|x - x^{(0)}(t)\| < \delta$$

then conditions (76.7) and (76.8) guarantee that all terms of the sequence $x^{(i)}$ belong to B. From the defining equation (76.5) we conclude that

$$\|x^{(i+1)} - x^{(i)}\| \leq \|\Re\left(g(x^{(i)}, t) - g(x^{(i-1)}, t)\right)\|$$
$$\leq K\,\|g(x^{(i)}, t) - g(x^{(i-1)}, t)\|$$
$$\leq K L\,\|x^{(i)} - x^{(i-1)}\|.$$

If $KL < 1$ then the series

$$x^{(0)} + (x^{(1)} - x^{(0)}) + (x^{(2)} - x^{(1)}) + \cdots$$

converges absolutely and uniformly on B since it is majorized by the geometric series $\sum (KL)^k$; and this implies the convergence of the approximation process. The periodic solution can also be interpreted as a fixed point of the mapping

$$x(t) \to y(t) := \Re f + \Re\,g(x, t).$$

Condition (76.7), which says that $g(x, t)$ is "small", is essential. We frequently force this condition to hold by introducing a small parameter μ. We write (76.3) in the form

$$(76.9) \qquad \dot{x} = A(t)\,x + f(t) + \mu\,g(x, t; \mu)$$

to indicate that the function $g(x, t)$ may also depend on μ. The inequalities (76.6) and (76.7) must be uniform in μ on a certain interval $0 \le \mu \le \mu_0$, respectively, if negative values of μ are also admitted, $0 \le |\mu| \le \mu_0$. The conditions for the constants K and L are

$$(76.10) \qquad \mu_0 K M(\delta) < \delta, \quad \mu_0 K L < 1.$$

Summarizing, we obtain the following

Theorem 76.1. Let the nonlinear differential equation (76.9) be given. Let A, f, and g be periodic in t with period ω. Suppose that the homogeneous equation $\dot{x} = A(t)\,x$ does not possess a solution which is periodic with period ω and that the function $g(x, t; \mu)$ satisfies a Lipschitz condition (76.6) for all x in a certain domain B, which is uniform for $0 \le t \le \omega$ and $0 \le |\mu| \le \mu_0$. Then there exists a $\mu_1 \le \mu_0$ such that for $|\mu| \le \mu_1$, (76.9) has a uniquely determined periodic solution $\psi(t, \mu)$ which depends continuously on μ and which tends toward the periodic solution $\varphi(t)$ of the linear equation (76.1) as $\mu \to 0$. The solution $\psi(t, \mu)$ can be obtained by successive approximations. The bound for μ depends only on the matrix A and on the Lipschitz constant L, because of (76.10).

The approximation process has still another interpretation. The variable $x^{(i)}$ defined by (76.5) satisfies a differential equation

$$\dot{y} = A(t)\,y + f(t) + \mu\,g\big(x^{(i-1)}(t), t\big),$$

whose right side is known if the iterates $x^{(0)}, \ldots, x^{(i-1)}$ have already been determined; in fact $x^{(i)}$ is the only, and therefore unique periodic solution of this equation. Thus the approximation consists in finding the periodic solution of a sequence of linear nonhomogeneous differential equations.

If the function $g(x, t; \mu)$ is analytic in the parameter μ, *i.e.* if there exists a power series expansion

$$(76.11) \qquad g(x, t; \mu) = g^{(0)}(x, t) + \mu\,g^{(1)}(x, t) + \cdots$$

(the coefficients $g^{(j)}$ are vectors whose elements are periodic in t) then the periodic solution $\psi(t, \mu)$ is also analytic in μ. Having taken care of the existence question, we can in this case proceed formally and substitute the expression

$$(76.12) \qquad x = \psi(t, \mu) = \varphi(t) + \mu\,\psi^{(1)}(t) + \mu^2\,\psi^{(2)}(t) + \cdots$$

into the differential equation (76.9). We obtain

$$\dot{\varphi}(t) + \mu \dot{\psi}^{(1)}(t) + \mu^2 \dot{\psi}^{(2)}(t) + \cdots = A \varphi(t) + \mu A \psi^{(1)}(t)$$
$$+ \mu^2 A \psi^{(2)}(t) + \cdots + f(t) + \mu g^{(0)}(\varphi + \mu \psi^{(1)} + \cdots, t)$$
$$+ \mu^2 g^{(1)}(\varphi + \mu \psi^{(1)} + \cdots, t) + \cdots.$$

We then must expand, whenever possible, the nonlinear functions as powers of μ and equate the coefficients of equal powers. We obtain a sequence of equations

$$\dot{\varphi} = A \varphi + f,$$
$$\dot{\psi}^{(1)} = A \psi^{(1)} + g^{(0)}(\varphi, t),$$
$$(76.13) \qquad \dot{\psi}^{(2)} = A \psi^{(2)} + g^{(1)}(\varphi, t) + \frac{\partial g^{(0)}}{\partial x}\bigg|_{x=\varphi} \psi^{(1)},$$
$$\cdots\cdots\cdots\cdots\cdots\cdots\cdots\cdots\cdots\cdots\cdots\cdots$$

the first of which is identical with (76.1), and we again have to find each of the periodic solutions. It is already obvious from the third equation in (76.13) that this approach can be carried out only if $g(x, t)$ is analytic in x also.

In order to test the stability of the periodic solution so found, we form the equation of the perturbed motion: We set

$$x = z + \psi(t; \mu), \quad \dot{x} = \dot{z} + \frac{\partial}{\partial t} \psi(t, \mu) = \dot{z} + \dot{\psi}.$$

It follows that

$$\dot{z} + \dot{\psi} = A z + A \psi + f + \mu g(z + \psi, t; \mu)$$
$$= A z + A \psi + f + \mu \left(g(\psi, t; \mu) + \frac{\partial g}{\partial x}\bigg|_{x=\psi} z + \cdots \right),$$

and since ψ satisfies equation (76.9),

$$(76.14) \qquad \dot{z} = A z + \mu \frac{\partial g}{\partial x}\bigg|_{x=\psi} z + \text{terms of second order}.$$

The variational equation is

$$(76.15) \qquad \dot{z} = \left(A(t) + \mu \frac{\partial g}{\partial x}\bigg|_{x=\psi} \right) z.$$

This is a linear equation with periodic coefficients which can be studied by the methods of sec. 60.

Example 1[1]). The scalar equation

$$\ddot{x} + (a' + b' x^2 + c' x^4) \dot{x} + k^2 x = p \sin t$$

describes the action of a vacuum tube generator excited by an alternating voltage, if the constants are appropriately interpreted. In order to

---

[1]) MALKIN [5].

be able to apply the theory discussed above we introduce a small parameter by means of the substitution

$$a' = a\mu, \quad b' = b\mu, \quad c' = c\mu,$$

which changes the equation into

(76.16) $$\ddot{x} + k^2 x = p \sin t - \mu(a + bx^2 + cx^4)\,\dot{x}.$$

Since resonance is excluded k cannot be an integer. Because of the simplicity of the equation it is not necessary to write it in vector form. We work with the periodic scalar solution

(76.17) $$\varphi(t) = \frac{p}{k^2 - 1}\sin t =: q \sin t$$

of the linear equation and we substitute the expansions

$$x = \varphi(t) + \mu x_1(t) + \mu^2 x_2(t) + \cdots,$$
$$\dot{x} = \dot{\varphi}(t) + \mu \dot{x}_1(t) + \mu^2 \dot{x}_2(t) + \cdots,$$

into the equation. In analogy with (76.13) we obtain

$$\ddot{x}_1 + k^2 x_1 = (a + b\varphi^2 + c\varphi^4)\,\dot{\varphi},$$
$$\ddot{x}_2 + k^2 x_2 = (a + b\varphi^2 + c\varphi^4)\,\dot{x}_1 + (2b\varphi x_1 + 4c\varphi^3 x_1)\,\dot{\varphi}.$$

The solution procedure becomes more perspicuous if the right sides are expanded into Fourier series and the solution is obtained by superposition; (72.17) must then be used. In this manner we obtain for x_1,

$$\ddot{x}_1 + k^2 x_1 = aq \cos t + \frac{bq^3}{4}(\cos t - \cos 3t)$$
$$+ \frac{cq^5}{16}(\cos 5t + 5\cos 3t + 10\cos t),$$

whence

$$x_1(t) = \left(aq + \frac{bq^3}{4} + \frac{5cq^5}{8}\right)\frac{\cos t}{k^2 - 1} + \left(-\frac{bq^3}{4} + \frac{5cq^5}{16}\right)\frac{\cos 3t}{k^2 - 9} + \frac{cq^5}{16}\frac{\cos 5t}{k^2 - 25}.$$

Example 2. The scalar equation

(76.18) $$\ddot{x} + k^2 x + a' x^3 = p \sin t$$

describes the behavior of a periodically excited oscillator with a nonlinear restoring force. It is known as the *Duffing equation*. Again we set $a' = \mu a$ and obtain

$$\ddot{x} + k^2 x = p \sin t - \mu a x^3.$$

As in the preceding example, we start with the zeroth iterate (76.17) and obtain for x_1 the equation

$$\ddot{x}_1 + k^2 x_1 = aq^3 \sin^3 t = \frac{aq^3}{4}(3\sin t - \sin 3t)\left(q := \frac{p}{k^2 - 1}\right)$$

and from it

$$x_1 = \frac{a\,q^3}{4}\left(3\,\frac{\sin t}{k^2-1} - \frac{\sin 3t}{k^2-9}\right).$$

For this example we shall compute the various bounds which appear in the proof of Theorem 76.1. Let δ be given. The domain in the $(x, \dot{x})$-plane, denoted by B in the proof, is defined by

$$|x - q\sin t| < \delta, \quad |\dot{x} - q\cos t| < \delta.$$

It is an annulus with radii $q - \delta$ and $q + \delta$. The maximum of the function $g = a x^3$ on this domain is $a(q + \delta)^3 =: M(\delta)$. The Lipschitz constant (*cf.* (76.6)) is the maximum assumed by the fraction

$$a\,\frac{x^3 - y^3}{x - y} = a\,(x^2 + xy + y^2).$$

For points of B, this constant is $L = 3a(q + \delta)^2$. In order to estimate the norm K of the operator $\Re$ we utilize (72.17),

$$\Re \sin m t = \frac{\sin m t}{k^2 - m^2}$$

and find that

$$\|\Re \sin m t\| = \frac{1}{k^2 - m^2}\,\|\sin m t\|.$$

If we let η^2 denote the distance from k^2 to the nearest perfect square then clearly

$$\|\Re \sin m t\| \le \frac{1}{\eta^2}\,\|\sin m t\|$$

and the constant K in the estimates can be set equal to $1/\eta^2$. Thus the inequalities (76.10) have the form

$$(76.19) \qquad |\mu|\,\frac{1}{\eta^2}\,a(q + \delta)^3 < \delta, \quad |\mu|\,\frac{1}{\eta^2}\,3a(q + \delta)^2 < 1.$$

But this condition is only sufficient and it gives much too small a value for the radius of convergence of the power series in μ. Studying a concrete example, we would rarely fall back on the general formulas and, in particular, we would not work with the norm of the operator $\Re$ but rather with an estimate for $\Re z$. We can also study the series (76.12) directly; it clearly represents a periodic solution within its domain of convergence.

The approximation procedure (76.13) can also be applied to find the almost periodic solution of an equation

$$(76.20) \qquad \dot{x} = A x + f(t) + \mu g(x, t; \mu)$$

in which A is constant, $f(t)$ is almost periodic, and $g(x, t; \mu)$ is *uniformly almost periodic* with respect to t for $x \in B$, *i.e.* the numbers L and τ appearing in Def. 73.1 can be chosen, for any given ε, independent of the particular value of $x \in B$. The function g may furthermore depend on a small parameter μ. The assumption that A is constant and not periodic as in (76.9), does not make the argument less general since by Theorem 61.1 the periodic matrix can be transformed into a constant matrix. Under this transformation the other two terms of the equation go over into almost periodic, respectively uniformly almost periodic, functions. We first show again, as in sec. 66, that the problem is equivalent to solving an integral equation of the form (76.4). In this case

$$\Re z(t) := \int\limits_{-\infty}^{+\infty} G(t - u) \, z(u) \, du$$

and the Green's matrix $G(s)$ is defined by (66.9). In proving the equivalence we use the fact that $G(s)$ has a jump at 0. The successive approximations themselves follow as in the periodic case. The constants K and $M(\delta)$, as well as the Lipschitz constant L, must now be so chosen that the estimates are valid for $-\infty < t < +\infty$. It remains only to show that the resulting solution is almost periodic. For this purpose we form the difference

$$\begin{aligned}
x(t + \tau) - x(t) = {}& \Re f(t)\big|_{t+\tau} + \mu \Re g\big(x(t), t; \mu\big)\big|_{t+\tau} \\
& - \Re f(t) - \mu \Re g\big(x(t), t; \mu\big).
\end{aligned}$$

Now we have

$$\begin{aligned}
\Re z(t)\big|_{t+\tau} = \Re z(t + \tau) &= \int\limits_{-\infty}^{+\infty} G(t + \tau - u) \, z(u) \, du \\
&= \int\limits_{-\infty}^{+\infty} G(t - u) \, z(u + \tau) \, du
\end{aligned}$$

and hence

$$\begin{aligned}
(76.21) \qquad x(t + \tau) - x(t) = {}& \Re\big(f(t + \tau) - f(t)\big) \\
& + \mu \Re\big(g(x(t + \tau), t + \tau; \mu) - g(x(t), t; \mu)\big).
\end{aligned}$$

We replace the last argument by

$$g\big(x(t + \tau), t + \tau; \mu\big) - g\big(x(t + \tau), t; \mu\big) + g\big(x(t + \tau), t; \mu\big) - g\big(x(t), t; \mu\big).$$

If we introduce the abbreviation

$$(76.22) \qquad c := \sup \big|x(t + \tau) - x(t)\big|, \quad -\infty < t < \infty,$$

and observe the Lipschitz condition (76.6) for g, we obtain

$$\big|g\big(x(t + \tau), t; \mu\big) - g\big(x(t), t; \mu\big)\big| < L \, \big|x(t + \tau) - x(t)\big| \leq L c.$$

Now we set

$$\begin{aligned}
c_1 &:= \sup \big|g(x, t + \tau; \mu) - g(x, t; \mu)\big|, \quad -\infty < t < \infty, \; x \in B, \\
c_2 &:= \sup \big|f(t + \tau) - f(t)\big|, \quad -\infty < t < \infty.
\end{aligned}$$

Then (76.21) implies

$$|x(t + \tau) - x(t)| < K c_2 + K c_1 + \mu K L c$$

and because of (76.22),

$$c - \mu K L c \leq K(c_1 + c_2).$$

Since f and g are almost periodic we can make c_1 and c_2 smaller than any given ε by the proper choice of τ and we can do this uniformly for $x \in B$, so that

$$c < \frac{2 K \varepsilon}{1 - \mu K L}.$$

But the denominator is clearly different from zero since (76.10) must be satisfied. Hence $x(t)$ is almost periodic.

77. Perturbed Linear Equations for the Resonance Case

The case treated in the last section, in which resonance was excluded, is characterized by the fact that the unperturbed equation (76.1) has a unique periodic solution $\varphi(t)$ or, equivalently, that the homogeneous equation

$$\dot{x} = A(t) x \quad \text{or} \quad \dot{y} = - A(t)^T y$$

has no periodic solution with period ω. In that case the periodic solution $\psi(t) = \psi(t, \mu g)$ which depends on the disturbance μg, tends to the solution $\varphi(t)$ of the unperturbed equation as $\mu g \to 0$. In the resonance case the situation cannot be so simple. For if the linear equation (76.1) has periodic solutions at all, that is if $f(t)$ satisfies the solvability conditions, then we immediately have a whole family of periodic solutions. Now it may happen that the nonlinear equation also has a family of periodic solutions and that the first family goes over into the second family as $\mu \to 0$. If, however, the perturbed equation has a unique periodic solution for each fixed μ then this solution approaches a single solution of the unperturbed equation as $\mu \to 0$. We may consider this solution, let it be denoted by $\tilde{\varphi}(t)$, as determined by the initial values. The question arises, which initial values determine those periodic solutions $\psi(t, \mu g)$ which tend to $\tilde{\varphi}(t)$ as $\mu \to 0$. In other words, we wish to find the "proper" initial values and then, starting from $\tilde{\varphi}(t)$, construct the solutions $\psi(t, \mu g)$. A plan for this procedure is obtained from the equations (76.13) if this time we mean by $\tilde{\varphi}(t)$ not the unique solution but a general periodic solution depending on a parameter. The second equation in (76.13) has periodic solutions only if the term $g^{(0)}(\tilde{\varphi}, t)$ satisfies the solvability conditions. We use these conditions and obtain a system of equations which determine the parameter for $\tilde{\varphi}(t)$. Thus we have then obtained $\tilde{\varphi}(t)$; but the function $\psi^{(1)}$ is not uniquely defined and depends on the parameters. To compute these parameters we utilize the solvability conditions of the third equation (76.13), etc. It can be shown that this process actually leads to a periodic solution. For this purpose certain additional conditions must also be satisfied so that the systems of equations used in the process have unique solutions. We prefer, however, to take a different approach, which was suggested by HALE [4]. Hale has proved that the procedure of successive approximations can be adjusted so that it will also work in the resonance case. Some preparation is necessary here.

We assume that the equation has the form

$$(77.1) \qquad \dot{y} = A y + \bar{g}(y, t),$$

$\bar{g}(y, t)$ is continuous with respect to y, periodic with respect to t, and satisfies a Lipschitz condition of the form (76.6). The assumption that A is constant is, as we have repeatedly noted, no restriction of the generality, nor is the assumption that $f(t)$ does not appear explicitly, since we can always turn to the perturbed motion (76.14). Let A be transformed to the form $\mathrm{diag}(A_1, A_2)$; A_1 is of order p and its characteristic roots λ are the ones which belong to periodic solutions with period ω of the homogeneous equation. Let

$$B(t) := \mathrm{diag}\,(e^{A_1 t}, E), \quad E = E_{n-p}.$$

This matrix is periodic and we have

$$\dot{B}(t) = \mathrm{diag}\,(A_1, 0)\, B(t).$$

If we substitute

$$y(t) = B(t)\, z(t), \quad \dot{y}(t) = \dot{B}(t)\, z(t) + B(t)\, \dot{z}(t)$$

in (77.1), we obtain

$$A B(t)\, z + \bar{g}\big(B(t)\, z, t\big) = \dot{B}(t)\, z + B(t)\, \dot{z}$$

and since

$$A B(t) - \dot{B}(t) = \mathrm{diag}\,(0, A_2)\, B(t) = \mathrm{diag}\,(0, A_2),$$

we solve for $\dot{z}$ and obtain

$$\dot{z} = \mathrm{diag}\,(0, A_2)\, z + B(t)^I\, \bar{g}\big(B(t)\, z, t\big).$$

The nonlinear term is of the same kind as that of (77.1). The matrix A_2 has no further critical characteristic roots. We may therefore assume from the start that the equation is given in the standard form

$$(77.2) \qquad \dot{x} = A x + g(x, t), \quad A = \mathrm{diag}\,(0, B).$$

B is of order $n - p$ and has no critical characteristic roots. The linear homogeneous equation

$$(77.3) \qquad \dot{x} = A x$$

has a p-parameter family

(77.4) $$\mathbf{x} = \mathrm{col}\ (\mathbf{c},\ 0),\quad \text{where}\ \mathbf{c} = \mathrm{col}\ (c_1,\ \ldots,\ c_p),$$

of constant, *i.e.* in particular, of periodic solutions. We must determine the constants $c_1, \ldots, c_p$ so that they give rise to a periodic solution of (77.2) and may be chosen as the zeroth iterate in the approximation procedure. As in sec. 76, we change the differential equation into an integral equation, but the integral operator must be defined differently here since we do not have a Green's matrix available. For this purpose we consider the space Φ of periodic functions $\mathbf{f}$ with period ω and set $\mathbf{f} = \mathrm{col}(\mathbf{g},\ \mathbf{h})$, where

$$\mathbf{g}(t) : = \mathrm{col}\ \big(f_1(t),\ \ldots,\ f_p(t)\big),\quad \mathbf{h}(t) : = \mathrm{col}\ \big(f_{p+1}(t),\ \ldots,\ f_n(t)\big),$$

$$\mathbf{y} : = \mathrm{col}\ (x_1,\ \ldots,\ x_p),\quad \mathbf{z} : = \mathrm{col}\ (x_{p+1},\ \ldots,\ x_n).$$

We introduce an auxiliary operator $\mathfrak{P}$ which is defined on the space Φ by

(77.5) $$\mathfrak{P}\mathbf{f} : = \frac{1}{\omega} \int_0^\omega \mathrm{col}\ \big(f_1(u),\ \ldots,\ f_p(u),\ 0,\ \ldots,\ 0\big)\ du$$

and we define a second auxiliary operator $\mathfrak{Q}$ as follows: For $\mathbf{f} \in \Phi$, $\mathfrak{Q}\mathbf{f} = \mathbf{x}^*$ is the unique periodic solution of the nonhomogeneous equation

(77.6) $$\dot{\mathbf{x}} = A\,\mathbf{x} + \mathbf{f}(t),$$

which satisfies the condition

(77.7) $$\mathfrak{P}\,\mathbf{x}^* = 0.$$

We first show that the definition makes sense and we do so by solving (77.6) directly, respectively by solving the equivalent system

$$\dot{\mathbf{y}} = \mathbf{g}(t),\quad \dot{\mathbf{z}} = B\,\mathbf{z} + \mathbf{h}(t).$$

By (72.7), the periodic solution of the second equation which involves the last $n - p$ components of $\mathbf{x}^*$ is given by

(77.8) $$\mathbf{z}^*(t) : = (e^{-B\omega} - E)^I \int_0^\omega e^{-Bu}\,\mathbf{h}(t + u)\ du.$$

In order that the equation for $\mathbf{y}$ have a periodic solution the solvability conditions must be satisfied. Here they have the simple form

$$\int_0^\omega \mathbf{g}(u)\ du = 0.$$

Then the periodic solution is

$$\mathbf{y}^*(t) = \int_0^t \mathbf{g}(u)\ du + \mathbf{y}_0$$

and y_0 is determined by the additional requirement (77.7), *i.e.* by

$$(77.9) \qquad \int_0^\infty y^*(t)\, dt = 0 .$$

If the component g_j is real we can find a constant ξ_j, $0 < \xi_j < \omega$, such that

$$y_j^*(t) = \int_{\xi_j}^t g_j(u)\, du .$$

For if we integrate this equation it follows that

$$\int_0^\omega y_j^*(t)\, dt = \int_0^\omega \int_{\xi_j}^t g_j(u)\, du\, dt = \int_0^\omega \int_0^t g_j(u)\, du\, dt - \int_0^\omega \int_0^{\xi_j} g_j(u)\, du\, dt$$

and since the left side is zero by (77.9), we have

$$\int_0^\omega \int_0^t g_j(u)\, du\, dt = \int_0^\omega \int_0^{\xi_j} g_j(u)\, du\, dt = \omega \int_0^{\xi_j} g_j(u)\, du ,$$

so that ξ_j is uniquely determined. If g_j is complex-valued, corresponding representations for the real and imaginary parts are derived separately.

With the help of the periodic function $x^* = \mathrm{col}(y^*, z^*)$ which is uniquely determined by f we have defined the operator

$$\mathfrak{Q} f := x^* .$$

By (77.9) it satisfies the equation (77.7), *i.e.* we have

$$(77.10) \qquad \mathfrak{P} \mathfrak{Q} f = 0 .$$

The operator $\mathfrak{Q}$ is bounded. For the component y^* this follows from the estimate

$$|y_j^*(t)| \le \left| \int_0^t g_j(u)\, du \right| + |y_{0j}^*| , \quad j = 1, 2, \ldots, p ,$$

and for the component z^* it follows from (76.2). Therefore there exists a constant Q such that

$$(77.11) \qquad \| \mathfrak{Q} f \| \le Q \| f \| .$$

Let $S(a) \subset \Phi$ be the set of all periodic functions f with the following property:

1) The norm of f is no greater than a fixed positive number δ,

$$\| f \| := \sup_{0 \le t \le \omega} |f(t)| \le \delta .$$

2) $\mathfrak{P} f$ is equal to a given vector $\mathrm{col}(a, 0) := a^*$, and $|a|$ is smaller than a fixed given number $\beta < \delta$.

With the help of $\mathfrak{P}$ and $\mathfrak{Q}$ we define the operator $\mathfrak{F} x$, $x \in S(a)$ by

$$\mathfrak{F} x := a^* + \mathfrak{Q} g\big(x(t), t\big) - \mathfrak{Q} \mathfrak{P} g\big(x(t), t\big) .$$

We have

$$\|\mathfrak{F}\boldsymbol{x}\| \leq |\boldsymbol{a}^*| + Q\,\|\boldsymbol{g}(\boldsymbol{x}(t),\,t)\| + Q\,\|\mathfrak{P}\boldsymbol{g}(\boldsymbol{x}(t),\,t)\|\,.$$

Since obviously

$$\|\mathfrak{P}\boldsymbol{f}\| \leq \frac{1}{\omega}\int_0^\omega \|\boldsymbol{f}(t)\|\,dt = \|\boldsymbol{f}\|$$

it follows that

$$\|\mathfrak{F}\boldsymbol{x}\| \leq |\boldsymbol{a}^*| + 2\,Q\,\|\boldsymbol{g}(\boldsymbol{x}(t),\,t)\|\,.$$

The second summand on the right is smaller than a number β_1, which depends only on Q and the bounds for $\boldsymbol{g}(\boldsymbol{x},\,t)$. If $\|\boldsymbol{g}(\boldsymbol{x},\,t)\|$ is sufficiently small then $\|\mathfrak{F}\boldsymbol{x}\| < \beta + \beta_1 < \delta$. Therefore $\mathfrak{F}\boldsymbol{x}$ has the first defining property for elements of $S(\boldsymbol{a})$. Property 2) is, of course, also given and therefore $\mathfrak{F}\boldsymbol{x}$ belongs to $S(\boldsymbol{a})$ for all $\boldsymbol{x} \in S(\boldsymbol{a})$, i.e. $\mathfrak{F}S(\boldsymbol{a}) \subset S(\boldsymbol{a})$. For two elements $\boldsymbol{x},\,\boldsymbol{y} \in S(\boldsymbol{a})$ we have

$$\mathfrak{F}\boldsymbol{x} - \mathfrak{F}\boldsymbol{y} = \mathfrak{Q}\big(\boldsymbol{g}(\boldsymbol{x}(t),\,t) - \boldsymbol{g}(\boldsymbol{y}(t),\,t)\big) - \mathfrak{Q}\,\mathfrak{P}\big(\boldsymbol{g}(\boldsymbol{x}(t),\,t) - \boldsymbol{g}(\boldsymbol{y}(t),\,t)\big).$$

If we estimate the right side, invoking the Lipschitz condition on $\boldsymbol{g}$, we obtain

$$\|\mathfrak{F}\boldsymbol{x} - \mathfrak{F}\boldsymbol{y}\| \leq 2\,Q\,L\,\|\boldsymbol{x} - \boldsymbol{y}\|\,.$$

If Q is sufficiently small then $\varkappa := 2QL < 1$ and we see from the inequality

$$\|\mathfrak{F}\boldsymbol{x} - \mathfrak{F}\boldsymbol{y}\| \leq \varkappa\,\|\boldsymbol{x} - \boldsymbol{y}\|,\ \ \varkappa < 1,$$

that the operator $\mathfrak{F}$ is a contraction mapping on the set $S(\boldsymbol{a})$. The mapping $\boldsymbol{x} \to \mathfrak{F}\boldsymbol{x}$ has a unique fixed point corresponding to a solution $\tilde{\boldsymbol{x}}(t)$ of the integral equation

$$\boldsymbol{x} = \mathfrak{F}\boldsymbol{x} = \boldsymbol{a}^* + \mathfrak{Q}\boldsymbol{g}(\boldsymbol{x},\,t) - \mathfrak{Q}\,\mathfrak{P}\boldsymbol{g}(\boldsymbol{x},\,t)\,.$$

The sequence

$$\boldsymbol{x}^{(0)} = \boldsymbol{a}^*,$$

$$\boldsymbol{x}^{(i)} = \mathfrak{F}\boldsymbol{x}^{(i-1)},\ \ i = 1,\,2,\,\ldots,$$

converges uniformly in t to the function $\tilde{\boldsymbol{x}}(t)$ and the inequality

$$(77.12) \qquad \|\tilde{\boldsymbol{x}} - \boldsymbol{x}^{(i)}\| \leq \|\boldsymbol{x}^{(1)} - \boldsymbol{x}^{(0)}\|\,\frac{\varkappa^i}{1 - \varkappa}$$

holds. This can be proved in the same way as the convergence of (76.5). Furthermore, one has

$$\dot{\boldsymbol{x}}^{(i)} = A\big(\mathfrak{Q}\boldsymbol{g}(\boldsymbol{x}^{(i-1)},\,t) - \mathfrak{Q}\,\mathfrak{P}\boldsymbol{g}(\boldsymbol{x}^{(i-1)},\,t)\big) + \boldsymbol{g}(\boldsymbol{x}^{(i-1)},\,t) - \mathfrak{P}\boldsymbol{g}(\boldsymbol{x}^{(i-1)},\,t)$$

and since $\mathfrak{Q}\boldsymbol{g}(\boldsymbol{x}^{(i-1)},\,t) - \mathfrak{Q}\,\mathfrak{P}\boldsymbol{g}(\boldsymbol{x}^{(i-1)},\,t) = \mathfrak{F}\boldsymbol{x}^{(i-1)} - \boldsymbol{a}^*$ and $A\boldsymbol{a}^* = 0$,

$$\dot{\boldsymbol{x}}^{(i)} = A\boldsymbol{x}^{(i-1)} + \boldsymbol{g}(\boldsymbol{x}^{(i-1)},\,t) - \mathfrak{P}\boldsymbol{g}(\boldsymbol{x}^{(i-1)},\,t)\,.$$

As in sec. 76, we see that the sequence $\dot{x}^{(i)}$ converges uniformly and conclude that $\tilde{x}(t)$ is continuously differentiable with respect to t and satisfies the differential equation

$$\dot{x} = A\,x + g(x,\,t) - \mathfrak{P}\,g(x,\,t)\,.$$

All this holds under the assumption that the number Q, *i.e.* the function $g(x,\,t)$, is sufficiently small. To formulate the complete result we again introduce a small parameter.

Theorem 77.1. Let the differential equation

$$\text{(77.13)} \qquad \dot{x} = A\,x + \mu\,g(x,\,t;\,\mu)\,, \quad A = \text{diag}\,(0,\,B)\,,$$

be given. Let $g(x,\,t;\,\mu)$ be periodic in t, continuous in all variables, once continuously differentiable with respect to μ, and twice continuously differentiable with respect to x. Then for each constant vector $a^* = \text{col}\,(a,\,0)$ whose norm is below a fixed bound β, $|a^*| < \beta$, there exists a continuously differentiable function $\tilde{x} = x(t,\,a^*,\,\mu) \in S(a)$ such that $x(t,\,a^*,\,0) \equiv a^*$ and

$$\text{(77.14)} \qquad \dot{x} = A\,x + \mu\,g(x,\,t;\,\mu) - \mu\,\mathfrak{P}\,g(x,\,t;\,\mu)$$

provided that μ is sufficiently small, $0 \leq |\mu| \leq \mu_0$. The bound μ_0 depends on β since it results from the conditions $|\mu| < 2QL$ and $\beta_1 + \mu\beta_1 < \delta_1$.

Let L_1 be a Lipschitz constant and γ a bound for the first derivatives. Then the iterates $x^{(i)}$ have continuous first order partial derivatives with respect to a^*. For the first iterate this is clear. We have

$$\frac{\partial x^{(0)}}{\partial a^*} = E\,.$$

We now proceed by induction. We assume that the kth iterate $x^{(k)}$ has continuous partial derivatives with respect to a^*. Then the expression

$$\text{(77.15)} \qquad \frac{\partial x^{(k+1)}}{\partial a^*} = E + \mu\left(\mathfrak{Q}\,\frac{\partial g}{\partial x}\,\frac{\partial x^{(k)}}{\partial a^*} - \mathfrak{Q}\,\mathfrak{P}\,\frac{\partial g}{\partial x}\,\frac{\partial x^{(k)}}{\partial a^*} \right)$$

$$\left(\frac{\partial g}{\partial x} : = \frac{\partial g(x,\,t)}{\partial x}\bigg|_{x=x^{(k)}(t)} \right)$$

is clearly continuous and for sufficiently small μ, $0 < |\mu| \leq \mu_1 \leq \mu_0$, all the functions $\partial x_i^{(k)}/\partial a_j$ are uniformly bounded. Let M be a bound for them. By (77.15) the difference

$$\xi_j(k) : = \frac{\partial x^{(k+1)}}{\partial a_j} - \frac{\partial x^{(k)}}{\partial a_j}$$

can be written

$$\mu\,(\mathfrak{Q} - \mathfrak{Q}\,\mathfrak{P})\left[\left(\frac{\partial g(x^{(k)},\,t)}{\partial x} - \frac{\partial g(\tilde{x},\,t)}{\partial x} \right) \frac{\partial x^{(k)}}{\partial a_j} \right.$$

$$\left. + \frac{\partial g(\tilde{x},\,t)}{\partial x}\left(\frac{\partial x^{(k)}}{\partial a_j} - \frac{\partial x^{(k-1)}}{\partial a_j} \right) + \left(\frac{\partial g(\tilde{x},\,t)}{\partial x} - \frac{\partial g(x^{(k-1)},\,t)}{\partial x} \right) \frac{\partial x^{(k-1)}}{\partial a_j} \right]\,.$$

The terms involving the solution $\tilde{x}(t)$ cancel out. We can make an estimate, utilizing the constants defined above:

$$\|\xi_j(k)\| = \left\|\frac{\partial x^{(k+1)}}{\partial a_j} - \frac{\partial x^{(k)}}{\partial a_j}\right\|$$
$$\leq \mu\,2\,Q\,L\,M\,(\|\tilde{x} - x^{(k)}\|) + \|\tilde{x} - x^{(k-1)}\|)$$
$$+ \mu\,2\,Q\,\gamma\left\|\frac{\partial x^{(k)}}{\partial a_j} - \frac{\partial x^{(k-1)}}{\partial a_j}\right\|.$$

The first term can be estimated with the help of (77.12) and is smaller than

$$\mu\,\varkappa\,M\,\frac{\varkappa^k + \varkappa^{k-1}}{1 - \varkappa}\,\|x^{(1)} - x^{(0)}\|.$$

The resulting inequality for $\xi_j(k)$ can be written in the form

$$\|\xi_j(k)\| \leq \varrho\,\varkappa^{k-1} + \sigma\,\|\xi_j(k-1)\|,$$

respectively

$$\frac{\xi_j(k)}{\sigma^k} \leq \frac{\varrho}{\sigma}\left(\frac{\varkappa}{\sigma}\right)^{k-1} + \frac{\xi_j(k-1)}{\sigma^{k-1}}.$$

The solution of the difference equation

$$p(k) = \frac{\varrho}{\sigma}\left(\frac{\varkappa}{\sigma}\right)^{k-1} + p(k-1), \quad k = 1, 2, 3, \ldots,$$

is

$$p(k) = \frac{\varrho}{\sigma}\,\frac{\left(\dfrac{\varkappa}{\sigma}\right)^k - 1}{\dfrac{\varkappa}{\sigma} - 1} = \frac{\varrho}{\sigma^k}\,\frac{\varkappa^k - \sigma^k}{\varkappa - \sigma}$$

and this implies

$$\|\xi_j(k)\| \leq \varrho\,\sigma^{k-1}\,\frac{\left(\dfrac{\varkappa}{\sigma}\right)^k - 1}{\dfrac{\varkappa}{\sigma} - 1} = \varrho\,\varkappa^{k-1}\,\frac{\left(\dfrac{\sigma}{\varkappa}\right)^k - 1}{\dfrac{\sigma}{\varkappa} - 1}.$$

If we choose the parameter μ so small that $\sigma < 1$, $\varkappa \neq \sigma$, then it follows that $\lim\limits_{k\to\infty} \xi_j(k) = 0$ (we have $\varkappa < 1$), and this implies the uniform convergence of the sequence $\partial x^{(k)}/\partial a_j$. The limit is a continuous function and must agree with the derivative $\partial \tilde{x}(t, a^*)/\partial a_j$. Hence the derivatives exist and are continuous.

Theorem 77.1 implies that the parameter a of a periodic solution of (77.13) must satisfy the condition

$$(77.16) \qquad \mathfrak{P}\,g(x(t, a, \mu), t; \mu) = 0, \quad |a| \leq \beta,$$

for all sufficiently small μ. Thus a is to be considered as a function of μ, $a = a(\mu)$. For, if $\bar{x}(t)$ is a periodic solution of (77.13) then

$$\ddot{\bar{x}} - A\,\bar{x} - \mu\,g(\bar{x}, t; \mu) = 0,$$

and hence also

$$(E - \mathfrak{P}) \left(\dot{\bar{x}} - A\,\bar{x} - \mu g(\bar{x}, t; \mu) \right) = 0$$

and

$$\mathfrak{P} \left(\dot{\bar{x}} - A\,\bar{x} - \mu g(\bar{x}, t; \mu) \right) = 0$$

and from the last two equations we obtain

(77.17)
$$\dot{x} = A\,\bar{x} + \mu(E - \mathfrak{P})\, g(\bar{x}, t; \mu)\,,$$
$$\mathfrak{P}\, g(\bar{x}, t; \mu) = 0\,.$$

If we now set

$$\mathfrak{P}\,\bar{x}(t) = : \operatorname{col}(a,\, 0)\,,$$

where a is defined as a function of μ, and start with this vector as the zeroth iterate then the approximation process yields exactly one periodic function satisfying equation (77.17) and hence agreeing with $\bar{x}(t)$ for sufficiently small μ. It is obvious that, conversely, a function which satisfies (77.14) and (77.16) is a periodic solution of (77.13).

Condition (77.16) can be considered as a defining equation for the parameters of the periodic solution and is equivalent to the solvability conditions (sec. 72): If (77.13) is to have a periodic solution $\bar{x}$ then the perturbation function $g(\bar{x}(t), t; \mu)$ must satisfy the conditions (72.12) which, because of the nature of A, can be written in form (77.16). If we set $g =: \operatorname{col}(q,\, h)$ then we obtain

$$\mathfrak{P}\, g(x(t), t; \mu) = \operatorname{col}\left(\frac{1}{\omega} \int_0^{\omega} q(x(t), t; \mu)\, dt,\, 0,\, \ldots,\, 0 \right)$$

and (77.16) assumes the form

(77.18)
$$\frac{1}{\omega} \int_0^{\omega} q(x(t, a, \mu), t; \mu)\, dt = 0\,.$$

The mean value of $q(x, t; \mu)$ with respect to t is the vector

$$q_0(x, \mu) : = \frac{1}{\omega} \int_0^{\omega} q(x, t; \mu)\, dt\,.$$

For $\mu = 0$, (77.18) becomes

(77.19)
$$\frac{1}{\omega} \int_0^{\omega} q(a(0), t; 0) = q_0(a_0, 0) = 0\,,$$
$$a_0 : = a(0)\,.$$

By well known theorems on implicit functions the system of equations (77.16), resp. (77.19), has a solution $a(\mu)$ for sufficiently small values of μ, such that $a(0) = a_0$, if the Jacobian determinant is different from zero,

(77.20)
$$\det\left(\frac{\partial q_0(x, 0)}{\partial x} \Big|_{x=a_0} \right) \neq 0\,.$$

The existence of the Jacobian matrix follows from the existence of the derivatives with respect to $\boldsymbol{a}$ which was shown above. In this manner we obtain

Theorem 77.2. If conditions (77.19) and (77.20) are satisfied then (77.13) possesses for sufficiently small values of μ a periodic solution which for $\mu = 0$ becomes a constant solution $\mathrm{col}\,(\boldsymbol{a}_0, 0)$.

We emphasize that Theorem 77.2 does not solve the problem of finding a periodic solution of (77.13). For, in order to write down the defining equation (77.19) we already have to know something about the periodic solution. We can, however, see from the proof how approximations for the defining equations can be obtained. We need only interrupt the successive approximation at the appropriate place.

In a practical case, of course, the procedure of computation would be adapted to the particular properties of the equation at hand, once the existence of a periodic solution is guaranteed by Theorem 77.2. In most cases the method of expansion in terms of powers of a small parameter is used (see below). We now wish to point out two further pecularities of the proof. First of all, we nowhere use the fact that ω is the smallest period of $\boldsymbol{g}(\boldsymbol{x}, t; \mu)$. ω can be any integer multiple of the smallest period. Now it can happen that (77.13) has a periodic solution whose smallest period is an integer multiple of the smallest period of $\boldsymbol{g}(\boldsymbol{x}, t; \mu)$. Such a solution is called *subharmonic*. If we argue in such a way that ω is the smallest period of the subharmonic solution then we see that Theorem 77.2 is also valid for subharmonic solutions.

Subharmonic oscillations are typically nonlinear phenomena. To make this clearer we write the equation in the somewhat special form

$$\dot{\boldsymbol{x}} = \boldsymbol{g}(\boldsymbol{x}) + \boldsymbol{f}(t)$$

and we assume that this equation describes a physical system subjected to a periodic external force $\boldsymbol{f}(t)$. If $\boldsymbol{g}(\boldsymbol{x}) = A\boldsymbol{x}$ is linear then only two cases are possible: Either there exist no periodic solutions at all or all the periodic solutions have the period of the external force. In technical language: For linear systems all forced oscillations have the frequency of the external force. In the nonlinear case, however, there may exist forced oscillations whose basic frequency is a fraction of the external frequency.

The second remark pertains to the fact that nowhere in the proof the special shape of $\mu\boldsymbol{g}(\boldsymbol{x}, t; \mu)$ was used. The smallness of this expression was caused by μ and all arguments remain valid if $\boldsymbol{g}(\boldsymbol{x}, t; \mu)$ contains linear terms or even is linear. This implies that for nonlinear equations

$$\dot{\boldsymbol{x}} = A(t)\,\boldsymbol{x} + \mu\boldsymbol{h}(\boldsymbol{x}, t; \mu),$$

resonance phenomena may appear even if the homogeneous equation does not have a periodic solution with period ω, provided there exists a solution whose period ω' differs from ω only by terms of the order of magnitude of μ.

This can be seen, for example, in the estimates (76.19). The norm computed there can be arbitrarily large if the number η is small. This number η which denotes the difference between k^2 and the square nearest to it is a measure for the difference between the external period and the period effective in the linear part. We outline the computation of the periodic solution for equation (76.18) in case $k^2 = 1 + x\mu$. We must treat

$$(77.21) \qquad \ddot{x} + x = \mu p \sin t - \mu(\alpha x + a x^3)$$

by the procedure which we have modified for the resonance case. It is necessary to introduce the factor μ in the first term on the right since otherwise the equation does not have a periodic solution for $\mu = 0$. We point out that (77.21) is not exactly of the form (77.13) since the matrix of the linear part has not been changed to the form (77.2). The computation proceeds as follows. We write the desired solution as a power series in μ,

$$(77.22) \qquad x = x_0 + \mu x_1 + \mu^2 x_2 + \cdots,$$

and obtain by substituting in the differential equation and then comparing coefficients, a system of second order differential equations with periodic coefficients,

$$\ddot{x}_0 + x_0 = 0,$$

$$\ddot{x}_1 + x_1 = p \sin t - \alpha x_0 - a x_0^3,$$

$$\ddot{x}_2 + x_2 = - \alpha x_1 - 3 a x_0^2 x_1,$$

$$\cdots\cdots\cdots\cdots\cdots\cdots\cdots\cdots\cdots$$

Since the matrix of the linear part is not of the special form (77.2), the zeroth iterate is not a constant, but rather we find from the first equation that

$$x_0 = c_1 \cos t + c_2 \sin t.$$

To determine the constants c_i we use the solvability conditions for the second equation

$$\int_0^{2\pi} (p \sin t - \alpha(c_1 \cos t + c_2 \sin t) - a(c_1 \cos t + c_2 \sin t)^3) \begin{Bmatrix} \cos t \\ \sin t \end{Bmatrix} dt = 0,$$

which lead to the equations

$$- \alpha c_1 - \frac{3}{4} a c_1 (c_1^2 + c_2^2) = 0 ,$$

$$p - \alpha c_2 - \frac{3}{4} a c_2 (c_1^2 + c_2^2) = 0 .$$

These are satisfied for $c_1 = 0$, $c_2 = \lambda$; λ is the real root of the cubic

$$h(\lambda) : = p - \alpha \lambda - \frac{3}{4} a \lambda^3 = 0 .$$

The solution $(0, \lambda)$ must be an isolated solution. Hence the determinant

$$\left(\alpha + \frac{3}{4} a \lambda^2 \right) \left(\alpha + \frac{9}{4} a \lambda^2 \right)$$

must not vanish; this is the Jacobian determinant of the left side with respect to $\mathfrak{c}(c_1, c_2)$ at the point $(0, \lambda)$. It can vanish only if either λ is a double root of the cubic or in case $p = 0, \lambda^2 = - \dfrac{4\alpha}{3a}$. In the second case there is no external force. The exceptional first case shall be excluded. One real root always exists. For certain parameter values all three roots are real and then there exist three distinct zeroth iterates of the form $x_0 = \lambda \sin t$. The other iterates contain only sine terms since equation (77.21) remains invariant under the substitution $t \to - t$, $x \to - x$. For the first iterate we obtain

$$x_1 = - \frac{a \lambda^3}{32} \sin 3t + c_4 \sin t$$

$(c_3 = 0)$, and to determine c_4 we use the sine-solvability condition of the differential equation for x_2, which leads to the equation

$$\left(\alpha + \frac{9}{4} a \lambda^3 \right) c_4 - \frac{3 \lambda^5}{128} a^2 = 0 .$$

We can continue in this manner. As for the stability of the constructed solutions *cf.* sec. 82.

To study nth order subharmonic solutions of the Duffing equation (76.18) we set $k^2 = \dfrac{1}{n^2} + \mu$ and $t = \tau n$. We obtain

$$\frac{d^2 x}{d \tau^2} + x = n^2 p \sin n\tau - \mu n^2 (x + a x^3) .$$

The term which does not contain x need not involve a small parameter since the equation of the zeroth iterate always has a two-parameter periodic solution

$$x_0 = \frac{n^2 p}{1 - n^2} \sin n\tau + c_1 \cos \tau + c_2 \sin \tau .$$

The first iterate x_1 satisfies the differential equation

$$\frac{d^2 x_1}{d\tau^2} + x_1 = - n^2 (x_0 + a\, x_0^3).$$

The Fourier expansion of the right side involves sine and cosine terms with arguments

$$\tau,\ 3\tau,\ (n-2)\,\tau,\ n\tau,\ (n+2)\,\tau,\ (2n-1)\,\tau,\ (2n+1)\,\tau,\ 3n\tau.$$

In case $n \neq 3$ we set the coefficients of $\sin \tau$ and $\cos \tau$ equal to zero and obtain equations for c_1 and c_2. If, on the other hand, $n = 3$ then we must also take into consideration the terms involving $\cos(n-2)\,\tau$ and $\sin(n-2)\,\tau$. It turns out that in the first case $(n \neq 3)$ the equations have a vanishing Jacobian determinant so that the method fails. In case $n = 3$ however, $c_1 = 0$ and c_2 are uniquely determined and we can proceed to the next iterate. For the details of the computation see, for example, MALKIN [5].

We also note the following. Even though at each step of the procedure which we discussed we must solve nonlinear equations for the parameters of the kth iterate we need not be concerned with the Jacobian determinant of these equations. It can vanish only for the first step and is different from zero for all further steps. This follows from Theorem 77.2. The existence of the solution is assured if the initial values of the zeroth iterate can be determined. Explicit calculation also shows that the Jacobian determinants of the subsequent steps are not equal to zero if that of the zeroth iterate does not vanish.

We briefly mention the question how the solutions behave if in the case of resonance the solvability conditions are not satisfied. For linear equations all solutions are unbounded in this case (sec. 69). For nonlinear equations only a very modest counterpart for this theorem is known.
Theorem 77.3.[1]) Let the equation

$$(77.23) \qquad\qquad \dot{x} = A(t)\,x + \mu f(t) + g(x, t)$$

be periodic with period ω. Let $g(x, t) = O(|x|^2)$. Let the equation

$$(77.24) \qquad\qquad \dot{y} = - A^T(t)\,y,$$

which is the adjoint of the linear part of (77.23) have a periodic solution $\bar{y}(t)$ with period ω, such that

$$(77.25) \qquad\qquad \int_0^{\omega} \bar{y}(u)^T f(u)\, du = c \neq 0.$$

Then each solution with sufficiently small initial values again and again assumes values of the order of magnitude $\sqrt{|\mu|}$, as t increases.

---

[1]) KARIM [1].

Proof. Let x be restricted to the domain $|x| \leq a$. There exists a constant b such that for such x the inequality $|g(x, t)| < b\,|x|^2$ holds. Let the periodic solution of (77.24) be normalized so that $|\bar{y}(t)| \leq 1$ for all t. From (77.23) and (77.24) we find that

$$\frac{d}{dt}\left(\bar{y}^T x\right) = \mu\,\bar{y}^T f(t) + \bar{y}^T g(x, t).$$

We integrate from t to $t + m\omega$ and, observing (77.25), obtain

$$\bar{y}(t)^T\left(x(t + m\omega) - x(t)\right) = c\,m\,\mu + \int_t^{t+m\omega} \bar{y}(u)^T g(x, u), u)\,du.$$

In this formula $x(t)$ is an arbitrary solution of (77.23). If $\sup |x(t)| = o\left(\sqrt{\mu}\right)$ we can choose a constant α arbitrarily small such that $\sup |x(t)| < \alpha \sqrt{|\mu|}$. Then the integral is smaller than $a\,|\mu|\,mb\omega x^2$ in absolute value and the last equation implies

$$c\,m\,|\mu| < m\,\omega\,b\,\alpha^2\,|\mu| + 2\alpha\sqrt{|\mu|}$$

and

$$(c - \omega\,b\,\alpha^2)\sqrt{|\mu|} < \frac{2\alpha}{m}.$$

If we choose $\alpha^2 < c/b\,\omega$ then the left side is positive and for large values of m we obtain a contradiction. Hence $|x|$ again and again becomes larger than $\alpha\sqrt{|\mu|}$.

In the case of nonresonance there exist solutions which consistently have the order of magnitude μ, for example periodic solutions.

78. Periodic Solutions of Autonomous Equations

For a periodic solution of an autonomous equation of the form

$$(78.1) \qquad \dot{x} = A\,x + \mu\,g(x; \mu)$$

the period is unknown a priori; in general it depends on μ. But the periodic solution can be constructed by the same process which was utilized in the preceding section. The function $g(x; \mu)$ is continuous in a certain domain G and once continuously differentiable. The matrix $A = \operatorname{diag}(A_1, A_2)$ is not given in the standard form (77.2) but of such a nature that the characteristic roots of A_1 have vanishing real parts and belong to simple elementary divisors, and the non-zero ones among them are rational multiples of each other. We may assume without loss of generality that A_1 is partitioned into submatrices

$$(0) \qquad \text{or} \qquad \begin{pmatrix} 0 & r_j\theta \\ -r_j\theta & 0 \end{pmatrix}, \quad r_j \text{ integer and positive.}$$

Then to each characteristic root of A_1 there corresponds a periodic solution of a certain period $T(\theta) = 2\pi k/\theta$, k an integer. The dependence

26*

of the matrix A_1 on θ is expressed by the notation $A_1 = A_1(\theta)$. We seek that solution of the nonlinear equation which for $\mu = 0$ becomes the solution with period $T(\theta)$. It is useful to express the dependence on the period by replacing θ^2 by $\theta_1^2 = \theta^2 + \mu\gamma(\mu)$. The function $\gamma(\mu)$ is to be determined later. We introduce a new variable by means of the substitution (*cf.* sec. 77)

$$\boldsymbol{x} = B(t)\,\boldsymbol{z}, \quad B(t) := \operatorname{diag}\left(e^{A_1(\theta_1)t},\ E\right).$$

The matrix of the linear part of the transformed equation is

$$\operatorname{diag}\left(e^{-A_1(\theta_1)t}\left(A_1(\theta) - A_1(\theta_1)\right)e^{A_1(\theta_1)t},\ A_2\right).$$

The first submatrix is independent of t and of the form μA_3, and the corresponding term can be considered as one of the perturbation terms. The nonlinear term of the transformed equation is

$$\mu \operatorname{diag}\left(e^{-A_1(\theta_1)t},\ E\right)\boldsymbol{g}\left(B(t)\,\boldsymbol{z};\ \mu\right).$$

It is periodic with period $T_1 := T\left(\sqrt{\theta^2 + \mu\gamma(\mu)}\right)$. The transformed equation is now in the standard form (77.2). We write it in the original notation

$$(78.2) \qquad \dot{\boldsymbol{z}} = A\,\boldsymbol{z} + \mu\,\boldsymbol{g}\left(\boldsymbol{z},\,t;\ \mu,\ \gamma(\mu)\right); \quad A = \operatorname{diag}\left(0,\ A_2\right).$$

The fourth argument of the nonlinear function expresses the fact that $\boldsymbol{g}$ depends on the function $\gamma(\mu)$, still unknown. We now consider the period T_1 as fixed, construct the periodic solution for a fixed μ by the procedure of sec. 77, and then reverse the transformation of variables. For this purpose we need the equations (77.19) which can be written in the form

$$(78.3) \qquad\qquad \boldsymbol{F}\left(\boldsymbol{a}(\mu),\ \mu,\ \gamma(\mu)\right) = 0.$$

$\boldsymbol{a}$ is the initial vector, $\gamma(\mu)$ the still unknown disturbance of the period. The left side of (78.3) can be written explicitly, and it is our task to determine $\boldsymbol{a}(\mu)$ and $\gamma(\mu)$ as functions of μ in such a way that (78.3) is valid for all sufficiently small μ. The vector equation (78.3) consists of p scalar equations for the $p + 1$ unknown functions $a_1(\mu), \ldots, a_p(\mu), \gamma(\mu)$. This is in agreement with the fact that the variables which determine the periodic solution of an autonomous differential equation depend on a parameter (*cf.* also sec. 68). For purely computational purposes we can also utilize the dependence of the period on the parameter μ in a different way. For this purpose we set

$$T_1 := T(\theta)\left(1 + \mu\,\delta(\mu)\right),$$

where $\delta(\mu)$ is again an unknown function, and we introduce a new independent variable τ by

$$(78.4) \qquad\qquad t = \tau\left(1 + \mu\,\delta(\mu)\right).$$

Then to a periodic solution of (78.1) with period T_1 there corresponds a solution of the transformed equation

$$\frac{d\boldsymbol{x}}{d\tau} = A\,\boldsymbol{x} + \mu\,\delta(\mu)\,A\,\boldsymbol{x} + \{1 + \mu\,\delta(\mu)\}\,\mu\,\boldsymbol{g}(\boldsymbol{x};\mu)$$

with period $T(\theta)$. This equation is put into standard form and a periodic solution having the known period T is constructed. Again the solvability conditions lead to defining equations for the initial vector $\boldsymbol{a}$ and the function $\delta(\mu)$.

Example. Consider the equation of *van der Pol*

$$(78.5) \qquad\qquad \ddot{x} + x = \mu(a + b\,x^2)\,\dot{x}.$$

In this case $T(\theta) = 2\pi$. The transformed equation is

$$\frac{d^2x}{d\tau^2} + x\{1 + \mu\,\delta(\mu)\}^2 = \mu(a + b\,x^2)\,\{1 + \mu\,\delta(\mu)\}\,\frac{dx}{d\tau}.$$

We may assume that $\delta(\mu)$ is analytic for sufficiently small μ and admits a power series expansion

$$\delta(\mu) = \alpha_0 + \alpha_1\mu + \alpha_2\mu^2 + \cdots.$$

For the solution we choose the expression (77.22). We compare coefficients and obtain with $\dfrac{dx}{d\tau} = x'$

$$x_0'' + x_0 = 0,$$

$$x_1'' + x_1 = (a + b\,x_0^2)\,x_0' - 2\alpha_0\,x_0$$

$$\cdots\;\cdots\;\cdots\;\cdots\;\cdots\;\cdots\;\cdots$$

In order to diminish the number of parameters involved we again use the fact that one component of the initial vector can be chosen arbitrarily, and set $x'(0) = 0$. Then $x_i'(0) = 0$ for all i. The first iterate is $x_0 = c_0 \cos\tau$. The right side of the equation for x_1 is

$$-\left(a\,c_0 + \frac{b}{4}\,c_0^3\right)\sin\tau - 2\alpha_0\,c_0 \cos\tau - \frac{b}{4}\,c_0^3\sin 3\tau$$

The solvability conditions are

$$2\alpha_0\,c_0 = 0, \quad \left(a + \frac{b}{4}\,c_0^2\right)c_0 = 0,$$

which implies $\alpha_0 = 0$, $c_0^2 = -4a/b$. We therefore must have sgn $a\,b = -1$. This fixes x_0 and we can take care of the equation for x_1. It involves two parameters. The initial condition $x_1' = 0$ and the two solvability conditions of the equation for x_2 yield three equations to determine the parameters of x_1 and the coefficient α_1. We may continue in this manner.

This method of solution is also known as the *Lindtstedt procedure* (*cf.* also sec. 79). The stability of the periodic solution will be dealt with in sec. 82.

79. Critical Cases of Second Order Autonomous Systems

In this section we will not use vector notation. x, y, etc. are scalar variables.

In the critical case a second order autonomous system of equations can be put into one of the three forms

$$\text{(79.1)} \qquad \begin{aligned} \dot{x} &= -y + p(x, y), \\ \dot{y} &= x + q(x, y). \end{aligned}$$

$$\text{(79.2)} \qquad \begin{aligned} \dot{x} &= p(x, y), \\ \dot{y} &= x + q(x, y). \end{aligned}$$

$$\text{(79.3)} \qquad \begin{aligned} \dot{x} &= p(x, y), \\ \dot{y} &= q(x, y). \end{aligned}$$

This was explained in sec. 70. The functions $p(x, y)$ and $q(x, y)$ have Taylor expansions starting with terms of at least second degree. Collecting the terms of the same degree in x, y we write

$$p(x, y) = p_2(x, y) + p_3(x, y) + \cdots,$$
$$q(x, y) = q_2(x, y) + q_3(x, y) + \cdots.$$

a) The origin of equations (79.1) is a singularity of the type of a center or of a focus, for a saddle point or a node is insensitive by sec. 21, and the singular point of the reduced auxiliary system

$$\text{(79.4)} \qquad \dot{x} = -y, \quad \dot{y} = x$$

is a center.

In order to determine whether the origin is a center or a focus we attempt to construct a periodic solution by the methods of sec. 78. If even for arbitrarily small initial values (x_0, y_0) it is impossible to find a periodic solution then the origin cannot be a center. It must then be a focus. The same conclusion is reached if there exists a limit cycle, that is an isolated periodic solution whose interior contains no further closed phase curves. By sec. 18, the origin has to be a focus in this case. This is the case, for instance, for equation (78.5).

In constructing the periodic solution for (79.1) we use the method of the small parameter to advantage. Since the equation does not involve such a parameter we must introduce it into the procedure. We denote the initial point with $(\varrho, 0)$, that is we choose as the initial point the

x-intercept of the curve and we regard the abscissa ϱ as the small parameter. The trivial solution corresponds to the value $\varrho = 0$. In analogy
to (78.4), we introduce a time τ depending on ϱ by

$$t = \tau(1 + \alpha_1 \varrho + \alpha_2 \varrho^2 + \cdots),$$

so that (79.1) becomes

$$
\begin{aligned}
\frac{dx}{d\tau} &= \left(-y + p_2(x, y) + \cdots\right)\left(1 + \alpha_1 \varrho + \alpha_2 \varrho^2 + \cdots\right), \\
\frac{dy}{d\tau} &= \left(x + q_2(x, y) + \cdots\right)\left(1 + \alpha_1 \varrho + \alpha_2 \varrho^2 + \cdots\right).
\end{aligned}
$$

(79.5)

Then we seek for sufficiently small ϱ a periodic solution of the form

$$
\begin{aligned}
x(\tau) &= \varrho\, x_1(\tau) + \varrho^2 x_2(\tau) + \cdots, \\
y(\tau) &= \varrho\, y_1(\tau) + \varrho^2 y_2(\tau) + \cdots,
\end{aligned}
$$

(79.6)

which has period 2π and is such that

$$x(0) = \varrho, \quad y(0) = 0.$$

We have used the fact that by hypothesis the solutions are analytic
functions of the initial values. If (79.6) is periodic then the coefficients
$x_i(\tau)$, $y_i(\tau)$ are periodic functions of period 2π and we have

$$
\begin{aligned}
x_1(0) &= 1, \quad x_i(0) = 0 \ \ (i = 2, 3, \ldots), \\
y_i(0) &= 0 \quad (i = 1, 2, \ldots).
\end{aligned}
$$

(79.7)

We substitute the expression (79.6) in the equations (79.5), expand on the
right by powers of ϱ, and equate the coefficients of equal powers. Then
we obtain a sequence of systems of differential equations for $x_i(\tau)$,
$y_i(\tau)$, $i = 1, 2, \ldots$, for which periodic solutions with the initial values
(79.7) must be found. Therefore for each individual system the two
solvability conditions must be satisfied. One of these conditions is used
at each step to determine the constants α_i. The other condition is superfluous in a certain sense and in general it will be satisfied only for finitely
many i. If it is not satisfied for a certain value i then there exists no
periodic solution for sufficiently small ϱ, and the origin is a focus.

The detailed computations follow. For the first iterate we have (the
prime denotes the derivative with respect to τ)

$$x_1' = -y_1, \quad y_1' = x_1,$$

and because of (79.7),

(79.8)
$$x_1 = \cos \tau, \quad y_1 = \sin \tau.$$

The second iterate satisfies

$$(79.9) \qquad \begin{aligned} x_2' &= -y_2 - \alpha_1 y_1 + p_2(x_1, y_1), \\ y_2' &= x_2 + \alpha_1 x_1 + q_2(x_1, y_1). \end{aligned}$$

The solvability conditions are

$$\int_0^{2\pi} \left((-\alpha_1 y_1 + p_2(x_1, y_1)) \cos \tau + (\alpha_1 x_1 + q_2(x_1, y_1)) \sin \tau \right) d\tau = 0,$$

$$\int_0^{2\pi} \left((\alpha_1 x_1 + q_2(x_1, y_1)) \cos \tau - (-\alpha_1 y_1 + p_2(x_1, y_1)) \sin \tau \right) d\tau = 0.$$

x_1 and y_1 satisfy (79.8). Since p_2 and q_2 are trigonometric polynomials of even degree and since α_1 cancels out of the first equation, it follows that $2\pi\alpha_1 = 0$, i.e. $\alpha_1 = 0$. Therefore the equations (79.9) contain no resonance term, and the periodic solutions can be found, taking (79.7) into consideration. They each contain a term with $\cos \tau$, respectively $\sin \tau$. Therefore the system for the third iterate has the form

$$(79.10) \qquad \begin{aligned} x_3' &= -y_3 - \alpha_2 \sin \tau + P_3, \\ y_3' &= x_3 + \alpha_2 \cos \tau + Q_3. \end{aligned}$$

P_3 and Q_3 are trigonometric polynomials depending on x_1, y_1, x_2, y_2. This time the solvability conditions are

$$(79.11) \qquad \begin{aligned} 2\pi\alpha_2 + \int_0^{2\pi} (Q_3 \cos \tau - P_3 \sin \tau) \, d\tau &= 0, \\ \int_0^{2\pi} (P_3 \cos \tau + Q_3 \sin \tau) \, d\tau &= 0. \end{aligned}$$

The first one is used to determine α_2. If the second one is not satisfied the origin is a focus. Otherwise we can find a pair of periodic functions for the third iterate and proceed to the fourth iterate. The equations and the solvability conditions appear in the form (79.10), resp. (79.11), each time. The constant α_k appears only in one of the conditions, and does so linearly. Only if the second solvability condition is satisfied every time the origin can be a center. In principle therefore, the given procedure does not yield a positive decision that the origin is a center. For a focus, on the other hand, a positive decision is possible. The origin is a focus if the second solvability condition of the kth step is not satisfied, *i.e.* if the integral

$$(79.12) \qquad g := \frac{1}{2\pi} \int_0^{2\pi} (P_k \cos \tau + Q_k \sin \tau) \, d\tau$$

is nonzero. In that case the kth iterate is no longer periodic but we have

$$x_k(\tau) = c_1 \cos \tau + c_2 \sin \tau + g\tau \cos \tau + \tilde{x}_k(\tau),$$

$$y_k(\tau) = c_1 \sin \tau - c_2 \cos \tau + g\tau \sin \tau + \tilde{y}_k(\tau),$$

where $\tilde{x}_k$, $\tilde{y}_k$ denote periodic functions known from the previous steps. c_1 and c_2 are computed from the initial conditions:

$$c_1 = -\tilde{x}_k(0), \quad c_2 = -\tilde{y}_k(0).$$

The first solvability condition, which involves α_{k-1}, has been used. If we substitute the iterates in (79.6) and observe (79.7) as well as the fact that $x_1, y_1, \ldots, x_{k-1}, y_{k-1}$, and hence $\tilde{x}_k$, $\tilde{y}_k$, are periodic then we obtain $x_k(2\pi) = 2\pi g$, $y_k(2\pi) = 0$ and

(79.13)
$$x\big|_{\tau=2\pi} = \varrho + \varrho^k\, 2\pi g + O(\varrho^{k+1}),$$
$$y\big|_{\tau=2\pi} = \quad 0 \qquad + O(\varrho^{k+1}).$$

The values on the left are the values of the functions $x(t)$, $y(t)$ at the period T which depends on the initial value ϱ,

$$T := 2\pi(1 + \alpha_1 \varrho + \alpha_2 \varrho^2 + \cdots)$$

(*cf.* the remark following (78.4)). The distance of the point of intersection of the trajectory with the positive x-axis from the x value in (79.13) is of the order of magnitude ϱ^{k+1}. We can therefore replace this point of intersection by the point $(\varrho + 2\pi g\varrho^k, 0)$ which is closer to or further from the origin than the initial point $(\varrho, 0)$, depending on the sign of g. Accordingly we have

Theorem 79.1. The origin is a stable or unstable focus according as the number g defined by (79.12) is negative or positive. The number g is also called the *Liapunov number*.

Since the polynomials P_k and Q_k in (79.12) are of degree k the number k must be odd.

Example (MALKIN [3]).
     We consider

(79.14) $\dot{x} = -y, \quad \dot{y} = x - bx^2 - cxy^2 + ay^3,$

and set

$$t = \tau(1 + \alpha_2 \varrho^2 + \cdots)$$

since $\alpha_1 = 0$, and we find the equations

$$x_2' = -y_2, \quad y_2' = x_2 - b\cos^2 \tau = x_2 - \frac{b}{2}(1 + \cos 2\tau).$$

From them and the initial conditions we find the periodic solution

$$x_2 = \frac{b}{2} - \frac{b}{3} \cos \tau - \frac{b}{6} \cos 2\tau,$$

$$y_2 = -\frac{b}{3} \sin \tau - \frac{b}{3} \sin 2\tau.$$

The next system of equations is

$$x_3' = - y_3 - \alpha_2 \sin \tau,$$

$$y_3' = x_3 + \alpha_2 \cos \tau - 2 b x_2 \cos \tau - c \sin^2 \tau \cos \tau + a \sin^3 \tau.$$

The second solvability condition is not satisfied for $a \neq 0$ and we have

$$g = \frac{a}{2\pi} \int_0^{2\pi} \sin^4 \tau \, d\tau.$$

For the stability behavior the sign of a is thus decisive: $a < 0$ implies stability.

The criterion of Theorem 79.1 is also obtained if we introduce polar coordinates $x = r \cos \varphi$, $y = r \sin \varphi$ into (79.1), then write the equation for $dr/d\varphi$ and attempt to solve it by an expression

$$(79.15) \qquad\qquad r = \varrho \, r_1(\varphi) + \varrho^2 r_2(\varphi) + \cdots.$$

The origin can be a center only if all the functions $r_i(\varphi)$ are periodic with period 2π. It is a focus if in computating the coefticients $r_i(\varphi)$ by Theorem 19.1, a nonperiodic function of the form

$$g\varphi + \text{periodic function}$$

is obtained at some step. We can show that this computation leads to the same numbers k and g as that following (79.5) (*cf.* MALKIN [3]). The individual steps of the computation are largely similar in the two procedures.

A procedure given by Liapunov rests on quite a different basis.[1] It starts from the fact that the function

$$(79.16) \qquad\qquad v_2 = x^2 + y^2$$

is a Liapunov function for the system of equations (79.4). We consider (79.4) as a system of approximations for (79.1) and simultaneously we consider (79.16) as an approximation of a Liapunov function

$$(79.17) \qquad\qquad v = v_2 + v_3(x, y) + v_4(x, y) + \cdots$$

---

[1] For the details see MALKIN [3].

for (79.1). The terms $v_3, v_4, \ldots$ are homogeneous polynomials in x, y and must be determined so that the derivative of (79.17) for (79.1) is negative definite. This derivative is

$$\frac{dv}{dt} = \left(2x + \frac{\partial v_3}{\partial x} + \frac{\partial v_4}{\partial x} + \cdots\right)\left(-y + p_2(x, y) + \cdots\right)$$
$$+ \left(2y + \frac{\partial v_3}{\partial y} + \frac{\partial v_4}{\partial y} + \cdots\right)\left(x + q_2(x, y) + \cdots\right).$$

It can be definite only if it starts with terms of even degree. Therefore the third degree terms must vanish. This leads to the equation

$$x\frac{\partial v_3}{\partial y} - y\frac{\partial v_3}{\partial x} = -2x\,p_2(x, y) - 2y\,q_2(x, y),$$

which must be solved by a homogeneous polynomial v_3 of third degree. Now we can show that the partial differential equation

$$x\frac{\partial v_m}{\partial y} - y\frac{\partial v_m}{\partial x} = \text{given homogeneous polynomial of degree } m$$

can always be solved for odd m. The fourth degree component of the derivative dv/dt is

$$x\frac{\partial v_4}{\partial y} - y\frac{\partial v_4}{\partial x} + p_2\frac{\partial v_3}{\partial x} + q_2\frac{\partial v_3}{\partial y} + 2x\,p_3 + 2y\,q_3.$$

We seek to determine v_4 so that this term becomes

$$G_4(x^2 + y^2)^2, \quad G_4 \text{ constant}.$$

Now an equation

$$x\frac{\partial v_{2m}}{\partial y} - y\frac{\partial v_{2m}}{\partial x} + F_{2m}(x, y) = G_{2m}(x^2 + y^2)^m,$$

where the homogeneous polynomial $F_{2m}(x, y)$ of degree $2m$ is given, is solvable only if the constant G_{2m} satisfies the relation

$$2\pi G_{2m} = \int_0^{2\pi} F_{2m}(\cos\varphi, \sin\varphi)\,d\varphi.$$

G_4 must therefore be chosen accordingly in this relation. If G_4 is negative or positive then the origin is a stable, respectively unstable focus. If $G_4 = 0$ then we must compute the fifth and sixth iterates, etc. If all the numbers G_{2m} vanish (which can of course never be determined by direct computation) then the origin is a center.

Example.[1]) If $p(x, y)$ and $q(x, y)$ are third degree polynomials,

$$p(x, y) = a_0 x^3 + a_1 x^2 y + a_2 x y^2 + a_3 y^3,$$

$$q(x, y) = b_0 x^3 + b_1 x^2 y + b_2 x y^2 + b_3 y^3,$$

---

[1]) SAKHARNIKOV [1]. The example is worked out completely there.

then $v_3 = 0$ and

$$F_4(x, y) = 2\big(x\,p(x, y) + y\,q(x, y)\big)$$
$$= 2\big(a_0 x^4 + (a_1 + b_0)\,x^3 y + (a_2 + b_1)\,x^2 y^2 + (a_3 + b_2)\,xy^3 + b_3 y^4\big).$$

Hence

$$G_4 = \frac{1}{2\pi} \int\limits_0^{2\pi} F_4(\cos\varphi,\,\sin\varphi)\,d\varphi$$
$$= \frac{2}{2\pi}\left((a_0 + b_3)\,\frac{3\pi}{4} + (a_2 + b_1)\,\frac{\pi}{4}\right) = \frac{1}{4}\,(3\,a_0 + a_2 + b_1 + 3\,b_3).$$

The condition $G_4 = 0$ is necessary for the origin to be a center.

In the case of equation (79.14) the defining equation for v_3 is

$$x\,\frac{\partial v_3}{\partial y} - y\,\frac{\partial v_3}{\partial x} - 2\,b\,x^2 y = 0$$

and we see immediately that v_3 depends on x only and is, in fact, equal to $-\dfrac{2}{3}\,b x^3$. The form F_4 is simply

$$2\,y\,q_3 = 2\,y\,(-\,c\,x\,y^2 + a\,y^3),$$

so that

$$G_4 = \frac{a}{\pi} \int\limits_0^{2\pi} \sin^4\varphi\,d\varphi.$$

If $a < 0$ then the focus is stable.

b) The discussion of the system of equations (79.2) is considerably more complicated. In addition to a center and a focus more complicated types of singularities can occur. Liapunov has settled the stability question (the results have been compiled in HAHN [4]). He constructs, similar to the manner explained above, suitable Liapunov functions and uses for this purpose essentially the special functions $Cs\,\theta$, $Sn\,\theta$ mentioned in sec. 22, a).

If the expansion of $p(x, y)$ begins with an odd degree term $p_{2k-1}(x, y) = -y^{2k-1}$ we introduce new variables r, θ by

$$x = r^{2k-2}\,Sn\,\theta, \qquad y = r\,Cs\,\theta,$$

and attempt to solve the equation for $dr/d\theta$, using an expression analogous to (79.15). Even for simple cases the computations are quite involved. For the special system

$$\dot{x} = -y^3 + \alpha\,y^2 x + \beta\,y\,x^2 + \gamma\,x^3,$$
$$\dot{y} = x + a\,y^3 + b\,y^2 x + c\,y\,x^2 + d\,x^3,$$

the conditions

$$2\,a\,(b + \beta) - (c + 3\,\gamma) = 0,$$
$$(b + \beta)\,(2\,a^3 + a\,\beta - \gamma) = 0,$$

are necessary and sufficient for the origin to be a center. If they are not satisfied then the origin is a focus.[1]

c) The system of equations (79.3) presents us with even greater difficulties. If the lowest terms in the expansion of p and q are of the same degree k then we use the auxiliary system

$$(79.18) \qquad \dot{x} = p_k(x,y), \quad \dot{y} = q_k(x, y),$$

and apply to it the procedure sketched in sec. 19. We arrive at a result which is valid for the complete system as well. If the complete system has stability properties with respect to the kth approximation (*cf.* sec. 68) then those are the stability properties of (79.18). This is, however, not the case if the origin of (79.18) is a center. In this case the system (79.3) must be studied directly, for example by introducing polar coordinates and using the expression (79.15) for the solution. If the lowest degree terms are not of equal degree then no general rule for the solution of the stability problem can be given.

The preceding considerations remain valid, to a certain extent, for nonautonomous equations also, namely in those cases in which the stability problem can be solved by a finite number of terms independent of time and for which the terms which depend on time are of higher degree in x and y.[2]

80. The Associated Coordinate System of a Periodic Solution

We can interpret the transition to the equation of the perturbed motion (sec. 35) as a transformation of coordinates. We introduce a new Cartesian coordinate system whose axes are parallel to those of the old system and whose origin coincides with the phase point which moves along the trajectory. The time scale remains unchanged. If, for example, we relate the curve with equations

$$(80.1) \qquad \begin{aligned} y_1 &= (1 + a\varrho) \cos (1 + \varrho)t, \\ y_2 &= (1 + b\varrho) \sin (1 + \varrho)t \end{aligned}$$

to the unit circle

$$\varphi_1 = \cos t, \quad \varphi_2 = \sin t,$$

then the coordinates in the system of the perturbed equation are

$$\xi_1 = y_1 - \varphi_1 = -2 \sin \frac{\varrho}{2} t \sin \left(1 + \frac{\varrho}{2}\right) t + a\varrho \cos (1 + \varrho)t,$$

$$\xi_2 = y_2 - \varphi_2 = 2 \sin \frac{\varrho}{2} t \cos \left(1 + \frac{\varrho}{2}\right) t + b\varrho \sin (1 + \varrho)t.$$

---

[1] ANDREEV [1].
[2] *cf.* also MALKIN [3].

In the (ξ_1, ξ_2)-plane they describe a curve of the character of an epicycloid. ϱ is the parameter of the family of curves. The (old) origin corresponds to the value $\varrho = 0$. It is obviously unstable since the square of the distances is

$$\xi_1^2 + \xi_2^2 = 4 \sin^2 \frac{\varrho}{2} t + O(\varrho)$$

and is not of the form $\chi(\varrho)$, for some $\chi \in K$.

In discussing a periodic solution we use yet another coordinate system to advantage, the so-called *associated* system. We consider an $(n-1)$-dimensional hyperplane which is orthogonal to the periodic solution and which moves with the phase point along the trajectory γ of the periodic motion. For the first $n-1$ coordinates of a neighboring curve we take the coordinates of the point at which it pierces the hyperplane, with respect to a Cartesian system whose origin lies on γ. The nth coordinate is the parameter value of the point at which the hyperplane is pierced, normalized by the arc length of the closed curve. Occasionally the nth axis is in the direction of the tangent. In the example (80.1) considered above, the "hyperplane" lies in the direction of the radius. The distance of the point of intersection from the unit circle is

$$\eta := \varrho \sqrt{a^2 \cos^2 t + b^2 \sin^2 t}.$$

The second coordinate is t. (The parameters of the two curves stand in the proportion $(1 + \varrho):1$.) Since $\eta = O(\varrho)$, the origin is stable. This implies that the periodic solution is orbitally stable (*cf.* below).

We next study the associated coordinate system for the periodic solution of an autonomous differential equation

$$(80.2) \qquad\qquad \dot{y} = f(y),$$

using the approach of ZUBOV [4]. We assume that the function $f(y)$ is continuous in a certain domain B and has continuous partial derivatives of first and second order. The closed trajectory of the periodic solution $y = \varphi(t)$, $0 \leq t \leq T$, is denoted by

$$(80.3) \qquad\qquad \gamma := \{\varphi(t) : 0 \leq t \leq T\}.$$

The period T is known. $S(\gamma, \delta)$ is a δ-neighborhood of γ which lies entirely in B.

We consider a hyperplane P_t which, for fixed t, passes through the point $p := \varphi(t) \in \gamma$ and which is perpendicular to γ. In order to write its equation in the y-coordinate system of R_n, we use the abbreviation $f(\varphi(t)) = g(t)$. Then the hyperplane is given by

$$(80.4) \qquad\qquad (y - \varphi(t))^T g(t) = 0.$$

Furthermore let Q_t be the disc which $S(\gamma, \delta)$ cuts from P_t,

$$Q_t : = S(\gamma, \delta) \cap P_t.$$

The number δ is chosen so small that the sets $Q_t^{\bullet}$ are pairwise disjoint for $0 \leq t \leq T$. This is possible since the curvature of the trajectory γ is bounded because of the hypothesis on $f(y)$.

Let p_0 be a point of Q_0. The trajectory $p(t, p_0)$ which begins at p_0 at time $t_0 = 0$ meets the set Q_t for the first time at a time τ which depends on t and p_0 (cf. Theorem 16.4),

$$(80.5) \qquad\qquad \tau = \tau(t, p_0).$$

If the distance $\varrho(p_0, \varphi(0))$ is sufficiently small then the vector

$$(80.6) \qquad\qquad z(t) : = p(\tau, p_0) - \varphi(t)$$

lies within Q_t. (80.4) implies

$$(80.7) \qquad\qquad z^T(t)\, g(t) = 0.$$

From (80.7) we obtain

$$(80.8) \qquad\qquad \dot{z}^T g + z^T \dot{g} = 0,$$

where

$$\dot{g}(t) = \frac{\partial f}{\partial y}\bigg|_{y=\varphi(t)} \frac{d\varphi}{dt}$$

and since by (80.6),

$$\dot{z}(t) = \frac{d p(\tau, p_0)}{d\tau} \frac{d\tau}{dt} - \dot{\varphi}(t) = f(p(\tau, p_0))) \frac{d\tau}{dt} - g(t)$$

we have

$$\left(f(p(\tau, p_0)) \frac{d\tau}{dt} - g(t) \right)^T g(t) + z(t)^T \dot{g}(t) = 0,$$

$$\frac{d\tau}{dt} f(p(\tau, p_0))^T g(t) - |g(t)|^2 = - z(t)^T \dot{g}(t),$$

and finally

$$(80.9) \qquad\qquad \frac{d\tau}{dt} = \frac{|g(t)|^2 - z(t)^T \dot{g}(t)}{f(z(t) + \varphi(t))^T g(t)}$$

and

$$(80.10) \qquad \frac{dz}{dt} = \frac{|g(t)|^2 - z(t)^T \dot{g}(t)}{f(z(t) + \varphi(t))^T g(t)} f(z(t) + \varphi(t)) - g(t).$$

These equations are identically valid in the vector z of (80.6). In addition, we consider them as differential equations defining the unknown functions $z(t)$ and $\tau(t)$.

A new Cartesian coordinate system is introduced as follows. For the origin we choose the point $\varphi(t)$ (t is fixed at first), in the hyperplane P_t.

We determine $n-1$ pairwise orthogonal unit vectors $b_1(t), \ldots, b_{n-1}(t)$ and supplement this system with a unit vector $b_n(t)$ in the direction of the tangent to the trajectory γ. We have

$$b_n(t) = \frac{f(\varphi(t))}{|f(\varphi(t))|} = \frac{g(t)}{|g(t)|}.$$

The column vectors $b_1, \ldots, b_n$ form an orthogonal matrix B. An arbitrary vector expressed in terms of the new basis $\{b_i(t)\}$, is denoted by $y = \mathrm{col}(y_1, \ldots, y_n)$. In the hyperplane P_t the old coordinates x and the new coordinates y are related by the equation

$$y = \varphi(t) + B(t) x.$$

Since $B(t)$ is orthogonal the inverse relation can be written in the form

(80.11) $$x = B(t)^T (y - \varphi(t)).$$

If in (80.9) and (80.10) we replace the vector z by $y - \varphi(t) = B(t)\,x$, then it follows that

(80.12) $$\frac{d\tau}{dt} = \frac{|g(t)|^2 - x^T B(t)^T \dot g(t)}{f(B(t)\,x + \varphi(t))^T g(t)},$$

and

$$\frac{dx}{dt} = -B(t)^T B(t) x$$

(80.13) $$+ B(t)^T \frac{|g(t)|^2 - x^T B(t)^T \dot g(t)}{f(B(t)\,x + \varphi(t))^T g(t)} f(B(t)\,x + \varphi(t)) - B(t)^T g(t).$$

The differential equations (80.10) and (80.13) are equivalent. Their integration is based on

Theorem 80.1. The scalar function

$$h(z, t) := z(t)^T g(t)$$

is a first integral of (80.10) which vanishes along each solution $p(t, z_0, t_0)$ of this differential equation. Hence the point $y_0 = z_0 + \varphi(t_0)$ lies in the set Q_{t_0}.

Proof. The derivative of $h(z, t)$ for (80.10) is equal to

$$\frac{d}{dt} h(z, t) = \dot z^T g + z^T \dot g = \frac{|g|^2 - z^T \dot g}{f(z + \varphi)^T g} f(z + \varphi)^T g - g^T g + z^T \dot g = 0.$$

Therefore $h(z, t)$ is constant along a solution of (80.10). If we choose $y_0 \in Q_{t_0}$ then $z_0^T g(t_0) = (y_0 - \varphi(t_0))^T g(t_0) = 0$. The constant is therefore zero●

The nth component of the vector x is obtained from (80.11),

$$x_n = b_n^T z = \frac{z^T g}{|g|} = \frac{h(z, t)}{|g|}.$$

From this we obtain, since $h(z, t)$ is constant,

$$x_n(t) = x_n(t_0) \cdot \frac{|g(t_0)|}{|g(t)|}.$$

For those solutions which pass through Q_0 at time $t = 0$, the component $x_n(t) = 0$. Therefore $x_n(t)$ must vanish identically.

In place of τ we introduce in (80.12) the variable $\theta = \tau - t$ and write

$$(80.14a) \qquad \frac{d\theta}{dt} = \frac{d\tau}{dt} - 1 = k_0(x, t).$$

For the right side of (80.13) we write $k(x, t) = \operatorname{col}(k_1, \ldots, k_n)$ so that

$$(80.14b) \qquad \frac{dx}{dt} = k(x, t).$$

As was shown above, $k_n(x, t) = 0$. $k(x, t)$ is periodic in t and vanishes for $x == 0$ as is easily seen by direct computation. The connection between the solutions of the original equation $\dot{y} = f(y)$ and those of the equations (80.14) is as follows. Let $y = p(t, y^0, 0)$ be a solution of (80.2) whose initial point y^0 lies in Q_0, and as in (80.5), let $\tau = \tau(t, y^0)$ be the time at which the trajectory $p(t, y^0, 0)$ meets the set Q_t for the first time. Then (80.11) defines functions

$$(80.15) \qquad x_i(t, x^0) = \sum_{j=1}^{n} b_{ji}(t)\left(p_j\left(\tau(t, y^0), y^0, 0\right) - \varphi_j(t)\right),$$
$$i = 1, 2, \ldots, n - 1,$$

which satisfy the differential equation (80.14b) and the initial conditions

$$x_i(0, x^0) = x_i^0 := \sum_{j=1}^{n} b_{ji}(0)\left(y_j^0 - \varphi_j(0)\right), \quad i = 1, 2, \ldots, n.$$

The function $\theta = \tau(t, x_0) - t$ is a solution of the equation (80.14a) with initial value $\theta = 0$ for $t = 0$. Let the functions

$$(80.16) \qquad \begin{aligned} \theta &= \theta(t, x^0, 0), \\ x_i &= x_i(t, x^0, 0), \quad i = 1, 2, \ldots, n, \end{aligned}$$

be those solutions of (80.14) which assume the value of 0, resp. $x^0 := \operatorname{col}(x_1, \ldots, x_n)$, for $t = 0$, and let $\tau = \theta + t$. Then

$$(80.17) \quad y_j(\tau, x^0) = \varphi_j(t) + \sum_{j=1}^{n=1} b_{ji}(t) x_i(t, x^0, 0), \quad j = 1, 2, \ldots, n,$$

is a solution of the system (80.3) if we consider τ as a function of t. The initial values are

$$(80.18) \qquad y_j^0 = \sum_{i=1}^{n-1} b_{ji}(0) x_i^0 + \varphi_j(0), \qquad j = 1, 2, \ldots, n,$$

and the point $\mathbf{y}^0$ lies in Q_0.

The relations (80.15) through (80.18) establish the desired connection between the original coordinates and those of the associated coordinate system.

The family of motions (80.1) satisfies the differential equations

$$\dot{y}_1 = -(1 + \varrho)\frac{1 + a\varrho}{1 + b\varrho} y_2, \quad \dot{y}_2 = (1 + \varrho)\frac{1 + b\varrho}{1 + a\varrho} y_1,$$

where ϱ is expressed in terms of y_1 and y_2 by the equation

$$\frac{y_1^2}{(1 + a\varrho)^2} + \frac{y_2^2}{(1 + b\varrho)^2} = 1.$$

The function $\varrho =: k(y_1, y_2)$ is a first integral of the differential equation. The periodic solution $\boldsymbol{\varphi} = \mathrm{col}(\cos t, \sin t)$ belongs to the value $\varrho = 0$. Furthermore $\mathbf{g} = \mathrm{col}(-\sin t, \cos t)$ and

$$B = \begin{pmatrix} \cos t & -\sin t \\ \sin t & \cos t \end{pmatrix}.$$

If we observe that $\mathbf{x}$ has only one component different from zero (since $x_2(t) \equiv 0$) we see that

$$B\mathbf{x} + \boldsymbol{\varphi} = \mathrm{col}\left((x_1 + 1)\cos t, (x_1 + 1)\sin t\right).$$

Furthermore

$$\mathbf{f}(B\mathbf{x} + \boldsymbol{\psi}) =$$
$$\mathrm{col}\left(-(1 + \bar{\varrho})\frac{1 + a\bar{\varrho}}{1 + b\bar{\varrho}}(x_1 + 1)\sin t, \ (1 + \bar{\varrho})\frac{1 + b\bar{\varrho}}{1 + a\bar{\varrho}}(x_1 + 1)\cos t\right).$$

In this formula

$$\bar{\varrho} := k\left((x_1 + 1)\cos t, (x_1 + 1)\sin t\right),$$

i.e. we have

$$(x_1 + 1)^2 \left(\frac{\cos^2 t}{(1 + a\bar{\varrho})^2} + \frac{\sin^2 t}{(1 + b\bar{\varrho})^2}\right) = 1;$$

$\bar{\varrho}$ is in general not constant. If we substitute the above expressions in equations (80.14) we obtain

$$\frac{d\theta}{dt} = \frac{(1 + x_1)^2}{(1 + \bar{\varrho})(1 + a\bar{\varrho})(1 + b\bar{\varrho})},$$

$$\frac{dx_1}{dt} = (1 + x_1)^2 \frac{(1 + b\bar{\varrho})^2 - (1 + a\bar{\varrho})^2}{(1 + a\bar{\varrho})(1 + b\bar{\varrho})} \cos t \sin t.$$

In the special case $a = b$, x_1 becomes constant as we have already pointed out above.

81. Stability Properties of a Periodic Solution

We consider an autonomous differential equation in R_n,

$$(81.1) \qquad \dot{x} = f(x), \quad f \in E,$$

and a solution $\varphi(t)$ which is periodic with period T. First of all we repeat the definitions of orbital and orbitally asymptotic stability (sec. 36) for the special case of the periodic solution. Let γ be defined by (80.3) and $p(t, x_0)$ be the general solution. The parameter t_0 can be set equal to zero without loss of generality. The periodic solution is called *orbitally stable* if there exists a function ψ of class K such that

$$\varrho\left(p(t, x_0), \gamma\right) < \psi\left(\varrho(x_0, \gamma)\right).$$

It is called *orbitally attractive* if there exists an estimate of the form

$$\varrho\left(p(t, x_0), \gamma\right) < \sigma(t, x_0), \quad \sigma \in L,$$

provided x_0 belongs to a certain δ-neighborhood $S(\delta, \gamma)$ of γ. It is called *orbitally asymptotically stable* if it is both orbitally stable and orbitally attractive.

Since the parameter t_0 does not occur in the definitions the orbital stability and the orbital attractivity are uniform with regard to the initial time. But if $\varphi(t)$ is orbitally stable its attractivity is also uniform with respect to x_0 which can be shown in analogy to the proof of Theorem 38.5. Let $\bar{x}_0$ be given and choose $\tau = \tau(\bar{x}_0)$ such that

$$(81.2) \qquad \sigma(t, \bar{x}_0) < \frac{1}{2}\, \psi\left(\varrho(\bar{x}_0, \gamma)\right) \quad \text{for} \quad t > \tau(\bar{x}_0).$$

Take a spherical neighborhood $K(\bar{x}_0)$ of $\bar{x}_0$ such that

$$\varrho\left(p(t, x_0'), \gamma\right) < \psi\left(\varrho(\bar{x}_0, \gamma)\right) \quad \text{for} \quad x_0' \in K(\bar{x}_0), \quad 0 < t < \tau(\bar{x}_0).$$

Let H be a closed subdomain of the domain of attraction of γ which contains γ in its interior and construct the balls $K(x_0)$ for all points $x_0 \in H$. Since H is closed the corresponding numbers $\tau(x_0)$ are bounded, $\tau(x_0) \leq \tau$. The subdomain H is covered by a finite number of these balls. Let their centers be $x_1, x_2, \ldots, x_N$. Then we have

$$\varrho\left(p(t, x_0), \gamma\right) < \psi\left(\max_i \varrho(x_i, \gamma)\right)$$

for $x_0 \in H$ and $0 < t \leq \tilde{\tau}$. Furthermore

$$\varrho\left(p(t, x_0), \gamma\right) < \sigma(t, 2\hat{x})$$

holds for $t > \tilde{\tau}$ by construction (*cf.* (81.2)). Here $\hat{x}$ is defined by

$$\psi\left(\varrho(\hat{x}, \gamma)\right) = \psi\left(\max_i \varrho(x_i, \gamma)\right).$$

The right side of the last inequality is independent of $x_0 \in H$. Therefore, the attractivity is uniform with respect to $x_0 \in H$.

27*

We now prove

Theorem 81.1. A nonconstant periodic solution of the autonomous differential equation (81.1) can never be attractive in the sense of Def. 35.6. Therefore, it can never be asymptotically stable in the sense of Liapunov.

Proof. The proof is indirect and we assume the existence of an estimate

$$\varrho\left(\boldsymbol{p}\left(t,\,\boldsymbol{x}_0\right),\,\boldsymbol{\varphi}\left(t\right)\right) < \sigma\left(t\right).$$

This estimate is valid for the particular solution $\boldsymbol{p} = \boldsymbol{\varphi}\left(t + c\right)$, provided c is sufficiently small. Therefore

$$\varrho\left(\boldsymbol{\varphi}\left(t + c\right),\,\boldsymbol{\varphi}\left(t\right)\right) < \sigma\left(t\right),$$

and this is a contradiction because the right side tends to zero whereas the left side is periodic in t and has a minimum different from zero.

If the solution $\boldsymbol{\varphi}\left(t\right)$ belongs to a family of periodic solutions then the theorem is immediately intuitive. Such a family exists, for instance, if the origin is a center in the phase plane or if a linear system of equations has a matrix with critical characteristic roots.

In the following, we assume that $\boldsymbol{f}\left(\boldsymbol{x}\right)$ satisfies the condition of (80.2). Then, the associated coordinate system exists. Its importance is seen in

Theorem 81.2. In order that the periodic solution $\boldsymbol{y} = \boldsymbol{\varphi}\left(t\right)$ of the autonomous differential equation (80.2) is orbitally stable, respectively orbitally asymptotically stable, it is necessary and sufficient that the trivial solution $\boldsymbol{x} = 0$ of the differential equation (80.14b) is stable, respectively asymptotically stable.

For the proof, we use (80.11) and the notations of sec. 80. The matrix B is orthogonal. Hence the distance $\varrho_1 := \left|\boldsymbol{x}\left(t,\,\boldsymbol{x}^0,\,0\right)\right|$ from the origin of the associated coordinate system is equal to the distance $\varrho\left(\boldsymbol{p}\left(t,\,\boldsymbol{y}^0\right),\boldsymbol{\varphi}\left(t\right)\right)$, and this distance is no smaller than $\varrho_2 := \varrho\left(\boldsymbol{p}\left(t,\,\boldsymbol{y}^0\right),\,\gamma\right)$. An estimate for ϱ_1 therefore implies a corresponding estimate for ϱ_2. If, on the other hand, the stability behavior of the periodic solution $\boldsymbol{\varphi}$ is known then there exists an estimate for ϱ_2: We have

$$\varrho_2 < \psi\left(\varrho\left(\boldsymbol{y}^0,\,\gamma\right)\right), \quad \text{respectively} \quad \varrho_2 < \psi\left(\varrho\left(\boldsymbol{y}^0,\,\gamma\right)\right)\sigma\left(t\right).$$

The time $\tau = \tau\left(t,\,\boldsymbol{y}^0\right)$ has been constructed so that (cf. (80.5))

$$\varrho\left(\boldsymbol{p}\left(\tau,\,\boldsymbol{y}^0\right),\,\gamma\right) = \varrho\left(\boldsymbol{p}\left(\tau,\,\boldsymbol{y}^0\right),\,\boldsymbol{\varphi}\left(t\right)\right) = \left|\boldsymbol{x}\left(t,\,\boldsymbol{x}^0,\,0\right)\right|.$$

Hence in case of orbital stability the estimate

$$\left|\boldsymbol{x}\left(t,\,\boldsymbol{x}^0,\,0\right)\right| < \psi\left(\varrho\left(\boldsymbol{y}^0,\,\gamma\right)\right) \leq \psi\left(\varrho\left(\boldsymbol{y}^0,\,\boldsymbol{\varphi}\left(0\right)\right)\right) = \psi\left(\left|\boldsymbol{x}^0\right|\right)$$

is valid and in case of orbitally asymptotic stability we have

$$\left|\boldsymbol{x}\left(t,\,\boldsymbol{x}^0,\,0\right)\right| < \psi\left(\left|\boldsymbol{x}^0\right|\right)\sigma\left(\tau\right) \leq \psi\left(\left|\boldsymbol{x}^0\right|\right)\sigma\left(t\right).$$

This implies stability, respectively asymptotic stability, in the sense of Liapunov for the solution $\boldsymbol{x} = 0$ of (80.14b).

Since the differential equation (80.14b) has periodic coefficients the stability, respectively asymptotic stability, of its equilibrium is always uniform.

The distinction made between orbital stability and stability in the sense of Liapunov enables us to assert

Theorem 81.3. The periodic solution $\varphi(t)$ of the differential equation (80.2) is stable in the sense of Liapunov if and only if the trivial solution $\theta = 0$, $x = 0$ of equations (80.14) is stable in the sense of Liapunov.
Proof. A) Let the solution $\theta = 0$, $x = 0$ of (80.14) be stable. Then we can make the difference $\theta = \tau - t$ arbitrarily small by choosing $|x^0|$ and θ_0 sufficiently small. We can then also make the distance $|p(t, y^0) - p(\tau, y^0)|$ arbitrarily small. But we have

$$\varrho\big(p(t, y^0), \varphi(t)\big) \leq \varrho\big(p(t, y^0), p(\tau, y^0)\big) + \varrho\big(p(\tau, y^0), \varphi(t)\big).$$

The first term on the right becomes arbitrarily small, as we have just noticed, and the second term becomes arbitrarily small because of the stability orbital which follows from Theorem 81.2.

B) The converse is seen as follows. We have

$$\varrho\big(p(t, y^0), p(\tau, y^0)\big) \leq \varrho\big(p(\tau, y^0), \varphi(t)\big) + \varrho\big(p(t, y^0), \varphi(t)\big).$$

The two terms on the right can be made arbitrarily small, since the periodic solution is stable and orbitally stable. This implies, however, that the difference $\theta = \tau - t$ can be made arbitrarily small by a suitable choice of the initial values.

By Theorems 81.2 and 81.3 the stability problem for the periodic solution has been reduced to a study of equations (80.14). Equation (80.14b) is of degree $n - 1$ and periodic. Its trivial solution is clearly asymptotically stable if the $n - 1$ characteristic multipliers of the variational equation are all smaller in absolute value than one. Solutions of equation (80.14b) even fade exponentially provided the initial values are sufficiently small and the solution of the first equation is of the form

$$\theta(t) = \theta(0) + O(|x^0|).$$

But then the equilibrium $\theta = 0$, $x = 0$ of equations (80.14) is stable in the sense of Liapunov, and by Theorems 81.2 and 81.3 the periodic solution $\varphi(t)$ is stable in the sense of Liapunov and orbitally stable. On the other hand, the linear part of equation (80.14b) agrees, up to a linear transformation, with the linear part of the equation of the perturbed motion for $\varphi(t)$; for we had $y = \varphi(t) + B(t)\,x$. The stability behavior of equation (80.14b) is therefore the same as that of the variational equation for the periodic solution. All this implies

Theorem 81.4 (ANDRONOV and WITT [1]). A periodic solution $\varphi(t)$ of an autonomous equation is stable in the sense of Liapunov and orbitally asymptotically stable, if the variational equation has $n-1$ characteristic multipliers of absolute value smaller than one.

The nth multiplier is always equal to one. This is seen as follows. The equation of the perturbed motion for the periodic solution $\varphi(t)$ of equation (80.2) has the form

$$\dot{z} = f\big(z + \varphi(t)\big) - \dot{\varphi}(t)$$

and the variational equation is

$$(81.3) \qquad \dot{\eta} = \frac{\partial f}{\partial y}\Big|_{y=\varphi} \eta.$$

Since the original equation is autonomous the equation

$$\dot{\varphi}(t+h) = f\big(\varphi(t+h)\big)$$

is valid identically in h. If we differentiate with respect to h and then set $h = 0$ we obtain

$$\frac{d}{dt}\,\dot{\varphi}(t) = \frac{\partial f}{\partial y}\Big|_{y=\varphi}\,\dot{\varphi}(t).$$

Therefore the derivative $\dot{\varphi}(t)$ is a solution of the variational equation, in fact a periodic solution. Equation (81.3) therefore has at least one characteristic exponent with zero real part, respectively a characteristic multiplier equal to one. Theorem 81.4 permits us to treat this critical case.

With the help of Theorem 81.4 we can easily prove a stability criterion due to Poincaré.

Theorem 81.5. Let the scalar system of equations

$$\dot{x} = f(x, y), \quad \dot{y} = g(x, y),$$

have a periodic solution $x = \varphi(t)$, $y = \psi(t)$ with period T. This solution is orbitally asymptotically stable if

$$h := \int_0^T \big(f_x(\varphi(t), \psi(t)) + g_y(\varphi(t), \psi(t))\big)\, dt < 0.$$

Proof. By (60.4) the product of the characteristic multipliers for the variational equation is equal to e^h. The condition of the theorem therefore implies that the multiplier which is different from one is smaller than one, and this means orbital asymptotic stability of the periodic solution by Theorem 81.4.

82. Examples: Testing for Stability

The stability of a periodic solution can be tested by investigating the stability of an equilibrium, the equilibrium of the equation of the perturbed motion or the equilibrium of the equivalent differential equation in the associated coordinate system. The latter method which was the content of Theorems 81.1 and 81.2, however, is mainly of theoretical importance. In practice (see for instance the example at the end of sec. 80) very little can be done with it and in most cases we will have to consider the equation of the perturbed motion, in fact a linear approximation of this equation, the variational equation, and carry out the test on the basis of the theorems of secs. 60 and 62. This is illustrated below in a few simple examples in which we limit our attention to scalar equations of the second order.

We first consider a perturbed conservative equation with periodic external force

$$(82.1) \qquad \ddot{x} + k^2 x = \mu\, g(x, \dot{x}, t; \mu) + f(t).$$

f and g are scalar functions periodic in t with period 2π. We assume that g has all the partial derivatives required in the following computations. We assume that a periodic scalar solution of period 2π is known and denote it by $\psi(t)$. The equation of the perturbed motion is

$$\ddot{y} + k^2 y + \ddot{\psi} + k^2 \psi = \mu\, g(y + \psi, \dot{y} + \dot{\psi}, t; \mu) + f(t)$$

and the variational equation is

$$(82.2) \qquad \ddot{y} + k^2 y = \mu\big(y\, g_x(\psi, \dot{\psi}, t; \mu) + \dot{y}\, g_{\dot{x}}(\psi, \dot{\psi}, t; \mu)\big).$$

The method of the small parameter furnishes a periodic solution of the form

$$\psi(t) = x_0(t) + \mu\, x_1(t) + \dots,$$

where $x_0(t)$ is the periodic solution of (82.1) belonging to the value $\mu = 0$, that is, the solution of

$$(82.3) \qquad \ddot{x}_0 + k^2 x_0 = f(t).$$

If we substitute the expansion of $\psi(t)$ in (82.2) and expand the right side by powers of μ we obtain

$$\ddot{y} + k^2 y = \mu\big(y\, g_x(x_0, \dot{x}_0, t; 0) + \dot{y}\, g_{\dot{x}}(x_0, \dot{x}_0, t; 0)\big) + O(\mu^2).$$

Now we usually work with the approximative equation

$$(82.4) \qquad \ddot{y} + k^2 y = \mu\big(y\, g_x(x_0, \dot{x}_0, t; 0) + \dot{y}\, g_{\dot{x}}(x_0, \dot{x}_0, t; 0)\big).$$

We are justified in doing so because the equilibrium of (82.4) has the same stability behavior as that of (82.2) in case the second equation has no critical case and in case $|\mu|$ is sufficiently small.

Equation (82.4) is of the type (60.5). It can be changed into a Hill's equation and then treated as in sec. 62. However, we prefer to apply the method of sec. 62 directly to (82.4) and therefore write this equation in the form

$$\ddot{y} - \mu\, r(t)\, \dot{y} + \left(k^2 - \mu\, s(t)\right) y = 0 ,$$

where

(82.5)
$$r(t) = g_{\dot{x}}(x_0, \dot{x}_0, t; 0), \quad s(t) = g_x(x_0, \dot{x}_0, t; 0) .$$

Its characteristic equation (*cf.* (62.2)) is

$$\lambda^2 + 2A\,\lambda + B = 0 .$$

The coefficients A and B can be expressed with the aid of a particular fundamental system $g(t)$, $h(t)$ of (82.4); g and h are so chosen that the initial conditions

$$g(0) = 1,\ \dot{g}(0) = 0,\ h(0) = 0,\ \dot{h}(0) = 1 ,$$

are satisfied. Then we have

(82.6) $\quad A = \dfrac{1}{2}\left(g(2\pi) + \dot{h}(2\pi)\right),\ \ B = g(2\pi)\,\dot{h}(2\pi) - \dot{g}(2\pi)\,h(2\pi),$

and in addition,

(82.7)
$$B = \exp\left(\mu \int\limits_{0}^{2\pi} r(t)\, dt\right).$$

The equilibrium of (82.4) is asymptotically stable if both roots of the characteristic equation are smaller than one in absolute value, that is if

$$\left| A \pm \sqrt{A^2 - B}\right| < 1.$$

These two inequalities are equivalent to the following conditions:

(82.8)
$$|B| < 1\ \text{ if }\ 0 < A^2 < B ,$$
$$|B| < 1\ \text{ and }\ 0 < 1 + B - 2|A|\ \text{ if }\ B < A^2 .$$

As in sec. 62 the fundamental system $g(t)$, $h(t)$ is obtained as a power series

$$g(t) = g_0(t) + \mu g_1(t) + \mu^2 g_2(t) + \cdots,$$
$$h(t) = h_0(t) + \mu h_1(t) + \mu^2 h_2(t) + \cdots,$$

subject to the defining equations

(82.9)
$$\ddot{g}_0 + k^2 g_0 = 0 ,$$
$$\ddot{h}_0 + k^2 h_0 = 0 ,$$
$$\ddot{g}_1 + k^2 g_1 = s(t)\, g_0 + r(t)\, \dot{g}_0 ,$$
$$\ddot{h}_1 + k^2 h_1 = s(t)\, h_0 + r(t)\, \dot{h}_0 ,$$

$$\cdots\cdots\cdots\cdots\cdots\cdots\cdots$$

and the initial conditions

$$g_0(0) = 1, \quad \dot{g}_0(0) = 0, \quad h_0(0) = 0, \quad \dot{h}_0(0) = 1.$$

All further initial values are equal to zero. Therefore

$$g_0(t) = \cos kt, \quad h_0(t) = \frac{1}{k} \sin kt,$$

and

$$A = \cos 2k\pi + O(\mu).$$

Furthermore by (82.7),

$$B = 1 + \mu \int\limits_0^{2\pi} r(t)\, dt + O(\mu^2).$$

First we assume that we do not have resonance, i.e. that k differs significantly from an integer, not only by an amount of the order of magnitude μ. Then $\cos 2k\pi$ is different from ± 1. The condition $|B| < 1$ is clearly satisfied if

$$(82.10) \qquad \mu \int\limits_0^{2\pi} r(t)\, dt = \mu \int\limits_0^{2\pi} g_{\dot{x}}(x_0, \dot{x}_0, t; 0)\, dt < 0.$$

For sufficiently small μ the conditions on A in (82.8) are satisfied so that (82.10) constitutes the sole stability condition in the nonresonance case.

Example. In equation (76.16) we have

$$g(x, \dot{x}) = -(a + bx^2 + cx^4)\,\dot{x}, \quad f(t) = p \sin t,$$

and the periodic solution of (82.3) becomes

$$x_0(t) = q \sin t, \quad q := \frac{p}{k^2 - 1}.$$

Condition (82.10) becomes

$$\mu \int\limits_0^{2\pi} (a + bq^2 \sin^2 t + cq^4 \sin^4 t)\, dt = \mu\pi \left(2a + bq^2 + \frac{3}{4} cq^4\right) > 0.$$

It is sufficient for the asymptotic stability of the periodic solution for sufficiently small μ.

In the case of resonance the situation is somewhat different. In that case k differs only by very little, *i.e.* by an amount of the order of magnitude of μ, from an integer m, $k^2 = m^2 + \alpha\mu$, and since the term $\mu\alpha x$ can be included in the expression for g (*cf.* example (77.21)) we can set $k = m$ throughout. The condition (82.10) remains; but since $\cos 2m\pi = \pm 1$ the expression $1 + B - 2|A|$ in (82.8) is $O(\mu)$. Now we have $g_0(2\pi) = 1$, $\dot{h}_0(2\pi) = 1$, and therefore

$$2A = 2 + \mu\left(g_1(2\pi) + \dot{h}_1(2\pi)\right) + \mu^2\left(g_2(2\pi) + \dot{h}_2(2\pi)\right) + \cdots,$$

$$B = 1 + \mu\left(g_1(2\pi) + \dot{h}_1(2\pi)\right) + \mu^2\left(g_2(2\pi) + \dot{h}_2(2\pi)\right)$$
$$+ \mu^2\left(g_1(2\pi)\,\dot{h}_1(2\pi) - \dot{g}_1(2\pi)\,h_1(2\pi)\right) + \cdots,$$

and

$$-2A + B + 1 = \mu^2\left(g_1(2\pi)\,\dot{h}_1(2\pi) - \dot{g}_1(2\pi)\,h_1(2\pi)\right) + O(\mu^3),$$

and the stability condition

$$(82.11) \qquad g_1(2\pi)\,\dot{h}_1(2\pi) - \dot{g}_1(2\pi)\,h_1(2\pi) > 0$$

follows, which must be satisfied in addition to (82.10).

We can change the form of the left side of (82.11). From the third equation (82.9) we obtain

$$g_1(t) = \frac{1}{m}\int_0^t \left(s(u)\cos mu - mr(u)\sin mu\right)\sin\left(m(t-u)\right) du$$

and

$$g_1(2\pi) = -\frac{1}{m}\int_0^{2\pi} \left(s(u)\cos mu - mr(u)\sin mu\right)\sin mu\,du.$$

Analogously it follows that

$$\dot{g}_1(2\pi) = \int_0^{2\pi} \left(s(u)\cos mu - mr(u)\sin mu\right)\cos mu\,du,$$

$$h_1(2\pi) = -\frac{1}{m^2}\int_0^{2\pi} \left(s(u)\sin mu + mr(u)\cos mu\right)\sin mu\,du,$$

$$\dot{h}_1(2\pi) = \frac{1}{m}\int_0^{2\pi} \left(s(u)\sin mu + mr(u)\cos mu\right)\cos mu\,du.$$

The periodic solution of (82.3) contains two arbitrary constants. Therefore we substitute in (82.4), etc., the expression

$$\varphi(t) = x_0(t) + c_1\cos mt + c_2\sin mt$$

in place of x_0 and determine the constants c_1, c_2 from the solvability conditions

$$G(c_1, c_2) := \int_0^{2\pi} g(\varphi, \dot{\varphi}, t; 0)\sin mt\,dt = 0,$$

$$H(c_1, c_2) := \int_0^{2\pi} g(\varphi, \dot{\varphi}, t; 0)\cos mt\,dt = 0.$$

We recognize that the four functional values $g_1(2\pi), \ldots, \dot{h}_1(2\pi)$ can be written in terms of the derivatives of G and H. By (82.5), we have

$$g_1(2\pi) = -\frac{1}{m}\frac{\partial G}{\partial c_1}, \qquad \dot{g}_1(2\pi) = \frac{\partial H}{\partial c_1},$$

$$h_1(2\pi) = -\frac{1}{m^2}\frac{\partial G}{\partial c_2}, \qquad \dot{h}_1(2\pi) = \frac{1}{m}\frac{\partial H}{\partial c_2},$$

and the inequality (82.11) assumes the form

$$\frac{\partial(G, H)}{\partial(c_1, c_2)} < 0.$$

For (82.10) we can also write (assuming $\mu > 0$)

$$-\frac{\partial G}{\partial c_1} + \frac{\partial H}{\partial c_2} < 0.$$

The stability conditions appear similar to the conditions for asymptotic stability of the equilibrium of an autonomous system of two equations: They involve positivity conditions for the trace and the determinant of a certain matrix. For an explanation of this similarity, which is more than formal, we direct the reader to a presentation of the Poincaré Theory in MINORSKY [1].

The above discussion refers to simple resonance for which the period of the periodic solution corresponds to the *smallest* period of the external force. If we wish to examine the stability of subharmonic oscillations we proceed as above. We must only observe that the solvability conditions for rth order resonance have the form

$$\int_0^{2\pi r} g(\varphi, \dot{\varphi}, t; 0)\cos\frac{t}{r}\,dt = 0, \qquad \int_0^{2\pi r} g(\varphi, \dot{\varphi}, t; 0)\sin\frac{t}{r}\,dt = 0,$$

since we must change the time scale (*cf.* the example in sec. 27).

Example. For the equation (77.21)

$$\ddot{x} + x = \mu p \sin t - \mu(\alpha x + a x^3),$$

we have (*cf.* sec. 77)

$$G(c_1, c_2) = p - \alpha c_2 - \frac{3}{4}a c_2(c_1^2 + c_2^2),$$

$$H(c_1, c_2) = \quad -\alpha c_1 - \frac{3}{4}a c_1(c_1^2 + c_2^2),$$

and

$$-\frac{\partial G}{\partial c_1} + \frac{\partial H}{\partial c_2} = 0,$$

$$\frac{\partial(G, H)}{\partial(c_1, c_2)} = -\left(\alpha + \frac{3}{4}a(c_1^2 + c_2^2)\right)\left(\alpha + \frac{9}{4}a(c_1^2 + c_2^2)\right).$$

We must set $c_1 = 0$; c_2 satisfies the equation

$$h(\lambda) := p - \alpha\lambda - \frac{3}{4}a\lambda^3 = 0.$$

The second condition can therefore be written in the form

$$(82.12) \qquad\qquad \frac{p}{\lambda}\,h'(\lambda) < 0.$$

The first condition cannot be checked by means of the first approximation.

The roots of the equation $h(\lambda) = 0$ are most simply found as points of intersection of the cubic with the straight line $\frac{4}{3a}(p - \alpha\lambda)$. We recognize that if we have three points of intersection then condition (82.12) is not satisfied for one of the two outside points. To this value of λ there corresponds therefore an unstable periodic motion. We also see that the absolute value of the "unstable" value λ lies between the absolute values of the two "stable" ones. If for a fixed amplitude of excitation p, we graph the variable $|\lambda|$, *i.e.* the amplitude of the forced vibration, as a function of $k^2 = 1 + \mu\alpha$ (α measures the deviation of the characteristic frequency of the system from the external frequency) then we obtain the well known amplitude-frequency-diagram (Figure 82.1). In this interpretation we must keep in mind that the construction fails for large values of $|\lambda|$; for in a concrete system the amplitudes are always bounded because of the ever present damping effect which was ignored in (77.21). If in a concrete system we change the frequency, *i.e.*

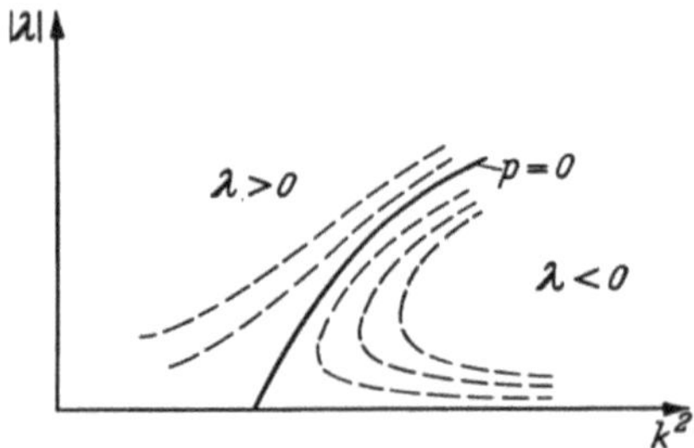

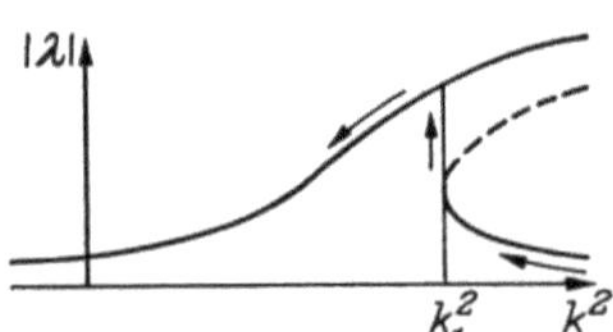

Fig. 82.1. **Response curve for the Duffing equation**   Fig. 82.2. Jump phenomenon. $k^2 = 1 + \alpha$

the variable α in the diagram, then the amplitude $|\lambda|$ changes. If we choose an initial state corresponding to a point on the lower stable branch (*cf.* the figure 82.2) curve and let α decrease then the amplitude "jumps" as α reaches the value α_1. This jump phenomenon can be found experimentally. As α increases, a downward jump occurs at a value α_2 which in general is larger than α_1. The theory developed here does not adequately explain this jump.[1])

---

[1]) *cf.* Minorsky [1], Stoker [1].

The stability of the periodic solution of an autonomous equation

$$(82.13) \qquad \ddot{x} + k^2 x = \mu g(x, \dot{x}; \mu)$$

can also be investigated with the help of the variational equation or its approximation

$$\ddot{y} + k^2 y = \mu \left(y g_x(x_0, \dot{x}_0; 0) + \dot{y} g_{\dot{x}}(x_0, \dot{x}_0, 0) \right)$$

which corresponds to (82.4) and can be written in the form (82.5). This time we must take account of the fact that one of the characteristic multipliers of this equation is always equal to one and that by Theorem 81.4 the periodic solution is orbitally asymptotically stable if the other multiplier is smaller than one. It suffices therefore to show that the quantity

$$B = \exp\left(\mu \int\limits_0^T r(t)\, dt \right)$$

is smaller than one, respectively that (in case $\mu > 0$)

$$(82.14) \qquad \int\limits_0^T g_{\dot{x}}(x_0, \dot{x}_0; 0)\, dt < 0.$$

We can change the form of this condition as we did in the nonautonomous case. The only solvability condition for (82.13) results ($cf.$ sec. 78) if we introduce the zeroth iterate $x_0 = c_0 \cos kt$ into the equation for the first iterate

$$\ddot{x}_1 + k^2 x_1 = g(x_0, \dot{x}_0; 0).$$

The condition has the form

$$G(c_0) := \int\limits_0^{\frac{2\pi}{k}} g(c_0 \cos kt, -kc_0 \sin kt; 0) \sin kt\, dt = 0.$$

Integrating by parts and substituting $kt = u$, we change the expression $G(c_0)$ into

$$-\frac{1}{k} c_0 \int\limits_0^{2\pi} (g_x \sin u + k g_{\dot{x}} \cos u) \cos u\, du$$

(the arguments of g_x and $g_{\dot{x}}$ are $c_0 \cos u$ and $-kc_0 \sin u$) and we further obtain

$$k G(c_0) = -k c_0 \int\limits_0^{2\pi} g_{\dot{x}}\, du - c_0 \int\limits_0^{2\pi} (g_x \cos u - k g_{\dot{x}} \sin u) \sin u\, du$$

$$= -k c_0 \int\limits_0^{2\pi} g_{\dot{x}}\, du - k c_0 \frac{d}{dc_0} G(c_0).$$

The first term on the right differs from the left side of (82.14) only by a quantity of order of magnitude μ since the period T differs by that much from $2\pi/k$. Since $G(c_0) = 0$ it follows that

$$c_0\, G'(c_0) = -\frac{1}{k}\, c_0 \int_0^{2\pi} g_{\dot{x}}(x_0,\, \dot{x}_0;\, 0)\, du,$$

and the stability condition is simply

$$(82.15) \qquad\qquad\qquad G'(c_0) > 0.$$

Example. In the case of equation (78.5) we have $k = 1$,

$$G(c_0) = -\pi\left(a + \frac{b}{4}\, c_0^2\right) c_0, \quad G'(c_0) = -\pi\left(a + \frac{3b}{4}\, c_0^2\right).$$

Since $c_0^2 = -\dfrac{4a}{b}$, we have $G'(c_0) = 2\alpha\pi$ and the periodic solution for $a > 0$ is orbitally asymptotically stable.

If the equation and hence the function $G(c_0)$ depends on a parameter then branch phenomena may occur similarly as for conservative systems of the second order (*cf.* the discussion in sec. 22). This is illustrated in a simple example. Let the equation have the form

$$(82.16) \qquad\qquad \ddot{x} + x = \mu(a\dot{x} + b\dot{x}^3 + c\dot{x}^5)$$

so that

$$G(c_0) = -\pi c_0\left(a + \frac{3b}{4}\, c_0^2 + \frac{5c}{8}\, c_0^4\right).$$

We assume that the coefficients depend linearly on a parameter λ; in fact let

$$(82.17) \qquad h(c_0,\, \lambda) := -c_0\pi(\lambda - \alpha - \lambda\beta c_0^2 + \lambda\gamma c_0^4).$$

Equation (82.16) can be considered as the equation of a vacuum tube generator and λ as the coefficient of a variable inductivity. We interpret the relation $h(c_0,\, \lambda) = 0$ in a $(\lambda,\, c_0^2)$-plane. By the criterion (82.15), those parts of the curve which lie above the region $h < 0$ correspond to stable states. We first consider the case $\alpha > 0$, $\beta > 0$, $\gamma = 0$ (figure 82.3). For small values of λ the origin is a stable focus. As λ approaches

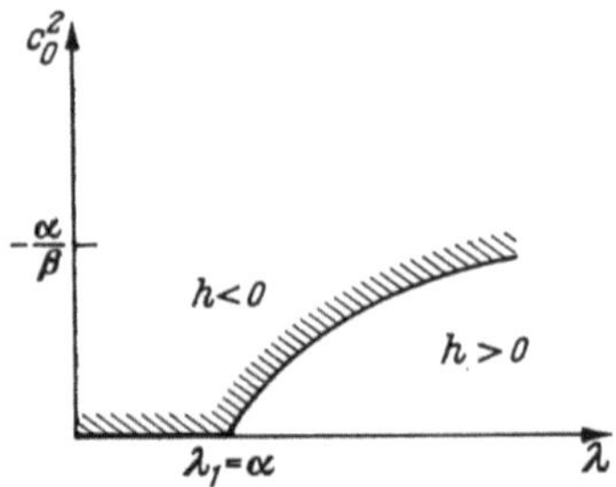

Fig. 82.3. $\gamma = 0$ in (82.17). Hatched parts: stable systems

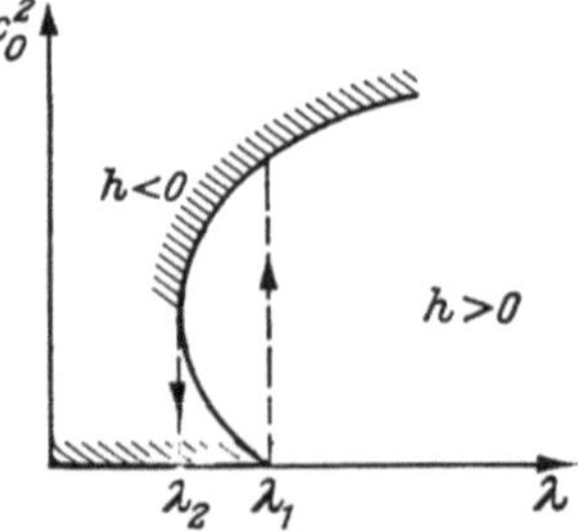

Fig. 82.4. $\gamma < 0$ in (82.17) Hatched parts: stable systems

the branch value λ_1 a stable limit cycle emerges and the origin becomes unstable. The amplitude of the periodic solution increases, starting from zero. If on the other hand the coefficient γ in (82.17) is different from zero, and in particular is negative, then the curve has the form of figure 82.4. Then there exists a second branch value $\lambda_2 < \lambda_1$ which is defined by $d\lambda/d(c_0^2) = 0$. For $\lambda = \lambda_2$ there exists a limit cycle of an amplitude different from zero and for $\lambda > \lambda_2$ there exist a stable and an unstable limit cycle. As λ surpasses the value λ_1, the origin becomes unstable and there remains a single stable limit cycle. The concrete system behaves as follows: The origin remains asymptotically stable while λ increases from zero to λ_1. Then suddenly the stable limit cycle belonging to $\lambda = \lambda_1$ appears, *i.e.* the system begins to oscillate with an amplitude different from zero.

For decreasing λ the oscillation ceases only when $\lambda = \lambda_2$; the jump takes place at a different place. This jump phenomenon can also be shown experimentally.

A survey on various branching phenomena is found in ANDRONOVA-LEONTOVICH and BELYUSTINA [1], further in the book of ANDRONOV, WITT, and KHAIKIN [1].

Bibliography [1]

AIZERMAN, A. M.
[1] On a problem concerning the stability "in the large" of dynamical systems. Usp. mat. Nauk **4**, 4, 187—188 (1949) (R).
[2] Lectures on the theory of automatic control. Moscow 1956 (R).
[3] Bedingungen für die Strukturstabilität gewisser Klassen von Regelsystemen. Regelungstechnik. Moderne Theorien. Tagung Heidelberg 1956, 93—99.

AIZERMAN, A. M., and F. R. GANTMAHER
[1] Stability in the first approximation for a periodic solution of a system of differential equations with discontinuous right hand sides. Prikl. Mat. Meh. **21**, 658—699 (1957) (R).
[2] Absolute stability of control systems. Moskow 1963 (R), transl. San Francisco 1964, Munich 1965.

ANDRÉ, J., and P. SEIBERT
[1] The local theory of piecewise continuous differential equations. Contr. nonl. Oscill. **5**, 225—255 (1960).

ANDREEV, A. F.
[1] Solution of the problem of the center and the focus in one case. Prikl. Mat. Meh. **17**, 333—338 (1953) (R).

ANDRONOV, A. A., and A. WITT
[1] Zur Stabilität nach Liapounow. Physikal. Z. Sowjetunion **4**, 606—608 (1933).

ANDRONOV, A. A., A. A. WITT and S. E. KHAIKIN
[1] Theorie der Schwingungen. Moskow 1959 (R), transl. Berlin 1965.

ANDRONOVA-LEONTOVICH, E. A., and L. N. BELYUSTINA
[1] The theory of bifurcations of second order dynamical systems and its applications to the investigation of the nonlinear problems of the oscillation theory. Trudy Simp. nelin. koleb. Kiev 1961, II, 7—28. (R, Engl. summary).

ANTOSIEWICZ, H. A.
[1] A survey of Lyapunov's second method. Contr. nonl Oscill. **4**, 141—166 (1958).

ARSCOTT, F. M.
[1] Periodic differential equations. Oxford 1964.

BAILEY, F. N.
[1] The application of Lyapunov's second method to interconnected systems. J. SIAM Control A 3, 443—462 (1965).

BARBASHIN, E. A., and N. N. KRASOVSKII
[1] On the stability of a motion in the large. Doklady Akad. Nauk SSSR **86**, 453—456 (1952) (R).

---

[1] (R) : = in Russian

[2] On the existence of Liapunov functions in the case of asymptotic stability in the large. Prikl. Mat. Meh. **18**, 345—350 (1954) (R).

BELLMAN, R.
[1] Stability theory of differential equations. New York 1953.
[2] Introduction to matrix analysis. New York 1960.

BELLMAN, R., and K. L. COOKE
[1] Differential-difference equations. New York 1963.

BERGEN, A. R., and I. J. WILLIAMS
[1] Verification of Ajzerman's conjecture for a class of third order systems. IRE Trans. AC-7, **3**, 42—46 (1962).

BESICOVITCH, A. S.
[1] Almost periodic functions. Cambridge 1931.

BHATIA, N. P.
[1] Anwendung der direkten Methode von Ljapunow zum Nachweis der Beschränktheit und der Stabilität der Lösungen einer Klasse nichtlinearer Differentialgleichungen zweiter Ordnung. Abh. Deutsche Akad. Wiss. Kl. Math. Phys. Techn. **1961**, Nr. 5 (1962).

BRAYTON, R. K., and W. L. MIRANKER
[1] A stability theory for nonlinear mixed initial value problems. Arch. rat. Mech. Analysis **17**, 358—376 (1964).

CAUGHEY, T. K., and A. H. GRAY jr.
[1] On the almost sure stability of linear dynamic systems with stochastic coefficients. J. appl. Mech. **1965**, 365—372 (1965).

CESARI, L.
[1] Asymptotic behavior and stability problems in ordinary differential equations. (Ergeb. Math. Grenzgeb. 16). Heidelberg 1963.

CHANG, S. S. L.
[1] Kinetic Lyapunov function for stability analysis of nonlinear control systems Trans. AIEE **83**, 91—94 (1961).

CHETAEV, N. G.
[1] Un théorème sur l'instabilité. C. R. (Doklady) Acad. Sci. URSS 1934, I 529—531 (1934).
[2] Stability of motion. Moskow 1955 (R).

CODDINGTON, E. A., and N. LEVINSON
[1] Theory of ordinary differential equations. New York 1955.

COLEMAN, C.
[1] Asymptotic stability in 3-space. Contr. nonl. Oscill. **5**, 257—268 (1960).

CONLEY, C. C., and R. K. MILLER
Asymptotic stability without uniform stability: almost periodic coefficients. J. Diff. Equ. **1**, 333—336 (1965).

COPPEL, W. A.
[1] On the stability of ordinary differential equations. J. London math. Soc. **38**, 255—260 (1963).
[2] Stability and asymptotic behavior of differential equations. Boston 1965.

CORDUNEANU, C.
[1] Sur la stabilité asymptotique. An. Şti. Univ. Iaşi Sect. I, **5**, 37—40 (1959).
[2] Sur la stabilité asymptotique. II. Rev. Math. pur. appl. **6**, 573—576 (1960).
[3] Application des inégalites différentielles à la théorie de la stabilité. An. Şti. Univ. Iaşi Sect. I, **6**, 46—58 (1960) (R, French summary).
[4] Complément au travail "Application des inégalites différéntielles à la

théorie de la stabilité." An. Şti. Univ. Iaşi Sect. I, 7, 247—252 (1961) (R, French summary).

[5] Sur la construction des fonctions de Liapunoff. Bull. Acad. Polonaise Sci. Ser. Sci. math. **10**, 559—563 (1963).

[6] Sur la stabilité partielle. Rev. Roumaine Math. pur. appl. **9**, 229—236 (1964).

CREMER, H., and E. H. EFFERTZ

[1] Über die algebraischen Kriterien für die Stabilität von Regelungssystemen. Math. Ann. **137**, 328—350 (1959).

DEBAGGIS, H. F.

[1] Dynamical systems with stable structures. Contr. nonl. Oscill. **2**, 37—59 (1952).

DILIBERTO, S. P.

[1] On systems of ordinary differential equations. Contr. nonl. Oscill. **1**, 1—38 (1950).

DOETSCH, G.

[1] Einführung in die Theorie und Anwendung der Laplacetransformation. Basel 1958.

DRIVER, R. D.

[1] Existence and stability of solutions of a delay-differential system. Arch. rat. Mech. Analysis **10**, 401—426 (1962).

EFENDIEV, A. R.

[1] On the region of influence of a singular point of higher order. Vestnik Moskov. Univ. I **1963**, 1, 14—25. (R, Engl. summary).

ERUGIN, N. P.

[1] On a problem of stability of systems of automatic regulation. Prikl. Mat. Meh. **16**, 620—628 (1952) (R).

EZEILO, J. O. C.

[1] Some results for the solutions of a certain system of differential equations. J. math. Analysis Appl. **6**, 387—393 (1963).

FILIPPOV, A. F.

[1] Differential equations with discontinuous right hand members. Mat. Sbornik **51** (93), 99—128 (1960) (R).

FLÜGGE-LOTZ, I.

[1] Discontinuous automatic control. Princeton 1953.

[2] Discontinuous automatic control. Rev. appl. Mech. **14**, 581—584 (1961).

FOSTER, H.

[1] Über das Verhalten der Integralkurven einer gewöhnlichen Differentialgleichung erster Ordnung in der Umgebung eines singulären Punktes. Math. Z. **43**, 271—320 (1937).

FROMMER, M.

[2] Über das Auftreten von Wirbeln und Strudeln (geschlossener und spiraliger Integralkurven) in der Umgebung rationaler Unbestimmtheitsstellen. Math. Ann. **109**, 395—424 (1934).

HAHN, W.

[1] Über Stabilität bei nichtlinearen Systemen. Z. angew. Math. Mech. **35**, 459—462 (1955).

[2] Eine Bemerkung zur zweiten Methode von Ljapunov. Math. Nachr. **14**, 349—354 (1956).

[3] Über Differential-Differenzengleichungen mit anomalen Lösungen. Math. Ann. **133**, 251—255 (1957).

[4] Theorie und Anwendung der direkten Methode von Ljapunov. (Ergebn. Math. Grenzgeb. 22.) Heidelberg 1959. Engl. transl. Englewood Cliffs 1963.
[5] Über Differentialgleichungen erster Ordnung mit homogenen rechten Seiten. Z. angew. Math. Mech. **46**, 357—361 (1966).
[6] Über Typen des Stabilitätsverhaltens. Mh. Math. **71**, 7—13 (1967).

HALANAY, A.
[1] Differential equations. Stability. Oscillations. Time lag. New York 1966.

HALE, J. K.
[1] Functional differential equations with parameters. Contr. diff. equ. **1**, 401—423 (1962).
[2] Linear functional differential equations with constant coefficients. Contr. diff. equ. **2**, 291—317 (1963).
[3] A stability theorem for functional-differential equations. Proc. nat. Acad. Sci. **50**, 942—946 (1963).
[4] Oscillations in nonlinear systems. New York 1963.

INGWERSON, D. R.
[1] A modified Lyapunov method for nonlinear stability analysis. IRE Trans. AC-**6**, 199—210 (1961).

KALMAN, R. E.
[1] Lyapunov functions for the problem of Lur'e in automatic control. Proc. nat. Acad. Sci. **49**, 201—205 (1963).
[2] Mathematical description of linear dynamical systems. J. SIAM Control A **1**, 152—192 (1964).

KALMAN, R. E., and J. E. BERTRAM
[1] Control system analysis and design via the "second method" of Lyapunov. I. II. Trans. ASME J. basic Engin. **82** (1962), 371—393; 394—400.

KAMENKOV, G. V.
[1] On stability of motion over a finite interval of time. Prikl. Mat. Meh. **17**, 529—540 (1953) (R).

KAMKE, E.
[1] Differentialgleichungen reeller Funktionen. Leipzig 1945.

KAPLAN, W.
[1] Operational methods for linear systems. Reading, Mass. 1962.

KARIM, R. I. I. A.
[1] Über den Resonanzfall bei Systemen von n nichtlinearen gewöhnlichen Differentialgleichungen erster Ordnung. Arch. rat. Mech. Analysis **7**, 21—28 (1961).

KAUDERER, H.
[1] Nichtlineare Mechanik. Heidelberg 1958.

KEIL, K. A.
[1] Das qualitative Verhalten der Integralkurven einer gewöhnlichen Differentialgleichung erster Ordnung in der Umgebung eines singulären Punktes. J.-Ber. Deutsche Math.-Verein. **57**, 111—132 (1955).

KODAMA, S.
[1] Stability analysis of a class of discrete control systems containing a nonlinear gain element. IRE Trans. AC-**7**, 5, 103—109 (1962).

KOVAL', P. I.
[1] Reducible systems of difference equations and the stability of their solutions. Usp. mat. Nauk **12**, 6 (78), 143—146 (1957) (R).

28*

Krasovskii, N. N.
[1] Theorems on the stability of motions determined by a system of two equations. Prikl. Mat. Meh. **16**, 546—554 (1952) (R).
[2] On the stability of the solutions of a system of two differential equations. Prikl. Mat. Meh. **17**, 651—672 (1953) (R).
[3] On the application of Lyapunov's second method to equations with retarded argument. Prikl. Mat. Meh. **20**, 315—327 (1956) (R).
[4] Certain problems of the theory of stability of motion. Moskow 1959 (R), transl. Stanford, Cal. 1963.

Ku, Y. H., and N. N. Puri
[1] On Liapunov functions of higher order nonlinear systems. J. Franklin Inst. **276**, 349—364 (1963).

Kudaev, M. B.
[1] Liapunov's function for the region of influence of a single singular point of higher order. Vestnik Moskov. Univ. I **1963**, No. 1, 3—13 (R, Engl. summary).

Kurzweil, J.
[1] On the reversibility of the first theorem of Lyapunov concerning the stability of motion. Czechoslov. math. Zhurn. **5** (80), 382—398 (1955) (R, Engl. summay).
[2] The converse second Lyapunov's theorem concerning the stability of motion. Czechoslov. math. Zhurn. **6** (81), 217—259; 455—473 (1956) (R, Engl. summary).

Kushner, H. J.
[1] On the stability of stochastic dynamical systems. Proc. nat. Acad. Sci. **53**, 8—12 (1965).

Lakshmikantham, V.
[1] On the stability and boundedness of differential systems. Proc. Cambridge phil. Soc. **58**, 492—496 (1962).
[2] Differential systems and extension of Lyapunov's method. Michigan math. J. **9**, 311—320 (1962).
[3] Notes on a variety of problems of differential systems. Arch. rat. Mech. Analysis **10**, 119—126 (1962).
[4] Differential equations in Banach spaces and the extension of Lyapunov's method. Proc. Cambridge phil. Soc. **59**, 373—381 (1963).
[5] Functional differential systems and extension of Lyapunov's method. J. math. Analysis Appl. **8**, 392—405 (1964).

LaSalle, J. P.
[1] A study on synchronous stability. Ann. of Math. (2) **65**, 571—581 (1957).
[2] The extent of asymptotic stability. Proc. nat. Acad. Sci. **48**, 363—365 (1960).
[3] Some extensions of Liapunov's second method. IRE Trans. CT-**7**, 520 to 527 (1960).
[4] Asymptotic stability criteria. Proc. Symp. appl. Math. **13**, 299—307 (1962).

LaSalle, J., and S. Lefschetz
[1] Stability by Liapunov's direct method with applications. New York 1961.

LaSalle, J. P., and R. J. Rath
[1] Eventual stability. Proc. II. Congr. IFAC Basle 1963, 556—560.

Lebedev, A. A.
[1] The problem of stability in a finite interval of time. Prikl. Mat. Meh. **18**, 75—94 (1954) (R).

[2] On stability of motion over a given interval of time. Prikl. Mat. Meh. **18**, 139—148 (1954) (R).

LEFSCHETZ, S.
[1] Differential equations. Geometric Theory. New York 1961.
[2] Stability of nonlinear control systems. New York 1965.

LEHNIGK, S. H.
[1] Die strukturelle Stabilität der Bewegung von Flugzeugen. Z. Flugwiss. **6**, 110—115 (1958).
[2] On Liapunov's second method with parameter dependent quadratic forms in the case of autonomous nonlinear equations which have a linear part. I. mezhd. kongr. IFAC avt. uprav. Moskow 1960.
[3] Stability theorems for linear motions with an introduction to Liapunov's direct method. Englewood Cliffs 1966.

LEIGHTON, W.
[1] On the construction of Liapunov functions for certain autonomous nonlinear differential equations. Contr. diff. equ. 2, 367—383 (1964).

LETOV, A. M.
[1] Inherently unstable control systems. Prikl. Mat. Meh. **14**, 183—192 (1950) (R).
[2] Stability in nonlinear control systems. Moskow 1955 (R), transl. Princeton 1961.

LIAPOUNOFF (LYAPUNOV), M. A.
[1] Problème générale de la stabilité de mouvement. Ann. Fac. Sci. Toulouse **9**, 203—474 (1907). (Translation of a paper published in Comm. Soc. math. Kharkow 1893, reprinted in Ann. math. Studies **17**, Princeton 1949.)
[2] Investigation of a singular case of the problem of stability of motion. Mat. Sbornik **17**, 252—333 (1893) (R).
[3] On a problem of stability of motion. Comm. Soc. math. Kharkow (2) III, 265—272 (1893) (R).
[4] Stability of motion. New York 1966. (Engl. transl. of [1] through [3] and of PLISS [4]).

LIVARTOVSKII, I. V.
[1] Certain questions of stability in the first approximation for differential equations with discontinuous right hand sides. Moskovsk. fiz-tehn. Inst. Isledov. Meh. prikl. Mat. Trudy 3, 47—63 (1959) (R).

LUR'E, A. I.
[1] Einige nichtlineare Probleme aus der Theorie der automatischen Regelung. Moskow 1951 (R), transl. Berlin 1957.

MAGNUS, K.
[1] Über ein Verfahren zur Untersuchung nichtlinearer Schwingungs- und Regelungssysteme. VDI-Forschungsh. 451 (1955), 32 S.

MALKIN, I. G.
[1] Sur un théorème d'existence de Poincaré-Liapounoff. C. R. (Doklady) Acad. Sci. URSS (2) **27**, 307—310 (1940).
[2] On a problem of the theory of stability of automatic control. Prikl. Mat. Meh. **16**, 365—368 (1952) (R).
[3] Theory of stability of motion. Moskow 1952 (R), transl. AEC-Transl. 3352 (in German: Berlin and Munich 1959).
[4] On the question of reversibility of Lyapunov's theorems on asymptotic stability. Prikl. Mat. Meh. **18**, 129—138 (1954) (R).

[5] Some problems in the theory of nonlinear oscillations. Moskow 1955 (R), trans. AEC-Transl. 3766 (1959).

MARKASHOV, L. M.
[1] On the characteristic exponents of the solutions of a second order linear differential equation with periodic coefficients. Prikl. Mat. Meh. **23** 1065—1075 (1959) (R, transl. Appl. Math. Mech. **23**, 1525—1535).

MASSERA, J. L.
[1] On Liapunoff's condition of stability. Ann. of Math. (2) **50**, 705—721 (1949).
[2] Contribution to stability theory. Ann. of Math. (2) **64**, 182—206 (1956); Erratum. Ann. of Math. (2) **68**, 202 (1958).
[3] On the existence of Lyapunov functions. Publ. Inst. Mat. Estad. Fac. Ing. Agrimens. Montevideo 3, No. 4, 111—120 (1960).
[4] A criterion for the existence of almost periodic solutions of certain systems of almost periodic differential equations. Publ. Inst. Mat. Estad. Fac. Ing. Agrimens. Montevideo 3, No. 3, 99—103 (1958) (Spanish, Engl. summary).
[5] Converse theorems of Lyapunov's second method. Symp. intern. Ecuac. diff. ordin. Mexico 1961, 158—163 (1962).
[6] The meaning of stability. Publ. Inst. Mat. Estad. Fac. Ing. Agrimens. Montevideo 4, No. 1, 23—47 (1964).

MASSERA, J. L., and J. J. SCHÄFFER
[1] Linear differential equations and functional analysis. I. Ann. of Math. (2) **67**, 517—573 (1958).

MATROSOV, V. M.
[1] On the stability of motion. Prikl. Mat. Meh. **26**, 885—895 (1962) (R, transl. Appl. Math. Mech. **26**, 1337—1353).

MINORSKY, N.
[1] Nonlinear oscillations. Princeton 1962.

MOVCHAN, A. A.
[1] The direct method of Lyapunov in stability problems of elastic systems. Prikl. Mat. Meh. **23**, 483—493 (1959) (R, transl. Appl. Math. Mech. **23**, 686—700).
[2] Stability of processes with respect to two metrics. Prikl. Mat. Meh. **24**, 988—1001 (1960) (R, transl. Appl. Math. Mech. **24**, 1506—1524).

MÜLLER, W.
[1] Qualitative Untersuchung der Lösungen nichtlinearer Differentialgleichungen zweiter Ordnung nach der direkten Methode von Ljapunov. Abh. Deutsch. Akad. Wiss. Kl. Math. Phys. Techn. **1965**, No. 4, 71 p.
[2] Über die Beschränktheit der Lösungen der Cartwright-Littlewoodschen Gleichung. Math. Nachr. **29**, 25—40 (1965).

NEMYTSKII, V. V., and V. V. STEPANOV
[1] Qualitative theory of differential equations. Moskow 1947 (R), transl. Princeton 1960.

NEYMARK, YU. I.
[1] The method of point mappings in the theory of nonlinear oscillations. Trudy Simp. nelin. Koleb. Kiev 1961, II, 268—307 (1964) (R, Engl. summary).

OBMORSHEV, A. N.
[1] Investigation of phase trajectories at infinity. Prikl. Mat. Meh. **14**, 383 to 390 (1950) (R).

Orlando, L.

[1] Sul problema di Hurwitz relative alle parti reali delle radii di un'equazione algebrica. Math. Ann. **71**, 233—245 (1911).

O'Shea, R. P.

[1] The extension of Zubov's method to sampled data control systems described by nonlinear autonomous difference equations. IEEE Trans. AC-**9**, 62—70 (1964).

Panov, A. M.

[1] On the behavior of solutions of systems of difference equations in the neighborhood of a fixed point. Izv. vyssh. ucheb. zaved. Mat. **1959**, 5 (12), 174—183 (R).

[2] Qualitative behavior of the trajectories of difference equations in the neighborhood of a fixed point. Izv. vyssh. ucheb. zaved, Mat. **1960**, 1 (14), 166—174 (R).

Peixoto, M. C., and M. M. Peixoto

[1] Structural stability in the plane with enlarged boundary conditions. An. Acad. Brasil. Ci. **31**, 135—160 (1959).

Perron, O.

[1] Über die Gestalt der Integralkurven einer Differentialgleichung erster Ordnung in der Umgebung eines singulären Punktes. I. II. Math. Z. **15**, 121—146; 273—295 (1922).

[2] Die Ordnungszahlen linearer Differentialgleichungssysteme. Math. Z. **31**, 748—766 (1929).

Persidskii, K. P.

[1] Über die Stabilität nach der ersten Näherung. Mat. Sbornik **40**, 284—293 (1933) (R, German summary).

Persidskii, S. K.

[1] On Liapunov's second method. Prikl. Math. Meh. **25**, 17—23 (1961) (R).

Petrov, V. V., and G. M. Ulanov

[1] Theory of two simple relay systems of selfregulation. Avt. Telemeh. **11**, 289—299 (1950) (R).

Pinney, E.

[1] Ordinary difference-differential equations. Berkeley 1959.

Pliss, V. A.

[1] A qualitative picture of the integral curves in the large and the construction with arbitrary accuracy of the region of stability of a certain system of two differential equations. Prikl. Mat. Meh. **17**, 541—554 (1954) (R).

[2] Some problems of the theory of stability of motion in the whole. Leningrad 1958 (R), transl. NASA TT F-250 (1965).

[3] The principle of reduction in the theory of stability of motion. Izv. Akad. Nauk SSSR Ser. mat. **28**, 1297—1324 (1964) (R).

[4] Investigation of a transcendental case of the theory of stability of motion. Izv. Akad. Nauk SSSR Ser. mat. **30**, 911—924 (1964) (R). Engl. transl. in Lyapunov [4].

Pontryagin, L. S.

[1] On the zeros of certain elementary transcendental functions. Izv. Acad. Nauk SSSR Ser. mat. **6**, 115—134 (1942) (R, transl. Amer. math. Soc. Transl. (2) **1**, 95—110).

[2] Ordinary differential equations. Moskow 1960 (R), transl. London 1962, Berlin 1965.

Popov, V.-M.

[1] Absolute stability of nonlinear systems of automatic control. Avt. Tele-

meh. **22**, 961—979 (1962) (R, transl. Autom. Remote Control **22**, 857 to 875).

REGHIŞ, M.
[1] On nonuniformly asymptotic stability. Prikl. Mat. Meh. **27**, 231—243 (1963) (R, transl. Appl. Mat. Mech. **27**, 344—362).

REISS, R., and G. GEISS
[1] The construction of Liapunov functions. IEEE Trans. AC-**8**, 382—383 (1963).

REISSIG, R.
[1] Erzwungene Schwingungen mit zäher Dämpfung und starker Gleitreibung. I. II. Math. Nachr. **11**, 231—238; **12**, 119—128 (1954).
[2] Erzwungene Schwingungen mit zäher und trockener Reibung. Math. Nachr. **11**, 345—384 (1954).
[3] Kriterien für die Zugehörigkeit dynamischer Systeme zur Klasse D. Math. Nachr. **20**, 67—72 (1959).
[4] Stabilitätsprobleme in der qualitativen Theorie der Differentialgleichungen. J.-Ber. Deutsche Math.-Verein. **63**, 97—116 (1960).

REISSIG, R., G. SANSONE and R. CONTI
[1] Qualitative Theorie nichtlinearer Differentialgleichungen. Roma 1963.

REKASIUS, Z. V., and J. E. GIBSON
[1] Stability analysis of nonlinear control systems by the second method of Liapunov. IRE Trans. AC-**7**, 3—15 (1962).

ROSEAU, M.
[1] Vibrations non linéaires et théorie de la stabilité. Berlin 1966.

SAHARNIKOV, N. A.
[1] Solution of the problem of the center and the focus in one case. Prikl. Mat. Meh. **14**, 651—658 (1950) (R).

SANSONE, G., and R. CONTI
[1] Non-linear differential equations. Oxford 1964.

SCHMEIDLER, W.
[1] Vorträge über Determinanten und Matrizen mit Anwendungen in Physik und Technik. Berlin 1949.

SCHULTZ, D. G., and J. E. GIBSON
[1] The variable gradient method for generating Liapunov functions. Trans. Amer. Inst. Electr. Eng. **81**, 203—209 (1962).

SCHWARZ, H. R.
[1] Ein Verfahren zur Stabilitätsfrage bei Matrizen-Eigenwertproblemen. Z. angew. Math. Phys. **7**, 473—500 (1956).

SKROWRONSKI, S., and S. ZIEMBA
[1] The boundedness of the motion of mechanical systems. Proc. Vibration Problems, Polon. Acad. Sci. **1**, 69—80 (1959).

SOLODOWNIKOW, W. W.
[1] Grundlagen der selbsttätigen Regelungen. I. II. Moskow 1952 (R) transl. Berlin u. München 1959.

STARZHINSKII, V. M.
[1] Survey of works on conditions of stability of the trivial solution of a system of linear differential equations with periodic solutions. Prikl. Mat. Meh. **18**, 469—510 (1954) (R).

STEPANOFF, V.
[1] Zur Definition der Stabilitätswahrscheinlichkeit. C. R. (Doklady) Acad. Sci. URSS **18**, 151—154 (1938).

STOKER, J. J.
[1] Nonlinear vibrations in mechanical and electrical systems. New York 1950.

STRAUSS, A.
[1] Liapunov functions and L^p-solutions of differential equations. Trans. Amer. math. Soc. **119**, 37—50 (1965).

SZEGÖ, G. P.
[1] Methods for constructing Liapunov functions for linear nonstationary systems. J. Electronics Control **13**, 69—88 (1962).
[2] On a new partial differential equation for the stability analysis of time invariant control systems. J. SIAM Contr. Ser. A **1**, 63—75 (1962).
[3] A contribution to Liapunov's second method: nonlinear autonomous systems. Trans. ASME J. basic. Eng. **84**, 571—578 (1964).

SZEGÖ, G. P., and G. R. GEISS
[1] A remark on "A new partial differential equation for the stability analysis of time invariant control systems". J. SIAM Control Ser. A **1**, 369 to 376 (1963).

VINOGRAD, R. E.
[1] On a criterion of instability in the sense of Lyapunov for the solutions of a linear system of ordinary differential equations. Doklady Akad. Nauk SSSR **84**, 201—204 (1952) (R).
[2] New proof of a theorem of Perron and some properties of regular systems. Usp. mat. Nauk **9**, (60) 2, 129—136 (1954) (R).
[3] The instability of the smallest characteristic exponent of a regular system. Doklady Akad. Nauk SSSR **103**, 541—544 (1955) (R).
[4] The inadequacy of the method of characteristic exponents for the study of nonlinear differential equations. Mat. Sbornik **41** (83), 431—438 (1957) (R).

VRKOČ, I.
[1] On the inverse theorem of Chetaev. Czechoslov. mat. Zhurn. **5** (80), 451—461 (1955) (R, Engl. summary).
[2] Integral stability. Czechoslov. mat. Zhurn. **9** (84), 71—129 (1959) (R, Engl. summary).

WANG, P. K. C., and F. TUNG
[1] Optimum control of distributed parameter systems. Trans. ASME J. basic. Eng. **84**, 67—79 (1964).

YOSHIZAWA, T.
[1] Liapunovs' functions and boundedness of solutions. Funkc. Eqvacioj **2**, 95—142 (1959).
[2] Stability and boundedness of solutions. Arch. rat. Mech. Analysis **6**, 409—421 (1960).
[3] Stability of sets and perturbed systems. Funkc. Eqvacioj **5**, 31—69 (1963).
[4] Some notes on stability of sets and perturbed systems. Funkc. Eqvacioj **6**, 1—11 (1964).

ZUBOV, V. I.
[1] Questions of the theory of Lyapunov's second method. Construction of

the general solution in the region of asymptotic stability. Prikl. Mat. Meh.
19, 179—210 (1955) (R).
[2] Methods of A. M. Lyapunov and their application. Leningrad 1957
(R), transl. Groningen 1964.
[3] Mathematical methods for the study of automatic control systems.
Leningrad 1959 (R), transl. New York 1962.
[4] Vibrations in nonlinear and controlled systems. Leningrad 1962 (R).

ZYPKIN, JA. S.
[1] Theorie der Relaissysteme der automatischen Regelung. Moskau 1955
(R), transl. München 1958.

Author Index

Subject Index

Die Grundlehren der mathematischen Wissenschaften
in Einzeldarstellungen
mit besonderer Berücksichtigung der Anwendungsgebiete

94. Funk: Variationsrechnung und ihre Anwendung in Physik und Technik. DM 98,—; US $ 24.50
95. Maeda: Kontinuierliche Geometrien. DM 39,—; US $ 9.75
97. Greub: Linear Algebra. DM 39,20; US $ 9.80
98. Saxer: Versicherungsmathematik. 2. Teil. DM 48,60; US $ 12.15
99. Cassels: An Introduction to the Geometry of Numbers. DM 69,—; US $ 17.25
100. Koppenfels/Stallmann: Praxis der konformen Abbildung. DM 69.—; US $ 17.25
101. Rund: The Differential Geometry of Finsler Spaces. DM 59,60; US $ 14.90
103. Schütte: Beweistheorie. DM 48,—; US $ 12.00
104. Chung: Markow Chains with Stationary Transition Probabilities. DM 56,—; US $ 14.00
105. Rinow: Die innere Geometrie der metrischen Räume. DM 83,—; US $ 20.75
106. Scholz/Hasenjaeger: Grundzüge der mathematischen Logik. DM 98,—; US $ 24.50
107. Köthe: Topologische Lineare Räume I. DM 78,—; US $ 19.50
108. Dynkin: Die Grundlagen der Theorie der Markoffschen Prozesse. DM 33,80; US $ 8.45
109. Hermes: Aufzählbarkeit, Entscheidbarkeit, Berechenbarkeit. DM 49,80; US $ 12.45
110. Dinghas: Vorlesungen über Funktionentheorie. DM 69,—; US $ 17.25
111. Lions: Equations différentielles opérationnelles et problèmes aux limites. DM 64,—; US $ 16.00
112. Morgenstern/Szabó: Vorlesungen über theoretische Mechanik. DM 69,—; US $ 17.25
113. Meschkowski: Hilbertsche Räume mit Kernfunktion. DM 58,—; US $ 14.50
114. MacLane: Homology. DM 62,—; US $ 15.50
115. Hewitt/Ross: Abstract Harmonic Analysis. Vol. 1: Structure of Topological Groups. Integration Theory. Group Representations. DM 76,—; US $ 19.00
116. Hörmander: Linear Partial Differential Operators. DM 42,—; US $ 10.50
117. O'Meara: Introduction to Quadratic Forms. DM 48,—; US $ 12.00
118. Schäfke: Einführung in die Theorie der speziellen Funktionen der mathematischen Physik. DM 49,40; US $ 12.35
119. Harris: The Theory of Branching Processes. DM 36,—; US $ 9.00
120. Collatz: Funktionalanalysis und numerische Mathematik. DM 58,—; US $ 14.50
121.
122. } Dynkin: Markov Processes. DM 96,—; US $ 24.00
123. Yosida: Functional Analysis DM 66,—; US $ 16.50
124. Morgenstern: Einführung in die Wahrscheinlichkeitsrechnung und mathematische Statistik. DM 34,50; US $ 8.60
125. Itô/McKean: Diffusion Processes and Their Sample Paths. DM 58—; US $ 14.50
126. Lehto/Virtanen: Quasikonforme Abbildungen. DM 38,—; US $ 9.50
127. Hermes: Enumerability, Decidability, Computability. DM 39,—; US $ 9.75
128. Braun/Koecher: Jordan-Algebren. DM 48,—; US $ 12.00
129. Nikodým: The Mathematical Apparatus for Quantum-Theories. DM 144,—; US $ 36.00
130. Morrey: Multiple Integrals in the Calculus of Variations. DM 78,—; US $ 19.50
131. Hirzebruch: Topological Methods in Algebraic Geometry. DM 38,—; US $ 9.50
132. Kato: Perturbation theory for linear operators. DM 79,20; US $ 19.80
133. Haupt/Künneth: Geometrische Ordnungen. DM 68,—; US $ 17.00
135. Handbook for Automatic Computation. Vol. 1/Part a: Rutishauser: Description of ALGOL 60. Approx. DM 58,—; approx. US $ 14.50
136. Greub: Multilinear Algebra. DM 32,—; US $ 8.00
137. Handbook for Automatic Computation. Vol 1/Part b: Grau/Hill/Langmaack: Translation of ALGOL 60. Approx. DM 64,—; approx. US $ 16.00
138. Hahn: Stability of Motion. DM 72,—; US $ 18.00
139. Mathematische Hilfsmittel des Ingenieurs. Herausgeber: Sauer/Szabó. 1. Teil. DM 88,—; US $ 22.00
143. Schur/Grunsky: Vorlesungen über Invariantentheorie. DM 28,—; US $ 7.00
144. Weil: Basic Number Theory. DM 48,—; US $ 12.00
146. Treves: Locally Covex Spaces and Linear Partial Differential Equations. Approx. DM 36,—; approx. US $ 9.00